White Skin, Black Fuel

The Zetkin Collective is a group of scholars, activists and students working on the political ecology of the far right. It was formed around the human ecology division at Lund University in the summer of 2018. There it organised the first international conference on Political Ecologies of the Far Right in November 2019, with some four hundred academics and activists from across the world (co-organised with CEFORCED and Action Group Hedvig). This is the first publication of the Collective. The members, who have all contributed to this book, are: Irma Allen, Anna Bartfai, Bernadette Barth, Lise Benoist, Julia Bittencourt Costa Moreira, Dounia Boukaouit, Clàudia Custodio, Philipa Oliva Dige, Ilaria di Meo, George Edwards, Morten Hesselbjerg, Ståle Holgersen, Claire Lagier, Line Skovlund Larsen, Andreas Malm, Sonja Pietiläinen, Daria Rivin, Luzia Strasser, Laudy van den Heuvel, Meike Vedder and Anoushka Eloise Zoob Carter. Andreas Malm has coordinated the writing.

White Skin, Black Fuel

On the Danger of Fossil Fascism

Andreas Malm
and the Zetkin Collective

London • New York

First published by Verso 2021
© Andreas Malm, the Zetkin Collective 2021

All rights reserved

The moral rights of the authors have been asserted

1 3 5 7 9 10 8 6 4 2

Verso
UK: 6 Meard Street, London W1F 0EG
US: 20 Jay Street, Suite 1010, Brooklyn, NY 11201
versobooks.com

Verso is the imprint of New Left Books

ISBN-13: 978-1-83976-174-4
ISBN-13: 978-1-83976-176-8 (US EBK)
ISBN-13: 978-1-83976-175-1 (UK EBK)

British Library Cataloguing in Publication Data
A catalogue record for this book is available from the British Library

Library of Congress Cataloging-in-Publication Data

Names: Malm, Andreas, 1977– author. | Zetkin Collective, author.
Title: White skin, black fuel : on the danger of fossil fascism / Andreas
   Malm and the Zetkin Collective.
Description: London ; New York : Verso, 2021. | Includes bibliographical
   references and index. | Summary: 'This is the first study of the far
   right's role in the climate crisis, presenting an eye-opening sweep of a
   novel political constellation, revealing its deep historical roots' –
   Provided by publisher.
Identifiers: LCCN 2020054666 (print) | LCCN 2020054667 (ebook) | ISBN
   9781839761744 (paperback) | ISBN 9781839761768 (US ebk) | ISBN
   9781839761751 (UK ebk)
Subjects: LCSH: Climatic changes – Government policy – Case studies. | Fossil
   fuels – Government policy – Case studies. | Right-wing extremists – Case
   studies. | Political ecology – Case studies.
Classification: LCC QC903 .M346 2021 (print) | LCC QC903 (ebook) | DDC
   304.2/8 – dc23
LC record available at https://lccn.loc.gov/2020054666
LC ebook record available at https://lccn.loc.gov/2020054667

Typeset in Minion by Hewer Text UK Ltd, Edinburgh
Printed and bound by CPI Group (UK) Ltd, Croydon CR0 4YY

Contents

Introduction ... vii

Part I

1. The Fortunes of Denial ... 3
2. Fear of a Muslim Planet ... 39
3. Fossil Fuels Are the Future ... 83
4. The Energy Wealth of Nations ... 103
5. Ecology Is the Border ... 133
6. White Presidents of the Americas ... 181

Part II

7. Towards Fossil Fascism? ... 223
8. Mythical Energies of the Far Right ... 255
9. Skin and Fuel ... 315
10. For the Love of the Machine ... 399
11. Death Holds the Steering Wheel ... 479

Coda: Rebel for Life ... 507
Postscript: A Strange Year in the Elevator ... 511

Acknowledgements ... 541
Index ... 543

Introduction

And [they] have brought humanity to the edge of oblivion: because they think they are white.
  James Baldwin, 'On Being White . . . and Other Lies', 1984

Klimaschutz und Antifa geht Hand in Hand, das ist doch klar.
  Chant at the Ende Gelände march to block the
  infrastructure of the Hambach coal mine, October 2018

In 2014, the party then known as the True Finns published a cartoon featuring a black man. He is dressed only in a grass skirt, his belly protruding over the belt. His nose is pierced with an animal bone. Eyes dilated, a wide-open mouth flashing absurdly large teeth, hysterically waving his left hand, he holds in his right a wooden bowl, in which four more animal bones jump up and down. At the top of his lungs, he screams: 'Even though the climate has not warmed since 1997, with this computer, I predict that the climate will warm by one hundred degrees, the moon will melt and the surface of the ocean will rise at least six hundred kilometres!! By the next week!!' To his right are two smaller figures, a man and a woman, white of skin. They look frightened, paralysed, cowardly as they stare at the black man's bowl. Professionally clad, they manage the climate institute of Finland. The woman exclaims: 'Ooh!! We have to spend more on wind turbines that function for only three days in a year!!' Satisfied, the witch doctor of climate science offers nothing of value in return: 'Great idea! I will give you a consultation.' *True* Finns, of course, would never cave in in such a ridiculous and despicable manner. 'So-called "climate science"', the party explains in the caption, 'has not been able to prove that human activity is the cause for the 1 degree rise in temperatures. Nevertheless, the climate directives force you to pay extra tax.'[1] True Finns would resist the extortion. They would refuse to believe in the fable, stop the pointless bleeding of resources and stand up for their own kind of energy.

Ever since climate change became a cause of concern, it has been widely assumed that people and policymakers will deal with it rationally. If they are informed about the danger, they will mend their ways. If only they realise how hard life will be on a planet that warms by 6 or 2 or even just 1.5°C, they will make an effort to emit less of the harmful gases and stake out a path towards quitting completely. If – having missed the previous warnings – they see the world actually starting to catch fire around them, surely they must then wake up and spring into action: this has been the premise for communication between the climate research community and the rest of society. The former passes on knowledge of how things are unfolding on earth and expects the latter to act in response, much as when a doctor gives an adult patient a diagnosis and prescribes a medication available at the nearest pharmacy. The condition is dire, but treatment guaranteed to have an effect. Ever the loyal crew of doctors, climate scientists have kept knocking on the doors of governments and delivering their messages – for instance, about how severe the consequences would be of a rise in average temperatures above 1.5°C, as laid out in a report from the Intergovernmental Panel on Climate Change (IPCC) in October 2018 – and waiting for at least some minimally

---

1 For a reproduction of the cartoon, see Johannes Kotkavirta, '*Perussuomalaiset julkaisi erikoisen poppamies-pilukuvan – "Talla varmasti kalastellaan ääniä*",' *Ilta-Sanomat*, 19 May, 2014.

adequate reaction. The same assumption of rationality has underpinned the expectation that a shift from fossil to renewable energy will happen when the price of the latter has fallen, or that well-informed consumers will choose the least damaging commodities, or that the international community will come to an agreement, or that modern civilisation and the human enterprise will once again demonstrate their problem-solving ingenuity and press on with the improvement of life on earth.

This assumption has been taking a beating for some time. Few, however, would have thought that a 1°C increase in average temperatures, an ever-rising tide of extreme weather events, an unhinging of the climate system observable to the naked eye in virtually every corner of the world would coincide with the surge of a political force that just flatly denies it all. The far right has not figured in any climate models. Variables of whiteness and race and nationalism have not been included. No IPCC scenario has counted on the possibility that deep into the early stages of global warming, just as the urgency of slashing emissions ought to be at its most overwhelming, state apparatuses in Europe and the Americas would be increasingly occupied by parties and presidents professionally clad and white of skin and eager to show the whole issue the door. In another conjuncture, the True Finns cartoon could be shrugged off as the bad joke of a good-for-nothing party on the European fringe; in the late 2010s, however, it plotted the inclinations of a far right storming into offices and chambers from Berlin to Brasília. Two trends now seemed to intersect: rapidly rising temperatures and rapid advances of the far right. There was no easily discernible end to either. Little, if anything, suggested that they would flatten or go into reverse of their own accord. So what happens when they meet?

The rise of the far right has, of course, been extensively and anxiously commented upon, but rarely as a trend rooted in a certain material base and growing into the atmosphere. In the authoritative *Oxford Handbook of the Radical Right*, published in 2018, we find 'chapters covering all major theoretical and methodological strands in this literature': discussions of religion, media, gender, violence, youth, charisma, euroscepticism, globalisation and plenty of other factors, but nothing on ecology.[2] One widely recognised expert, Cas Mudde, put out a global survey of *The Far Right Today* in 2019 and passed by the issue with complete silence.[3] The 'surprising dearth' of research on the climatic dimension has been noted.[4] It creates a picture of the far right as rising somewhere else than in a rapidly warming world. But 'from now on, every issue is a climate issue', writes Alyssa Battistoni, formulating a theorem bound to become truer with time.[5] Far-right politics in the 1930s or 1980s could perhaps be studied outside of the natural environment. In the 2010s or 2040s, one cannot understand what it is doing in and to the world if that context is bracketed out: here we propose to put it front and centre.

What follows is the first systematic inquiry into the political ecology of the far right in the climate crisis.[6] We have investigated what the main far-right parties have said, written and done on climate and energy in thirteen European countries: Austria, Denmark, Finland, France, Germany, Hungary, Italy, the Netherlands, Norway, Poland, Spain, Sweden and the United Kingdom. Europe is the continent that gifted the world with both the fossil economy and fascism. On the other hand, some parts of it – particularly in the north – have, until recently, enjoyed a reputation as the most enlightened forerunners in climate policy as well as the most humane hosts of refugees. We focus on Europe, but we also look at two countries in the Americas – the United States and Brazil – that have long been recognised for their outsized impact on the climate system and that both, only two years apart, fell under the rule of presidents at the far-right end of the spectrum and on a rampage against nature.

2   Jens Rydgren, 'The Radical Right: An Introduction', in Jens Rydgren (ed.), *The Oxford Handbook of the Radical Right* (Oxford: Oxford University Press, 2018), 1.

3   Cas Mudde, *The Far Right Today* (Medford, MA: Polity, 2019). The closest Mudde comes to mentioning ecology is a reference in passing to 'so-called "eco-terrorism"' on page 132.

4   Matthew Lockwood, 'Right-Wing Populism and the Climate Change Agenda: Exploring the Linkages', *Environmental Politics* 27 (2018): 713.

5   Alyssa Battistoni, 'Within and Against Capitalism', *Jacobin*, 15 August 2017.

6   The one volume published on the topic is Bernhard Forchtner (ed.), *The Far Right and the Environment: Politics, Discourse and Communication* (Abingdon: Routledge, 2020). Empirically rich, this collection of case studies has analytical weaknesses (see further below) and lacks a focus on climate.

One of them, Donald Trump, was, of course, the ubiquitous face of anti-climate politics in the second half of the 2010s. He has now lost the White House. The time has come to take stock of the phenomenon he represented and ponder in what guise it might reappear. Was he a four-year-long American nightmare that has finally ended – a freak of the local culture, unlikely to come back to haunt an even warmer world? Can we breathe a sigh of relief that at least we won't have to deal any more with this kind of insanity? Our prognosis is less upbeat. In fact, as we shall show, the phenomenon Trump represented – precisely insofar as fossil fuels and whiteness came together in his character – extends far beyond US borders. Only by subjecting it to comparative study, drawing in countries not known for giving their middle fingers to climate, can we catch sight of it as something more than a Republican eccentricity or even a personal Trumpian idiosyncrasy – namely, as a systemic tendency, emerging at a particular moment in the history of the capitalist mode of production. If indeed it has that status, it needs to be known and fought as such.

Part I presents the main findings. It offers a history of the conjuncture of climate change and nationalist politics. How has the far right dealt with global heating and its drivers so far? We look back on developments in recent decades, with an emphasis on the second decade of the twenty-first century. We trace the evolution of a set of ideas about climate and nation, energy and race, from the earliest organisation of denial to the stances of the party family that has shaken up European politics. Those ideas are neither set in stone nor uniform across the parties. To the contrary, the far right is in flux and adopts varying positions in different countries and will continue to adapt to shifting circumstances. But the repertoire of far-right climate politics is not infinitely malleable. It will not expand much beyond some basic standpoints worked out in passionate devotion to the far right's universal object of love: its own ethnically pure nation. In the settings we examine, this means, to all intents and purposes, the white nation. So what does it mean to defend the white nation in a climate emergency? Some on the far right have inverted the logic of the Finns' cartoon and decided that the emergency is in fact real and the white nation the best shield against it. While this position might seem antithetical to climate denial, there is, we shall argue, less to separate them than first meets the eye.

Part II tries to make some sense of all this. How is it possible for the anti-climate politics of the far right to come to prominence at this late hour? What would it mean to live in a world both hotter and further to the right than now? Here we engage in what might be called political climate modelling: taking the trends of the recent past and following them into the future, extrapolating and speculating on possible scenarios.[7] How should the phenomenon be designated and defined? In a pathbreaking essay, Cara Daggett has proposed the term 'fossil fascism': we consider its meaning and contrast it with classical fascism and compare the contemporary far right with that of interwar Europe.[8] Part II thus sketches a deeper history of the nexus. It traces lineages of resurfacing ideas and contends that white skin and black fuel have been coupled for a long time – indeed, machines powered by fossil fuels were infused with racism from the very first moment of their global deployment. The European incubator for skin and fuel was an empire. Any exploration of it must begin with Frantz Fanon and continue with others who saw the onward march of metropolitan technology from the receiving end. It is our contention that one cannot understand recent developments, or their possible continuation and aggravation, without such a longer view.

But colonial history is only one source of the problem we are dealing with. We shall have to attack it from multiple angles. From what sources does the far right pump its fantasies about defending the nation and fighting conspiracies and arming itself with superior energy for the tasks? What is the more profound significance – cultural, psychological – of the phenomenon in this very late capitalism? Not the least important, what is its relation to the regions of bourgeois civilisation that would abjure any association with the far right? Those who think that the mainstream way of dealing with the climate stands in absolute, irreconcilable opposition to that of the far

---

7 An exercise pioneered by Joel Wainwright and Geoff Mann, *Climate Leviathan: A Political Theory of Our Planetary Future* (London: Verso, 2018).

8 Cara Daggett, 'Petro-masculinity: Fossil Fuels and Authoritarian Desire', *Millennium: Journal of International Studies* (2018), online first, 3.

right will need to think again. The latter is not some *deus ex machina* that descended on earth just as the problem was about to be sorted out. We shall see how the one bleeds into the other. Or, to paraphrase Max Horkheimer: she who does not wish to speak of fossil capital and the liberal ideology that has sustained it should also be silent about fossil fascism and its prefigurations. One of our central arguments is that the anti-climate politics of the far right has risen in conjunction with some pressing material interests of the dominant classes. The tactics for protecting those interests have varied: they exist on a continuum, where the main thrust easily glides into the extreme.

What we will not do, however, is stack up a million footnotes to substantiate the reality of climate breakdown. We will presume knowledge of its ABCs. A superabundance of scientific evidence is always one click away; often it is enough merely to open the window. Whether it is a firestorm colouring the sky a hazy yellow or the snow that never fell this winter, the fingerprints of the crisis cover more and more of everyday life (which evidently does not mean it cannot be denied: a paradox we must probe). Sometimes, people gasp at the sights and say, 'Oh, so this is what climate change looks like', but they tend to forget that it is a cumulative process, the effects progressively magnified by the total amount of greenhouse gases emitted: and more are emitted each year, each week, every minute. A taste of global heating is only ever a foretaste. Ten years of business-as-usual from now, this year's bushfires or mild winters might be remembered as rather pleasant by comparison: it's as though we're caught in an escalator, heading up, up towards temperatures of a 'severity that makes ordinary human society impossible'.[9]

But the metaphor of an escalator is too deterministic. It is not the case that, once humans entered this process, destination and speed were fixed. Imagine, instead, a curious kind of elevator: a large company of people is invited into it by a clique of men who promise mind-blowing views from above. The elevator can only rise one

---

9 Henry Shue, 'Mitigation Gambles: Uncertainty, Urgency and the Last Gamble Possible', *Philosophical Transactions of the Royal Society A: Mathematical, Physical and Engineering Sciences* 376 (2018): 4.

floor at a time. At every floor, before advancing to the next one, the travellers must decide whether to push the 'up' button. They can choose instead to stop, start the descent and get out. Now imagine that, after some time, a fire alarm goes off. For every floor, it rings higher and blinks brighter; soon smoke starts seeping in. Arguments break out about whether to continue. Clearly, this metaphor is a little contrived and partial – every metaphor of the climate crisis fails to do justice to its object – but it captures one aspect the escalator misses: each moment of sustained business-as-usual *is the outcome of conflict.*[10]

This book studies the behaviour of some people inside the elevator: the first part after the alarm has gone off, the second mostly before that moment. In the first, we present a contemporary history, but it is one that will reverberate for a long time to come. People around the world are already suffering the consequences of decisions made in the 1990s and early 2000s; in the next decade, they will start receiving the fallout from the 2010s. It takes about a decade for most of the warming from one pulse of CO_2 to materialise; then the heat stays on practically forever, so that people in, say, 2030 will live in the heat from what was emitted up to 2020.[11] Documentary records of the previous decade might then be informative. There are people who should be held to account.

Before the alarm, there was, of course, no innocent harmony, no evenly shared rewards from the panorama – to the contrary, those who insisted most forcefully on pushing the 'up' button employed a great deal of brutality.[12] But only under the sirens does the full significance of their acts become legible. This also applies to the forebears of the contemporary far right, namely the classical fascists, who shared with

---

10   Eventually, of course, the elevator risks becoming more like an escalator, when further heating is unavoidable and positive feedback mechanisms take over.

11   Katherine L. Ricke and Ken Caldeira, 'Maximum Warming Occurs about One Decade after a Carbon Dioxide Emission', *Environmental Research Letters* 9, no. 12 (2014): 1–8; Kirsten Zickfeld and Tyler Herrington, 'The Time Lag between a Carbon Dioxide Emission and Maximum Warming Increases with the Size of the Emission', *Environmental Research Letter* 10 (2015): 1–3.

12   And, to stretch the metaphor further, most people were not invited but coerced into the elevator and denied gas masks and other protective gear.

it the defining pursuit of the pure white nation.[13] How did they deal with fossil fuels and their technologies? While fascism has been inspected from most points of view, its love affair with those particular productive-destructive forces have largely escaped attention as such: now is the time to revisit it. The prehistory of fossil fascism holds a key to the positions of the far right today, and it is part of what brought us into this mess in the first place. But fascism also has a history of love for nature, which is staging its own comeback. Where could it lead us?

While this is a big book that tries to catch up with a sorely understudied topic, we make no pretence of an exhaustive or conclusive inquiry: this is a first essay. Many of our interpretations are tentative, in the nature of hypotheses. We deal with the Old World and two of its offshoots in the New, leaving out some – notably Canada and Australia – that have their own distinct articulations of energy and race. Nor do we deal with the far right in any country in the global South besides Brazil; India is a major omission. We offer no richly textured ethnography of the lifeworlds of the people who might sympathise with the parties and their climate and energy policies. One member of The Zetkin Collective, Irma Allen, is doing just this work among Polish coal miners; another, Ståle Holgersen, is planning the same among oil workers in Norway. We concentrate on climate, paying scant attention to other aspects of the ecological crisis – the sixth mass extinction, the collapsing insect populations, the plastic and air pollution, the land depletion . . . Certain factors of class and gender deserve more in-depth treatment than we give them here. We home in on race and racism, the far right and fascism in the past and present, without capturing more than a fraction of their determinants; we cannot provide a total overview of the variables that have conspired to breathe new life into them, nor of the political content of the parties we study.[14]

---

13 Cf. Jens Rydgren, 'The Sociology of the Radical Right', *Annual Review of Sociology* 33 (2007): 246; Richard Saull, Alexander Anievas, Neil Davidson and Adam Fabry, 'The *Longue Durée* of the Far-Right: An Introduction', in Richard Saull, Alexander Anievas, Neil Davidson and Adam Fabry (eds), *The Longue Durée of the Far-Right: An International Historical Sociology* (London: Routledge, 2015), 5–7.

14 Thus, we do not, for instance, deal with the differences in the economic, social and family policies of European far-right parties.

Our object is both overdetermined and contradictory, and we reflect on some of the many nuances.

We are, furthermore, aiming at a moving target. The climate system and the political systems of the world are drifting towards pronounced volatility – in the former case, without precedent in the annals of human history – and many of the faces and names in our story might soon sink into oblivion. Trump is a man of the past. During 2019, when most of this book was written, three far-right parties were ejected from government: the Austrian FPÖ, the Lega Nord of Italy and the Danish People's Party.[15] If global heating goes one way, the rate of CO_2 ever rising – two or three more parts per million each year – the advances of the far right are distinctly less unilinear. They have proven rather easier to reverse with resistance. By the late 2010s, one formation of the European far right that only years prior appeared among the most formidable and fearsome had come to an inglorious end: the Golden Dawn of Greece (another case here left out). But as such, the far right seems here to stay for a while. Its forces may look different tomorrow, but they are less likely to vanish overnight than to evolve and gather force and put their imprint on any transition away from fossil fuels, if such a thing ever transpires. We are trapped inside the elevator with them, and we need to have an idea of where they come from, what they do, how they think, what steps they might take next.

Because those seeking to go higher and burn more fossil fuels have never ceased to be victorious, we are now in a situation where full breakdown can be averted only with the most herculean redirection and restructuring of the world economy. Every barrel of extracted oil, every container of coal, every cubic foot of gas: every tonne of carbon released into the air speeds up the rush. But, conversely, every piece of fossil fuel left untouched limits the hazards. Every emission avoided relieves suffering. Every step to decarbonise our economies – fully and immediately freeing them of fossil fuels and starting the hard work of undoing the damage – counts. These are the parameters within which a difference is made, now and in the near future. Sustained business-as-usual is, more

---

15 The latter lost its status as main parliamentary prop for the government, of which it was not formally a part.

than ever, the outcome of conflict: during 2019, the world saw the greatest popular mobilisations around the climate issue so far in history. This book deals with the opposite side, which no climate movement can wish away. Progress has a tendency to provoke furious reaction, and this movement has not been an exception. Nor will anti-fascists and anti-racists be able to ignore this context. Rather, their old struggle against the far right is taking on a novel aspect. It is increasingly difficult to tell it apart from the struggle to preserve the conditions in which human and other life can thrive on this planet.

After Clara Zetkin had written the first essay to ever engage in depth with fascism from within the workers' movement, months after Mussolini's march on Rome, in early 1923, she was tasked with drafting a resolution on the topic for the Comintern, still not under the full control of Stalin. She called for 'a special structure to lead the struggle against fascism, made up of workers' parties and organizations of every viewpoint' and listed six outstanding tasks. Number one: 'Collecting facts on the fascist movement in every country.' (Number five on a subsequent list: 'Refuse to ship coal to Italy.')[16] It is in this spirit we submit the following study: our contribution to the resistance, the product of a collective project that we hope will be of some use in other collective projects.

If nothing else, the anti-climate politics of the far right should shatter any remaining illusion that fossil fuels can be relinquished through some kind of smooth, reasoned transition with everyone on board. Climate is reputed to have a unique ability to inspire fraternisation and 'post-political' consensus: because it concerns humanity as a whole, people of all loyalties and persuasions should be able to agree on a safety plan.[17] But a transition will happen through intense polarisation and confrontation, or it will not happen at all. Things might well get ugly. Indeed, they already are.

*

16   Clara Zetkin, *Fighting Fascism: How to Struggle and How to Win* (Chicago: Haymarket, 2017), 72, 74.
17   'Post-political': see Erik Swyngedow, 'Apocalypse Forever? Post-Political Populism and the Spectre of Climate Change', *Theory, Culture and Society* 27 (2010): 213–32; the symposium on Swyngedow's argument in *Capitalism Nature Socialism* 24 (2013): 6–48; and postscript below.

The manuscript for this book was originally completed in late January 2020. Weeks later, contemporary politics underwent the caesura known as Covid-19. As so much else, publication was put on hold, while the jagged upwards curve of the far right – if not of global heating – turned downwards in some places, went into prolonged fall or quickly rebounded, in a world now suffering from two emergencies (if not more). We have left the manuscript all but unchanged. Instead, we have added a postscript that surveys the scene of 2020, the year when the overheated world became officially sick, another year of continued mutations on the far right.

The Zetkin Collective
November 2020

PART I

1
The Fortunes of Denial

Climate science has produced three fundamental insights about the state of the world. There is a secular trend for average temperatures to rise. It can only be attributed to human emissions of greenhouse gases, predominantly carbon dioxide from the burning of fossil fuels. The impacts on ecosystems and societies are negative, indeed potentially catastrophic if emissions continue unabated. Taken together, these three observations mark out a need for mitigation, or the closing of the sources of those gases. This, in turn, implies that the capitalist world economy has to shed its energetic foundation – fossil fuels – for humans and other species to be able to live well: but that is unlikely to happen naturally, as when a snake moults its skin because the old has had its day. It would rather require some disruptive intervention. Ever since climate science solidified in a series of milestones that happened to coincide with the collapse of actually existing socialism around 1989 – James Hansen's testimony to the US Senate in 1988, the founding of the IPCC in the same year, the first assessment report from that panel in 1990, the second in 1995 – these basics have, consequently, been disputed. If there is no warming trend, or if it cannot be attributed to humans, or if the impacts are harmless or even beneficial, then there is no need for any action.[1]

---

1 This basic taxonomy of climate scepticism was first established by Stefan Rahmstorf, 'The Climate Sceptics', Munich Re Group, pik-potsdam.de, 2004; see further

Such is the inverted ABC of what is often referred to as 'climate scepticism', but as scholars who study it have pointed out, 'scepticism' is an undeservedly generous term. It suggests that proponents are animated by the rational virtue of being sceptical about general assertions, engaged in noble scientific methodology, inclined to ask critical questions and open to unexpected answers. But that is not how they comport themselves. No matter how large the mountains of evidence dumped on them, they stand firm in their beliefs – the antithesis of a rational and scientific disposition. They are no more sceptical about climate change than an incorrigible Holocaust-denier is sceptical about the facticity of the Holocaust. This is so because, at the bottom of their hearts, they are motivated by 'an abiding *faith* in industrial science and technology, free enterprise, and those great institutions' of modern capitalism endangered by the pursuit of climate action. Hence, Peter J. Jacques argues, 'denial' is the more appropriate term.[2] In the case of the far right, however, another object of faith is superimposed on these institutions. It consists of *a racially defined nation powered by fossil fuels.*

All European far-right parties of political significance in the early twenty-first century expressed climate denial. Some moved away from it, as we shall see, but it remained the default position. A prototypical case was the Alternative for Germany, or Alternative für Deutschland (AfD), whose entry into the Bundestag in 2017

Wouter Poortinga, Alexa Spence, Lorraine Whitmarsh et al., 'Uncertain Climate: An Investigation into Public Scepticism about Anthropogenic Climate Change', *Global Environmental Change* 21 (2011): 1015–24. Cf. the slightly different model of 'stages of denial' in Michael E. Mann and Tom Toles, *The Madhouse Effect: How Climate Change Denial Is Threatening Our Planet, Destroying Our Politics, and Driving Us Crazy* (New York: Columbia University Press, 2016), 53–67.

2   Peter J. Jacques, 'A General Theory of Climate Denial', *Global Environmental Politics* 12 (2012): 9, 11. Emphasis in original. Cf. James Lawrence Powell, *The Inquisition of Climate Science* (New York: Columbia University Press, 2011), 4; Spencer Weart, 'Global Warming: How Skepticism Became Denial', *Bulletin of the Atomic Scientists* 67 (2011): 46–7; and on why Holocaust 'revisionism' should be called 'denial', Deborah Lipstadt, *Denying the Holocaust: The Growing Assault on Truth and Memory* (London: Penguin, 2016 [1993]), 23–4. On the relation between climate and Holocaust denial, see further below. On climate denial as defence of capitalist modernity, see also Aaron McCright and Riley E. Dunlap, 'Anti-reflexivity: The American Conservative Movement's Success in Undermining Climate Science and Policy', *Theory, Culture & Society* 27 (2010): 100–33.

– the first time a far-right party achieved that feat in half a century – was greeted by pundits as a sensation: this was 'a watershed moment', 'a political earthquake', 'a seismic shock'.[3] The AfD became the third largest party. After the conservative CDU and the social-democratic SPD formed a new coalition government, it took up the role as the leading opposition. The *Grundsatzprogramm* of the AfD adopted an unequivocal line on climate change: it has 'been going on for as long as the earth has existed'. Our planet has always alternated between cold and warm temperatures, and the current warmth is no more unnatural than that of the Middle Ages or the Roman Empire. 'The IPCC tries to prove that human CO_2 emissions cause global warming with severe consequences for mankind', but these attempts are based on 'hypothetical models' and 'not backed by quantitative data or measured observations.' The AfD could establish that no rise in temperatures has occurred since the end of the 1990s, despite increasing emissions. The additional CO_2 should be greeted as a godsend: 'Carbon dioxide is not a pollutant but an indispensable component of all life', a higher atmospheric concentration of which brings about more plentiful crops and abundant food.[4] Everything held true by the scientific consensus is, in this view, false. Denial of trend and attribution as well as impact – or, flat-out climate denial – was inscribed in the foundational document of the first truly successful far-right party in Germany since the Third Reich.[5]

After its founding in 2013 by a group of economists critical of the EU, the AfD turned ever further to the right, with denial becoming a firm party line. The main leaders reiterated it frequently. 'I do not believe that humans can contribute much to this', the party's federal spokesman Alexander Gauland reaffirmed in August 2018, when

---

3 Kate Connolly, 'Germany Faces First Far-Right Party in Parliament Since 1960s', *Guardian*, 22 September 2017; Jasmin Siri, 'The Alternative for Germany after the 2017 Election', *German Politics* 27 (2018): 141; Cas Mudde, 'What the Stunning Success of AfD Means for Germany and Europe', *Guardian*, 24 September 2017.

4 AfD, *Programm für Deutschland: Das Grundsatzprogramm der Alternative für Deutschland*, 2016, 156.

5 On this status of the party and general developments during its first five years, see Lars Rensmann, 'Radical Right-Wing Populists in Parliament: Examining the Alternative for Germany in European Context', *German Politics and Society* 36 (2018): 41–73.

an interviewer pressed him on recent heat records.[6] A stout, blond former policeman, environmental spokesman Karsten Hilse declared, like a general going into battle, that 'here and now, the AfD is fighting the false doctrine of man-made climate change', backing up his cause with the finding that '0.3 percent of [scientific] studies indicate that global warming is man-made'.[7] In a party motion to the Bundestag from June 2018, the AfD made a particularly ambitious effort to overhaul the collected wisdom of climate science: CO_2 is an all-too thinly disbursed 'gas of life'; the more we emit of it, the better; elevated concentrations fertilise plants, shrink deserts and green the planet, while no trace of anthropogenic warming can be found in the data. 'The entire climate problematic comes down to a non-problem.'[8]

In the usually staid Bundestag, the representatives of the AfD introduced a code of conduct unknown since the dying days of the Weimar Republic. When opponents spoke, they rained down insults – 'nonsense!', 'foolish!', 'impossible!' – or broke out in theatrical laughter; when a party member took the podium, they gave boisterous applause. 'We were elected by people who want us to tell the truth', one representative explained.[9] *Mut zur Wahrheit*, or 'courage to tell the truth,' was a party motto. On the parliamentary stage, it inspired histrionic displays and verbal aggression against other parties that presumably cheated citizens of their patrimony.[10] Thus,

---

6 Dominik Rzepka, 'ZDF-Sommerinterview: Gauland hält Klimapolitik für sinnlos', *ZDF*, zdf.de, 12 August 2018. Cf. for example *Deutschlandfunk Nova*, '"Ich würde nie . . ."' mit Alice Weidel', deutschlandfunknova.de, 20 September 2017; *Schwäbische Zeitung*, 'Weidel: Klimawandel nicht menschengemacht', schwaebische. de, 13 September 2017.

7 Reuters, 'AfD nennt Klimawandel "Irrlehre" und will Energiewende beenden', de.reuters.com, 23 March 2018; Fabian Schmidt, Marc Röhlig and Sebastian Maas, 'Mit diesen absurden "Bewiesen" leugnen zwei AfD-Redner den Klimawandel', *Bento*, 23 March 2018.

8 Deutscher Bundestag, 'Aufgabe der Energie- und Klimaschutz-Zwischenziele 2030 des Energiekonzeptes 2010 – Für eine faktenbasierte Klima- und Energiepolitik', Antrag der Abgeordneten Karsten Hilse, Dr. Heiko Wildberg, Dr. Rainer Kraft, Udo Theodor Hemmelgarn, Marc Bernhard und der Fraktion der AfD, Deutscher Bundestag Drucksache 19/2998, 27 June 2018.

9 Griff Witte and Luisa Beck, 'The Far Right Is Back in the Reichstag, Bringing with It Taunts and Insults', *New York Times*, 8 June 2018.

10 Rensmann, 'Radical', 45, 59.

in a Bundestag debate in January 2018, climate spokesman Rainer Kraft attacked all other parties for practising 'eco-populist voodoo' and, more particularly, the Greens for seeking to establish 'an eco-socialist planned economy' under the disguise of climate protection.[11] In the truth-telling crusade of the AfD, such denial became a prime rallying cry. It was brought forward as the front banner in 2019, when the climate crisis took centre stage in German politics. By now, the country had the most dynamic climate movement in Europe, if not the world. The school strikes and youth-led demonstrations known as Fridays for Future attracted greater crowds than anywhere else, a regular show of anger and anxiety in squares from Munich to Leipzig that instilled the sense of emergency in public opinion. Masses of activists from Ende Gelände, roughly 'here and no further', continued to break into coal mines and their adjoining infrastructure and shut them down. Actions under the brand of Extinction Rebellion disrupted the normally tranquil district of ministerial buildings in Berlin, and, by late summer 2019, Germans ranked climate as their single greatest concern; immigration – recently number one – had slipped down.[12] How did the AfD respond?

It launched agitated attacks on all three branches of the movement. The AfD referred to the school strikes as the utterly illegitimate 'No Education Friday'.[13] When the second-largest trade union federation, Verdi, threw the weight of its 2 million members behind them, the AfD reviled it as a gravedigger of German industry and 'traitor to the workers'.[14] Ende Gelände was a band of 'eco-terrorists' guilty of trampling down carrots on their way to coal mines, while Extinction Rebellion was classified a 'religious esoteric sect'; all in all, the AfD saw reason to fear the imminent end of German capitalism. 'Will there soon be anarchism, eco-socialism – will there soon be the third

---

11   Deutscher Bundestag, Plenarprotokoll 19/7, 18 January 2018, 577–8.

12   *Die Zeit*, 'Klimawandel ist für die meisten Deutschen das größte Problem', zeit. de, 19 August 2019.

13   Karsten Hilse in Deutscher Bundestag, Plenarprotokoll 19/108, 28 June 2019, 13447.

14   *AfD Kompakt*, 'Frank Bsirskes klimastreikende Gewerkschafter verraten Arbeitnehmer', afdkompakt.de, 6 August 2019.

socialist dictatorship on German soil? The AfD is working against this development with all its might', explained Karsten Hilse in another blazing Bundestag speech.[15]

In parallel with the climate mobilisations, the party doubled down on denial. Swearing itself to private property and the liberty to 'take profitable opportunities', it read the issue through a lens of an Argus-eyed anti-socialism.[16] Kraft charged the other parties with 'threatening the end of the world and stirring up mass hysteria, so that people will accept that more and more of their property and their freedom are stolen', while Hilse had another few cracks at the entirety of the science: thirty years of research had failed to come up with 'one single piece of evidence' for CO_2 affecting climate.[17] Any climatic fluctuations – a priori of a natural kind – should be dealt with through adaptation. And even if they were anthropogenic, the AfD liked to argue, Germany only accounted for 2 per cent of current emissions, so that its decarbonisation would reduce global warming by a farcical 0.000284 or 0.000653 degrees Celsius (Hilse's calculations). There would be a green version of the old East Germany with nothing to show for it.

Other far-right parties in Europe toed a similar line. Although less passionate about the issue than the AfD, the Dutch Party for Freedom, or Partij voor de Vrijheid (PVV), led by the flamboyant Geert Wilders, consistently expressed its contempt for the science. In an interview just before the 2010 general election, when the PVV reached its highest peak thus far with 15 per cent of the votes, environmental spokesperson Richard de Mos made clear that it would 'not go along with the climate hype'. Demanding a national investigation on 'whether CO_2 is really a problem', rhetorically asking for evidence that sea levels are on the rise, he branded 'the climate story a scientifically outdated money-wasting hobby of the elite'.[18] In 2017,

---

15 Deutscher Bundestag, Plenarprotokoll 19/118, 17 October 2019, 14423–4. The first socialist dictatorship, in this counting, would have been the Third Reich.

16 Ralf Havertz, 'Right-Wing Populism and Neoliberalism in Germany: The AfD's Embrace of Ordoliberalism', *New Political Economy* 24 (2019): 385–403.

17 Deutscher Bundestag, Plenarprotokoll 19/118, 17 October 2019, 14432, 14424; Plenarprotokoll 19/108, 28 June 2019, 13448.

18 Hajo Smit, 'Richard de Mos (PVV): "Met ons ligt er een kans"', climategate.nl, 4 June 2010.

one PVV senator went on another rant of denial, rehashing the claim that temperatures and sea levels have always fluctuated and exhorting his fellow senators to 'stop the climate hoax'.[19] The issue belonged in the PVV trash bin, one item among many to be thrown aside. With outstanding monomania, the party attacked Islam and cared for little else.

After 2017, however, the PVV was overtaken by a fresh force on the Dutch far right: the Forum for Democracy, or Forum voor Democratie (FvD). Positioning itself as a more urbane, culturally conservative alternative, it integrated Muslim-bashing into a broader diagnosis of the ills of the Western world. If Wilders's hair resembled a toupee and his manners those of an urchin, Thierry Baudet of the FvD dressed like a smart aristocrat, played the piano and quoted gloomy philosophers of the right. He sought to restore pride in white culture. He wanted Western civilisation to break free from 'Cultural Marxism' – a notion we shall return to in the second part of this book – to stop flagellating itself for supposed sins and remember that it used to 'spread with confidence to every corner of the world'. The FvD had been 'called to the front' to reinvigorate the West – or as Baudet also liked to put it, to save 'our boreal world'.[20] An archaic term for the north and the northern wind, 'boreal' alludes to the idea of Europeans as a people of Aryan and polar stock who should have their continent to themselves: a codeword for the white race, supposedly in existential danger.

In this view of the world, climate was not an issue to rush past. It was at the very centre of attention. Baudet, at first, merely repeated Wilders, calling climate change a 'hoax', questioning attribution and proclaiming that 'more CO_2 has a great positive effect on plant growth.'[21] Somewhat elastic on the issue, the FvD had on its website a

---

19 Hans Labohm, 'Maiden speech PVV-senator Dannij van der Sluijs inzake klimaat – feiten, feiten', climategate.*nl*, 29 June 2017.

20 Quotations from Joost de Vries, 'Meet Thierry Baudet, the Suave New Face of Dutch Rightwing Nationalism', *Guardian*, 3 April 2019; Thijs Kleinpaste, 'The New Dutch Disease Is White Nationalism', *Foreign Policy*, 20 March 2018.

21 Hanneke Keultjes, 'Alle ballen op Baudet in Haags-Amsterdams verkiezingsdebat', *Het Parool*, 9 February 2018; Thierry Baudet's Twitter account, 3 January 2018; RTL, 'Prinsjesdagdebat: Thierry Baudet versus Jesse Klaver', RTL Late Night, 19 September 2018.

section about the opportunities to profit from green technology, but, as the party geared up for provincial elections in 2019, that section was deleted along with every other admission of an ecological problem, the denial upgraded from scattered comments to a theme second only to immigration. Baudet took every chance to hammer away at 'the climate madness'. Emissions cuts were not only unnecessary and expensive, but a cloak for socialist regulation. In the elections in March 2019, the FvD received 15 per cent of the votes, making it the largest party in the Netherlands. Afterwards, Baudet explained the victory – part of a general 'awakening' in the West – with the FvD 'making opposition to climate policies our main electoral theme. The winning ticket is to say bluntly that we don't believe in their stuff anymore.'[22]

Arriving somewhat late to this scene, the Netherlands got its own Climate Intelligence Foundation, or CLINTEL, in 2019, a denialist think tank close to the FvD, funded by men with years in positions at Shell and airline companies on their CVs.[23] Hitting back at the spreading 'hysteria', CLINTEL published a so-called World Climate Declaration under the headline 'There Is No Climate Emergency': among the signatories, several Shell figures, aviation businessmen and Paul Cliteur, parliamentary group leader and theorist of the FvD.[24] Baudet himself seemed to take an ever greater interest in the issue. But PVV and Wilders were again neck and neck with him in 2019, the contest no longer one of abusing Islam only. The Dutch far right had entered the race against climate.

Over in Austria, the Freedom Party, or Freiheitliche Partei Österreichs (FPÖ), caused a minor brouhaha in 2015 when newly appointed environmental spokesperson Susanne Winter called

---

22 Urs Gehriger, 'There's a Proper Reawakening across Europe Going On', *Die Weltwoche*, 28 March 2019. Cf. the analysis of FvD claiming climate denial as its special policy 'file' in *Stuk Rood Vlees*, 'Rechtspopulisme en de media met Léonie de Jonge en Alyt Damstra', stukroodvlees.nl, 29 March 2019.

23 Notably Guus Berkhout (formerly of Shell) and real estate billionaire Niek Sandmann (formerly of the Dutch Antilles Express). See for example Richard Collett-White, 'Climate Science Deniers Planning European Misinformation Campaign, Document Leaked', *DeSmog UK*, 6 September 2019.

24 CLINTEL, 'World Climate Declaration', clintel.org, 27 November 2019.

climate change a 'religion' and 'a web of lies spun by the media that needs to be destroyed'.[25] Subsequently the party focused on attribution. In the long run-up to the 2017 election that made him vice-chancellor of Austria for one and a half years, chairman Heinz-Christian Strache littered his campaign with indictments of the sun: 'Nothing can be done about global warming, due to the solar flares and the heating of the sun', ran a typical line.[26] Similar statements were issued, ad nauseam, by the top echelons of the FPÖ, one of the two parties ruling Austria after the 2017 landslide of the right.[27]

One European country long appeared to be immune to this surge: Spain. In 2019, however, it became just like any other when Vox took it by storm. In the national election in April, it entered parliament as the fifth largest party, only to sail on a wave of popularity to third place in the November election, bamboozling the forecasts. Photogenic leader Santiago Abascal stated that 'climate has changed as long as the earth has existed'. Jumbling denial of trend and attribution, Vox pinned any change on the sun, the moon, the rotation of the earth, volcanoes and naturally occurring atmospheric phenomena but absolutely not on CO_2 emitted by humans. It would, said Abascal, be 'very arrogant' to believe that humans could alter the climate. It would be 'even more arrogant' to think that the alteration could be rectified by 'coercive laws and taxes'. Twisting *cambio climático* to *camelo climático*, Vox spread the Spanish version of the keyword 'hoax', further specified this as 'the biggest scam in history' and, in somewhat inquisitorial fashion, fulminated against the false 'climate religion'. 'New religions are imposed on us, be it the female [i.e. feminist] or the climatic, telling us the new commandments: you

---

25 *Der Standard*, 'FPÖ-Umweltsprecherin Winter nennt Klimawandel "Lügengebäude"', derstandard.at, 9 July 2015.

26 *Heute*, 'Strache: Wir sind nicht schuld an Erderwärmung', heute.at, 3 June 2017. See for example Florian Freistetter, 'Strache und die Sonnenstürme: Wie der FPÖ-Chef mit falschen Behauptungen zum Klimawandel davonkommt', *Vice*, vice.com, 19 September 2017; *Kurier*, 'Strache und der Klimawandel: Alles verhandelbar', kurier.at, 7 September 2017; Philip Pramer, 'Wurde in Grönland Wein angebaut?', *Fakt = Fakt*, 5 June 2017.

27 Michael Bonvalot, 'Warum die FPÖ den Klimawandel in Frage stellt', *Vice*, 9 October 2017.

should not have children, not have a car and not eat meat', complained Abascal.[28]

With that late addition, the status of explicit denial as the default position of the far right was confirmed. Not many years earlier, such denial was regarded as a spent force, and nowhere more so than in Europe. It was widely believed to have beaten a sorry retreat to politically irrelevant margins. Obituaries were written for it. To understand this reversal of fortunes – nothing less than spectacular, as we shall see – we first need to go back to the years when the science came of age.

The Origins of Denial

In the summer of 1988, the United States experienced the worst heat waves and droughts since the Dust Bowl. Ominous images of burning forests, withering fields and sweltering cities filled the American press and elicited nervous suspicion: was this the work of the so-called greenhouse effect? Had the danger of which some scientists warned already arrived? It was amid this tense national atmosphere James Hansen intervened with his testimony to the Senate, in which he forthrightly asserted that 'we can ascribe with a high degree of confidence a cause and effect relationship between the greenhouse effect and observed warming'. The suspicions were sound: 'It is already happening now.' Describing the extreme summer as a taste of things to come, the report in the *New York Times* also noted that the testifying scientists 'said that planning must begin now for a sharp

---

28 Christian Palencia Solórzano, 'VOX contra la comunidad científica: Abascal afirma que el cambio climático es un invento', *Los Replicantes*, losreplicantes.com, 15 March 2019; D. Cruz, 'Santiago Abascal, líder de VOX, niega que el cambio climático exista', *Sport: Fuera de Juego*, 10 April 2019; Daniel Bellaco, 'Jóvenes y científicos de todo el mundo claman para que se detenga el cambio climático, Santiago Abascal lo niega', *Digital Sevilla*, digitalsevilla.com, 15 March 2019; Isidoro Eduardo Robaina, 'VOX y su argumentario contra el cambio climático: "No vamos a malgastar más dinero en esta estafa"', *La Marea*, 5 June 2019; Isidoro Tapia, 'Vox y su "camelo climático"', *El Confidencial*, 20 February 2019; Raquel Nogueira, 'Vistalegre plus ultra, nueva entrega en la cruzada negacionista de Vox', *Ethic*, 8 October 2019.

reduction in the burning of coal, oil and other fossil fuels that release carbon dioxide'.[29] *Planning* for a sharp reduction? The very notion injected panic in fossil capital.

With signs of trouble ahead multiplying, the IPCC was established in 1988, the United Nations began preparations for a concerted response and awareness of the problem spread across what was still referred to as 'the free world'. There was no time to lose. In 1989, several urgent counter-initiatives were launched: Exxon formulated an internal plan for how to drive home 'the uncertainty in scientific conclusions regarding the potential enhanced Greenhouse effect' and ran its first advertorial on the subject.[30] A suite of companies set up the Global Climate Coalition to contest the science. The key conservative think tank known as the George C. Marshall Institute published its first report attacking it. When more than a hundred heads of state gathered in Rio de Janeiro in the summer of 1992 and adopted the United Nations Framework Convention on Climate Change (UNFCCC), with its call to prevent 'dangerous anthropogenic interference with the climate system', further alarm was stoked.[31] Socialism appeared to be passing out of history, but at the very same moment, fossil capital had to gear up for a war to safeguard its freedom.

Rarely has a dominant class so swiftly, purposefully and effectively built up an ideological state apparatus (ISA) in an hour of need. Scholars of the entity refer to it as 'a denial machine', but it also fits

---

29 Philip Shabecoff, 'Global Warming Has Begun, Expert Tells Senate', *New York Times*, 24 June 1988. On this pivotal moment, see further Spencer Weart, *The Discovery of Global Warming*, rev. ed. (Cambridge, MA: Harvard University Press, 2008), 150–1; Naomi Oreskes and Erik M. Conway, *Merchants of Doubt: How a Handful of Scientists Obscured the Truth on Issues from Tobacco Smoke to Global Warming* (London: Bloomsbury, 2010), 183–4; Nathaniel Rich, *Losing Earth: The Decade We Could Have Stopped Climate Change* (London: Picador, 2019), 125–32, 132.

30 Plan quoted in Geoffrey Supran and Naomi Oreskes, 'Assessing ExxonMobil's Climate Communications (1977–2014)', *Environmental Research Letters* 12 (2017): 15.

31 Riley E. Dunlap and Aaron M. McCright, 'Organized Climate Change Denial', in John S. Dryzek, Richard B. Norgaard and David Schlosberg (eds), *The Oxford Handbook of Climate Change and Society* (Oxford: Oxford University Press, 2011), 150; Oreskes and Conway, *Merchants*, 186; Peter J. Jacques, Riley E. Dunlap and Mark Freeman, 'The Organisation of Denial: Conservative Think Tanks and Environmental Scepticism', *Environmental Politics* 17 (2008): 351, 362.

Louis Althusser's criteria of an ISA: 'a system of defined institutions, organizations, and the corresponding practices', which, through their day-to-day activities, uphold some elements of the dominant ideology.[32] A classic example of an ISA is the school. The teacher turns to his pupils and, swinging his pointer, asks them to provide the right answers. In a church, the priest invites the congregation to mass and offers everyone the body given up for you; in a television show, the host looks the audience in the eye and raises it to the level of participant; in a party, the leaders spur their members to canvass for the upcoming elections – in an ISA, the subjects are hailed or *interpellated* and, responding to the call, partake in some material practise by which the ideology is dispensed.[33] Now interpellations happen all over the place, whenever someone addresses someone else and seeks to purvey an idea or prompt a course of action. If a man shouts to his neighbours below that they too should hang the national flag from their balconies, he interpellates them, on his own and in the moment; in ISAs, such acts are organised over time. Their messages can compete and comingle in a cacophony of communication.

But why call these entities ideological *state* apparatuses? A federation of sports clubs or the museums of a town are not necessarily part of the state, as normally defined, but they are clearly capable of organising interpellations. Ideological apparatuses seem to be plural and fluid, located at the interface between civil society and state, more often than not existing outside government control. Some of them are even built to question elements of the dominant ideology – an LGBT organisation in Poland, a movement for immigrants' rights in Denmark. These deserve the label of '*counter*-apparatuses'. But for ideological apparatuses that reproduce the dominant ideology, we can retain Althusser's original term, the S for 'state' not a literal suggestion that a king or prime minister rules them like an embassy,

---

32 Riley E. Dunlap and Aaron McCright, 'Climate Change Denial: Sources, Actors and Strategies', in Constance Lever-Tracy (ed.), *Routledge Handbook of Climate Change and Society* (London: Routledge, 2010), 246; Dunlap and McCright, 'Organized', 144; Louis Althusser, *On the Reproduction of Capitalism: Ideology and Ideological State Apparatuses* (London: Verso, 2014 [1995]), 77.

33 See Louis Althusser, *Lenin and Philosophy and Other Essays* (New York: Monthly Review Press, 2001); Althusser, *On the Reproduction*.

but a sign precisely of that reproducing and cementing function.[34] On this account, the denial machine did indeed emerge as an ISA. It was formed to secure one element of the dominant ideology against the peril of climate science. The doctrine at stake – the credo and communion of fossil capital – can most simply be summarised as *fossil fuels are good for people*.

The basis of this doctrine was a particular material mode of accumulating capital, in ascendancy since the early nineteenth century: the generation of profit *through* extraction and combustion of fossil fuels. A fire that never goes out, capital here expands by taking coal and oil and gas out of the ground and burning them. When profits have been made, they are reinvested in the same cycle on a larger scale, so that ever-greater clouds of CO_2 are released in the process. This is what we refer to as 'fossil capital'.[35] It ties various brands of capitalists together in a dependence on fossil energy, the material substratum for any number of commodities: a car manufacturer needs steel for its factory and gasoline for the vehicles on the road. A steel producer uses coal to process iron ore. A software company runs on electricity from the nearest gas-fired power plant, and so on; throughout the capitalist mode of production, fossil fuels are consumed as an input. But for that to happen, there must also be someone who produces those same fuels as an output. This, of course, is the specialty of the coal and oil and gas corporations, the *raison d'être* of the capitalists who invest in mines and rigs and pipelines to pull up the stock of energy from its reservoirs. Karl Marx observed that for capital accumulation in general to commence, capital has to be concentrated on the one hand and workers possessing no other commodity than their labour-power amassed on the other; he termed

---

34 This draws on Nicos Poulantzas, *State, Power, Socialism* (London: New Left Books, 1978), 28–9; Göran Therborn, *The Ideology of Power and the Power of Ideology* (London: Verso, 1980), 78–9, 84–6; Frieder Otto Wolf, 'The Problem of Reproduction: Probing the Lacunae of Althusser's Theoretical Investigations of Ideology and Ideological State Apparatuses', in Katja Diefenbach, Sara R. Farris, Gal Kirn and Peter D. Thomas (eds), *Encountering Althusser: Politics and Materialism in Contemporary Radical Thought* (London: Bloomsbury, 2013), 252; Jan Rehmann, *Theories of Ideology: The Powers of Alienation and Subjection* (Chicago: Haymarket, 2014), 175.

35 This is based on Andreas Malm, *Fossil Capital: The Rise of Steam Power and the Roots of Global Warming* (London: Verso, 2016).

this process 'primitive accumulation', and so we can, analogously, speak of a *primitive accumulation of fossil capital*.[36]

An unfortunate English rendering of the German *ursprünglich*, 'primitive' has the connotation of something archaic and long ago superseded. The process should rather be understood as *primary*, a logical antecedent without which the whole thing would die down. If no one digs up the coal, the steel producer will have no coke for the furnaces, the car manufacturer no steel for the chassis. Only if the stock of energy is continuously hauled up and offered as discrete commodities can other capitalists purchase it and set it on fire, as part of their cycle of accumulation, ever intertwined with the cycle of profiting directly from the sale of fossil fuels. We can thus distinguish between *fossil capital in general* and *primitive accumulation of fossil capital* as two moments of *fossil capital as a totality*, much as we can tell the flames from the billets in a fire. The first term refers to capital for which fossil fuels are a necessary auxiliary in the production of other commodities, the second to the department known in the vernacular as 'the fossil fuel industry', the third to the two in their unity. When we use 'fossil capital' with no qualifier, it is the latter – the fire as a whole – we have in mind.

Now, from this base grows a political structure of a determinate character. The capitalists who preside over the primitive accumulation of fossil capital constitute a *class fraction*.[37] Given their role in the total metabolism and process of production, they make up a subcategory of the capitalist class, a bearer or agent of the special task of supplying

---

36   For elaborations, see ibid., 291, 320–6, 355–61; Andreas Malm, *The Progress of this Storm* (London: Verso, 2018), 102–5. It could be objected that this would be tantamount to the undue inflation of Marx's concept to cover the primitive accumulation of bauxite and oranges and any other raw material or input. But fossil fuels are the universal substratum for the production of surplus-value – not a material for this or that specific product, as bauxite for aluminium or oranges for juice, but a type of energy utilised across the spectrum of commodity production. It is this special status of fossil fuels in the total metabolism of capital that comes into view in the climate crisis. Conversely, we could say that the problem of ozone depletion was relatively easily managed because there did not exist any primitive accumulation of chlorofluorocarbon capital with which the rest of capital lived in symbiosis, and hence no capitalist class fraction with the capacity to sabotage the Montreal Protocol enacted in 1989.

37   The following draws on Nicos Poulantzas, *Political Power and Social Classes* (London: New Left Books, 1973), 79–81, 84–5, 233, 239, 299.

fossil fuels to the market; they glow with the drive to maximise profit from the selling of these and no other commodities. Delivering materials to the fire is what they do. Fossil capital in general, on the other hand, is no class fraction, because it is precisely the *generality* of capital, comprising auto and steel and computer companies and any other entities in the habit of expanding value by – among other conversions – turning fossil fuels into CO_2. It is a broad, not to say universal category, too amorphous and open-ended to constitute a fraction *sensu stricto*. Marx's 'primitive accumulation' was not executed by a particular class fraction – any merchant, landowner or slave trader could engage in it – but, in our case, it is the permanent mission of a subset of the capitalist class that we can simply refer to as *primitive fossil capital*. Located at the deepest material base, this fraction is also capable of operating at the highest political levels. It has a venerable history not only of fulfilling its economic task, but also of acting as a *political force*, using its narrow composition and centralised operations to bend governments to its will, or just whisper in their ears.

Under the threat of climate mitigation, the stakes are of a different order for primitive fossil capital. It faces an *existential crisis*, because the prevention of dangerous anthropogenic interference with the climate system ultimately requires that it ceases to exist. The lion's share of coal and oil and gas still in the ground must be left there for the duration, which means that this particular class fraction cannot continue to reproduce itself by extracting *more* of them to sell – but asking it to stop doing so is like asking a human being to stop breathing. There is no way around this contradiction. Primitive fossil capital has to be liquidated wholesale. For the rest of capital, however, climate mitigation rather represents a *structural crisis*. It would have to cease being fossil and might reinvent itself as *non-fossil* capital. A car manufacturer can potentially source its steel from a plant that reduces iron ore with something else than coke (such as hydrogen gas). A software company will be just as contented if the electricity comes from wind turbines. Since the transition would have to affect actually existing capitalism as the greatest totality of all, it might very well be painful, require large-scale destruction of fixed capital and induce serious losses for some. But capital as such may survive it. We cannot know this for a certainty, since a transition of this kind has never happened before – particularly not under such an ultra-tight schedule – but it is not a logical impossibility, not an axiomatic end as it is for primitive fossil capital. When the threat of climate mitigation first appeared in the late 1980s and early 1990s, the latter found itself questioned to the core. It then spared none of its capacity to act as a class fraction in the realm of politics to stave off an existential crisis and *thereby also protected fossil capital in general from a structural one*. This division of labour has remained operational into the time of this writing, with some peculiar political effects.

The first thing primitive fossil capital did was to set up the denial machine – or, a synonym, *the denialist ISA*. A plethora of think tanks sprung up to fight back against climate science. They employed professional denialists, hosted anti-IPCC conferences, organised symposiums for policymakers, testified in Congress, appeared on television and in radio debates, flooded media with advertisements and produced 'an endless flow of printed material' disseminating their beliefs.[38] From the start, the corporation then known as Exxon made critical contributions to the apparatus, through its own efforts as well as via uncountable think tanks, front groups, legislators, columnists and other generously funded proxies.[39] Exxon was one of the sponsors of the Global Climate Coalition, alongside fellow oil companies Shell, BP, Amoco and Texaco. They were joined by car manufacturers GM, Ford and Chrysler, chemical giant DuPont and business umbrellas such as the American Petroleum Institute, the US Chamber of Commerce, the National Association of Manufacturers and the American Highway Users Alliance, to name only some. A broad church for Anglo-American fossil capital, the Coalition is today largely forgotten, but in the early 1990s it was the largest pressure group in international climate negotiations and left an indelible mark on their trajectory.[40]

---

38 Aaron McCright and Riley E. Dunlap, 'Defeating Kyoto: The Conservative Movement's Impact on the US Climate Change Policy', *Social Problems* 50 (2003): 355–7, 362–3; Jacques et al., 'The Organisation', 355–7 (quotation from 355).

39 See for example Oreskes and Conway, *Merchants*, 246–7; Steve Coll, *Private Empire: ExxonMobil and American Power* (New York: Penguin, 2012), 184, 619–20; Powell, *The Inquisition*, 66, 98, 102–4, 112–6.

40 Powell, *The Inquisition*, 94–6; David Ciplet, J. Timmons Roberts and Mizan R. Khan, *Power in a Warming World: The New Global Politics of Climate Change and the Remaking of Environmental Inequality* (Cambridge, MA: MIT Press, 2015), 140; Peter

Exxon was the exemplary driving force of denial. The coal industry, however, was nearly as quick on the draw; in 1991, American coal interests set up the Information Council on the Environment to 'reposition global warming as a theory (not fact)'.[41] But in these early years, primitive fossil capital also gathered around itself fossil capital in general in the efforts to defend the doctrine of fossil fuels as a blessing.[42] All of this happened primarily on American soil. The US-born ISA then diffused a bundle of tropes in the public conversation, not always consistent with one another but united in political intent. One said that temperatures are not in fact on the rise. Another held that swings in the climate – including any perceptible warming – are caused by the sun and occur as part of a natural cycle.[43] Of particular interest for our purposes, however, is the trope of carbon dioxide as a gift to life, since it occasionally lifted the veil on some deeply ingrained associations between energy and race.

In a bid to influence the Rio summit in 1992, the Global Climate Coalition distributed a video claiming that more CO_2 in the atmosphere would fertilise crops and help feed the world. In 1998, the Western Fuels Association, a consortium of coal companies headquartered in Colorado, established the front group Greening Earth Society to further purvey the idea that excess CO_2 should be welcomed.[44] But the most famous composition from this genre came later, in 2006, when the Competitive Enterprise Institute – another key think tank in the apparatus, recipient of lavish Exxon funding – released a sixty-second commercial simply called 'Energy'.[45] In the opening scenes, happy people mill around in New York's Central

---

Newell and Matthew Paterson, *Climate Capitalism: Global Warming and the Transformation of the Global Economy* (Cambridge: Cambridge University Press, 2010), 37.

41 Kathy Mulvey and Seth Shulman, *The Climate Deception Dossiers*, Union of Concerned Scientists, 2015, 19–20.

42 Dunlap and McCright, 'Organized', 147–8.

43 For a survey and rebuttal of these and other 'arguments', see Mann and Toles, *The Madhouse*, 53–115.

44 Dick Russell, *Horsemen of the Apocalypse: The Men Who Are Destroying Life on Earth – and What It Means for Our Children* (New York: Hot Books, 2017), 23; Powell, *The Inquisition*, 94; Oreskes and Conway, *Merchants*, 203; Dunlap and McCright, 'Organized', 150.

45 James Hoggan, *Climate Cover-up: The Crusade to Deny Global Warming* (Vancouver: Greystone, 2009), 82–3.

Park. A blonde woman of model beauty blows soap bubbles; a group of equally blond children skip rope; another white woman jogs on a beach. A blonde girl blows on a dandelion, scattering its seeds. The voice-over says: 'There's something in those pictures you can't see. It's essential to life. We breath it out. Plants breath it in' – cut to an old-growth forest – and this miraculous invisible medium comes straight from 'the earth and the fuels we find in it. It's called carbon dioxide, CO_2.' Cuts to images of a refinery and an oil derrick, the voice-over continuing: 'The fuels that produce CO_2 have freed us from *a world of back-breaking labour*', the last five words spoken over the image of the only black person to appear in the clip. She raises her arms high to strike a heavy pestle into a wooden mortar, presumably pounding cassava or some other African crop. A thatched hut can be seen in the background. This black woman represents the world from which fossil fuels have freed us, 'lighting up our lives'. Then suddenly the pastoral piano melodies are broken up by a drone of sinister strings and the warning: 'Now some politicians want to label carbon dioxide a pollutant – imagine if they succeed. What would our lives be like then?' In the final scene, we are back at the blonde girl scattering the dandelion seeds, who gets to personify the slogan: 'Carbon dioxide – they call it pollution. We call it life.'[46]

Thanks to fossil fuels, white people have ascended the evolutionary ladder to the height of comfort and affluence. Black people have stayed behind in the fossil-free bottom to break their own backs. Now, imagine if CO_2 would be treated as a pollutant – what would *our* lives look like then? Primitive fossil capital clearly did not shy away from interpellating white people and framing mitigation as a threat to their life: the 'Energy' commercial inspired other think tanks to play up the trope of fossil fuels and their derivative gas as life-enhancing substances.[47] Anne Pasek has named this genre of denial 'carbon vitalism' and picked out seven beliefs that hold it together. CO_2 is only toxic at artificially high levels that can never be

---

46 As of autumn 2020, the commercial could still be watched on YouTube: search for 'Competitive Enterprise Institute global warming energy'.

47 In fact, there is evidence suggesting that GM and Ford financed this ad campaign. Kevin Grandia, 'Leaked Memo Claims That GM, Ford Financed Pro-CO_2 Ad Campaign', *DeSmog* UK, 27 July 2006.

reached in the atmosphere and so it cannot be a pollutant; it is essential for photosynthesis and thus beneficial to plants; it does not have the ability to alter the climate by trapping heat; current atmospheric levels are far below those that reigned on the luxurious earth of the dinosaurs – we still live in a CO_2 famine; a return to such geological heights should be the aim of energy policies; to burn fossil fuels is to render the biosphere a service; any measures to cap their use would be detrimental to life itself.[48] Whose life? The Competitive Enterprise Institute gave its answer, but other carbon vitalists would probably argue that everyone on earth would prosper in a CO_2-saturated atmosphere, black people included, their cassava growing better too. Some just have the burden to kindle that flame.

The denials of trend, attribution and impact were united by the overarching trope of the enormous and insurmountable uncertainty of the science: no firm conclusions can be drawn on any of the issues at hand; the methods of climate scientists are riddled with conjectures and outright counterfeits; springing to action on such slippery foundations would be foolhardy. Or, in the plain language of an advertorial from Exxon's future partner Mobil, printed in the *New York Times* in 1997: 'Let's face it: The science of climate change is too uncertain to mandate a plan of action that could plunge economies into turmoil.' From this followed the trope of scientists and activists as a bunch of alarmists and religious zealots. Or, in the words of another Mobil advertorial, published two years earlier: 'The sky is not falling' – 'Good news: The end of the Earth as we know it is not imminent.'[49] Those who dared to question the doxa of the doomsday were the brave sons of Galileo. They faced persecution from the guardians of 'the hoax' – another prominent trope, canonised by James Inhofe, a Republican senator funded by Exxon, who said in a speech that global warming is 'the greatest hoax ever perpetrated on

---

48   Anne Pasek, 'Fixing Carbon: Bodily Politics and Vitalist Discourses in Climate Denial', paper presented at the second annual Energy and the Left conference, New York University, 30 March 2018.

49   Quoted in Supran and Oreskes, 'Assessing', 6, 10. Cf. for example Dunlap and McCright, 'Climate', 251; Powell, *The Inquisition*, 16, 74, 79, 88; Heather W. Cann and Leigh Raymond, 'Does Climate Denialism Still Matter? The Prevalence of Alternative Frames in Opposition to Climate Policy', *Environmental Politics* 27 (2018): 436.

the American people' and then went on to publish a book-length study of *The Greatest Hoax: How the Global Warming Conspiracy Threatens Your Future*, the cover featuring a faceless leader of the conspiracy with his hands around a gleaming globe.[50] (But Inhofe is probably most famous for trying to disprove climate change by taking a snowball into the Senate and tossing it on the floor.) To this must be added the anti-communism so defining for the denialist ISA. We shall return to it in some depth later.

The denialist ISA interpellated a range of subjects: businessmen, car owners, Americans, rational agents; perhaps most importantly, everyone who identified themselves as a beneficiary of the free market. The Heartland Institute, perhaps *the* key think tank of the apparatus, in 2020 still trumpeted the mission 'to discover, develop, and promote free-market solutions to social and economic problems'.[51] It spoke to anyone who was already a subject of the free market. Broad in its appeal, the denialist ISA operated across other ISAs firmly entrenched in the American social formation – churches, schools, courts, trade unions, radio and television programmes, but most of all the Republican party – as a kind of transversal, single-issue apparatus. Because climate mitigation posed a threat to privileges tied to a whole range of subject positions, this apparatus – devoted to the literal denial of one problem – could combine several interpellative elements in a cohesive, if not exactly coherent, structure. It developed a most impressive capacity for public outreach, as well as for symbiotic existence with centrally located parts of the American state apparatus; put simply, it had a hotline to decision-makers. It hailed as much the citizens as the rulers of the American empire.

In one respect, however, the denialist ISA presents a challenge to Marxist theories of ideology. Ever since Georg Lukács and Antonio Gramsci, such theories have worked on the assumption that the most effective bourgeois ideology is the one least obvious and ostentatious in its class bias, inconspicuous enough to sink into popular

---

50 Dunlap and McCright, 'Organized', 153; Mann and Toles, *The Madhouse*, 95–8.
51 heartland.org, accessed 30 January 2019. For an updated analysis of the denialist output of the Heartland Institute, see Cann and Raymond, 'Does Climate'.

consciousness as the normal way of doing things. All have sought to loosen the strictures of the base/superstructure model. But original climate denialism looks as though someone had striven for the most overdrawn caricature of that model and staged a mock play of material interests paying for ideas. Massive trawling of the output from the American denialist ISA has documented that, in the period 1993 to 2013, agents with direct corporate funding were vastly more likely to spread the message that climate change is a natural cycle and CO_2 good.[52] Initially, the efforts to camouflage this base logic were minimal; as crude as any ideological campaign had ever been, the purposes were written on the foreheads of its priests and patrons. It was all about fending off regulations that would trim profits in the short term. United in their outspoken faith in the free market, the deniers were – so Jacques quotes Gramsci – a 'real, organic vanguard of the upper classes', and anyone with a modicum of critical instinct could see this.[53] In the longer term, it was all about ensuring the very reproduction of fossil capital. The snake feared for its head and secreted a venom of disinformation: as simple as that.

In spite of this transparency – or perhaps because of it, in the triumphalist mood after the Cold War – the denialist ISA was eminently successful on its home terrain. The American public had expressed high degrees of worry around the time of the Hansen testimony, in response to which Bush the elder vowed to counter 'the greenhouse effect with the White House effect', but by the mid-1990s, doubts were sown deep in the nation.[54] The Republican party had been swayed by the ISA. Representatives of the latter had achieved a status as legitimate authorities on the subject, leading to decades of

---

52 Justin Farrell, 'Corporate Funding and Ideological Polarization about Climate Change', *PNAS* 113 (2016): 92–7; cf. Justin Farrell, 'Network Structure and Influence of the Climate Change Counter-Movement', *Nature Climate Change* 6 (2015): 370–4.

53 Jacques, 'A General', 12. Cf. Peter Jacques, 'Ecology, Distribution, and Identity in the World Politics of Environmental Skepticism', *Capitalism Nature Socialism* 19 (2008): 9; Jacques et al., 'The Organisation', 354; Oreskes and Conway, *Merchants*, 248; Dunlap and McCright, 'Climate', 245, 250; Dunlap and McCright, 'Organized', 144–5; Powell, *The Inquisition*, 13; Andrew J. Hoffman, *How Culture Shapes the Climate Change Debate* (Stanford, CA: Stanford University Press, 2015), 40.

54 Bush quoted in Oreskes and Conway, *Merchants*, 185; on public worry, see Weart, *The Discovery*, 44–5.

'balance' in media reporting – one minute to someone who believes in global warming, one minute to someone who does not.[55] All the standard tropes of denial were in rapid circulation, within and beyond US borders. Most importantly, international climate politics had developed a determinant pattern it has retained ever since: the US, responsible for more CO_2 emissions than any other country, could not be counted on for even the mildest of action. But, at the same time, there were indications that the denialist ISA faced a kind of crisis.

The Crisis of Denial

We now know that primitive fossil capital possessed rudimentary knowledge of the problem since at least the 1960s. One early moment of dissemination occurred in 1959, when three hundred industry executives, government officials and researchers convened for a symposium in New York on the theme of 'Energy and Man'. It was meant to mark the centennial of the first discovery of oil in the US, but one scientist on the podium, physicist Edward Teller, spoiled the party by telling the audience that CO_2 blocks infrared radiation, and so continued emissions of that gas might well 'melt the icecap' and cause 'all the coastal cities' to be submerged.[56] To better understand the process, the oil industry turned to Stanford and other top universities for collaborative research. In 1965, the American Petroleum Institute, or API, the main trade association for oil and gas corporations active in the country, received a report from its president Frank Ikard on the findings so far. Speaking to the annual general meeting, he did not mince words:

> This report unquestionably will fan emotions, raise fears, and bring demands for action. The substance of the report is that there is still time to save the world's peoples from catastrophic consequences of pollution, but time is running out. One of the most important

---

55   See for example Maxwell T. Boykoff and Jules M. Boykoff, 'Balance as Bias: Global Warming and the US Prestige Press', *Global Environmental Change* 14 (2004): 125–36.

56   Benjamin Franta, 'On Its 100th Birthday in 1959, Edward Teller Warned the Oil Industry about Global Warming', *Guardian*, 1 January 2018.

predictions of the report is that carbon dioxide is being added to the earth's atmosphere by the burning of coal, oil, and natural gas at such a rate that by the year 2000 the heat balance will be so modified as possibly to cause marked changes in climate beyond local or even national efforts.[57]

– a resumé of the first IPCC reports, preceding them by three decades. In 1968, the same API received another report on state-of-the-art research on fossil fuels and climate, which again included the elementary insights – trend, attribution, impact – and concluded that 'there seems to be no doubt that the potential damage to our environment could be severe'.[58] And that was only a start.

No corporation was more proactive in instigating climate research than Exxon. As early as 1957, scientists working for what was then known as Humble Oil published peer-reviewed calculations of the atmospheric impacts of CO_2 from fossil fuels.[59] Two decades later, one senior in-house scientist informed top managers about a 'general scientific agreement' on the ensuing climate hazards, which would have to be swiftly addressed; he estimated 'a time window of five to ten years' before humanity must make the critical decisions.[60] Exxon now reacted by driving straight to the research front, for no altruistic reasons: it sniffed a danger to its business. Exactly how close was it? The corporation equipped one of its supertankers with a laboratory for investigating the share of CO_2 absorbed by the oceans, ran

---

57 Benjamin Franta, 'Early Oil Industry Knowledge of CO_2 and Global Warming', *Nature Climate Change* 8 (2018): 1024–5.

58 E. Robinson and R. C. Robbins, *Sources, Abundance and Fate of Gaseous Atmospheric Pollution*, Stanford Research Institute, Menlo Park, CA, report prepared for American Petroleum Institute, 1968 (excerpts available at smokeandfumes.org), 110. See further Neela Banerjee, Lisa Song and David Hasemyer, 'Exxon's Own Research Confirmed Fossil Fuels' Role in Global Warming Decades Ago', *Inside Climate News*, insideclimatenews.org, 16 September 2015; Neela Bannerjee, 'Oil Industry Group's Own Report Shows Early Knowledge of Climate Impacts', *Inside Climate News*, 5 February 2016.

59 H. R. Brannon Jr., A. C. Daughtry, D. Perry et al., 'Radiocarbon Evidence on the Dilution of Atmospheric and Oceanic Carbon by Carbon from Fossil Fuels', *Transactions of the American Geophysical Union* 38 (1957): 643–50.

60 James F. Black quoted in Neela Banerjee, John H. Cushman Jr., David Hasemyer and Lisa Song, *Exxon: The Road Not Taken* (n.p.: Inside Climate News, 2015), 1–2.

advanced climate models, perused the latest literature and predicted, anno 1982, that the atmospheric rate of CO_2 would reach 415 parts per million in 2019. It could not have been more spot on: in June 2019, the rate hit 415 parts per million.[61] An internal consensus formed in the early 1980s, as Exxon's researchers and managers stared a warmer world in the face: it was real; it called for action; fossil fuels would soon be in the cross-hairs.[62] Other corporations knew too – Shell, BP, GM, Exxon's future partner Mobil, coal giant Peabody, all keen to read up and attend symposiums and hearings on 'the greenhouse effect', as the problem was then known.[63] The basic knowledge continued to make its way through the circuits of primitive fossil capital, into the Global Climate Coalition, whose very own scientists in 1995 wrote a seventeen-page internal primer asseverating that 'the potential impact of human emissions of greenhouse gases such as CO_2 on climate is well established and *cannot be denied*'.[64]

And yet deny it they did. Primitive fossil capital established and kept the denialist ISA going against its better knowledge, deliberately misleading the subjects of its interpellations. We must correct Althusser on one point: 'The bourgeoisie has to believe in its own myth before it can convince others.'[65] The faith in denial, if not in

---

61 Lee Wasserman, 'Did Exxon Deceive Its Investors on Climate Change?', *New York Times*, 21 October 2019; Fiona Harvey, 'Latest Data Show Steep Rises in CO_2 for Seventh Year', *Guardian*, 4 June 2019. See further Banerjee et al., *Exxon*; Supran and Oreskes, 'Assessing', 6–10, 15; Coll, *Private*, 185; Rich, *Losing*, 47–8; Sybille van den Hove, Marc Le Menestrel and Henri-Claude de Bettignies, 'The Oil Industry and Climate Change: Strategies and Ethical Dilemmas', *Climate Policy* 2 (2002): 4; Suzanne Goldenberg, 'Exxon Knew of Climate Change in 1981, Email Says – but it Funded Deniers for 27 Years', *Guardian*, 8 July 2015.

62 Banerjee et al., *Exxon*, for example 4–5, 22, 55.

63 Mulvey and Shulman, *The Climate*, 4; Rich, *Losing*, 89, 144, 188; Marco Grasso, 'Oily Politics: A Critical Assessment of the Oil and Gas Industry's Contribution to Climate Change', *Energy Research and Social Science* 50 (2019): 109. On the development of climate science in the US in the 1980s and how oil corporations tracked it, see further Rich, *Losing*.

64 Copy of document included in Mulvey and Shulman, *The Climate*, 26. Emphasis added.

65 Louis Althusser, *For Marx* (London: Verso, 2005 [1965]), 201. But the bourgeoisie might well have managed to convince itself about the lie and believe in its own propaganda: cf. Stanley Cohen, *States of Denial: Knowing about Atrocities and Suffering* (Cambridge: Polity, 2001), 38.

capitalism itself, was only ever half-hearted, at the most. After Hansen's testimony and the creation of the IPCC, primitive fossil capital – led by Exxon and the API – launched into denial of something it had itself observed and counted on; thus in 1997, Lee Raymond, the CEO of Exxon, declared that 'the earth is cooler today than it was 20 years ago', due to 'natural fluctuations' that had nothing to do with fossil fuels.[66] The next year, a team at the API outlined a 'road map' for how to turn climate change into 'a non-issue'. Victory in this pursuit was defined as the moment when 'average citizens "understand" (recognize) uncertainties in climate science', such perceptions become 'part of the "conventional wisdom"' and 'media coverage reflects balance'.[67] The laymen's impression of a debate between researchers who believed in global warming and those who disputed it was completely manufactured by the class fraction that *knew*, before almost anyone else, that there was no reason to have such a debate, any more than one over heliocentrism or the laws of thermodynamics. The debate was a victorious trick, the denial but a tactic. Some of the early reports might have been buried deep in desks and archives, but the knowledge was updated and the duplicity renewed on a regular basis. Exxon, for instance, spoke with a consistently forked tongue over the years, saying one thing in internal documents and something entirely different in advertorials and other PR material.[68]

The apparatus was erected on lies, and it also suffered from contradictions in its external communication. It paid purportedly independent scientists to wage war on science. To gain credibility, the denialist ISA contracted some willing old white men with distinguished scientific careers – if only in peripheral disciplines – to debunk the elementary insights; most august among them were Richard Lindzen and Fred Singer. The apparatus raised the banner of reason and attacked climate scientists for being prone to myth. No corporation could afford to abandon its pretence to rationality: even

---

66 Quoted in Bannerjee, *Exxon*, 45; Russell, *Horsemen*, 24.
67 Document included and discussed in Mulvey and Shulman, *The Climate*, 12. The original revelation of the plan is in John H. Cushman, 'Industrial Group Plans to Battle Climate Treaty', *New York Times*, 26 April 1998.
68 As painstakingly documented by Supran and Oreskes, 'Assessing'.

at the height of its sponsorship of the ISA, ExxonMobil self-identified as 'a science- and technology-based company'.[69] As the evidence for human-induced and potentially catastrophic global warming accumulated relentlessly over the 1990s, cracks began to appear in this edifice. It came to be regarded, outside the community of believers, as the temple of an obscurantist faith-group that refused any contact with actual science and reason. The crudity was not necessarily a strength. From the start, it made the apparatus vulnerable to exposure: the Information Council on the Environment set up in 1991, for example, fell dead in the same year, after journalists revealed that its putative scientists simply fronted for coal companies.[70] When the second IPCC report in 1995 marshalled a new mountain of evidence in support of the conclusion – phrased in characteristically restrained terms – that there was 'a discernible human influence on global climate', the discrepancy between the consensus and the caucus became too glaring in the eyes of too many.[71]

Parts of fossil capital, including its primitive fraction, now realised that they caught more heat than they repelled by preaching overt denial. In 1997, BP broke ranks with the Global Climate Coalition. So did DuPont, and Shell in 1998, and Ford in 1999 – explaining that membership in the Coalition had 'become something of an impediment for Ford Motor Company to achieving our environmental objectives' – followed in 2000 by the defections of Texaco, GM and Chrysler.[72] All of these former deniers suddenly professed their acknowledgement of global warming and the need to do something about it. This was probably not due to a change of heart, given the decades-old knowledge; when the Coalition fractured in the late 1990s, a bolt of enlightenment could not have been the cause. Instead, it should be sought in the arrival of new strategies held together by what we might call *capitalist climate governance*.

---

69 Quoted in van den Hove et al., 'The Oil', 4. On the rationalist rhetoric, see further Powell, *The Inquisition*, for example 11–15, 99.
70 Mulvey and Shulman, *The Climate*, 21; Hoggan, *Climate*, 32–4.
71 Quoted in Oreskes and Conway, *Merchants*, 205.
72 Ford quoted in Lester R. Brown, 'The Rise and Fall of the Global Climate Coalition', Earth Policy Institute, earth-policy.org, 25 July 2000; on BP, see van den Hove et al., 'The Oil', 11–12.

This new dawn rose from Kyoto, home of the eponymous protocol. In the lead-up to the UNFCCC summit in that Japanese city in 1997, BP, DuPont and other early defectors lobbied hard for market mechanisms to be integrated in the agreement. Under pressure from the denialist ISA, the Clinton administration plumped for the idea, and the EU gave up its initial opposition. The Kyoto Protocol indeed came to centre on so-called flexible mechanisms: instead of cutting their own emissions, advanced capitalist countries could pay poor counterparts to do it for them.[73] This opened the floodgates to capitalist climate governance, or an array of mitigation measures that (1) postpone any showdown with fossil capital into the distant future, (2) impose no serious limits on accumulation and (3) open up novel opportunities for the generation of profit – or, in short, a form of climate governance that *harnesses the energies of capital.* Here global warming is accepted as a fact and capital repositioned as solution. After Kyoto, the world thus became awash in schemes for clean coal and clean oil, carbon capture and storage, gas as a bridge fuel, carbon trading, voluntary offsetting, climate derivatives, REDD+, a forest of other acronyms, company plans for climate neutrality and all the rest, with geoengineering as the outer horizon.[74] The Global Climate Coalition was supplemented by bodies with names such as World Business Council for Sustainable Development or Business for Innovative Climate and Energy Policy. Based in the US and EU, a dense network of institutions formed a parallel ISA, devoted to averting direct interventions from states and securing one element of the dominant ideology: let capital itself deal with the problem, for capital is good (even though it is fossil). This was the era of solutions, opportunities, win-win and advertisements steeped in verdant greenery.[75]

---

73   Ciplet et al., *Power*, 139–41.
74   For a comprehensive survey that advances a very generous interpretation of these and other initiatives, see Newell and Paterson, *Climate*.
75   For the parallel ISA, its organisations and activities, see the research by J. P. Sapinski, for example 'Constructing Climate Capitalism: Corporate Power and the Global Climate Policy-Planning Network', *Global Networks* 16 (2016): 89–111; 'Corporate Climate Policy-Planning in the Global Polity: A Network Analysis', *Critical Sociology* 45 (2019): 565–82. See also the excellent study by Christopher Wright and Daniel Nyberg, *Climate Change, Capitalism, and Corporations: Processes of Creative Self-Destruction* (Cambridge: Cambridge University Press, 2015).

In simple terms, after a decade of fairly unanimous denial, a chunk of the capitalist class opted for greenwashing as a more promising ideological strategy. It was not without contradictions either, obviously. There is an inverted symmetry between the two: in denial, fossil capital continues with business-as-usual and says untruthfully that no action is needed; in greenwashing, it continues with business-as-usual and falsely claims to undertake the needed action. Duplicity inheres in both. When a shift occurred, it was prompted not by conversion but calculation, as capitalist climate governance allowed fossil capital to expand as before but in apparent *alignment* with science and reason – the freedom to dress up in a greener skin, without the planning for sharp reductions originally envisaged.

The denialist ISA did not thereby go out of business. To the contrary, the years of Kyoto saw its first major victory: when Bush the younger withdrew the US from the Protocol in March 2001, the efforts of the Coalition, ExxonMobil and their partners had borne some epochal fruit.[76] In the same year, the Coalition disbanded. Its objective 'was to comment on the Kyoto Protocol and since the administration has decided to pursue another course, the work of the coalition is essentially done', one spokesman for the National Mining Association stated, a mission-accomplished line repeated – only partly disingenuously – by other representatives of the already thinned-out front.[77] The Coalition did make decisive contributions to the scuppering of Kyoto in the US, and ever since, a global agreement with binding emissions reductions has been perceived as off the table in that particular country.[78]

The Kyoto years, then, marked a turning point not so much by convincing capital *in toto* as by validating a diversity of tactics. The two ISAs operated in parallel, less in conflict than in sync; both denialism and capitalist climate governance aimed at forestalling any

---

76 McCright and Dunlap, 'Defeating'; cf. Oreskes and Conway, *Merchants*, 215; Powell, *The Inquisition*, 96, 112; Coll, *Private*, 85; Wright and Nyberg, *Climate*, 81.

77 Carl Raulston quoted in Pamela Najor, 'Global Climate Coalition Ends Its Work; Voice for Industry Opposed to Global Treaty', *The Heat Is Online*, 25 January 2002.

78 Rich, *Losing*, 186.

mitigation that might trammel self-expanding value, only they targeted different sides of the equation. One actor might pursue the former option in one moment and the latter the next, or both simultaneously. We shall see this diversity of tactics reappear at ever-higher stages. In the times of Kyoto, with a fillip from Bush, the denialist ISA charged ahead in its homeland. While BP rebranded itself 'Beyond Petroleum' in 2001, ExxonMobil persisted in embracing denial openly; in the early millennium, it generated more profit than any other company in the world, staking a claim to the most profitable capitalist enterprise in history, and so could afford some bad press. Peabody likewise stayed true to the cause.[79] As it expanded its holdings with hundreds of mines in Australia, Indonesia and other parts of Asia, feeding the emissions explosion underway in China, this company claimed to be fructifying the earth with more CO_2.[80] So did Murray Energy, another leading US coal company that continued to fund, among other think tanks, the Competitive Enterprise Institute, scattering the dandellion seeds.

But none approached the obstinacy and success of the Koch brothers. Inheritors of a fortune their father had drawn from pipelines and refineries, Charles and David Koch were in 2009 the sixth and seventh richest men in the world, respectively. They owned pipelines and refineries, factories producing fertilizers and coke, coal-fired power plants and swathes of Canadian tar sands, from which they exported more oil than any competitor: and they never backed a millimetre from denial. In the contest of base/superstructure kitsch, they poured more money – on one estimate, three times more – into the denialist ISA than ExxonMobil itself during the first decade of the millennium.[81] The think tanks they funded took no breaks, the Heartland Institute and the other dedicated bodies working as frantically as ever to spread the doubt. Obsessive preachers of denial like Marc Morano toured the world without budget constraints. If there was a period in the early millennium when these people suffered

---

79 Den Hove et al., 'The Oil', 8; Powell, *The Inquisition*, 106, 111–5; Coll, *Private*, 310–1, 326, 623–4; Hoffman, *How Culture*, 39.
80 See Russell, *Horsemen*, 107–22.
81 Jane Mayer, *Dark Money: The Hidden History of the Billionaires behind the Rise of the Radical Right* (New York: Anchor, 2017), 5, 264, 60, 251.

unpopularity and even ostracism, they merely, as we shall see, bided their time.

Some elements of this hard core, however, eventually found it prudent to officially distance themselves from what others perceived as institutionalised crackpottery. After the Union of Concerned Scientists released a dossier of damning revelations in 2007 – year of the fourth IPCC report, preceded by the Stern report and Al Gore's film *An Inconvenient Truth*, causing a temporary spike in climate awareness in the Western cultural sphere – ExxonMobil solemnly vowed to discontinue the funding of denialist groups. Yet the money continued to flow.[82] Some channels appear to have been made invisible, for just as ExxonMobil claimed to have closed its spigots, another mode of funding took off: anonymous foundations. Their business idea was to receive vast sums of money and transfer them to earmarked recipients while keeping the identities of the benefactors perfectly secret. Starting in 2003, the twin foundations DonorsTrust and the Donors Capital Fund became the main conduit of money to the US-based public faces of the apparatus – the Heartland Institute, the Competitive Enterprise Institute, the American Enterprise Institute, the Heritage Foundation and dozens more – and made sure to leave no traces of the original sources behind.[83] Here was a way of plastering over the contradictions of the ISA. Capitalists could now write anonymous checques to it and appear innocent of its excesses.

Even the supposedly most enlightened members of primitive fossil capital failed to extricate themselves from the apparatus. BP and Shell remained part of the API, even as it developed its road map for killing the climate issue.[84] At the same time, Shell led the World

---

82 Elliot Negin, 'Why Is ExxonMobil Still Funding Climate Science Denier Groups?', Union of Concerned Scientists, 31 August 2018.

83 Robert J. Brulle, 'Institutionalizing Delay: Foundation Funding and the Creation of US Climate Change Counter-Movement Organizations', *Climatic Change* 122 (2014): 681–94. Cf. Jason M. Breslow, 'Robert Brulle: Inside the Climate Change "Countermovement"', *Frontline*, pbs.org, 23 October 2012; Suzanne Goldenberg, 'Secret Funding Helped Build Vast Network of Climate Denial Thinktanks', *Guardian*, 14 February 2013; Mayer, *Dark*, 252–4.

84 Mulvey and Shulman, *The Climate*, 9–12; cf. Grasso, 'Oily', 106.

Business Council for Sustainable Development.[85] Individual capitalists evidently banked on both denial and greenwashing as tactics for reproducing themselves. The double duplicity reached new heights when Rupert Murdoch – himself a long-time 'sceptic' and owner of Fox News and the *Wall Street Journal*, two of the loudest megaphones of denial – in 2011 declared his company 'carbon-neutral across all of our global operations'.[86] The discord in the class was of little account. Across the board, whether they betted on denial or greenwashing or both, corporations in the business of producing fossil fuels systematically obstructed anything that might translate into forceful mitigation, with fossil capital in general backing them up.[87]

Over the course of the Bush reign, however, the balance shifted towards capitalist climate governance virtually everywhere – except for the parts of the US controlled by Republicans. These were increasingly viewed as isolated anomalies. 'Environmental scepticism', wrote a team of researchers in a seminal study in 2008, 'appears to be primarily a US phenomenon.'[88] When Barack Obama took the White House, it appeared to have lost even there. In 2012, Robert Brulle, a scholar of the denial machine and expert on the dark money of the Donors, offered the following prediction:

> I think the outright climate-denier component is probably atrophying, but you're going to sort of see more of the technology-can-fix-this approach rhetoric, and a voluntary action through technological innovation or some sort of things like that will replace the outright climate denial as that becomes less and less viable.[89]

Funding the deniers came with a rising cost to company reputation, and incognito transfers were not always a stable solution; they tended to leave a trail of scandal in its wake. Capital did seem

---

85 J. P. Sapinski, 'Climate Capitalism and the Global Corporate Elite Network', *Environmental Sociology* 1 (2015): 272, 274.
86 Quoted in Wright and Nyberg, *Climate*, 93.
87 Ciplet et al., *Power*, 143.
88 Jacques et al., 'The Organisation', 363.
89 Breslow, 'Robert'. Similar expectations are reflected in Cann and Raymond, 'Does Climate'.

to move in the opposite direction. In 2014, Google left the American Legislative Exchange Council (ALEC) in a public show of disgust: for many years, the Council, a business body specialising in drafting laws favourable to the free market, had given voice to denial. With its water-powered data centres and wind turbines flying on kites, Google wanted to be seen as part of the solution, 'so we should not be aligned with such people. They're just literally lying', an executive explained.[90] More than ever, the deniers looked like late Holocene fossils. In September 2015, *New York Magazine* ran a story with the headline 'Why Are Republicans the Only Climate-Science-Denying Party in the World?'[91] A few weeks later, capitalist climate governance experienced its second holy moment, when world leaders finally negotiated a successor to the Kyoto Protocol, which had centred on flexible mechanisms but retained the rule of mandatory emissions cuts; at COP21 in Paris, the latter was thrown overboard and replaced with the principle of *voluntary* emissions cuts, long championed by business fronts at the UN negotiations and, of course, by the US.[92] Now everyone was on board. Hands held high, faces gleaming, the leaders congratulated themselves on what UN secretary-general Ban Ki-moon called 'a peace pact with the planet'. 'Now you don't hear much about the sceptics', he sighed his relief, while the CEO of Unilever labelled them 'the only endangered species'.[93] The *Guardian* chimed in:

> The Paris agreement signals that deniers have lost the climate wars ... The whole world agreed, we need to stop delaying and start getting serious about preventing a climate crisis. We've turned the corner; climate

---

90 Mulvey and Shulman, *The Climate*, 24. But that did not prevent Google from funding the Competitive Enterprise Institute for various other lobbying services it performed on its behalf. Stephanie Kirchgaessner, 'Revealed: Google Made Large Contributions to Climate Deniers', *Guardian*, 11 October 2019.
91 Jonathan Chait, 'Why Are Republicans the Only Climate-Science-Denying Party in the World?', *New York Magazine*, 27 September 2015.
92 On the long path to voluntary emissions cuts in the UN negotiations, see Ciplet et al., *Power*.
93 Ban Ki-moon and Paul Polman quoted in Richard Valdmanis, 'Once a Fixture, Climate Sceptics Say They Are Being Stifled in Paris', *Star*, 8 December 2015.

denial is no longer being taken seriously. The world has moved on, and contrarians have become irrelevant relics of the fossil fuel age.[94]

One should never underestimate the tendency to overestimate the rationality of bourgeois civilisation.

The Revenge of Denial

When the early twenty-first-century far right denied climate change, it did not advance novel arguments or adduce any fresh evidence. It recycled tropes put in circulation by the denialist ISA in the years after the fall of the Berlin Wall. When the AfD said that no warming has been detected, models are hypothetical and CO_2 is the gas of life, this was the output of an ideological toxic waste landfill. In the 1990s, pioneers of denial such as Fred Singer spread a wisecrack that environmentalists were watermelons, green on the outside and red on the inside; Jean-Marie Le Pen of the Front National actually cut up a watermelon to demonstrate the point.[95] The far right did not keep purportedly independent scientists in its employ or publish its own phony research syntheses, but passed on what it had picked up from conservative think tanks and associated sources, with which they collaborated on occasion. If primitive fossil capital was the historical engine of the denial machine, the far right had become the exhaust pipe.

A far-right party, however, was no mere appendage of the denialist ISA. It was an apparatus in its own right, specialised in interpellating subjects as *members of an ethnically constituted nation*. It hailed people thought to be white. Denial here sat in a slightly different context. When the AfD abjured climate science, it was impelled by forces other than ExxonMobil in 2001: in itself it had no profits to protect, nor did it need to factor in the effects of climate change

---

94 Dana Nuccitelli, 'The Paris Agreement Signals That Deniers Have Lost the Climate Wars', *Guardian*, 14 December 2015.

95 Oreskes and Conway, *Merchants*, 248, 251–2; Jens Martin Skibsted, 'Foreign Climate: Why European Right-Wingers Should Be Tree Huggers', *Huffington Post*, 21 May 2014.

in long-term business planning. It might have been as rational as any other party, in the limited sense of striving to maximise political influence and deploying optimal methods for that goal, but there was nothing that tied it to biophysical realities, similar to the bonds that kept fossil fuel producers abreast of the progress in the field.[96] We are unlikely to discover internal documents from the AfD or the Front National or the Sweden Democrats that detail their basic knowledge of the science at an early date; conversely, it is entirely possible that a Gauland or a Baudet genuinely *believed* in what he said, no calculated deception at play. And if they merely pretended to believe, they were less vulnerable to shaming and correction. The interpellation of subjects of a white nation has never been constrained by what exists or not, and this released the far right from the contradiction that came to plague the denialist ISA. There were no limits to the falsehoods it could propagate, as long as its constituencies were receptive to them. Denial exacerbated the reputation of certain corporations as dishonest. It reinforced the image of a far-right party as honest. Whereas the former had pecuniary interests in broadcasting denial, the latter had not built their fortunes on derricks and mines, a circumstance which appeared to give them a greater potential to attract popular support for denial, the substance of which was set free.

Three decades after the maturation of climate science, with a clear view of the impacts of climate breakdown, one could indeed expect that literal, organised denial would have been fading away. But from Europe to the Americas, as of the late 2010s, *it wielded more political power than at any point in history*. This was a remarkable turn of events, one few if anyone could have predicted around the time of Paris. It was positively sensational in the case of Germany. While the denialist ISA strengthened its grip over the Republicans and thereby a good half of the US, Germany was regarded as virtually genetically free of denial. In 2013, four German scholars wrote in the journal *Global Environmental Change*:

---

96 On this rationality of far-right parties, cf. Jens Rydgren, 'Is Extreme Right-Wing Populism Contagious? Explaining the Emergence of a New Party Family', *European Journal of Political Research* 44 (2005): 416.

All major political parties represented in the national parliament relate in some positive way to environmental and climate protection goals. The political parties are extremely unlikely to launch campaigns in favor of climate-change skepticism, and climate-change skeptics have no strong political outlet in the current political landscape of Germany.[97]

Interestingly, the authors attributed the extreme contrast with the US to the factor of race. 'Public attitudes and political debates are not organized along racial divides in Germany so that "race" as a variable is typically not even addressed' – hence the absence of American-style denial. But after the publication of that article, racism, of course, made its most astounding comeback on the German political scene since the end of the Second World War.

As an integral component of the ethnonationalist agenda of the far right, climate denial travelled some distance from the headquarters of ExxonMobil and Peabody. When the same German scholars concluded that denial was 'a phenomenon of the Anglo-American cultural sphere rather than a worldwide trend', they gave expression to the prevailing Anglo-American exceptionalism in denial studies – a paradigm that must now be overturned, since the phenomenon by the late 2010s had become a more worldwide trend than even the Global Climate Coalition could have hoped for.[98] It had greater mass appeal, as measured in electoral support, than it ever achieved under the auspices of the denialist ISA, *on account of its migration into the ideology of the ethnic nation*. If it had thereby severed some of the direct links to fossil capital, it had lost none of its utility for it. Whenever the position 'there is no need for action' (AfD) affected policy, the interests of fossil capital as a totality were handsomely served.[99] We may then posit that the far right now *objectively worked as the defensive shield of fossil capital as a totality and primitive fossil capital in particular*, even if – or rather precisely because – it was not

---

97 Anita Engels, Otto Hüther, Mike Schäfer and Hermann Held, 'Public Climate-Change Scepticism, Energy Preferences and Political Participation', *Global Environmental Change* 23 (2013): 1019.
98 Ibid., 1025.
99 Deutscher Bundestag, 'Aufgabe'.

set up or financed by them; not a pawn in their game, like the Global Climate Coalition or the Information Council on the Environment. Insofar as denialism had become detached from the base by taking up residence in the swelling superstructure of far-right ethnonationalism, it *stabilised that base all the more effectively*. (A vindication, as it were, of the main lines in Western Marxist theories of ideology.)

Corporate funding never dried up completely, but it kept a lower profile after ExxonMobil made its vow in 2007. For a decade, conservative think tanks appeared to be the main motor of the machine, rendering it somewhat more purely ideological in character.[100] We might then outline a rough periodisation of denialism: 1989–2001: the early corporate phase, ending with the dissolution of the Coalition; 2001–07: the late corporate phase, ending with ExxonMobil's termination of public sponsorship; 2007–16: the conservative phase, ending with the election of Trump; 2016–: the far right phase, with more of a smooth continuation than a break between the phases. The third functioned as a bridge to full ideologisation in the fourth. Certainly, the original denialist ISA was still in operation in the fourth; in the late 2010s, the Heartland Institute, the Competitive Enterprise Institute and other key think tanks were – as we shall see – as active and energetic as ever. But as a phenomenon in world politics, the structure of denialism was transformed by its integration into the ISAs of far-right parties and presidencies. If the classical model of the apparatus had guardianship of the free market as its main vocation, the central interpellation of the far right was nationalist, and for far-right parties in Europe, this meant one thing above all else: hostility to immigration. It was on the back of such hostility that organised denial for the first time made inroads deep into European politics.

---

100 Dunlap and McCright, 'Climate', 246, 250.

2
Fear of a Muslim Planet

A rule has remained in force up to the moment of this writing: every time a European far-right party denies or downplays climate change, it makes a statement about immigration. It says: the problem facing our societies has nothing to do with climate – forget about that hoax – the real danger is the presence of too many non-white foreigners and, to be more precise, too many Muslims in our land. The corollary does not have to be uttered in words. It is there in every breath a European far-right party exhales, understood by addresser and addressee alike, because such a party has a totalising vision of the ethnically homogenous nation. It wants the dominant ethnic group – the white majority of Swedes or French or Italians – not only to rule the territory in question, but, ideally, to be the sole population living within its borders. By their very nature, aliens who have come to reside there constitute a threat to the nation.[1]

Such are the rudiments of the ethnonationalism that unites this party family. It might possibly be distinguished from more 'civic' forms of nationalism, for which citizenship or residence could be

---

[1] Rydgren, 'Introduction', 1–3; Tamir Bar-On, 'The Radical Right and Nationalism', in Rydgren, *The Oxford*, 17–18, 21–6; Cas Mudde, *Populist Radical Right Parties in Europe* (Cambridge: Cambridge University Press, 2007), 18–24; Matt Golder, 'Far Right Parties in Europe', *Annual Review of Political Science* 19 (2016): 480.

sufficient criteria for belonging.[2] The far right cares little for niceties and formalities. Its notion of the nation has genealogical depth, giving precedence to the body of people who trace their identity back to equally homegrown ancestors, in a chain stretching to some misty dawn and – it is hoped – into the future.[3] But the future will come about only after a change of course, a stop to the influx from the outside and, preferably, a reversal of the flow: a removal of the foreign elements. Elimination of their presence is the *telos*.[4] Jens Rydgren, leading specialist in the field and editor of *The Oxford Handbook of the Radical Right*, has deemed this ethnonationalism 'the master frame' of the party family, developed first by the Front National (FN) under Jean-Marie Le Pen in the mid-1980s and then exported across the continent. While some analysts would rather privilege anti-elite populism, Rydgren points out that for the far right, that stance is rather secondary: the elite is to be despised because it has opened the borders and invited the enemy. Ethnonationalism is the primary standpoint that subsumes all others, the beginning and the end of far-right politics.[5]

Whatever other issue it speaks about, then, a European far-right party is preoccupied with this one. Hostility to immigration is the one programmatic position shared by all members of the party family, the one sentiment that predisposes their supporters to vote for them, from the original breakthrough of the FN in 1983 to the Bundestag elections in 2017 and onwards.[6] This is the one problem that the parties want to confront with utmost urgency. A typical list of concrete demands includes a full halt to immigration from non-Western, non-white – above all Muslim – countries; termination of

---

2 But Bar-On rightly points out that 'civic' nationalism has its roots in, and is hard to separate from, ethnic variants. Bar-On, 'The Radical Right', 22–3.

3 Thus the AfD demanded the reinstatement of the old law that made German blood and ancestry the prerequisites for citizenship. Rensmann, 'Radical', 50–1.

4 Rydgren, 'The Sociology', 244; Rydgren, 'Introduction', 4; Bar-On, 'The Radical Right', 20–1, 28.

5 Rydgren, 'Is Extreme', 414–6, 426, 433; Rydgren, 'Introduction', 5–6.

6 Kai Arzheimer, 'Explaining Electoral Support for the Radical Right', in Rydgren, *The Oxford*, 150; on this as the determinant of AfD support, see Michael A. Hansen and Jonathan Olsen, 'Flesh of the Same Flesh: A Study of Voters for the Alternative for Germany (AfD) in the 2017 Federal Election', *German Politics* 28 (2019): 1–19; Rensmann, 'Radical', 55.

programmes permitting immigrants to settle on grounds of asylum or family reunification or employment; zero tolerance for anyone having entered the country illegally; expedited deportations: and the commencement of some process of repatriation. A reform package of that sort would solve a whole host of problems. Michelle Hale Williams has conceptualised immigration as the 'funnel issue' of the far right, the issue through which all others pass, on their way towards resolution in the cessation – and ultimately reversion – of the arrival of non-white foreigners. Thus unemployment is a symptom of immigration and would come to an end with it; the same with rampant crime, sexual violence, segregation, poverty, anomie and decay and any other malady society suffers from.[7] In Williams's model of the far-right worldview, every problem is a function of the ur-problem of immigration. But we may stretch her point and propose that when a far-right party formulates a position on any matter whatsoever, that position must pass through the funnel of anti-immigration *even if it concerns something the party concludes is a non-problem.* That is, it does not necessarily have to say (but it might, as we shall see) 'climate change is a symptom of immigration'. It can just as well say 'climate change is not a problem at all', and if it does so, it is because *that view has been productively related to the war on immigration.* This emphatically does not mean that a far-right party is a single-issue party – only that immigration is the narrow pipe through which all other aspects of the world must be guided. It is for the far right what profit is for ExxonMobil.

And being far-right in the world today involves a choice. Some say the problem that casts a shadow of imminent catastrophe over the coming decades is climate breakdown: those on the far right say – must say – that it is something else. A couple of decades passed after the invention of the master frame before the choice came to the fore. It was the historical exploit of Jean-Marie Le Pen to give a new lease of life to the European far right, after the coma of the post-war decades, by drawing inspiration from the ideas of the Nouvelle Droite, revamping ethnonationalism and directing

---

[7] Michelle Hale Williams, *The Impact of Radical Right-Wing Parties in West European Democracies* (Basingstoke: Palgrave Macmillan, 2006), 60–1.

political energies towards immigration; with that master frame in place, he then developed a position on climate with all the finesse of his watermelon spectacle. In a chat with readers of *Le Monde* before the presidential election in 2007, he got to speak on both topics. On immigration, he proposed the application of a 'suction pump' to rid France of 'all the people who are on our territory even if we did not wish them to come'. On climate change, he branded it a 'dogma' devised 'to terrorise people', claimed that any warming had its origins in the sun and reassured the readers of *Le Monde* that 'there are 15,000 scientists who are of my opinion'.[8] Three years later, the Front National organised its first party activity devoted to the second topic: a conference in Nanterre on the theme of 'Climate Change, Myth or Reality?' 'A scam', Le Pen adjudicated; a 'catastrophism' and 'crime' contrived by socialists and environmentalists to increase taxes and, more importantly, open the borders further still. The buzz around climate 'is meant to justify the reception of an increased number of refugees, since their situation was created by us who overconsume and have contributed to the destruction of the environment of these so-called climate refugees.'[9] We shall see this particular trope returning much later. Ever the firebrand, Le Pen followed up on the Nanterre conference by characterising global warming as a 'conspiracy' against 'the whites, the developed countries, who are held responsible for the misery of the world'.[10] The black witch doctor was on his way.

In Scandinavia, aspiring far-right parties had followed the model of the Front National since the 1980s and continued to do so – virtually by logical necessity – when the choice of the problem of the century was pressed upon them. In early 2008, Siv Jensen, leader of the Norwegian Progress Party, or Fremskrittspartiet (FrP), rallied her apparatus to the message that

---

8 Philippe Le Cœur, Alexandre Picquard and Christine Combeau, 'Jean-Marie Le Pen: "Je n'ai pas changé dans mes convictions"', *Le Monde*, 23 March 2007.

9 *Le Parisien*, 'Jean-Marie Le Pen dénonce "l'écologisme" et les "bobos gogos"', leparisien.fr, 30 January 2010.

10 AFP, 'Pour Jean-Marie Le Pen, les "crimes" communistes plus nombreux et plus lourds que ceux des nazis', *20 Minutes*, 26 November 2010.

global warming might as well be natural as man-made and confessed to the press that 'climate scepticism' could be a vote-winner: 'We might lose a bit in the short term, but in the long run I think the climate issue can be just as important [for us] as immigration.'[11] One week later, the party proposed that Norway's borders be closed to people from Muslim countries and the maximum number of admitted asylum-seekers cut to one hundred individuals per year.[12] This coincided with the moment when the world was preparing for the upcoming COP15, the decisive UNFCCC summit in Copenhagen – or *Hope*nhagen, as the city was renamed momentarily in honour of the gathering – where a worthy successor to the Kyoto Protocol was expected to be negotiated. In Denmark, the Danish People's Party, or Dansk Folkeparti (DF), prepared in its own fashion.

The day before COP15 began, on 6 December 2009, the DF held an 'apolitical climate conference' in the Danish capital. It was organised in collaboration with the True Finns, the United Kingdom Independence Party, Lega Nord and other family members from the European Parliament, so as to give a platform to 'independent scientists' ignored by the IPCC, among them Fred Singer. Pia Kjærsgaard, leader of the DF – as resoundingly successful in that capacity as Siv Jensen in Norway – publicised the findings of the conference: 'First, we cannot with any certainty say that there is a global warming. Second, any human influence on climate is highly doubtful'; the prevalent belief in trend and attribution is 'a form of mass psychosis'; the IPCC acts like the Inquisition; the talk of 'the world perishing in a climate Ragnarok' deserves only sarcasm; in more hands-on terms, 'all the billions spent on CO_2 reductions ought to be used for other purposes'.[13] So if climate was no problem whatsoever, what could there be to worry about?

'Europe will – maybe not in twenty, but rather thirty to forty years from now – have a Muslim majority population, if nothing is done.

---

11 Andreas Nielsen, 'Siv skal ta klima-bløfferne', *Verdens Gang*, 28 March 2008.
12 Kristoffer Rønneberg, 'Frp vil stenge grensen', *Aftenposten*, 7 April 2008.
13 Pia Kjærsgaard, 'Klimafantaster og klimarealister', Dansk Folkeparti, 7 December 2009.

That'll mean the end of our culture and the end of European civilization', said Morten Messerschmidt to *FrontPage Magazine* in 2006.[14] A rising star of the DF, leader of its contingent in the European Parliament, combining slickness and rowdiness in the best manner of the Danish far right, it was he who chaired proceedings at the anti-COP15 conference. He had made his choice abundantly clear. In 2001, Messerschmidt and other activists in the DF youth wing distributed a poster with two pictures: on the left side, three blonde girls over the caption 'Denmark today'; on the right, two persons in black masks waving copies of the Qur'an, blood splattered on their white martyr's robes, over the caption 'Denmark in 10 years'. The poster explicated: 'Mass rapes – brutal violence – insecurity – forced marriages – oppression of women – gang crime: this is what a multi-ethnic society offers us.' In the year when it was printed, the DF became the chief parliamentary pillar of support for the Danish centre-right government, a position it maintained for ten years straight and, after a social-democratic parenthesis, recaptured in 2015.[15]

A court gave Messerschmidt a conditional sentence of two weeks in prison for inciting racial hatred through the poster. That did not slow down his career, which came to involve dilating upon the party line on climate. In a long essay published in the run-up to COP15, he inspected scenarios such as lethal heat waves, stronger hurricanes, rising sea levels submerging land, a collapsing Gulf Stream – and found them to be nothing but 'bogeymen'. Climate change is a natural phenomenon that humans can do nothing to stop, but alas, debates on the issue have become 'ruled by doomsday prophecies'.[16] What about his own prophecies? In June 2018, seven years after the blonde girls of Denmark would have been replaced by Muslim blood rituals and multi-ethnic carnage, he looked back on the poster and the two weeks he never served in jail and argued that 'while we

14 Jamie Glazov, 'Europe's Suicide?', *FrontPage Magazine*, frontpagemag.com, 26 April 2006. On the extreme Islamophobia that defined the party as a force in Danish political affairs, see for example Jens Rydgren, 'Explaining the Emergence of Radical Right-Wing Populist Parties: The Case of Denmark', *West European Politics* 27 (2004): 474–502.
15 For more recent developments, see further below.
16 Morten Messerschmidt, 'Klimapolitik med fornuft', *180 Grader*, 27 May 2008.

dramatised the problem back then, we can now confirm that we understated it'. Happily, however, since 2001 the mood of the Danish people had turned in the direction of a 'will to self-defence. Better late than never.'[17]

Between 2001 and 2018, the DF did indeed become one of the most influential far-right parties – to wit, one of the most influential parties of any kind – in a European country. In the early years of this long boom, it distributed two other posters: one with the Danish parliament submerged by the rising sea over the caption 'Do you believe them?'; another with a blonde girl aged four or five, smiling trustingly into the camera, over the caption 'When she retires, there will be a Muslim majority in Denmark.' Concern over the demographic expansion of Muslim populations, from a baseline of immigrant communities towards a majority by means of breakneck procreation, is the touchstone of full racialisation of Muslims. It is the device that conclusively transforms Muslimness into a hereditary trait. Predictions of a Muslim majority refer not to the amount of people who will perform the *salat* or the *hajj* or observe the Ramadan fast thirty or forty years from now – confessional practice is beside the point – but to the number of progeny whose identity can be traced to ancestors from Muslim countries, carrying the threat within them. The life choices and tastes and affinities and personal quirks of all these future Muslims are inconsequential, or perhaps entirely predictable: their essence precedes their existence.[18] It is present in the body at the moment of birth. Homo sapiens they might look like on the surface, but these people are rendered members of a special Homo islamicus, as Edward Said observed in *Covering Islam: How the Media and the Experts Determine How We See the Rest of the World*, the ground-breaking study of what we now call Islamophobia – meaning that they are effectively construed as a race, in opposition to a West that is sometimes Christian, sometimes secular, but always

---

17 Morten Messerschmidt, 'Messerschmidt taler ud om sin dom – jeg fortryder intet', *Berlingske*, 25 June 2018.
18 This draws on Andreas Malm, *Hatet mot muslimer* (Stockholm: Atlas, 2009), 108–14.

white.[19] Whether recently arrived or reproduced over the fourth or sixth generation, the Muslim body was the quintessentially non-white body in the imaginary of the far right in early twenty-first-century Europe.

In the years spanning the fourth IPCC report and COP15, across the strait from Denmark, the Sweden Democrats, or Sverigedemokraterna (SD), readied for the leap into parliament. Unlike the DF and the FrP, this party had emerged straight out of the street-fighting neo-Nazi skinhead movement. The Danes and the Norwegians started out as neoliberal populists who wanted to bring down the Scandinavian welfare state and relieve themselves of taxes and then adopted the master frame of the French, but the Swedes came from an unbroken lineage of homebred Nazis and fascists, surviving in hibernation over the post-war decades and finally glimpsing hope in the 1980s. They had little luck in the beginning. Compared to their counterparts, their entrance into the mainstream seemed to be held off longer. By the first decade of the millennium, however, the boots and bomber jackets had been tucked away, replaced by suits donned by a leadership – the so-called gang of four – headed by Jimmie Åkesson.[20] Strategically savvy, this clique of young men saw the DF as the northern star to follow and patiently worked to import the Danish success.

On a typical day in 2008, Åkesson wanted to 'call attention to the mathematical fact that the Muslim minority in a not-too-distant future will constitute a majority in more and more parts of Sweden if ongoing developments continue'.[21] But the ideologist in the gang of four was Richard Jomshof, editor-in-chief of party journal *SD-Kuriren*. In those crucial years, he penned a manifesto of sorts printed over and over again in the pages of the journal, entitled 'The

---

19 Edward Said, *Covering Islam: How the Media and the Experts Determine How We See the Rest of the World* (London: Routledge & Kegan Paul, 1981), 57. Cf. Zygmunt Bauman, *Modernity and the Holocaust* (Cambridge: Polity, 2000 [1989]), 60.

20 The classical history of the party before entrance into parliament, co-authored by Stieg Larsson – best known for his *Millennium* trilogy, less so for his long-time activism in the Swedish section of the Fourth International – is Stieg Larsson and Mikael Ekman, *Sverigedemokraterna: Den nationella rörelsen* (Stockholm: Ordfront, 2001).

21 Jimmie Åkesson, 'Vi vill värna svensk kultur', *Dagens Nyheter*, 26 February 2008.

Islamisation of Sweden'. 'The Muslim population', it argued, 'grows ever faster, not only due to mass immigration but also through high fertility.' This 'creates a very dangerous situation' bound to 'threaten Swedish security and cohesion in the long term. We seriously risk a future when Sweden is transformed into a Muslim state; the unthinkable becomes thinkable.'[22] The text was illustrated with a drawing of a dark-skinned man holding up an architectural plan for how to rebuild the city hall of Stockholm into a mosque, the three crowns of the tower in the process of being taken down. The prolific Jomshof also came up with an exact year for when the balance would shift to a nationwide Muslim majority: 2032.[23] How could this disaster in the making possibly be averted? 'There is only one way out of this madness; continuous investments in our families, dramatically reduced immigration and increased repatriation.'[24] 'Or', in the words of a fellow activist who aimed higher, 'does Islam itself have to disappear for what is Western to remain?'[25]

Sinister forecasts of this kind were broadcast from all SD channels in the period leading up to the Swedish election of 2010. Like its European peers, the party here absorbed with special avidity the writings of Bat Ye'or, whose influence on the early twenty-firstcentury far right has yet to be fully appreciated. 'Bat Ye'or' is the penname of Gisèle Littman, an amateur historian who has spent most of her life in Britain. Her *pièce de résistance* is a book called *Eurabia: The Euro-Arab Axis*, published in 2005 with a minor American university press. It contains a stunning revelation. Contemporary history revolves around the year 1973, when Arab countries slapped an oil embargo on the West. European economies were brought to a stand-still, and the Arabs – a term interchangeable with 'the Muslims' in Ye'or's universe – controlled not only the oil, but also the banks on which these economies depended. To switch on the flow of oil and cash again, the Arab-Muslims demanded nothing less

---

22   Richard Jomshof, 'Islamiseringen av Sverige', *SD-Kuriren*, no. 69, August 2007.
23   Richard Jomshof, 'Gammeldanskar', *SD-Kuriren*, no. 53, June 2002.
24   Richard Jomshof, 'Allt fler muslimer i Europa', *SD-Kuriren* online, sdkuriren.se, 16 November 2006.
25   Jonas Karlsson, 'Kort kommentar om islam och västerländsk vänster', *SD Blogg*, sdblogg.se, 14 September 2007.

than the complete surrender of Europe to the rule of Islam. And Europe agreed. Since 1973, a secret organ, a cabal that few have heard of and fewer seen, an 'occult machinery' called the Euro-Arab Dialogue, or EAD, has governed the continent.[26] The institutions of the EU are merely facades for the EAD. In the reality Ye'or has uncovered, the EAD permanently controls universities and libraries, parties and parliament, media and the schools, at the beck and call of the sheiks of oil and finance who direct events from the Middle East and regularly remind their partners about the pain they will inflict unless Islamisation runs its course.

This is why Muslims are present in Europe. They did not come spontaneously, as migrants looking for jobs or as refugees fleeing conflict. They arrived as a result of a meticulous plan to Islamise the continent, for which Ye'or can provide an exact date and place: this 'Islamization was actually planned at a Euro-Arab Seminar that was held at the University of Venice from March 28 to 30, 1977.' At that seminar, the leaders of Europe – ever fearful of losing their oil and income – yielded to the Arab-Muslim demand for 'an implantation of homogenous ethnic communities', and so it began. But if immigration proceeded in accordance with a secret masterplan, how can Ye'or know about it? Since the meetings of the EAD 'were closed and the proceedings not published', she admits, the initiatives of that occult machinery 'can only be deduced from the fact of their subsequent implementation.'[27] Even if she has not seen the minutes, Ye'or can infer their content merely by looking around. Muslim immigrants

---

26 Bat Ye'or, *Eurabia: The Euro-Arab Axis* (Madison, NJ: Fairleigh Dickinson University Press, 2005). 'Occult machinery': 268. The creation of the EAD along those lines is described on 52–83. The best reading of Ye'or so far, which also begins the process of gauging her influence, is Reza Zia-Ebrahimi, 'When the Elders of Zion Relocated to Eurabia: Conspiratorial Racialization in Antisemitism and Islamophobia', *Patterns of Prejudice* 52 (2018): 314–37. See also for example Matt Carr, 'You Are Now Entering Eurabia', *Race and Class* 48 (2006): 1–22; Sindre Bangstad, *Anders Breivik and the Rise of Islamophobia* (London: Zed Books, 2014), ch. 5. For her influence, cf. Gabrillea Lazaridis, Marilou Polymeropoulou and Vasiliki Tsagkroni, 'Networks and Alliances against the Islamisation of Europe: The Case of the Counter-Jihad Movement', in Gabriella Lazaridis and Giovanna Campani (eds), *Understanding the Populist Shift: Othering in a Europe in Crisis* (London: Routledge, 2017), 70–103.

27 Ye'or, *Eurabia*, 91–5.

have been deployed as the infantry of a conquest. They are on a mission to physically supplant the natives. The mosques are their outposts. But the mightiest weapon in the Muslim arsenal is the womb itself, whose products swarm the world in general and Europe in particular, to the extent that the latter continent exists no longer: it is now 'Eurabia'.

In Ye'or's narrative, then, the apocalypse has already occurred. All was lost in 1973. Since then, ex-Europe has sunk into ever more abject subordination to its Arab-Muslim overlords, who call the shots without opposition – without anyone even noticing – siphon off 'poll taxes' from ordinary taxpayers, enforce sharia law and prohibit thoughts disagreeable to Islam. Oil is the modern bedrock of their power. But the roots of the evil go all the way back to the seventh century, when Islam was set in motion as a ceaseless campaign to enslave, tax, kidnap and massacre everyone who is not a Muslim – and 1,300 years later, it finally managed to overrun European borders. The hapless surviving natives are now relegated to the status of *dhimmi*, a term by which Ye'or means something like 'the passive and submissive slaves of the Muslims'. When a *dhimmi* sees a Muslim on the street, he walks in the gutter and accepts insults without responding. He does not fight back; he lets himself be humiliated and robbed. The people formerly known as Europeans have been reduced to 'a mass of anonymous *dhimmis*, a collective chattel without history and political rights', who always have to give the best jobs and houses to the Muslims spitting at them.[28]

Now, one could imagine that these protocols of the elders of *Eurabia* would be at most the object of satire, but it became the foundational work in a whole genre of books about the Islamisation of Europe.[29] For parties such as the SD, it was gospel. A rhapsodic review of Ye'or's book in the party press ended by asking, 'So what is the solution?' Answer: 'A comprehensive sending home

---

28 Quotation from Bat Ye'or, 'Eurabia: How Far Has It Gone?', presentation at the Counter-Jihad Conference, Brussels, 18 October 2017. See further Ye'or, *Eurabia*, for example 123, 198, 202–4, 243, 256, 269.

29 A comparison with the classical *Protocols* is expertly laid out in Zia-Ebrahimi, 'When the Elders'.

[*hemsändelse*] of Muslims settled in Europe.'[30] Ye'or might have been pessimistic about the future – barring some allusions to the possibility of reversing 'decades of policy' – but the far right gave her message an activist twist: for another few minutes, Eurabia can be avoided.[31] This thinking shaped the SD on the verge of its belated breakthrough. The text that did more than any other to finally catapult the SD into the parliament was an op-ed by Åkesson in the country's largest newspaper, laying out the whole diagram – mass immigration, high fertility rates, the timeless evil of the Islamic religion, the descent of Sweden into an abyss of mass rapes and pork bans and sharia laws and foreskin removal financed by native tax-payers, producing a sum total of impending loss of the country to the enemy majority – and rounded off with the promise to 'reverse the trend with all the might I can muster'.[32] It caused a sensation in the fall of 2009. Never before had the politics of the SD stood in such an intense national limelight. At exactly this moment, some eyes were turned towards COP15 in Copenhagen – but 'sit still in the boat, it has been rocked before', *SD-Kuriren* soothed its readers, for 'all ages have had their theories of the end of the world. There have been intimations about doomsday, Ragnarök, end of the times, hell and Armageddon' and now it's this climatic theory, even though 'planet earth has always undergone massive changes in climate. In fact even when humans and automotive technology did not exist.'[33] The party espoused the tenets of default denial. It had its choice rewarded by the electorate in 2010, when the SD flew past the parliamentary threshold, and again four years later, and again, as we shall soon see, after some serious incidents in Sweden another four years later.

It is noteworthy that this sort of apocalypticism shares some motifs with established climate change discourse. There is pollution that must

---

30 Harry Vinter, 'Eurabien – Europas integration med islam', *SD-Kuriren*, no. 63, April 2005. The author's name is almost certainly a pseudonym.
31 Ye'or, 'Eurabia: How Far'.
32 Jimmie Åkesson, 'Muslimerna är vårt största hot', *Aftonbladet*, 19 October 2009.
33 Margareta Sandstedt, 'Sitt still i båten, det har gungat förr', *SD-Kuriren* online, 28 April 2008.

be put under control. There are prognoses, projections, approaching deadlines, tipping points and irreversible change, foundations of our lives already now undergoing swift degradation, a duty to coming generations: a civilisation in existential danger. Fear and anxiety race through the hearts of the spectators. Such emotions are deflected from climate and directed towards another target, although not necessarily through the window of the Eurabia theory. In the 2010s, it was complemented by the theory of 'the Great Replacement', so called after a book by French writer Renaud Camus suggesting that white people of Europe are being systematically replaced by overbreeding non-whites – particularly Muslims – courtesy of the treacherous elites. A dystopia equalling Ye'or's, if not as rich in ornamental details, this theory likewise impressed a sense of urgency on its adherents: since we have for so long stood apathetically before the catastrophe rushing towards us, our only salvation now is 'remigration'.[34] When picking its apocalypse, the far right could draw on both resources, or just spit out its choice without reference.

Thus Geert Wilders in 2017 predicted that 'our women' will feel safe nowhere and sharia law be imposed everywhere any day, 'but not a single European government dares to address these existential questions. They worry about climate change. But they will soon be experiencing the Islamic winter.'[35] The same Susanne Winter of the FPÖ who called for the destruction of the 'web of lies' raised alarm over 'a Muslim immigration tsunami' and demanded Islam be 'thrown back to where it came from, the other side of the Mediterranean'.[36] The same Heinz-Christian Strache who blamed the sun for any climatic fluctuations lamented that the Great Replacement had already been effected in Austria but vowed to keep on fighting: 'We don't want to become a minority in our own country.'[37] The AfD spoke about *Volkstod* – death of

---

34   See Jacob Davey and Julia Ebner, *'The Great Replacement': The Violent Consequences of Mainstreamed Extremism*, Institute for Strategic Dialogue, isdglobal.org, 2019.
35   Quoted in Stella Schaller and Alexander Carius, *Convenient Truths: Mapping Climate Agendas of Right-Wing Populist Parties in Europe* (Berlin: adelphi, 2019), 91.
36   Klaus Höfler, 'Grazer FPÖ-Obfrau Winter: "Kinderschänder Mohammed"', *Die Presse*, 13 January 2008.
37   Philip Oltermann, 'Austrian Deputy Leader Endorses Far-Right Term

the *Volk*, meaning 'people' or 'race'. In the election manifesto that brought the party into the Bundestag, the demand for 'minus immigration' ranked alongside that for German withdrawal from all international climate agreements, a balance widely reported in the press.[38] Alexander Gauland, who scoffed at the notion of human influence on climate, could not stop banging on about the approaching day of doom – to sample some typical lines from one of his speeches: if

> our German nation state goes down the drain, there won't be a second chance. We cannot move to another Germany . . . Our children and grandchildren should live in this Germany . . . Our ancestors did not build this land so that our political elite would ruin it and let it go to the dogs. And that is why we are now called upon to preserve the inheritance of our fathers and forefathers . . . I do not want to live in a country where Muslims have the majority . . . This is, dear friends, a policy of human flooding.[39]

In the spring of 2019, the AfD plastered Berlin with a poster in its 'Learning from European History' series, this one reproducing a nineteenth-century painting in the kitschiest orientalist style. A group of men surround a naked white woman. The men are dark of skin, wrapped in turbans and busy inspecting the body of the woman, her neck and breasts, one of the prospective buyers putting his fingers into her mouth: it is a slave market. The text of the poster read *Damit aus Europe kein "Eurabien" wird!* – so that Europe does not become 'Eurabia'![40]

"Population Replacement"', *Guardian*, 29 April 2019; cf. Davey and Ebner, *The Great*, 16.

38 Schneider, Jens 2017, ' "Minuszuwanderung" und Ausstieg aus allen Klimaschutzorganisationen', *Süddeutsche Zeitung*, 9 March.

39 *Frankfurt Allgemeine Zeitung*, 'Gaulands Rede im Wortlaut', faz.net, 5 June 2016.

40 For example Alexander Fröhlich, 'Die nackte Frau und die bösen Turbanträger', *Der Tagesspiegel*, 12 April 2019. The painter of the original was Jean-Léon Gérôme. Another of his paintings, *The Snake Charmer*, has oftened served as the cover of Said's *Orientalism*.

The AfD had pored over the revelations of Bat Ye'or.[41] Her Amsterdam-based think tank International Center for Western Values listed Thierry Baudet as one of its 'friends'.[42] The FvD leader clearly has absorbed some of her thoughts; when asked why conservative parties do not dispute climate science, he snapped: 'They are *dhimmis*.' For him, it was but another apocalypticism. 'Mind you, it all comes down to a retelling of the Noah's Ark, with a flood as punishment for our sins, which we can prevent by repenting. I believe that around the year 1000 we were caught up in similar fantasies.'[43] Baudet had made his choice too. He desperately wanted Europe to remain 'predominantly white', claimed in characteristically pretentious terms that Dutch society was being 'diluted homeopathically' by non-white people and, fond of organic metaphors, warned of 'malicious, aggressive elements' being injected into its 'body'.[44]

All of this suggests a logic of far-right denial deviating from that of the original ISA. Whereas the latter said 'the free market is precious, all is well, climate change is not a real problem', the former said 'the nation is precious, it is going under, climate change is a non-problem'; where ExxonMobil had only the joys of the pump to offer, the far right had its own apocalypse. It proved itself superior in political potency. A wedge of denial was driven deep into the countries for decades perceived as the world's prime paragons of climate mitigation – countries like Austria and the Netherlands, but Denmark, Sweden and Germany above all.

But the far right did not just deselect climate as an item on the political agenda with the potential to distract from its mania. That could not account for the ferocity of the denial. There was nothing coolly indifferent about it. Similarly, the far right could just as well have resolved to follow its standard funnel methodology, accepted global warming as one of the many ills of the present and blamed it on immigration in general and Muslims in particular. That would not have

---

41 See for example posts from AfD Altötting-Mühldorf's Facebook page, 12 July 2017; AfD Hessen's Twitter account, 5 April 2018; AfD Havelland's Facebook page, 5 April 2018.
42 International Center for Western Values, icwv.org, n.d.
43 Gehriger, 'There's a Proper'.
44 Kleinpaste, 'The New Dutch'.

been any more far-fetched than treating unemployment or sexual violence in the same manner. Indeed, some voices on the far right did add climate to their list and squeeze it through the pipe, as we shall see. The fact that the overwhelming majority would not touch it suggests that *some form of investment was at stake*, if not of the immediately financial, ExxonMobil variety. Some of the very same values the far right found in the nation it appears to have located also in the fossil economy, so that defence of the former and of the latter became one and the same thing, deflection and aggression two valences of a single undertaking. If so, the climate politics of the far right – or rather its *anti*-climate politics – was related to the war on immigration as something much more central to the columns than the foiling of a diversion. But the effect was identical. To the extent that the far right moved forward, the issue of climate change fell behind. In this, the trend went far beyond the words and deeds of the parties as such.

A Syndrome of Selection

Only rarely capable of originality in the realm of ideas, the European far right could hear its choice of apocalypse reverberating through the mainstream right. In the first decade of the millennium, when the questions of climate and Islam crossed paths – the years not only of IPCC and Al Gore, but also of al-Qaida, the Muhammad cartoon crisis, Ayaan Hirsi Ali, the French veil bans and a series of other flare-ups – the Swedish association of employers published a magazine named after the prefix of neoliberalism and neoconservatism and a popular film character with superhuman powers. The first issue of *Neo* had the theme 'It's looking bright – and we have only just started' and celebrated the amazing progress of the world thanks to the free market. The second issue was themed 'It looks dark – and it has only just started' and brooded over Islam. Threats stemming from Muslim immigration became an abiding preoccupation of *Neo*, and in January 2007 it filled the cover with ridiculously terrified people running and screaming and madly trying to save themselves from 'greenhouse gases on the attack'. Editor Sofia Nerbrand, a leading intellectual of the Swedish capitalist class, added scare quotes to 'the climate threat' and contended that 'there is in fact no consensus among researchers about the reasons for the fluctuations in climate'.[45] The other house organ of the bourgeois intelligentsia, *Axess*, named after its owner Antonia Ax:son Johnson – as fourth-generation head of one of Scandinavia's greatest conglomerates, the richest woman in Sweden – and the general blessing of having access to things, painted the world in the same colours: bright for markets, dark for Muslims, blank for climate but occasionally dark for concerns about it. The Eurabia theory found its way into the output of the *Axess* sphere too.[46]

But it was, of course, the American right that led the way for the rest of the world. A figure who personified the convergence of reactions, in all his coarse and derisive style perfectly pitched to the zeitgeist, was Mark Steyn, trusted pundit at Fox News and *The Rush Limbaugh Show* and author of *America Alone: The End of the World As We Know It*, a bestseller lauded by such luminaries as Christopher Hitchens and Martin Amis and recommended by George W. Bush to the White House staff.[47] The book was Steyn's gift to the burgeoning Eurabia genre. White women in Europe give birth to perilously few children. The fast-breeding Muslims are poised to take over – the threshold of 50 per cent of the population will be crossed in the 2030s, that of 100 per cent sharia law in 2040 – making the natives an endangered population. Northern Europe is the ground zero for this ordained oblivion, but it will ultimately be a global event. 'We are living through a remarkable period: the self-extinction of the races who, for good or ill, shaped the modern world', Steyn opined in the *Wall Street Journal*.[48] The only solution that might just ward off the end of the world is the one practised during the war in Bosnia by Serbian nationalists, whose influence on this way of thinking we shall

---

45 Sofia Nerbrand, 'Håll huvudet kallt', *Neo*, January/February 2007.

46 See particularly Pierre Lellouche, 'Ett islamiskt hot mot Europe?', in Kurt Almqvist (ed.), *Den sekulära staten och islam i Europa* (Stockholm: Antlantis/Axel och Margareta Ax:son Johnsons stiftelse, 2007), 113–22.

47 Mark Steyn, *America Alone: The End of the World As We Know It* (Washington, DC: Regnery, 2006); Christopher Hitchens, 'Facing the Islamist Menace: Mark Steyn's New Book Is a Welcome Wake-Up Call', *City Journal*, Winter 2007; Martin Amis, 'The Decline of the West', *The Times*, thetimes.co.uk, 14 April 2007; Irwin M. Stelzer, 'Reader of the Free World', *Weekly Standard*, 12 March 2007.

48 Mark Steyn, 'It's the Demography, Stupid', *Wall Street Journal*, 4 January 2006.

return to. For Steyn, they knew and lived the truth: 'If you can't outbreed the enemy, cull 'em.'[49]

In the prologue to his tract, Steyn juxtaposed the two fears.

> Much of what we loosely call the Western world will not survive the twenty-first century, and much of it will effectively disappear within our lifetimes . . . and unlike the ecochondriacs' obsession with rising sea levels, this isn't something that might possibly conceivably hypothetically threaten the Maldive Islands circa the year 2500; the process is already well advanced as we speak.[50]

At the Heartland Institute's tenth International Conference on Climate Change in the summer of 2015, Steyn was a keynote speaker, up at the top of the bill with James Inhofe.[51] Taking a poke at COP21 in Paris, soon-to-be presidential candidate Ted Cruz invited Steyn in December of the same year to testify before the Senate.[52] By now a star in the firmament of the denialist ISA, the next year the same Steyn gave a talk at a conference titled 'The Climate Surprise: Why CO_2 Is Good for the Earth', hosted by the newly formed CO_2 Coalition, heir to the George C. Marshall Institute and one of ExxonMobil's main pre-2007 beneficiaries; here he elaborated on a theory only sketched in *America Alone*.[53] The world is threatened by 're-primitivization'. There are two agents of the process: the Muslims and the climate movement. Both hate the West. Allies in the struggle against modernity, they do, however, sometimes come into conflict.

> Earth Hour. You know Earth Hour? This ridiculous thing every year where they turn off the lights for an hour on a Saturday evening and you pretend you're doing something to save the planet by sitting in the dark for an hour. In Sweden, a Swedish town a couple of weeks ago, they

---

49   Steyn, *America*, 5.
50   Ibid., xiii.
51   ICCC-10: 'Speakers', climateconference.heartland.org/speakers.
52   US Senate Committee on Commerce, Science, and Transportation, 'Data or Dogma? Promoting Open Inquiry in the Debate over the Magnitude of Human Impact on Earth's Climate', commerce.senate.gov, 8 December 2015.
53   'The Climate Surprise: Why CO_2 Is Good for the Earth', CO2coalition.org, 25 April 2016; Steyn, *America*, xiii, 137.

had to cancel Earth Hour because they've taken in so many so-called refugees [laugh] that it's unsafe to switch the lights off for an hour. Because the hot Scandinavian blondes are going to be feeling a lot of unwanted attention as they're wandering around an entirely darkened Swedish town for 60 minutes. And that's fascinating to me.[54]

But action in unison was the rule. An interviewer from the CO_2 Coalition asked Steyn to break down the features shared by 'the environmentalist, pro-climate disruption [sic] crowd' and the 'Islamic supremacists'. Both, he explained, evince 'an actual hatred of humanity'. The former aims primarily at doing away with the humans who emit too much:

> If you're picking which two billion people are allowed to survive, they point out that someone who is born in Somalia has a carbon footprint that's only a twentieth of the size of someone who is born in Sweden or Canada. So therefore it's the Western world that needs to put itself out of business.[55]

The climate movement wants to extinguish white humanity and hand the planet over to the Somalis. This might count as the most advanced attempt yet to formulate a unified theory of an Islamic-climatic threat to the survival of the 'races' that 'shaped the modern world'.

During the weeks in 2016 when Trump established his lead over the other Republican presidential candidates, Steyn went on a booked-out tour in Australia. Advertised as 'the world's greatest conservative commentator and writer', he was invited by the think tank Institute of Public Affairs, an Australian node in the anglophone network of such bodies, likewise financed once upon a time by ExxonMobil and Shell.[56] Steyn contributed to the Institute's 2015

---

54 Mark Steyn, 'Mark Steyn's Stand against Climate Alarmism: In-Depth with the Climate Crybully Conniption-Inducer', *New Criterion*, 29 March 2016.
55 Ibid.
56 Institute for Public Affairs, 'Mark Steyn Australian Tour', ipa.org.au, February 2016; Brad Norrington, 'Think Tank Secrets', *Sydney Morning Herald*, smh.com.au, 12 August 2003; *The Age*, 'The Global Warming Sceptics', 27 November 2004.

anthology *Climate Change: The Facts* (also featuring Richard Lindzen). In the years following its appearance, this volume was the first to show up when typing 'climate change' into the search bar of Amazon's book section. Then in 2018, Steyn was back at Fox News to chastise Democrats and CNN for being more critical of white supremacists than of undocumented immigrants. 'The white supremacists are American citizens. The illegal immigrants are people who shouldn't be here.'[57]

If Mark Steyn was a lone sailor on these oceanic routes of convergence, he could be brushed aside. He was not. In his company were other notable contributors to the Eurabia genre, such as Bruce Bawer, author of *While Europe Slept: How Radical Islam Is Destroying the West from Within* and advocate of the 'mass expulsion' of Muslims from the continent, and Bruce Thornton, author of *Decline and Fall: Europe's Slow Motion Suicide* and discoverer of the continent's Muslims as the incarnation of the subterranean subhuman species 'the Morlocks' (first invented by H. G. Wells in *The Time Machine*).[58] Both published excited denialist screeds in *FrontPage Magazine*.[59] Operated by the David Horowitz Freedom Center, *FrontPage* mostly spewed out hatred of Muslims. Donors Capital funded the David Horowitz Freedom Center.[60] Horowitz was the largest donor to Wilders's PVV, and a featured speaker at an annual meeting of ALEC. The American Enterprise Institute enlisted Ayaan Hirsi Ali and teamed up with 'counterjihad' activists; the Heritage Foundation gave the stage to celebrities from the anti-Islam crusade; the Heartland Institute protested that the true history of 1,400 years of Islamic attacks on the West was being excised from European school

---

57   Judd Legum, 'The Unrepentant Racism of Tucker Carlson Tonight', Fox News, 18 January 2018.

58   Bruce Bawer, *While Europe Slept: How Radical Islam Is Destroying the West from Within* (New York: Doubleday/Random House, 2006), 230; Bruce Thornton, *Decline and Fall: Europe's Slow Motion Suicide* (New York: Encounter, 2007), 78–9. For an excellent reading of Bawer, see Sindre Bangstad, 'Eurabia Comes to Norway', *Islam and Christian-Muslim Relations* 24 (2013): 375–81.

59   Bruce Bawer, 'Inside the "Warmist" Faith', *FrontPage Magazine*, 14 August 2012; 'Junk Science and Leftist Folklore Have Set California Ablaze', *FrontPage Magazine*, 17 August 2018.

60   See the transactions tracked, as of November 2018, by *Conservative Transparency*: 'DonorsTrust & Donors Capital Fund', conservativetransparency.org.

curricula, and so on.⁶¹ In its fourth phase, the denialist ISA formed close relations with Islamophobic thinkers and their tanks – often referred to as 'the counterjihad movement' between roughly 2005 and 2015, when the term went out of fashion – the black box of billionaire money spreading its largesse over both.⁶² The love was reciprocal.

A more refined way of acting out the posture was to say that both climate change and Islamophobia had been blown out of all proportion. This was the argument of Pascal Bruckner, reliable 'philosopher' of the French right and author of two companion volumes: *The Fanaticism of the Apocalypse: Save the Earth, Punish Human Beings* followed by *An Imaginary Racism: Islamophobia and Guilt*. In the former, Bruckner used his broadest brush to paint a picture of concerns over climate change as a medley of self-flagellation, Gnosticism, the Prometheus myth, the Mayan calendar, disaster films, noble savages, the Genesis narrative of the fall from grace and one or two other ingredients. 'Ecologism' is but an updated catalogue of theological and mythological scare stories. The science of climate change 'eludes the test of verification'; it engages in 'marvellous logical acrobatics'; it is a 'magical skeleton' that requires of its practitioners chiefly 'a diploma in intimidation'.⁶³ It is out to punish us for the mere sake of it. Under the rule of 'commissars of carbon', there will be

---

61 Branko Marcetic, 'Populist Billionaires', *Jacobin*, 25 August 2017; Kanyakrit Vongkiatkajorn, 'Verizon Leaves Lobbying Group ALEC over Ties to Anti-Muslim Activist', *Mother Jones*, 15 September 2018; Nathan Lean, 'A Top Conservative Think Tank Embraces Islamophobia', *Truthdig*, 22 June 2012; Linda Sarsour, 'If the First Thing You Ask a Muslim Woman Is If She's American, You Might Be a Bigot', *Guardian*, 18 June 2014; Joy Pullmann, 'College Board Excises Islam From European History Curricula', Heartland Institute, 29 June 2016.

62 See Chloe Patten, 'Boston Bombings: Beware the Multi-Million Dollar Islamophobia Industry', *The Conversation*, 24 April 2013; *PressProgress*, 'Exclusive: Rebel Media's Ezra Levant Received Foreign Funding from "Anti-Muslim" Think Tank', pressprogress.ca, 17 August 2017; Arun Kundnani, *The Muslims Are Coming!: Islamophobia, Extremism, and the Domestic War on Terror* (London: Verso, 2013), 250; Nathan Chapman Lean, *The Islamophobia Industry: How the Right Manufactures Hatred of Muslims* (London: Pluto, 2017), 159.

63 Pascal Bruckner, *The Fanaticism of the Apocalypse: Save the Earth, Punish Human Beings* (Cambridge: Polity, 2013), 64, 78, 121, 122.

no more skiing, surfing, snowboarding, tobogganing; put away your skis and poles, and forget all-terrain vehicles and motorsports at the coast. You have to stop doing all that. Biking and organic food, nothing else. You used to have fun? Well, now you have to atone for it,

and switch off the refrigerator and give up toilet paper and eat nothing but gruel. Bruckner sees before him how the commissars will sterilise people. A special shudder runs down his spine when he comes across a report showing that men 'exhale a much higher amount of carbon than women do'. But worst of all is the idealisation of African poverty and the demand that everyone be reduced to a third-world standard of living. The end point of all of this can only be 'the slow extinction of the human race.'[64]

The critique of Islamophobia has the same epistemological status as climate science, in the eyes of Bruckner. It represents 'the fabrication of a new crime of opinion' and reinvents the notion of the original sin. But if Islamophobia is an imaginary problem, it turns out that Islam is a very real one indeed. It has become a culture of death, with any and all redeeming features gone, and now it makes demands on us, inside Europe itself – and so we must fight back against it, get properly armed and swap an Islamisation of Europe for a Europeanisation of Islam. 'We will not win this war solely with spies, tanks, and planes'; we will win it only if 'we persuade ourselves and the rest of the world of the eminent virtues of our civilization'. And 'there comes a time when we simply have to say: this is the way we live. Take it or leave it.'[65] This is the way we live, all-terrain vehicles and the rest of the package.

It was not so difficult to spot where climate science and the critique of Islamophobia touched the same raw nerve in Bruckner. Both came with an immanent questioning of the ways of the West – 'the disease of the end of the world is purely Western' (*The Fanaticism of the Apocalypse*); 'the liberal capitalistic and

---

64 Ibid., 147, 156, 154.
65 Pascal Bruckner, *An Imaginary Racism: Islamophobia and Guilt* (Cambridge: Polity, 2018), 24, 178, 155.

imperialist West is guilty of causing everything bad on Earth' (*An Imaginary Racism*).[66] It was just a slightly convoluted way of expressing Mark Steyn's grand theory of climate and Islam as two fronts of 're-primitivization'. If Bruckner represented the smugly sophisticated highbrow end of the right-wing spectrum, Anders Breivik was at the combatant pole. In the 1,500-page compendium he posted online to justify his massacre of social-democratic youth activists in Norway in 2011, called *2083: A European Declaration of Independence*, he not only copy-and-pasted the totality of Eurabia lore, but also endorsed the view of global warming as a fabrication cooked up to rob the West of its wealth.[67]

We are here dealing with a political syndrome of the early twenty-first century. In the historical moment when Europe and North America were requested to phase out the fossil fuels on which their economies had been built, the forces of the right made a careful selection: the problems we really should fret about have to do with immigration – with people, that is, who are not quite white like us, not ancestrally connected to the land, but who nevertheless have come to dwell here. If the selection could be read off with some precision from the writings of Mark Steyn or the transactions of Donors Capital, it could also, and more significantly, be measured in elections. The undergrowth of ideas shot up at election time, in Europe, in the United States and beyond: but perhaps in Europe in particular, a continent whose politics now became utterly overshadowed by the one issue of immigration. Just as Europe faced the climate crisis, it turned its gaze to the sons and daughters of non-European lands.

We can then revisit our earlier observation that every time a European far-right party denies or downplays the climate crisis, it makes a statement about immigration and note that by and large, the reverse is also true. When such a party rails against immigration, it de facto comments on global heating. An immensity of political energy has been thrown into the war on the former, and if future

---

66 Bruckner, *Fanaticism*, 181; Bruckner, *An Imaginary*, 60.
67 Anders Behring Breivik, *2083: A European Declaration of Independence*, 2011, 656–8. For analysis of this text, see Bangstad, *Anders*, ch. 3.

generations will have to sift through the reasons for why so little was done to prevent a climate catastrophe foretold, this will be one of them: the fictional problem of an immigrant invasion and a Muslim takeover supplanted the real thing. While some lived in biophysical reality, the far right – often the winner on the ground – levitated into an imaginary, hallucinatory, phantasmagorical sphere. Steyn and his lot would, of course, say exactly the opposite. And thus the forces were lined up.

A Break with Reality

One could imagine that the actual strike of a climatic disaster would turn the tables on the right, perhaps even bring it to its senses. That, however, cannot be taken for granted. The extreme summer of 2018 gives a hint of other possibilities. A heat dome covered the northern hemisphere for months. There was a La Niña phase, which ordinarily would convey cool water and air towards the north; instead, heat records were smashed like icicles with a sledgehammer. The highest overnight temperature ever recorded came in from the coast of Oman. The highest temperature ever recorded in Africa came in from the Algerian desert. But it was in the Arctic that the maxima were most brutally shattered, as wildfires made their way through the Siberian tundra – accelerating the melting of permafrost and the consequent release of methane – and engulfed forests in Sweden; within the Arctic Circle, the heat wave had no historical precedent, leaving climate scientists, as so often before, baffled by the speed and ferocity of the developments.[68]

Among the worst affected countries, one was just a few weeks away from an election. That country was Sweden. It received barely any rain at all between late May and August. Nights classified as tropical were measured in the far north. July became the hottest month ever recorded, averages reaching 5.5 degrees Celsius above normal, the ultra-red zones on the maps of weather anomalies looking like

---

68 For example Damian Carrington, 'Heatwave Made More Than Twice As Likely by Climate Change, Scientists Find', *Guardian*, 27 July 2018; Jonathan Watts, 'Heatwave Sees Record High Temperatures around the World This Week', *Guardian*, 13 July 2018.

patches of badly burnt skin. During the endless sun, harvests withered and farmers sent their cattle to mass slaughter for lack of fodder.[69] 'If we get this kind of summer for two or three years in a row, Swedish agriculture will be finished', said one expert to *Dagens Nyheter*, the leading daily.[70] The tinderbox inevitably ignited: wildfires roared through the plantation forests of the northern plains, turning chlorophyll into charred coal for miles on end, prompting evacuation of villages and orders to tens of thousands to stay inside their homes and forcing the Swedish state to call upon fighter jets to bomb the conflagrations, in an eerie war game no one had witnessed before. If they didn't burn, trees across the nation – normally at their lushest this time of the year – took on sickly red and brown hues. In the farthest north, the top summit of Kebnekaise, consisting of a cone-formed glacier in a drawn-out process of meltdown, calved chunks of ice. By the end of July, it had shrunk to the point where it was no longer Sweden's highest peak, a kind of geological closure.[71]

The nation contracted acute climate anxiety. *Dagens Nyheter* reported that the temperatures 'stand at levels so novel to Sweden that it's difficult to put them in any context', the joys of summertime 'under the shadow of a doomsday feeling'. This placed the Sweden Democrats in something of a quandary. They had planned for the election scheduled for early September to be about immigration, and now suddenly the country spoke about the climate crisis instead, the issue dominating headlines and talk shows and climbing rapidly among the top concerns of the electorate.[72] 'What does your party

---

69 Records were collated and maps constructed by the Swedish Meteorological Institute: SMHI: 'Juli 2018 – Långvarig hetta och svåra skogsbränder', smhi.se, 31 July 2018.

70 Johan Kuylenstierna quoted in Jannike Kihlberg, 'Klimatet i Skandinavien ändras fortare än man trott', *Dagens Nyheter*, 11 August 2018.

71 As reported in for example Oscar Magnusson, 'Kebnekaises sydtopp inte längre högst', *Norrländska Socialdemokraten*, 1 August 2018.

72 The Sweden Democrats must be forgiven for this expectation, since all other parties except one (the Left Party) had signed up to the view that 'integration' – euphemism for the presence of non-white immigrants in the country – was the central political problem of the times. For an excellent analysis of this and other aspects of the 2018 election, see Göran Therborn, 'Twilight of Swedish Social Democracy', *New Left Review* series 2, no. 113 (2018) – which, however, does not say a single word about the extreme weather and the climatic dimensions of the election. Such implicatory denial is still all too prevalent on the left. On implicatory denial, see further below.

want to do to cut emissions?' journalists asked, pressing their microphones into the chests of SD representatives who had memorised lines on mass immigration. Observers speculated that the summer might burn away some of their support.

In recent years, the party had softened the tone if not the substance of its position on climate. Leaked internal communication suggests that the leadership worried that pushing denial would reduce the party's chances to win over more female voters (the kind of limited rationality that might rein in a far-right party).[73] Even so, the SD's energy and climate agenda had been crafted under the influence of denialists from the Swedish business community, and the spokesmen for this policy portfolio had indeed continued to harp on the same old strings – 'elevated levels of CO_2 have so far been a blessing for our planet'; 'no one has been able to say that a couple of degrees would be that bad, and I don't think it will happen'; 'we don't participate in a competition about who is most worried about climate change'; 'the predictions have extremely large insecurities', and the rest of it.[74] Taking a cue from Trump, in January 2017 the party proposed to cut the budget of the Swedish Meteorological Institute as punishment for its 'propaganda'. The shadow budget for that year put 'the climate' within quotation marks, and the main online journal of the SD published a stream of denialist pieces.[75] But the leadership preferred to speak little on the subject. 'One shouldn't exaggerate the possible consequences for Sweden', environmental spokesman Martin Kinnunen claimed in June 2018, days before the dimensions of the drought and the heat wave became apparent, in a statement at the softer end of the scale. 'It will be developing countries that suffer, if any', he said, to explain the low priority of the issue.[76]

---

73 David Baas, 'SD-politik styrs dolt av klimatförnekare', *Expressen*, 19 October 2016.

74 Ibid.; Josef Fransson, 'Förhöjda halter av CO_2 är en välsignelse', *Bärgslagsbladet – Arboga Tidning*, bblat.se, 16 December 2015; Martin Kinnunen, 'Parisavtalet är en papperstiger', *Aftonbladet*, 9 June 2017; Johanna Alskog, 'SD: SMHI ägnar sig åt tendentiös opinionsbildning', *Altinget*, altinget.se, 9 January 2017; Jonathan Jeppsson, 'SD: Skär ner 8 miljarder på klimat och miljö', *Aftonbladet*, 14 January 2017.

75 Oscar Sjöstedt m. fl. (SD), Motion till Riksdagen 2016/17: Budgetpropositionen för 2017, 81; misc. articles on climate change in *Samtiden*.

76 Niklas Svensson, 'Miljön är inte högst upp på dagordningen', *Expressen*, 22 June 2018.

In a scorched country, such a line of disinterest didn't work any longer. The SD had to have something to say. In early August 2018, Jimmie Åkesson returned from vacation and kick-started his election campaign with a speech in his hometown, a sleepy village in the southern countryside, as surrounded by yellow lawns and dried-out fields as any other: and began by praising the weather of the past months. 'We have enjoyed an amazing summer, with sun and heat, warm seas and lakes and warm swims', he rejoiced. 'During the winter half year, this is the weather we are prepared to pay tens of thousands to find somewhere else – we shouldn't forget that.' Taunting those who connected the dot with climate change, Åkesson went on: 'Many conclude that this is the ultimate proof that the world is going under. Is it going under? I actually don't think so. To turn a single summer's weather into politics is simply not serious. It is the worst kind of populism', a charge he relished turning against his opponents. He did, however, acknowledge that the farmers in Sweden – so treasured by the party – were in some dire straits. What should be done to ease their burden? The SD leader submitted that taxes on diesel and gasoline must be slashed. Indeed, retroactive tax cuts must be decreed for those fossil fuels. The audience roared its approval. Then Åkesson returned to his favourite terrain and repeated that 'there is only one party putting Sweden and the needs of Swedes before mass immigration from all over the world, and that's us, and we intend to win this election'.[77]

Over the next weeks, the SD held firm to this line: lose no sleep over the summer; set the fossil fuels free from undue restrictions; deal harshly with the threats to Sweden. When the election manifesto was unveiled, the press reported cuts on fuel taxes as one of its main points.[78] The document also demanded the scrapping of environmental regulations for industry; the abolition of the recently instituted, very modest tax on aviation; the keeping of the contested Bromma airport in Stockholm, mostly used for domestic business flights; eased rules for the EPA tractors venerated by some of the

---

77 *Expressen*, 'Se hela Jimmie Åkessons sommartal', 11 August 2018.
78 Amanda Lindholm, 'SD vill hyra fängelseplatser utomlands', *Dagens Nyheter*, 29 August 2018.

party's rural constituencies.[79] Åkesson refused to concede anything to the evidence that the heat dome was linked to climate change and flaunted an attitude best characterised as wilful ignorance. 'I don't know how fast the climate is changing', but 'the earth's climate has always changed and always will', he said at one point; at another, 'I am no expert on weather, but people talk about there having been an El Niño and a heat wave it brought us this summer, and there's probably something to that.'[80] When a journalist asked him 'what does your climate anxiety look like?', Åkesson replied: 'I have anxiety, above all about the fact that a fourth of our country's women don't dare to go outside in the dark' – a reference, as everyone understood, to the immigrant rape gangs that supposedly roamed the streets of the nation.[81]

On immigration, as on climate, the SD had tempered its rhetoric a few notches upon entrance to parliament. Most importantly, the idea of 'repatriation' – that is, mass expulsion of people living in Sweden on the basis of some origins-related criterion of not belonging there – had been put on the shelf. After the government of social democrats and greens closed the borders to refugees in late 2015, however, there was little to separate the party from national policy. The Sweden Democrats gleefully welcomed its adversaries to the position that the house of the nation is full, and then dusted off the banner from their skinhead years and readied for 'the next great battle in migration policy': the removal of certain people to the countries 'where they should live' (Åkesson).[82] This was now referred to as 'remigration'. Who should be prompted to leave? Spokespersons vaguely and sweepingly pointed to 'the many who don't feel at home in this country and live in marginal areas' and suggested that the Swedish Migration Agency be converted into a Remigration Agency tasked

---

79 Sverigedemokraterna, *Sverigedemokraternas valmanifest 2018*, sd.se, August 2018, 11, 23–4, 26.

80 Karin Eriksson, 'Jimmie Åkesson (SD): Vi planerar för att det blir extraval nästa år', *Dagens Nyheter*, 16 August 2018; Malin Roos, 'Får jag sova tre timmar i sträck är jag nöjd', *Expressen*, 20 August 2018.

81 Roos, 'Får jag'.

82 TT (Tidningarnas Telegrambyrå), 'Åkesson vill inte göra SD radikalare nu', *Sydsvenskan*, 10 February 2018.

with the administration of departures.[83] As any good executor of ethnic cleansing, the party described the process as voluntary. The premise built into it, of course, was that recently arrived or long-settled Iraqis and Syrians, Afghanis and Iranians, Somalis and Palestinians, their children and grandchildren and others among the 2.5 million Swedes of foreign descent were *not* clamouring to evacuate, which was, presumably, why the SD existed as a party in the first place. Any outflow would have to be implemented through coercion. Yet the demand for 'remigration' – much like the 'minus immigration' of the AfD – was given pride of place in the 2018 election manifesto.[84]

On the way into the election campaign, Richard Jomshof, now secretary of the party, reiterated that 'Western Europe will perish because of Islamisation, if we do not dare tackle the problems we have' and proceeded to lay out the vision of remigration.[85] A third member of the gang of four, Vice Speaker of Parliament Björn Söder, repeated his belief that one cannot be a Jew and a Sami and at the same time a Swede, effectively drawing a line around the white-majority population, shading the party with ethnonationalism writ large.[86] Below the top branches, a dense vegetation of richer elaborations grew, some of which had to break through cracks in the cordon the leadership tried to maintain around it. 'Swedes are white and the country is ours', ran the refrain of a song shared online by a leading representative in the southern city of Växjö, who then confirmed her view that 'in order to be Swedish one has to be born Swedish and have Swedish parents'.[87] 'One CANNOT force different races to live together. Ethnicity is something you are born with and Swedish genes cannot be acquired, that's a fact', wrote a candidate in the southern

---

83   *Expressen*, 'SD-toppen vill starta skolor i Syrien – med svenska lärare', 13 February 2018.
84   Sverigedemokraterna, *Sverigedemokraternas*, 7, 20.
85   Karin Eriksson, '"Vi kan inte lämna över till en generation som förstör allt"', *Dagens Nyheter*, 6 April 2018.
86   For a commentary on the statement in English, see Peter Wolodarski, 'Sweden's Far Right May Loom Large, but I Am Confident It Won't Govern', *Guardian*, 8 August 2018.
87   *Aftonbladet*, 'SD-kvinnans länk på Facebook: "Svenskar är vita"', aftonbladet.se 19 August 2018.

city of Helsingborg, also hitting out at 'politicians and celebrities with Jewish blood' and Arabs who have 'lost their brain cells'.[88] Another candidate to a municipal assembly paid tribute to Adolf Hitler again and again. Another called for the prosecution of traitors to the nation. Another expressed her wish for all Muslims to drown in a cesspool. Yet another incited the burning of immigrants, and on it went, in an incontinence chronic to the party, the leaks impossible to plug up for the gang of four however hard they tried (if one believes their own protestations following every media scandal).[89] It did nothing to dent the support for the SD. Nor did the experiences of the summer.

In 2018, the Sweden Democrats cemented their position as the third largest party, with 17.5 per cent of votes, up by 4.6 per cent – the eighth consecutive election yielding a growing share. There were no signs that the party performed worse than average in areas most affected by the disasters. In fact, the reverse was true. In Ljusdal, where the largest wildfire in the modern history of Sweden raged, SD was the second most popular party in the election to Riksdagen, with 19.9 per cent of the votes; on the dried-out island of Öland – its landscape shrivelled, farmers desperate, basic water supplies maintained by trucks from the mainland – it reached above 20. No hard evidence suggested that immigrants or Muslims or Jews threatened the well-being of those places. But some fantasies die hard: on the day after the election, Mattias Karlsson, fourth man in the gang and leader of the party's parliamentary group, posted the following update on Facebook, accompanied by a picture of the knee-bending soldiers of imperial king Carolus Rex:

> Sweden is in really bad trouble. So it has been many times before in history ... Yet we have always stood up, yet we have always prevailed in spite of the odds and we have survived. Thanks to the unyielding steadfastness, conviction, will to sacrifice and leadership of a small cohort of patriots. We have to be that cohort now.

---

88 David Baas, 'SD-politiker förnedrar offren för Förintelsen', *Expressen*, 31 August 2018.

89 Ibid.; David Baas, 'SD-politikern: "Bara att hoppas att färjan förliser"', *Expressen*, 6 September 2018.

Fate has appointed us this time. There is no time to rest or grieve lost illusions and hopes. We have not chosen this, but our opponents have for real forced us into an existential battle for the survival of our culture and our nation. There are only two options, victory or death.[90]

The far right rode on a break with the reality of climate breakdown.

A Way to Exit

In the UK, the syndrome of selection grew into the Brexit project. One could hear Katie Hopkins's heart flutter. A celebrity worthy of the early twenty-first-century culture industry – star on *The Apprentice* show, friend of Donald Trump, social media shocker, sometime columnist and businesswoman – she would go on to become one of the more boorish Brexiteers. In 2015, Hopkins told her readers in the *Sun* that refugees trying to cross the Mediterranean were 'cockroaches'. She had a solution for them: 'Bring on the gunships, force migrants back to their shores and burn the boats.' She believed that 'white Christians' like herself were being pushed out of Europe and that 'black extremists' were clearing out whites from South Africa. Whereas white genocide was urgently real for this reality bigot, the climate crisis was 'clap trap'; commenting on COP21, she wondered 'if there aren't more pressing matters at our door'. 'Why are 150 world leaders, 40,000 delegates and all the usual luvvies fiddling around with climate "change" while ISIS makes the world burn?'[91]

In the second decade of the century, the trend coursed through the British superstructure: the *Sun*, *Daily Mail*, *Daily Express*, and

---

90 Mattias Karlsson's Facebook page, 10 September 2018.
91 James Martin, 'Katie Hopkins Wrote This in the Sun about Migrants and Now Everyone Is Really Angry', *Huffington Post*, 18 April 2015; Natalie Martinez, 'The White Nationalist Influencers of Instagram', *Media Matters*, mediamatters.org, 8 April 2019; Fiona Day, 'Katie Hopkins Refers to Climate Change as "Clap Trap"', *Closer*, 27 September 2014; Katie Hopkins, 'Why are 150 world leaders, 40,000 delegates and all the usual luvvies fiddling around with climate "change" while ISIS makes the world burn?', *Daily Mail*, 30 November 2015.

the *Spectator* alternated between sorties against Muslims and sallies against climate science.[92] So did a journalist by the name of Boris Johnson. He installed himself as a loudspeaker for one of the crankier denialists on the British scene, Piers Corbyn – amateur weather forecaster and astrophysicist, elder brother of Jeremy – who dedicated his life to proving that it was all about the sun. 'Global temperature depends not on concentrations of CO_2 but on the mood of our celestial orb,' Johnson transmitted in the *Telegraph* in 2013. Corbyn–Johnson possessed the knowledge that, since the sun is losing its spots, 'we are in for a prolonged cold period. Indeed, we could have 30 years of general cooling.'[93] The prime minister *in spe* was back at it two years later, in time for the Paris Agreement, which he understood as 'driven by a primitive fear that the present ambient warm weather is somehow caused by humanity', a fear 'without foundation'. Here he belted out one of the slier arguments from the denialist depot: that humans are arrogant to think that they could alter the climate. What are we against the sun? Affecting modesty and erudition, Johnson described climate science as hubris against our solar 'governor and creator' and compared it to the folly of Agamemnon, who thought it was his own fault that no wind blew him to Troy – 'it was all about him, him, him.'[94] Humble humans understand that their actions matter little and can carry on with a light heart.

But 'the problem is Islam. Islam is the problem', Johnson pounded after the London bombings in 2005.[95] 'The real problem with the Islamic world is Islam', he repeated in an essay called 'And Then

---

92   Melanie Phillips was an early practitioner: see for example *Londonistan* (New York: Encounter, 2006); 'Blame the Trees!', *Daily Mail*, 13 January 2006; 'Our Political Landscape', *Spectator*, 28 September 2008.

93   Boris Johnson, 'It's Snowing, and It Really Feels Like the Start of a Mini Ice Age', *Telegraph*, telegraph.co.uk, 20 January 2013. The second is a quote from Corbyn. For a profile of him, see Tim Adams, 'Piers Corbyn: The Other Rebel in the Family', *Guardian*, 24 January 2016. For an earlier accolade to Corbyn signed by Johnson, see Boris Johnson, 'The Man Who Repeatedly Beats the Met Office at Its Own Game', *Telegraph*, 19 December 2010.

94   Boris Johnson, 'I Can't Stand This December Heat, but It Has Nothing to Do with Global Warming', *Telegraph*, 20 December 2015. Arrogance is also a theme in Johnson, 'It's Snowing'; cf. Santiago Abascal above.

95   Boris Johnson, 'Just Don't Call It War', *Spectator*, 16 July 2005.

Came the Muslims'; moreover, there is 'no stronger retrograde force' on the planet. It represents the absolute negation of progress. 'There must be something about Islam that indeed helps to explain why there was no rise of the bourgeoisie, no liberal capitalism and therefore no spread of democracy in the Muslim world.'[96] How could the rule of this trinity be assured? Journalist Johnson also turned to the dark continent. 'The best fate for Africa', he wrote in the *Spectator*, 'would be if the old colonial powers, or their citizens, scrambled once again in her direction; on the understanding that this time they will not be asked to feel guilty'.[97] As with greenhouse gases, empire was no reason to have a bad conscience. Rather there ought to be more of it.

These were some of the ideas circulating through the constellation of right-wing forces that would pull off Brexit. In another conjuncture, months before the election that brought Margaret Thatcher to power in 1979, Stuart Hall observed a 'dialectic between the radical-respectable and the radical-rough forces of the Right'. The neofascists of the National Front marched in the streets; the Tories 'reworked' their hobbyhorses on race and immigration *sotto voce*. The 2010s saw another act of this 'great moving right show', as Hall called it, with half of the cast renewed, the rough part now played by *inter alia* the United Kingdom Independence Party (UKIP).[98] It led the Tories towards Brexit. Or did it follow? The identity of the leader was uncertain, for the Tories and UKIP lived in an 'essentially symbiotic relationship', the partners stimulating one another.[99] But the tugging of UKIP was clearly decisive in pushing the Tories to announce a referendum on EU membership, and it is

---

96 Quotations from Frances Perraudin, 'Boris Johnson Claimed Islam Put Muslim World "Centuries Behind"', *Guardian*, 15 July 2019. 'No stronger retrograde force' was a formulation Johnson borrowed from his oracle Winston Churchill.

97 Boris Johnson, 'Cancel the Guilt Trip', *Spectator*, 2 February 2002.

98 Stuart Hall, 'The Great Moving Right Show', *Marxism Today*, January 1979, 19–20. Note that we are not here attempting a complete explanation of Brexit, any more than of the Swedish election of 2018 in the previous section; we are instead looking for trends.

99 Tim Bale, 'Who Leads and Who Follows? The Symbiotic Relationship between UKIP and the Conservatives – and Populism and Euroscepticism', *Politics* 38 (2018): 264.

for this the former party looks likely to be remembered. Owing to the first-past-the-post system, it never won more than one seat in the House of Commons – despite taking 12.6 per cent of the votes in 2015 – a considerable underachievement compared to the rest of the European far right. But UKIP helped to precipitate the 'departure from the European Union of one of its biggest member states. That, by anyone's standards, constitutes impact – and on a truly historic scale', even if it was the Tories who reworked the themes and ultimately reaped the rewards, much like under Thatcher.[100] Here is another peculiarity of the English: the far right is repeatedly reconstituted *inside the main conservative party*, which, to an unusual degree, excites and then sweeps up the rougher elements. The SD outgrew the Moderates, the AfD challenged Angela Merkel, but in the UK the far right ended up chasing Boris Johnson.

During the lead-up to Brexit, UKIP was the voice of denial in British politics. In 2013, party leader Nigel Farage parroted Boris Johnson – or was it the other way around? – by brandishing before the European Parliament two print-outs from NASA purportedly showing how the Arctic ice cap had grown by 60 per cent in one year. 'We are now', he asserted, 'going into a period of between 15 and 30 years of global cooling. We may have made one of the biggest, stupidest collective mistakes in history by getting so worried about global warming.' But the mistake was not haphazard: its purpose was to foist a world government on gullible nations.[101] Gerard Batten, who took over UKIP in 2018, deemed the worry 'a scam to milk the masses'. (He also deemed Islam 'a death cult'.)[102] When the IPCC released its 1.5°C report in the same year, the party's energy spokesperson Roger Helmer fired off a fusillade of angry tweets: 'How can they get away with crying wolf for decades, when nothing actually materializes?', and 'Why should people around the world spend billions on

---

100   Ibid., 274. The British far-right share in the European Parliament, however, was uncommonly large: see Danny Dorling and Sally Tomlinson, *Rule Britannia: Brexit and the End of Empire* (London: Biteback, 2019), 298.

101   Farage speech on UKIP MEP's YouTube channel, 11 September 2013; Farage interview on Alex Jones's 'American TV', YouTube, 16 December 2009.

102   Gerard Batten's Twitter account, 16 December 2017; Peter Walker, 'Ukip's Gerard Batten Reiterates His Belief that Islam Is a "Death Cult"', *Guardian*, 18 February 2018.

reducing CO_2 emissions when the climate is behaving as if we had already done so?'[103]

The same viewpoints could be found among figures of the Tory right, such as Jacob Rees-Mogg, hypergentleman and sidekick of Johnson. He believed that the effect of CO_2 on climate 'remains much debated'. Unease over global warming was the result of psychologically archetypal 'eschatological fears', no different from the ancient Roman expectation that the world would end in 634 BC, the foundation of which was 'a prophecy involving twelve eagles'. Thanks to this sort of groundless 'environmentalist obsession', British citizens suffered high electricity prices.[104] Jacob Rees-Mogg directed a hedge fund with millions of pounds invested in oil and gas. He was referred to the parliamentary watchdog for failing to disclose these interests when intervening in debates about, for example, deep sea mining: 'There may be endless supplies of gas. There may be oil spurting out', an 'enormous wealth' that ought to be 'sucked out of the earth'.[105] Jacob had a sister named Annunziata, who would defect from the Tories to join the Brexit Party – Nigel Farage's new outfit, after leaving the Batten-led UKIP – only to return to the fold, a little shuttle to confirm the symbiosis. In the hot spring of 2016, she tweeted: 'I'm really enjoying this global warming. Simply tropical for late April in London.'[106] Some years earlier, she had shared her tips for how to get rich from Canadian tar sands, or the 'black gold mine'.[107]

There was a clearinghouse for these ideas, located on 55 Tufton Street in Westminster. A stone's throw from Whitehall, this building – four storeys, an inconspicuous brick facade – was

---

103 Roger Helmer's Twitter account, 8 and 15 October 2018. This is just a pick. On Helmer's views on climate science, see further Matthew Reed, ' "This Loopy Idea" ': An Analysis of UKIP's Social Media Discourse in Relation to Rurality and Climate Change', *Space and Polity* 20 (2016): 234.

104 Jacob Rees-Mogg, 'Climate Change Alarmism Caused Our High Energy Prices', *Telegraph*, 23 October 2013.

105 Jane Merrick, 'Leading Tory Backbench MP Jacob Rees-Mogg "Failed to Declare Interests" ', *Independent*, 14 December 2014.

106 Annunziata Rees-Mogg's Twitter account, 27 April 2016.

107 Annunziata Rees-Mogg, 'How to Invest in Canada's Black Gold Mine', *Money Week*, 16 October 2006.

owned by aerospace businessman Richard Smith. It housed the offices of the Global Warming Policy Foundation, Britain's prime denialist think tank. Set up before COP15 to tip the balance towards inaction, the Foundation was the creation of Nigel Lawson, officially Lord Lawson of Blaby, who served as Thatcher's chancellor, in which capacity he privatised North Sea oil. He was also the chair of a consulting firm that helped oil giants – Shell, Total, BP – do business in Eastern Europe and advised coal companies in Poland (of which more shortly).[108] Not niggardly about office space, the building housed a dozen other think tanks and subagencies, several of which combined hostility to the EU with initiatives to deny or downplay the climate crisis: thus the European Foundation published a dossier with no fewer than '100 Reasons Why Global Warming Is Natural'. The Foundation was directed by Richard Smith himself and had Roger Helmer of UKIP on its advisory board.[109] By far the most influential, however, was the Institute of Economic Affairs: recipient of annual bounties from BP since 1967, it pushed carbon vitalism and other types of denial and in 2013 celebrated 'twenty years of denouncing the eco-militants', a persistency that had contributed to preventing 'wholesale political change on climate issues'. The Institute maintained that 'evidence of climate impact is still hard to prove, and harm even more difficult to establish'.[110] It promiscuously shared staff and ideas with the Global Warming Policy Foundation.

---

108   Tom Bawden, 'The Address Where Eurosceptics and Climate Change Sceptics Rub Shoulders', *Independent*, 10 February 2016; *DeSmog UK*, '55 Tufton Street', n.d.; *Little Sis*, '55 Tufton Street', n.d., littlesis.org; Damian Carrington, 'Lord Lawson's Links to Europe's Colossal Coal Polluter', *Guardian*, 6 March 2012. For extensive documentation of the activities of the Foundation, see *DeSmog UK*, 'Global Warming Policy Foundation', n.d.

109   Charlotte Meredith, '100 Reasons Why Global Warming Is Natural', *Daily Express*, express.co.uk, 20 November 2012; *DeSmog UK*, '55 Tufton'.

110   Lawrence Carter and Alice Ross, 'Revealed: BP and Gambling Interests Fund Secretive Free Market Think Tank', *Unearthed*, 30 July 2018; David Pegg and Rob Evans, 'Revealed: Top UK Thinktank Spent Decades Undermining Climate Science', *Guardian*, 10 October 2019; Roger Bate, 'Twenty Years Denouncing the Eco-Militants', Institute of Economic Affairs, iea.org.uk, 10 October 2013. The IEA was based at another address, a two minutes' walk from 55 Tufton Street.

Amid the nearly all-white, nearly all-male inbreeding going on at 55 Tufton Street, the éminence grise was Matthew Elliott, a wizard of Whitehall lobbying who had links to the Koch brothers.[111] He conjugated the Tufton organisations and had one of his own, called the TaxPayers' Alliance. It did gigs with the Competitive Enterprise Institute.[112] (In 2007, Elliott expressed rage over the waste of tax money on conveniences for Muslim prisoners.)[113] If the Tufton network aimed to operate as the UK wing of the denialist ISA, however, it had a voice modulated for British conditions and a particular axe to grind: getting out of the EU.[114] Smith, Lawson, Helmer, Elliott and their peers wished to see the UK cut the European red tape that suffocated business and instead consummate the Special Relationship with a free trade agreement. Some of the abhorrent regulations were justified by references to the climate. In his 2009 book *The Great European Rip-off: How the Corrupt, Wasteful EU Is Taking Control of Our Lives*, Elliott and his co-author David Craig accused the EU of finagling citizens into expressing concern about this issue and using it to 'make a huge power-grab'. Indeed, 'our euro-leaders were so fired up with their new climate change crusade that some started talking about a 60 per cent or even an 80 per cent reduction in emissions by 2050' – aims so preposterous that their sole rationale must be to rip off 'the rest of us'.[115] From this deceit and robbery, the UK had to break free. The same hankering for freedom – from high fuel prices, green taxes, trading regulations, the EU – was expressed by five top Tories and Tufton fellows in *Britannia Unchained: Global Lessons for Growth and Prosperity*. Shackled in

---

111 See Chloe Farand and Mat Hope, 'Matthew and Sarah Elliott: How a UK Power Couple Links US Libertarians and Fossil Fuel Lobbyists to Brexit', *DeSmog UK*, 19 November 2018.

112 Victoria Seabrook, 'Climate Science Denying US Neocon Groups Promote "Simple" and "Freer" Trade with UK at Tory Party Conference', *DeSmog UK*, 5 October 2016.

113 Matthew Elliott, 'Prison Spends £17,000 on Footbaths and Toilets for Muslims', *Telegraph*, 27 August 2009.

114 See further Felicity Lawrence, Rob Evans, David Pegg et al., 'How the Right's Radical Thinktanks Reshaped the Conservative Party', *Guardian*, 29 November 2019.

115 David Craig and Matthew Elliott, *The Great European Rip-off: How the Corrupt, Wasteful EU Is Taking Control of Our Lives* (London: Random House, 2009), 135, 246. Manipulation of surveys to produce climate concern: 171–4.

chains, the former empire had to regain its deserved 'share of the world's wealth and resources'.[116]

When 'Vote Leave' was founded as the official campaign before the referendum, it established its first headquarters at 55 Tufton Street. Elliott set it up; Lord Lawson served as chairman.[117] By his side Elliott had Dominic Cummings, later the special advisor to Boris Johnson, who in turn became the face of Vote Leave. Sure of its priorities, this crew danced to the same tune as UKIP: the real problem facing our society is immigration. Although the UK stood outside the Schengen area, and so was untouched by the arrival of refugees in the EU in 2015, UKIP depicted membership as a sluice that would inevitably allow those people to flood the nation. Just as the British had had to welcome migrants from member states in Eastern Europe, so they would have to accommodate untold numbers from the Middle East (the former horror a starting point for the Brexit project). For this storytelling campaign, UKIP enjoyed a support wave – in the year leading to the referendum, the plurality of voters rated immigration the number one issue and UKIP the 'best party' on it – on which Farage rose to become John the Baptist of Brexit.[118] A defining moment, to match his NASA print-outs, was his unveiling of a poster with a caravan of refugees marching under the heading 'BREAKING POINT: The EU Has Failed Us All'.[119]

If that was the rough edge of the leave campaign, Elliott and Cummings vouched for the respectability. But telling the one from the other was not always easy. Before Farage's caravan, Vote Leave published its own poster with the heading 'The EU Is Letting in More and More Countries' and a map of the European mainland in grey, Britain in blue and Turkey, Syria and Iraq in red and orange, colours of the danger zone. Turkey had 76 million inhabitants, the poster

---

116   Kwasi Kwarteng, Priti Patel, Dominic Raab, Chris Skidmore and Elizabeth Truss, *Britannia Unchained: Global Lessons for Growth and Prosperity* (Basingstoke: Palgrave Macmillan, 2012), 2. High fuel prices: 62.

117   See for example Bawden, 'The Address'.

118   Matthew Goodwin and Caitlin Milazzo, 'Taking Back Control? Investigating the Role of Immigration in the 2016 Vote for Brexit', *British Journal of Politics and International Relations* 19 (2017): 451, 455–6, 460; Bale, 'Who Leads', 270, 273.

119   See Media Mole, 'Nigel Farage's Anti-EU Poster Depicting Migrants Resembles Nazi Propaganda', *New Statesman*, 16 June 2016.

pointed out. It was 'lined up to join' the EU, and located right next to it: the numberless masses of Syria and Iraq. A thick arrow shot from the Levant across the grey area into the heart of England, as though all those hundreds of millions were about to leap over the continent and squeeze in among the Joneses.[120] Boris Johnson put his signature to a statement on how continued membership would perpetuate 'uncontrolled immigration' and invite the populations of Syria and Iraq.[121] Needless to say, this was all make-believe. Turkey was not about to be admitted to the EU; there were no freeways connecting Aleppo to Berlin to Calais. And yet 'the campaign was always talking about immigration. The most proud moment for many of Vote Leave's staff was how well the Turkey leaflet did', one insider source later revealed.[122] The crescendo continued until the day of the referendum.[123]

When Prime Minister Boris Johnson in 2019 asked for the mandate to finally effectuate the Brexit the nation had chosen three years prior, he no longer trafficked in rough climate denial. On his road to 10 Downing Street, he had served as the mayor of London, introducing bike sharing programmes and retrofitting buildings to cut emissions, but also shrinking congestion zones. As MP, he voted against every single climate measure brought before the House – decarbonisation targets, onshore wind farms, vehicle emissions taxes – but verbally questioned the expansion of the Heathrow airport. As foreign secretary, he claimed to be lobbying the US to rejoin the Paris Agreement, but axed the staff of climate diplomats.[124] Forked tongue? Flip-flopping? From 10 Downing Street, he likewise muted the rhetoric

---

120   Alexander Kent, 'Political Cartography: From Bertin to Brexit', *Cartography Journal* 53 (2016): 199–201.

121   Boris Johnson, 'The Only Way to Take Back Control of Immigration Is to Vote Leave on 23 June', Vote Leave, 26 May 2016.

122   Carole Cadwalladr, 'Shahmir Shanni: "Nobody Was Called to Account. But I Lost Almost Everything"', *Guardian*, 21 July 2018.

123   For a thorough analysis of the role of the immigration issue in the referendum, see Goodwin and Milazzo, 'Taking'.

124   Nathanial Gronewold, 'Boris Johnsons Stance on Climate Change Has Flip-flopped', *Science*, 25 July 2019; Matthew Weaver, 'Prof David King Says Would-Be PM Oversaw "Devastating" Cuts in Efforts to Tackle Crisis', *Guardian*, 18 June 2019; Jonathan Watts and Pamela Duncan, 'MPs and the Oil Industry: Who Gave What to Whom?', *Guardian*, 11 October 2019.

on Islam and immigration, often reminding audiences of his Turkish great-grandfather and apologising for the hurt his remarks had caused Muslims. For him, 2019 was respectability time.

In the election in December of that year, Johnson squared off against Labour under Jeremy Corbyn, who had spent a lifetime on anti-imperialist and anti-racist campaigning and now pushed the most comprehensive climate programme ever embraced by a major party in an advanced capitalist country: zero emissions in 2030, eviction of incompliant companies from the stock exchange, nationalisation of energy suppliers, windfall taxes on oil and gas companies to pay for the transition to clean energy, job guarantees for workers in those industries, a permanent ban on fracking, massive rollout of renewables. Climate topped the Labour election manifesto.[125] It was given a place in the Conservative manifesto too, on the third-to-last page. There the Tories vowed to use 'free markets, innovation and prosperity' to 'lead the global fight against climate change'. Long before that, they reminded voters what Brexit was all about: 'taking back control of our borders'.[126] Now was the time to 'get Brexit done' and 'unleash Britain's potential', the two Tory slogans of the election, seen as the most important in a generation or two. The climate merited no bywords.

Thus when Channel 4 arranged an election debate on the topic of the climate emergency, two party leaders abstained themselves: Boris Johnson and Nigel Farage. On the empty seat of the former was placed an ice sculpture, slowly melting in the studio. A fortnight later, Johnson got to choose his own image of himself: on a photoshoot visit to a factory producing industrial machinery, he had a Styrofoam wall decorated with the word 'GRIDLOCK' and a digger wrapped in the Union Jack. He got behind the wheel of the 'incredible digger' – his words (the fuel powering the vehicle needed no mention) – put his foot down, launched the vehicle through the mock bricks and emerged on the other side like an impressive winner. The bulldozer PM, smashing fetters with his national machine:

---

125 Labour Party, *It's Time for Real Change: The Labour Party Manifesto 2019*, labour.org.uk, November 2019, 9–24.

126 Conservative and Unionist Party, *Get Brexit Done: Unleash Britain's Potential*, vote.conservatives.com, November 2019, 55, 20.

business-as-usual plus the spectacle.[127] Despite polls suggesting that climate was a bigger concern for voters than in any previous election, this was the line that won.[128]

One of many queuing to congratulate Johnson for thrashing Corbyn was the Global Warming Policy Foundation. It praised the PM for achieving the mandate for a climate and energy policy that 'won't undermine Britain's international competitiveness' or 'hurt businesses'.[129] The Foundation expected a return on investment: when Johnson ran for Tory leadership, its director gave him £25,000 (he donated four times the amount to Vote Leave).[130] Johnson rather maintained a tepid commitment to capitalist climate governance, as did some of the other Tufton players, such as the Institute of Economic Affairs and TaxPayers' Alliance. When Extinction Rebellion hit London, the latter warned of a 'leap into eco-socialism' and patiently explained that only unfettered capitalism could reduce emissions, the proof being Starbucks' adoption of green standards, so 'forget the protest, pack up and go and get a Starbucks'.[131] In this division of labour, the semi-brainy magazine Johnson used to edit, the *Spectator*, continued to publish in-your-face denial of the basic science, claiming that sea levels were falling and small islands growing and heat records fake under some not-so-subtle headlines: 'Don't Blame Oil and Coal Companies for Climate Change'; 'Don't Blame the West for the Climate Emergency'; 'The Naked Socialist Agenda behind the

---

127 See for example Lizzy Buchan, 'Boris Johnson Union Flag Branded JCB "Brexit" Digger through Piles of Boxes in Bizarre Campaign Event', *Independent*, 10 December 2019.

128 See for example Damian Carrington, 'Climate Crisis Affects How Majority Will Vote in UK Election – Poll', *Guardian*, 30 October 2019.

129 Global Warming Policy Foundation, 'GWPF Congratulates Boris Johnson on Election Victory', thegwpf.com, 13 December 2019.

130 Peter Geoghegan, 'Revealed: Climate Change Denier Makes Big Donations to Boris Johnson and Jeremy Hunt', *OpenDemocracy*, 12 June 2019; Richard Collett-White, 'Major Tory Donor and Johnson Backer to Lead UK Climate Science Denial Group', *DeSmog UK*, 18 December 2019. The director in question was Terence Mordaunt, a successor of Lord Lawson. Mordaunt was the owner of the Bristol Port and a relative of Penny Mordaunt, who, as defence minister, was highly active on the anti-Turkish front in the referendum campaign. See for example Daniel Boffey and Toby Helm, 'Vote Leave Embroiled in Race Row over Turkey Security Threat Claims', *Guardian*, 22 May 2016; *DeSmog UK*, 'Terence Mordaunt', n. d.

131 Darwin Friend [sic], 'TPA Responds to Extinction Rebellion', Taxpayers' Alliance, taxpayersalliance.com, 9 October 2019.

Global Warming Hysteria'.[132] If there was a household magazine for the Brexit bourgeoisie, it was the *Spectator*.

Katie Hopkins also partied on election night. 'Nationalism is back. British people first', she tweeted.[133] Her friend Tommy Robinson was so exultant that he joined the ruling party. This man began his political career in the British National Party, founded the English Defence League to physically attack Muslims and became the poster child of the radical-rough right; in 2018, he joined Batten as a special advisor to UKIP; when Boris Johnson 'thwarted our country becoming a socialist dump', he found a better home. (His announcement included a special shout-out to Dominic Cummings.)[134] The legs of the Brexit right and the street-fighting right were tangled throughout this conjuncture. Just a week before the referendum, on the same day that Farage posed with his caravan poster, a neo-Nazi fatally shot and stabbed Labour MP Jo Cox – guilty of supporting immigration and the EU – while shouting 'Britain first'.[135] The outcome of the referendum set off a cascade of racist violence, with more than 2,300 incidents reported in the first thirty-eight days.[136]

In parallel with Tory attempts to get Brexit done in 2018 and 2019, far-right marches in the streets of England attracted the greatest numbers since the 1930s. Then this force fell in line. So happy were they with Johnson's 'hardline approach' to Islam that five thousand members of Britain First – the group Cox's murderer had been involved in and got his slogan from – collectively abandoned their organisation after the election and instead became members of the Tories.[137] Johnson did not, of course, pick his cabinet from this flank. He got them from 55 Tufton Street. The Institute of Economic Affairs

---

132 Respectively: Ross Clark, 10 October 2019; Lionel Shriver, 17 August 2019; Maurice Newman, 31 August 2019.

133 Katie Hopkins's Twitter account, 12 December 2019.

134 His Telegram message quoted in Marcus Barnett, 'Anti-Islam Activist "Tommy Robinson" Announces He's Joined the Tories', *Morning Star*, 13 December 2019.

135 For the views of the man who killed Cox, Thomas Mair, see Kester Aspden, 'The Making of a Bedsit Nazi: Who Was the Man Who Killed Jo Cox?', *Guardian*, 6 December 2019.

136 For data and analysis, see Jon Burnett, 'Racial Violence and the Brexit State', *Race and Class* 58 (2017): 85–97.

137 Mark Townsend, 'Britain First Says 5,000 of Its Members Have Joined Tories', *Guardian*, 28 December 2019.

boasted that fourteen ministers were its own 'alumni'.[138] All five authors of *Britannia Unchained* were given minister titles. Jacob Rees-Mogg was there, as Leader of the Commons.[139] Thus the Tories consolidated their status as the fully inclusive denomination of the British right: a peculiar way of institutionalising a most general convergence, in which the syndrome of selection was law.

---

138   Lawrence et al., 'How the Right'.
139   For a detailed examination, see Mat Hope and Richard Collett-White, 'Is Boris Johnson's Cabinet the Most Anti-climate Ever?', *DeSmog UK*, 25 July 2019.

3
Fossil Fuels Are the Future

In the second decade of the millennium, around the time of the Paris Agreement, the far right tended to sit like a protective belt around fossil capital in Europe and rebuff any attempts to encroach upon its freedom. This often translated into resistance against even capitalist climate governance, however feeble it might be. The Paris Agreement had its most vociferous – perhaps even sole – European foe in the far right. A case in point was the party of Perussuomalaiset (PS), previously known in English as the True Finns – a better translation of the adjective would be 'ordinary', 'typical' or 'regular'; but the party eventually preferred to call itself merely 'the Finns' in English-language channels. It considered the agreement 'catastrophic' to the domestic economy and demanded that the yoke be lifted off industry and taxpayers.[1] In Sweden, the only party to vote against ratification and then call for withdrawal was, of course, the SD; in the Netherlands, the PVV expressed its desire to emulate Trump's pullout because 'it is all nonsense that only creates extra burdens for our people'; in Britain and Germany, UKIP and the AfD pressed for a hard exit from this

---

1   PS, 'Järkevää ilmastopolitiikkaa olisi tuoda Suomeen teollisuutta muista maista', *Perussuomalainen*, December 2017, 3; Matti Matikainen, 'Presidenttiehdokas Huhtasaari teollisuuden ulosliputtamisesta: "Kyllä tulee hintaa päättäjiemme moraaliselle ylemmyydentunnolle"', *Suomen Uutiset*, 17 December 2017.

and other climate agreements signed by their countries.[2] In Austria, the FPÖ called Paris a *Mogelpackung*, or 'sham'.[3] Upon entering the government, however, this party shelved its earlier demand for withdrawal and nominally accepted the obligations, and we shall soon encounter similar deviations from the line.

But why, we must ask, would any far-right party feel the need to get out of such a permissive construction as Paris? A common explanation has long been that nationalists oppose supranational treaties in the defence of sovereignty – they do not want a global entity to order about their nation – but that line of reasoning is marred by two problems.[4] First, the Paris Agreement did no such thing. It preserved the nation as the sacrosanct unit of decision-making, free to choose its own ambitions for emissions cuts and free to live up to them or not, without any system of allocation, oversight, compliance or sanctions. If the resistance of the far right was inspired by the perceived need to safeguard national sovereignty, it would presumably have waned with the turn from Kyoto to Paris. Instead, the rise of the nation-state as the primary locus of mitigation efforts in the post-Kyoto era – itself an extension of capitalist climate governance, with its ideal of voluntary measures – occurred *in tandem with the swell of nationalism*.[5] Fragmentation along national lines and nationalist insistence on withdrawal appeared to be two moments of the same process, the latter merely carrying it to the extreme: our nation needs to do literally nothing.

Second, the standard explanation begs the question of why the far right would judge the interests of the nation best served by

---

2 Mats J. Larsson, 'Åkesson: Bättre om Sverige lämnar klimatavtalet', *Dagens Nyheter*, 7 July 2017; Michael van der Galien, 'PVV steunt Trump: "Klimaatakkoord Parijs levert alleen maar lastenverzwaringen op, wij willen zijn voorbeeld volgen"', *De Dagelijkse Standaard*, 1 June 2017; UKIP, *Britain Together: UKIP 2017 Manifesto*, ukip.org, 56.

3 FPÖ Salzburg, 'FPÖ-Rauch: Weltklimavertag ist Mogelpackung und nicht nachhaltig', fpoe-salzburg.at, 4 October 2016. On the consistent far-right opposition to Paris, see further Schaller and Carius, *Convenient*, 21, 63–4.

4 See for example Bernhard Forchtner and Christoffer Kølvraa, 'The Nature of Nationalism: Populist Radical Right Parties on Countryside and Climate', *Nature and Culture* 10 (2015): 200, 213–19.

5 The outstanding chronicle of this process is Ciplet et al., *Power*. On the hollowness of Paris, see for example Jen Iris Allen, 'Dangerous Incrementalism of the Paris Agreement', *Global Environmental Politics* 19 (2019): 4–11.

unmitigated climate breakdown. Much as one could imagine the parties blaming it on immigration, one could picture them arguing that it ruins the most priceless heirlooms of the homeland – the top summit of a mountain chain, for instance – and lashing out against the foreigners responsible for most of the CO_2 accumulated in the atmosphere. Why not send the soldiers of Carolus Rex to fight that existential battle? Or why not, at the least, demand an international treaty to do such work? Far-right agonies over the impacts of global heating on the national heritage – Alpine glaciers, the plazas and cathedrals of Venice, Polish farms – were conspicuous by their absence.[6] Where it could, the far right instead boosted the burning of fossil fuels at home, exploiting national sovereignty to set as much of them on fire as possible, without any perceptible resentment against other countries that did the same. Again, this suggests some sort of tie between the ethnonationalist project and the energetic base any hint at climate action must perforce call into question.

As much as the Paris Agreement respected the three precepts of capitalist climate governance we identified above – most importantly, the text never mentions 'fossil fuels', a silence that could not be louder – it upheld the *idea* of mitigation. Most famously, it committed the signatories to 'holding the increase in the global average temperature to well below 2°C above pre-industrial levels and pursuing efforts to limit the temperature increase to 1.5°C' (the latter a concession to a long-standing demand from countries in the global South: their one diplomatic gain at COP21).[7] Much of the far right could not stand even that. Here too, it drove capitalist climate governance beyond itself – not only can we postpone the showdown with fossil capital, we should jettison the very notion of ever having to do so – and returned, full circle, to the original rejectionism of ExxonMobil et al.

At most, the defence of sovereignty is a partial but insufficient explanation for the far-right tendency to abhor something as lax as Paris. Moreover, formal protection of self-determination could be entirely compatible with a push for renewable energy at home. So could a quest for economic self-sufficiency. But far-right parties very

---

6 As noted by Schaller and Carius, *Convenient*, 24, 41.
7 United Nations 2015, *Paris Agreement*, unfccc.int, 5.

rarely advocated a massive scale-up of renewable energy on their own territories and their own initiatives: they tended to be lukewarm to the idea at best, with a few aberrant cases of some ambiguous nominal support. More often they detested the very thought. They reserved a particular hatred for wind power.

Of Minarets and Mills

The 2010s saw a sharpening of the far-right agitation against wind as a source of energy. Roger Helmer of UKIP admitted to disliking large solar farms, but 'most of all, I hate wind turbines because they're symbols of monstrous, pointless waste, and futile political correctness.'[8] His party liked to think of itself as the avant-garde of resistance to wind in rural English communities.[9] Marine Le Pen, heir to her father's party, said she was fine with solar and biogas but wanted an immediate moratorium on the erection of turbines, which 'are hideous and do not function', 'costly and monstrous'.[10] In the spring of 2019, the party launched a campaign specifically to stop turbines, defined as 'visual and sound pollution' and 'a disaster for the environment'.[11] In the motherland of wind, the Danish People's Party announced in 2018 that it would henceforth oppose all construction of wind farms on land and only accept offshore projects. The demand was immediately placated, when the government and the DF agreed to reduce the number of onshore turbines from the current 4,300 to 1,850 in 2030, while three new offshore farms were tabled – in other words, more than half of standing wind turbines were to be *taken down* in the next decade, a rare case of dismantling

---

8   Roger Helmer's Twitter account, July 1, 2012.
9   See Reed, ' "This Loopy Idea" '.
10 *Terra Eco*, 'Marine Le Pen: "Je suis plus cohérente que les Verts"', terraeco.net, 21 February 2012; Elisabeth Jeffries, 'Nationalist Advance', *Nature Climate Change* 7 (2017): 469–71. Pascal Bruckner expresses a similar sentiment: 'As for the much-vaunted wind turbines, they produce about as much noise as a railway switching yard, as well as tinnitus and pressure on eardrums.' Bruckner, *Fanaticism*, 113.
11 *Europe 1*, 'Les éoliennes sont "un drame pour l'environnement", estime Jordan Bardella', europe1.fr, 7 April 2019; Paul Vermeulen, 'Le Rassemblement National lance la campagne "Stop éoliennes" ', *Novopress*, 9 April 2019.

renewable energy infrastructure.[12] The AfD was working for the same turnaround in Germany. In the parlance of the party, 'so-called renewable energy' isn't really that – it's just unreliable, inadequate, expensive and, in the case of wind, outright awful, since turbines 'destroy the image of our cultural landscapes and pose a lethal danger to birds'.[13] 'I'm very much against windmills. I want all the windmills out of the Netherlands. Except for the old ones, of course', said Thierry Baudet and chuckled.[14]

Another European country with promising potentials for wind power – rolling hills, mountains, steppes – is Hungary. But in 2010, the Fidesz party swept the elections and began to establish a stranglehold on the nation. Six years later, the office of Prime Minister Viktor Orbán announced that 'the wind only blows in afternoons and evenings' and issued a list of restrictions on construction: no turbines would be allowed within twelve kilometres of an existing building, within forty kilometres of a radar installation or within fifteen kilometres of a military airfield, and if any site could be identified outside of those limits, *all* neighbours within one kilometre would have to approve of the project. This amounted to an effective nationwide ban.[15] In 2016, when the PS was part of the government, new subsidies to wind power in Finland were stopped, largely due to the efforts of agriculture and forestry lobbyists close to the party.[16] During that year, the party waged a national campaign against the scourge. It began with a press conference featuring a man crying over the pain the nearby turbines had caused him and his family – a result, the Finns explained, of the 'undetectable ultrasound' or 'infrasonic waves' from the installations. Ten per cent of the population was at risk of contracting

---

12 Berlingske, 'DF glæder sig over færre vindmøller på land fremover', 29 June 2018. When this book was being finalised, Denmark had had a social-democratic government for half a year, but this decision had not been revoked.
13 AfD, *Programm*, 158.
14 Gehriger, '"There's a Proper"'.
15 Géza Németh, 'Így nyírta ki a kormány a szélerőműveket', *Napi.hu*, napi.hu, 23 December 2016; Eszter Vass, 'Wind Power Utilisation Made Impossible in Hungary', *DailyNews Hungary*, 23 September 2016.
16 Jarmo Liski, 'Näin tuulivoimaloita vastaan lobataan – perussuomalaisille tukea MTK:lta ja keskustavaikuttajalta', *Yle*, 11 October 2016.

various diseases. To top it off, the turbines endangered wildlife. 'They make bats' insides explode.'[17]

Why the animus? Wind turbines were the most conspicuous symbol of a policy – or even just a pretence – of climate mitigation. They were eyesores for principled opponents of the latter, banners flying the idea that renewable energy might one day take over, hoisted inside the space these people considered their own. But there was also another dimension to it. There were striking similarities to the hatred of minarets, mosques and calls to prayers.[18] A minaret and a mill shared some physical properties: immovable structures, they could reach soaring heights and tower above their surroundings; they had a potential for making wailing sounds, from the *adhan* or the blades; they permanently occupied public space. Local campaigns against both kinds of projects were often couched in technical objections. There will, detractors said, be noise, spoiled views, fractures in the rural or urban landscape; but below the surface, there was a sentiment of intrusion or even invasion. Campaigns tended to fall back on a high-pitched rhetoric of 'this does not belong here', targeting projected Islamic places of worship and wind energy converters as monuments of a nefarious foreign power: Muslims overrunning natives, environmentalists overhauling everyday life. Why should our community be sacrificed for these outsiders? Resistance to external imposition became a leitmotif in spatial conflicts over both Islam and wind in early

---

17   Niko Hatakka and Matti Välimäki, 'The Allure of Exploding Bats: The Finns Party's Populist Environmental Communication and the Media', in Forchtner, *The Far*, 142–3; Schaller and Carius, *Convenient*, 17.

18   This paragraph draws on a comparative reading of John Barry, Geraint Ellis and Clive Robinson, 'Cool Rationalities and Hot Air: A Rhetorical Approach to Understanding Debates on Renewable Energy', *Global Environmental Politics* 8 (2008): 67–98; Susana Batel and Patrick Devine-Wright, 'Populism, Identities and Responses to Energy Infrastructures at Different Scales in the United Kingdom: A Post-Brexit Reflection', *Energy Research and Social Science* 43 (2018): 41–7; Jocelyne Cesari, 'Mosque Conflicts in European Cities: An Introduction', *Journal of Ethnic and Migration Studies* 31 (2005): 1015–24; Todd H. Green, 'The Resistance to Minarets in Europe', *Journal of Church and State* 52 (2010): 619–43; Uriya Shavit and Fabian Spengler, 'For Whom the Bell Tolls? Contesting the *Adhans* in Majority Non-Muslim Societies', *Journal of Muslim Minority Affairs* 36 (2016): 447–64. The associative links to resistance against immigration in campaigns against wind are noted in Barry et al., 'Cool Rationalities', 78.

twenty-first-century Europe, a posture reverberating in deep affective wells: the feeling of one's own life being a plaything in the hand of unintelligible distant forces; the experience of having one's own life of privilege questioned by transitions, away from a homogenously white population or, if only potentially, away from an economy based on fossil fuels.

'Migrants are like wind turbines, everyone agrees to have them but no one wants them in their backyard', explained Marine Le Pen, revealing less about 'everyone' than about how the *far right* perceived the two.[19] The parties fought on both fronts as though they were one. In its one-page manifesto for the period 2017–21, the PVV presented eleven steps for the 'de-islamisation' and reclamation of the Netherlands: the first included the closing down of all mosques; the seventh an end to all public investment in wind power.[20] 'The AFD rejects the minaret and the cry of the *muezzin* as icons of Islamic supremacy'.[21] SD activists led the struggle to thwart nineteen attempts by the Muslim community in the city of Borlänge to build a place of worship, as well as the (less successful) struggle to stop the expansion of nearby wind farms.[22] Representatives in municipalities across Sweden – the middle layer responsible for most of the racist expletives that reached the media – were engaged in numerous drawn-out campaigns against both provocations. (As for Hungary, the last minaret was built five hundred years ago.)[23] Under the towers, real or imagined, existing or planned, activists of the far right stoked resentment against a world turned upside down.

---

19   *Le Point*, 'Marine Le Pen compare les migrants à des éoliennes', lepoint.fr, 14 April 2019.

20   PVV, 'Concept – Verkiezingsprogramma PVV 2017–2021: Nederland weer van ons!', pvv.nl, 2017.

21   AfD, *Programm*, 98.

22   Roger Helmer's Twitter account, 29 September and 8 August 2018; Peter Kadhammar, 'SD satte stopp för den nya moskén', *Aftonbladet*, 23 January 2018; Anders Nordin, 'SD vill stoppa utbyggnad av vindkraft', *Dalarnas Tidningar*, 4 September 2013.

23   Elżbieta M. Goździak and Péter Márton, 'Where the Wild Things Are: Fear of Islam and the Anti-refugee Rhetoric in Hungary and in Poland', *Central and Eastern European Migration Review* 7 (2018): 131.

In Defence of Fossil Capital in General

Not only wind power but *all* domestic mitigation policies tended to be the object of far-right dissent. It was here that the ministration to fossil capital became most obvious. In Upper Austria, the far right maintained close ties to steel and automotive corporations, not the least through the figure of Manfred Haimbuchner, deputy provincial governor and vice chairman of the FPÖ. He was convinced that actual mitigation would cause the 'deindustrialisation of the world and Upper Austria'.[24] He also headed the party think tank Atterseekreis, which in 2015 published an anthology with contributions from various personalities of the country's capitalist class, called *Courage to Speak Out*. The foreword made the outlook plain: 'Through high taxes, high environmental standards and absurd climate protection legislation, we endanger our industrial base, which is also the foundation for jobs and prosperity' – very much a mindset of the bourgeois mainstream, elevated to national protocol when the FPÖ took over the helm together with the ÖVP in 2017.[25]

Across the border in Hungary, the Fidesz government assiduously catered to the needs of one industry in particular: the manufacturing of cars. Soon after coming to power in 2010, Orbán turned to German auto companies and wooed them deeper into his country with an array of sweet location advantages. Hungary cut its corporate tax to the lowest rate in the EU, at 9 per cent, further reduced by means of various allowances to a negligible 3.6 per cent for the largest multinationals. A company moving its car factory to Hungary would benefit from a platform next door to the main markets in Western Europe, with weak unions, long hours and low wages – in 2016, Hungarians put in 1,740 hours per year, compared to 1,613 in Austria, but received half the pay – and a standing offer to help rewrite labour laws.[26] In late 2018, this deference to foreign – read: German – auto

---

24   Bonvalot, 'Warum'.
25   Manfred Haimbuchner and Alois Gradauer, 'Mut zur Wahrheit: Ser geschätzter Leser, sehr geschätzte Leserin!', in Manfred Haimbuchner and Alois Gradauer (eds), *Mut zur Wahrheit* (Linz: Freiheitlicher Arbeitskreis Attersee, 2015), 15.
26   Valerie Hopkins, 'Hungary Ties Growth to Bumper of German Manufacturers', *Financial Times*, ft.com, 22 November 2018; Nathan Johnsson, 'Car Industry Remains

producers reached a new level when Fidesz rammed through a set of statutes that became colloquially known as the 'Slave Law'. Companies were now entitled to request four hundred hours of overtime (up from 250) and wait for three whole years (rather than one) before remunerating their employees for such work.[27] In the same year, BMW announced the construction of a new plant in Hungary, which was expected to raise the share of cars in the country's export from one-third to nearly one-half.[28]

How, then, did the Fidesz government respond to the idea of cutting CO_2 emissions? After such far-reaching efforts to create a choice business environment, it came as no surprise that it hedged every nominal commitment to Paris and EU targets with caveats about the need to protect the car industry.[29] In regard to EU ambitions to cut emissions by 40 per cent (from 1990 levels) and expand renewable energy to a 27 per cent share by 2030, the Fidesz government was openly resistant in solidarity with its main beneficiary: 'Hungary will never accept a proposal from Brussels that would hurt the German, more precisely the Bavarian, auto industry', Foreign Minister Péter Szijjártó stated in 2018, claiming total unity of interests between Fidesz and BMW and Audi.[30] The minister had made personal assurances to the owners of these two companies never to abide by such directives.[31] If Hungarian climate policies were

---

Vital to Hungary: Is Hungary Vital to Car Makers?', *Budapest Business Journal*, 14 December 2018; *Deutsche Welle*, 'Hungary Rolls Out Red Carpet for German Carmakers', dw.com, 7 August 2018; Peter Wilkin, 'The Rise of "Illiberal" Democracy: The Organization of Hungarian Political Culture', *Journal of World-Systems Research* 24 (2018): 7, 20–1; but see especially Agnes Gagyi and Tamas Gerocs, 'The Political Economy of Hungary's New "Slave Law"', *Left East*, 1 January 2019.

27 Palko Karasz and Patrick Kingsley, 'What Is Hungary's "Slave Law," and Why Has it Provoked Opposition?', *New York Times*, 22 December 2018.

28 Edward Taylor and Krisztina Than, 'BMW to Build 1 Billion Euro Car Factory in Hungary', Reuters, 31 July 2018.

29 See for example Péter Kaderják in 'European Council: Environment Council, Public Session', video.consilium.europa.eu, 9 October 2018.

30 Sándor Czinkóczi, 'Szijjártó inkább a bajor autóipart támogatja az uniós klímacélok helyett', *444*, 444.hu, 21 September 2018; MTI (Magyar Távirati Iroda), 'Szijjártó szerint a klímacéloknál fontosabb az autóipar versenyképessége', *HVG*, 21 September 2018.

31 Miklós Halász-Szabó, 'Hungarian Govt Sides with Car Industry against Environmental Targets', *Hungary Today*, 21 September 2018.

designed to maintain the appeal for foreign capital, however, they also had some application to the project of nurturing a national bourgeoisie in the fold of Fidesz. Plutocrat Lörinc Mészáros, a frontman for Orbán, had by 2018 risen to a position as the eighth-richest man in the country, winner of 83 public tenders and owner of 121 companies; by 2019, he was also poised to take over 100 per cent of the shares in the country's only coal-fired power plant.[32] That plant was supplied by coal from local mines, but Hungary had limited fossil fuel extraction. The climate policies of one of the most notorious far-right governments in Europe were, in other words, primarily geared to fossil capital in general. They were as yet rarely couched in terms of denial, but rather hidden in official indifference to the issue, paired with the all-consuming passion.

For the 2010s, the Fidesz government must count as one of the most consistent in focusing on the problem of Muslim immigration, its rhetoric spiced up with the incessant demonisation of George Soros. A financial investor and liberal philanthropist, Soros was best known in his native Hungary as a rich Jew. After the so-called refugee crisis of 2015, the Fidesz government turned itself into a loudspeaker for the theory that Soros – 'the international speculator' (Orbán) – manoeuvred in the background to hand Europe over to a 'Muslim invasion'.[33] It was his money that worked to open the borders. In October 2017, the Fidesz government mailed a 'national consultation' to all eligible voters, asking them to voice their opinions about what it called the 'Soros Plan' for settling 'at least one million immigrants from Africa and the Middle East annually on the territory of the European Union, including Hungary'; furthermore, the state informed its citizens that Soros planned to force Hungary to give these aliens lenient sentences for any crimes, serve them lavishly with welfare, subordinate the national language and culture to their

---

32 Katalin Erdélyi, 'Túlnyomórészt uniós forrásokból gazdagodtak Mészáros Lőrinc családi cégei az elmúlt hét évben', *Átlátszó*, 15 January 2018; Péter S. Föld, 'Mészáros Lőrinc nem stróman, csak időnként vesz ezt-azt Orbán Viktornak', *Fühü*, fuhu.hu, 30 December 2018; MTI, 'Megszólalt Mészáros Lőrinc a Mátrai Erőműről', *24.hu*, 15 December 2018.

33 Griff Witte, 'Once-Fringe Soros Conspiracy Theory Takes Center Stage in Hungarian Election', *Washington Post*, 17 March 2018.

preferences and punish anyone in opposition to such measures.[34] Around the same time, Fidesz filled Hungary with advertisements portraying a Soros smiling maliciously under the caption '99 percent oppose illegal immigration – don't let Soros have the last laugh.'[35] Seldom had the figures of the subversive Muslim and the subversive Jew, unified in their contempt of national borders, fused to such spectacular effect.

In spring 2018, Hungary headed to the polls again. Fidesz beat the drums against Muslims and the rich Jew behind them ever more ferociously; at one rally, a party supporter explained that 'Soros is the mastermind behind the Islamisation of Europe'.[36] Such propaganda handed Fidesz another victory preserving the two-thirds majority in parliament it had enjoyed since 2010. It also drowned out all other themes. A search on the official, Hungarian-language YouTube channel of Fidesz in late 2018 yielded 42 hits for 'immigration' (*bevándorlók*) and 122 for 'Soros'. For 'climate politics' (*klimapolitika*) and 'environmental protection' (*környezetvédelmi*), the result was zero.[37] But in response to a question fielded by an oppositional member of parliament in October 2018 regarding the recent IPCC report, one minister crossed the line: 'The degree to which human activity impacts climate is up for debate.'[38] For the Hungarian government, this marked the sudden awakening of a passion about the climate. We shall examine it further below. Protecting the car industry, resisting emissions cuts, ignoring climate change, vilifying Muslims and Jews and eventually falling in line with denial: the early twenty-first-century European far right in power.

---

34 For the content of this 'national consultation' and Soros's own rebuttal of it, see George Soros, 'Rebuttal of the October 9 National Consultation in Hungary', georgesoros.com, 20 November 2017.

35 See for example Zack Beauchamp, 'Hungary Just Passed a "Stop Soros" Law That Makes it Illegal to Help Undocumented Migrants', *Vox*, 22 June 2018. For more on the place of Soros in the state ideology of Fidesz, see Paul Ledvai, *Orbán: Europe's New Strongman* (London: C. Hurst and Co, 2017); on the Soros conspiracy theory in general, Stephen J. Whitfield, 'The Persistence of the *Protocols*', *Society* 55 (2018): 417–21.

36 Gyongyi Horgasz quoted in Witte, 'Once-Fringe'.

37 Fidesz YouTube search conducted on 26 December 2018.

38 Csaba Tibor Tóth, 'A fideszes államtitkár szerint "vitatható", hogy az ember okozza-e a klímaváltozást', *Mérce*, 30 October 2018. The statement was made by Péter Cseresnyés, deputy minister of innovation and technology.

It has been suggested that far-right parties in countries that import practically all their fossil fuels would be keen to seek out alternatives, in the interest of self-reliance.[39] That hypothesis is gainsaid by Hungary, and by Spain, and by Sweden, which have few or no domestic resources of oil or coal or gas. 'There are no good alternatives to fossil fuels', according to Martin Kinnunen, the SD climate spokesman; renewables are uneconomical and technically inferior. Sweden will have to import gas and coal to power plants for the foreseeable future. All targets for renewables in the Swedish energy mix, all ambitions for energy efficiency and a carbon-free fleet vehicle ought to be scrapped – 'we don't want sectoral goals, we don't want national goals', Åkesson stressed after the extreme summer of 2018.[40] Existing laws, budgets and investments in emissions reductions should be annulled, because they turn off companies and undercut Swedish competitiveness – a constant refrain in SD anticlimate agitation. If cuts have to be made, it sometimes argued, in the moments when it claimed to abide by the scientific consensus, then measures should be undertaken *anywhere but in Sweden*, although this nation could potentially contribute through rekindled flexible mechanisms.

There is an analogy here to the logic of ethnopluralism. Other ethnicities and cultures are acceptable in their own homelands *but not in our country*; similarly, a far-right party such as the SD could occasionally suspend its general denial and accept the need for mitigation – if only in the most abstract sense – somewhere far away.[41] In other words: white Swedish culture and fossil energy might not be a universal ideal, but it is what *we* are going to have on the patch of the earth we get to preside over. Hence the SD could fantasise about El Niño one day and praise carbon trading in the next. Here was another point where capitalist climate governance and organised denial overlapped, opening up for a Rupert Murdoch–like juggling of rhetorical tactics. The bottom line was to

---

39   Lockwood, 'Right-Wing', 724.
40   Kinnunen in *SR*, 'SD om klimatet: Det är på global nivå utsläppen spelar roll', sr.se, 25 June 2018; Åkesson in *SVT*, 'Utfrågningen – Jimmie Åkesson', svt.se, 2 September 2018.
41   See for example Sverigedemokraterna, *Sverigedemokraternas*, 23.

make sure nothing would be done. With no fossil fuels to dig up from Sweden, it was their burning the SD haloed; much as in Austria and Hungary, the defence and adulation concerned fossil capital in general.

In Defence of Primitive Fossil Capital

In countries with considerable production of fossil fuels, on the other hand, the far right did indeed tend to be the most ardently supportive of all actors. Finland has its own peculiar energy source: peat, consisting of moss, shrubs and other plant remains that have sunk into wetlands, the water preventing them from fully decomposing, so that organic matter is packed in dense layers similar to the early stages of coal formation. Finnish bogs hold peat deposited over thousands of years. When they are drained of water, several metres of sediments can be harvested and carted off to power plants for burning. The process generates CO_2 emissions comparable to the dirtiest lignite coal.[42] The fuel is semi-fossil, erroneously thought of as a thing of the pre-industrial past. In the 2010s, Finland, having scaled up extraction massively since the 1980s, became the world's largest producer of peat and depended on it as a buffer against spikes in energy demand.[43] Any serious attempt to reduce Finnish greenhouse gas emissions would have to bring extraction levels close to zero. Nothing of the sort was in the cards, and the PS, the most loyal defender of peat, wanted to go further and exempt it from paying any and all taxes. Probably not coincidentally, the party had its stronghold in the west of the country, the heartland of the industry: out of the fifteen municipalities with the highest electoral support for the PS

---

[42] A comprehensive survey of peat in Finland, including its climatic impacts, is Arvo Leinonen (ed.), *Turpeen tuotanto ja käyttö: Yhteenveto selvityksistä* (Helsinki: VTT, 2010); see further BirdLife Suomi, Greenpeace, Luonto-Liito, Maan ystävät, Natur och Miljö, Suomen luonnonsuojeluliitto and WWF, *Puolueiden vastaukset vaalikyselyyn – kysymyksittäin*, report retrieved from epressi.com, 2015.

[43] Tilastokeskus, 'Energian kokonaiskulutus nousi 4 prosenttia tammi–kesäkuussa', stat.fi, 27 September 2018.

in 2015, fourteen produced peat.[44] Another energy source also had a special friend in the party: a bill to ban the use of (imported) coal by 2029 was opposed as 'unrealistic'.[45] These were the fuels by which the Finns would ward off the approaching witch doctor.

'Fossil fuels are the future!' exclaimed the headline of an article authored by Roger Helmer. Bring it on and burn it: 'If Europe is to have a future (other than as a depressed third world continent) it's time to come to terms with the fossil fuel revolution' and gratefully receive the promises of fracking, cheap coal and glutted oil markets.[46] UKIP trotted out its love for all fossil fuels and its wish to see as much of them as possible extracted in the independent homeland. A new life should be given to the coal industry, emblem of a great past wasted away by those who sold the nation to foreigners: 'Our enemy number one is the ugly ghastly windmills. Our policy is to reopen the mines. We have at least enough coal reserved for 200 more years', Helmer fleshed out the line, adding that his party did not 'regard CO_2 as a pollutant'.[47]

Somewhat ironically given Margaret Thatcher's relation to coal, echoes could, again, be heard on the Tory right, where Jacob Rees-Mogg noted that 'coal is plentiful and provides the least expensive electricity per megawatt', but 'unfortunately, coal-fired power stations are being shut down because of European Union regulations'.[48] In July 2018, just before he stepped into the public eye as the herald of the Tory far right, Rees-Mogg voted to reduce support for the partial or complete conversion of coal-power plants to biomass, as did his party: the purity of coal must be protected against government interference.[49] Like him, UKIP also wanted to see a thousand fracking

---

44 Calculated from Tilastokeskus, 'Eduskuntavaalit 2015 vaalikarttapalvelu', stat.fi (election results from 30 April 2015), and Geologian Tutkimuskeskus, 'Turtevarojen tilinpito', gtk.fi (distribution of peat resources).

45 BirdLife suomi et al., *Puolueiden*; PS, 'Pienituloisilla'.

46 Roger Helmer, 'Fossil Fuels are the Future!', rogerhelmermep.wordpress.com, 7 November 2012.

47 Catherine Thorleifsson, 'From Coal to Ukip: The Struggle over Identity in Post-Industrial Doncaster, *History and Anthropology* 27 (2016): 566.

48 Rees-Mogg, 'Climate'.

49 For this and other parts of his voting record on 'low carbon electricity generation', see the entrance on Jacob Rees-Mogg at *TheyWorkForYou*.

rigs bloom in Britain.[50] So did Boris Johnson the journalist, calling fracking 'glorious news for humanity' and wanting 'no stone unturned'.[51] But the North Sea was a more dependable frontier. In November 2019, the Johnson government issued a round of licences for oil and gas exploration, state and industry uniting behind the goal of increasing extraction from 5.7 to 20 billion barrels, a near quadrupling. The preceding Tory government of Theresa May promised to 'ensure that we extract every last drop of oil and gas that it is economic to extract'.[52] How Johnson could beat that policy remained to be seen.

'Eradicating Islam should be the primary target of Dutch foreign policy', proclaimed the PVV before its electoral success in 2010; it also came out in favour of new coal-power plants.[53] Its instinct was to disapprove of all caps on fossil fuel use. In the late 2010s, wrangling over the most controversial fossil fuel production in the country came to a head: for decades, gas extraction had rattled the province of Groningen, with frequent earthquakes damaging homes and other buildings, to the exasperation of the local population. When the government promised that operations would be fully phased out in 2030 – far too distant a date in the eyes of many – Wilders took to Twitter, alleging that reductions in gas production would increase electricity prices for the Dutch people while 'billions of euros' were spent on Africa and the EU.[54] The PVV argued that gas extraction should continue, since wind and sun offered no real alternatives.[55] But, as in the rest of the country, it was displaced by the FvD, which had more success campaigning for continued extraction. In Denmark, the DF in 2017 backed the North Sea Agreement, giving tax breaks to

---

50 See for example UKIP, *Britain*, 57.
51 Boris Johnson, 'Ignore the Doom Merchants, Britain Should Get Fracking', *Telegraph*, 9 December 2012.
52 Chancellor Philip Hammond in Greg Muttitt, Anna Markova and Matthew Crighton, *Sea Change: Climate Emergency, Jobs and Managing the Phase-Out of UK Oil and Gas Extraction*, Platform, Oil Change International and Friends of the Earth Scotland, platformlondon.org, May 2019, 22. For UK oil and gas policies, see further this report.
53 Bar-On, 'The Radical Right', 31; Smit, 'Richard'.
54 Geert Wilders's Twitter account, 1 July 2018.
55 Tristan Braakman and Willem van Reijendam, 'Verkiezingen: wat zeggen de partijen over de gaswinning?', *RTV Noord*, 8 February 2017.

companies drilling for oil and gas in the country's territorial waters in the North Sea and initiating the largest-ever investment into its fossil fuel industry (but so did several other parties, including the social democrats).[56] Finnish peat, British fracking, Dutch gas and Danish oil were, however, heaps of fine gravel compared to the mountain chains of German lignite.

Powerhouse of European capitalism, Germany was the world's sixth largest emitter of CO_2 from fossil fuel combustion as of 2018. Out of the twenty-eight EU member states, it accounted for 21 per cent of emissions.[57] Much stemmed from lignite, also known as brown coal, a soft, moist, low-rank variety, at the next step of sedimentation after peat. Unlike hard black coal, lignite is not extracted in pits, but in open-cast mines that can stretch for miles; not traded and transported across oceans, but burnt in power plants in the vicinity. German lignite complexes have generated plentiful profit for their private owners.[58] Forests, fields and several hundred of villages have been razed to make way for brownish basins occupying as much space as middle-sized cities. On their floors, so-called diggers – huge coal excavators, looking somewhat like horizontal Eiffel towers on wheels, the largest mobile machines on the planet – slowly chew up the ground and spit out fuel to the nearby plants, whose concave grey chimneys belch out perpetual clouds into the atmosphere. In the 2010s, this country was the main producer of lignite in the world. Out of the ten largest point sources of CO_2 in the EU, as of 2016, seven were German lignite plants.[59] Nothing would yield deep emissions cuts in Europe faster than their total closure. Nothing less was required, and thus lignite formed the front line in the battle over German energy, split over two

---

56 Alexander Hecklen, 'Forstå sagen: Derfor bruger Mærsk 21 milliarder kroner på Tyra', *DR*, dr.dk, 1 December 2017; Regeringen, *Aftale mellem regeringen, Socialdemokratiet, Dansk Folkeparti, Det Radikale Venstre og SF*, 22 March 2017.

57 Joint Research Centre (European Commission), 'Fossil CO_2 Emissions of All World Countries: 2018 Report', edgar.jrc.ec.europa.eu, 23 November 2018; Eurostat, 'Greenhouse Gas Emission Statistics', ec.europa.eu, 25 June 2019.

58 For the basics of German lignite, see Kerstine Appunn, 'Factsheet: Coal in Germany', *Clean Energy Wire*, 5 December 2017.

59 *Sandbag*, 'New Data: European Coal Emissions Plummet by 11% in 2016', sandbag.org.uk, 1 April 2017; Anne Hähnig, 'Es lebe der Bagger', *Die Zeit*, 8 November 2018.

geographical sectors: the Rhineland in the west and Lusatia in the east, the latter of which also happened to be the stronghold of the AfD.

In the 2017 election, more than 30 per cent of voters in the region of Lusatia cast their ballot for the AfD. Nowhere else in the country did the party register such robust support. Pro-coal, anti-climate messages were central to the AfD sweeping Lusatia, alongside the usual fare of xenophobia.[60] When the *Große Koalition* mooted the closure of the mines in 2017, the AfD castigated Angela Merkel's Christian Democrats for 'laying our country to waste not only by a disastrous asylum policy' but also by means of 'a left-green ideologised climate policy'.[61] Pointing to the relative poverty of Lusatia, the jobs in the coal industry – a few thousands in the region; fifteen to twenty thousand in the nation as a whole, to be compared with some 340,000 in the renewable energy sector – and the business interests involved, the AfD took up the cudgels for indispensable mining. It posed as the outlet for the popular feeling of 'being abandoned by the old parties for the sake of ideological climate goals', 'ideological' here meaning baseless and fraudulent, the defence of coal merging with denial.[62] Karsten Hilse – himself from Lusatia – came out to speak at a rally for the mines.[63]

The AfD was not, of course, the sole actor in this drama. Ende Gelände organised a series of mass actions after 2015 on both the western and eastern fronts, with thousands of activists streaming into the lignite mines, occupying the diggers and blocking the railway tracks conveying the coal to the plants. The pressure built up to the extent that the government in the summer of 2018 launched a 'coal exit commission' tasked with discussing and proposing a complete phase-out at some point.[64] Composed of industrialists,

---

60 Christian Taubertund and Simone Wendler, 'Warum die AfD die Lausitz überrollt', *Lausitzer Rundschau*, lr-online.de, 26 September 2017.

61 AfD-Fraktion im Sächsischen Landtag, 'CDU-Politik gefährdet tausende Braunkohle-Jobs in der Lausitz', *AfD Kompakt*, 20 October 2017.

62 AfD-Fraktion im Sächsischen Landtag, 'AfD: Ohne gleichwertige Arbeitsplätze kein Kohle-Ausstieg!', afd-fraktion-sachsen.de, 29 November 2018.

63 DPA, 'Verbalattacken bei AfD-Kundgebung für Arbeit in der Lausitz', *Berliner Morgenpost*, 10 March 2018.

64 Benjamin Wehrmann, 'Germany's Coal Exit Commission', *Clean Energy Wire*, 24 July 2018.

unionists, mayors, mainstream environmentalists and climate scientists, the body was lambasted by the AfD for capitulating to activists and 'disregarding law and order': the AfD demanded its immediate dissolution.[65] Voices around the commission, meanwhile, warned that a rapid phase-out would drive the east even deeper into the arms of the AfD.[66] This was one factor behind the commission's decision to set the end date for coal at 2038, two decades into the future.[67] Thus the far right lent new weight to the prolongation of business-as-usual at the sites of the largest CO_2 sources in Europe.

But that was not the end of the struggle over German coal: the AfD marched on with the demand to rip up the 2038 decision. In late 2019, the eastern lignite states of Brandenburg and Saxony went to the polls, the future of coal seemingly in the balance. The top AfD candidates from the two states met up at their shared border and, in the company of Karsten Hilse, ceremoniously swore that lignite will be extracted for another thousand years, a number with a certain resonance in German nationalism.[68] In both, the AfD soared to second place and came within reach of the first. Ninety per cent of its voters applauded the line on climate.[69] The results strengthened the party's confidence in arrant anti-climate politics, which included not only coal extraction *in perpetuum*: the party sought the total abolition of *Klimaschutz* as a political endeavour.[70] It wanted to terminate the *Energiewende*, or 'energy transition', the acclaimed German programme for shifting to renewable energy which had, however,

---

65 Steffen Kotré, 'Bundesregierung muss Kohlekommission auflösen', *AfD Kompakt*, 26 September 2018; cf. Tino Chrupalla, 'Kohle-Kommission gefährdet Arbeitsplätze in der Lausitz', *AfD Kompakt*, 19 September 2018.

66 Thorsten Metzner, Hans Monath and Susanne Ehldering, 'Ministerpräsident Woidke: "Ein schneller Kohleausstieg stärkt die AfD" ', *Der Tagesspiegel*, 1 September 2018; Hähnig, 'Es lebe'.

67 Adam Vaughan, 'Germany Agrees to End Reliance on Coal Stations by 2038', *Guardian*, 26 January, 2019.

68 Benjamin Konietzny, ' "Braunkohle reicht 1000 Jahre": Die AfD tröstet die Lausitzer Seele', *NTV*, ntv.de, 7 August 2019. The top candidates were Jörg Urban and Andreas Kalbitz, the latter Höcke's sidekick in Der Flügel.

69 Johannes Hillje, 'Wider der Spaltung beim Klimaschutz!', *Die Zeit*, 3 October 2019.

70 See for example Deutscher Bundestag, 'AfD-Antrag zur Aufgabe der Klimaschutz-Ziele stößt auf breite Kritik', 17 October 2019; Matthias Kamann, 'AfD will Politik zum Klimaschutz in Deutschland komplett stoppen', *Die Welt*, 15 October 2019.

run out of steam in the late 2010s, precisely because of long-lived lignite. The *Energiewende* 'hangs on our competitiveness like a millstone around a neck'.[71] A startling announcement seemed to come naturally: opposition to *Klimaschutz* would henceforth be *the* most important issue for the AfD, succeeding hostility to the EU, immigration and Islam. 'We have a unique selling point here', Gauland explained.[72] With this shift in emphasis, the party held a congress in December 2019, elected a new leadership and, heartened by the recent polls, set itself a goal for the next national election: gaining executive power.[73] It could no longer be dismissed as a chimera.

Around the time of Paris, few would have expected German politics to soon be overshadowed by a party of this complexion. But a transition in the powerhouse was now less of a foregone conclusion than ever. With the AfD evolving into one of the strongest nodes of the far right, and with equivalents across the industrial core of northwestern Europe, from the bogs of Finland to the fields of Groningen, one might begin to question the immigration-as-funnel metaphor. Does it do justice to the centrality of energy in contemporary far-right politics? Were fossil fuels merely ancillary material dragged through the pipe, or related to the war on immigration in a more symmetrical, reciprocal fashion? Were they even, as the turn of the AfD suggested, paramount? To study the relation in higher resolution, one might turn to two other fossil fuel producers: Poland and Norway.

71   Deutscher Bundestag, Plenarprotokoll 19/7, 18 January 2018, 577. See further AfD, *Programm*, 157–63; cf. Alice Wiedel's Twitter account, 28 September 2018.

72   AFP, 'Gauland: Widerstand gegen Klimaschutzmaßnahmen neues Hauptthema für die AfD', *Die Welt*, 29 September 2019; Markus Wehner, 'Warum die AfD aufs Klima-Thema setzt', *Frankfurter Allgemeine Zeitung*, 2 October 2019.

73   Cecilia Reible, 'AfD-Bundesparteidag: Neuer Vortstand will Regierungsfähigkeit', *Tagesschau*, 1 December 2019.

4

The Energy Wealth of Nations

When one boat after another capsized and sent hundreds of migrants into the depths of the Mediterranean, the bodies of seventy-one suffocated Syrian refugees were found in a lorry abandoned on a motorway in Austria, some ten thousand destitute people made the trek from the Budapest railway station to northern Europe and the body of Alan Kurdi washed up on a Turkish beach, one country, barely touched by the influx, was heading for the voting booths: Poland, ruled for the past eight years by the liberal-conservative Civic Platform. Now the party called Law and Justice, or Prawo i Sprawiedliwosc (PiS), seized the moment. Immigrants in general and Muslims in particular were on their way to flooding Poland. In response to requests from the EU, the government of Civic Platform had agreed to accept seven thousand refugees over two years, all thoroughly checked for any threats to security, divided into small groups and settled on the condition that invidiuals had a clear intention to integrate into Polish society.[1] The PiS considered it a death sentence to the nation that had to be revoked. On 16 September 2015, Jarosław Kaczyński, founder and leader of the PiS, gave a speech to the parliament on how the migrants would inevitably take control, through

---

1 Krzysztof Jaskulowski, *The Everyday Politics of Migration Crisis in Poland: Between Nationalism, Fear and Empathy* (New York: Palgrave, 2019), 37.

a process that will more or less look like this: first the number of foreigners suddenly increases, then they do not obey – do not want to obey, they declare they do not want to obey – our customs . . . and then or even simultaneously they impose their sensitivity and their claims in the public space in different spheres of life, and they do so in a very aggressive and violent way.

This had already happened in other European countries, Kaczyński averred.

If somebody says all of this is not true then have a look around Europe, let's take Sweden. There are 45 zones there governed by sharia law, there is no control of the state . . . Or what is going on in Italy? Churches have been taken over and are often treated as toilets. What is going on in France? Non-stop arguments, sharia introduced, even patrols which check if sharia is observed . . . So do you want that all of this becomes reality in Poland, that we stop feeling at home in our own country? Is that what you want?[2]

The speech struck a chord with the nation. The next week, deputy chairman of the PiS Antoni Macierewicz backed it up with an online video, in which he presented a map of the European continent with the names of the countries written in Arabic and thick arrows running up from the Balkans to pierce Poland. According to Macierewicz, the refugees of the Middle East had made a coordinated decision to move their route onto Polish territory, for a very determinate purpose: they 'openly say that they will be combating Polish civilisation and culture'.[3] It should perhaps be pointed out that the map was a fraud. But resistance to any EU quotas and emergency measures and reception of asylum-seekers had by now become a PiS battle cry that reached ever dizzier heights. Closer to the election, on 12 October, Kaczyński campaigned in a region just north of Warsaw and spoke at a rally:

---

2 Michał Krzyżanowski, 'Discursive Shifts in Ethno-Nationalist Politics: On Politicization and Mediatization of the "Refugee Crisis" in Poland', *Journal of Immigrant and Refugee Studies* 16 (2018): 86.
3 Ibid., 88–90.

That information about getting 100,000 Muslims into Poland, is that true? Well the minister of health should respond to that because all of this is related to various dangers in that area. We already have symptoms of very serious diseases, not seen in Europe for long: cholera on the Greek islands, dysentery in Vienna, various parasites and protozoans which are not dangerous in those people's organisms but can be dangerous here.[4]

On the very same day, Kaczyński also paid a symbolic visit to Ostroleka, where the government three years earlier had halted the construction of a new coal-fired power plant. Through his visit he sought to demonstrate how the Civic Platform had treated coal as a resource to be squandered. 'We will throw out that way of thinking', he told the media: 'We know that coal is needed, because it is our only serious energy resource, and so we must have coal and we must have coal-fired power plants.'[5] The Ostroleka plant would be promptly built on his party's watch.

Two weeks later, the PiS won a landslide victory in the election. For the first time since 1989, a single party took an outright parliamentary majority. After seizing power, it stood by its basic promises and ratcheted up the rhetoric on both immigration and coal. On the latter, Prime Minister Beata Szydło, another party dignitary, used the occasion of St. Barbara's Day – a centuries-old festivity, named after the patron saint of miners, who dress in smart uniforms and feathered hats and drink in beer halls – to lay out the philosophy of the black rock as a synonym of development and modernity. 'There will', she stressed, be no strong Polish economy without a strong mining industry.' Antoni Macierewicz – the man with the map, by now the minister of defence – chose to visit the Bełchatów power plant, number one point source of CO_2 in the EU, on some counts the largest lignite complex in the world, a forest of chimneys surrounded by a black-brown wasteland even more massive than any of the

---

4 Ibid., 90.
5 *Onet Biznes*, 'Kaczyński: musimy mieć i węgiel, i elektrownie węglowe', biznes.onet.pl, 12 October 2015.

German giants. 'Poland stands on coal', he proclaimed, 'and this will not change.'[6]

According to the PiS, coal was the past, present and boundless future of the nation. And the fuel did have a distinguished history in Poland: going back to the fifteenth century, coal production took on a new significance after the end of the Second World War, when the nation rose out of the ashes of Nazi occupation and found itself blessed with largely intact mines in Silesia. These were used as a springboard for rapid, forced industrialisation under the Polish People's Republic, which blared out propaganda tidbits like 'fight for coal', 'the great battle for coal' and, indeed, 'Poland stands on coal' (*Polska węglem stoi*). Electricity output from coal grew tenfold in the first two decades after inclusion in the Eastern bloc; by the 1970s, more than five hundred mines were in operation and Poland one of the world's main exporters of hard coal (mostly to other parts of the bloc), making the fuel the backbone not only of the country's energy system but of its economic development and stability as such. Here was 'our black gold', the regime blazed, proof of Poland's ability to sustain itself and vault into a future of prosperity.[7]

After the collapse of the Eastern bloc, the coal industry went through a chaotic process of privatisation and restructuring of uncompetitive mines. Several hundreds of thousands of jobs were bled out. Yet by the time of the 2015 election, Poland still derived 84 per cent of its electricity from coal, rendering the nation a place in the EU as the second largest emitter of CO_2 from power generation. During its time in government, the Civic Platform tried to maintain a delicate balance of staying open to the EU and defending domestic coal. In 2009, it enshrined a long future for the fuel in a plan for the period up to 2030, singling it out as 'a major factor stabilising Poland's energy security'. But for the PiS, this was far from enough. Just as the Civic Platform had betrayed the nation by signing up to receive a

---

6 Both quotations from Magdalena Kuchler and Gavin Bridge, 'Down the Black Hole: Sustaining National Socio-Technical Imaginaries of Coal in Poland', *Energy Research and Social Science* 41 (2018): 136.

7 Kuchler and Bridge, 'Down', 140–3; Paul Josephson, *Would Trotsky Wear a Bluetooth? Technological Utopianism under Socialism, 1917–1989* (Baltimore: Johns Hopkins University Press, 2010), 103–5.

couple of thousand refugees, so it had sold out the black gold by failing to invest properly in its future and by overseeing the closure of mines in traitorous deference to the EU.[8] The PiS revived the status of the rock in the national imaginary and italicised coal nationalism, fusing nation and fuel into one ethno-material body, by which Poland would rise like a phoenix once again – this time not from Nazi occupation, but from the dismal period of liberal, EU-friendly decomposition.[9]

The PiS, then, flew into office on the wings of promises to resurrect coal and drove the policies of its predecessor one step further, into a more assertive, sentimental agenda. In government, the party extended the lifetime of the fuel into the second half of the century: by 2050, it should still provide 50 per cent of electricity in Poland. Coal also formed the basis of an Electromobility Development Plan, according to which urban pollution would be combatted by means of electrical cars, fuelled by Ostroleka and Bełchatów and other such facilities. The PiS embraced the vision of 'clean coal' – the ultimate oxymoron of capitalist climate governance – and advanced Poland as a global leader in the field, equipped with research hubs and engineering programmes for washing coal to a shiny green.[10]

On the ground, however, all that glittered in black was not gold. In the years after the 2015 election, disappointment spread among the mining communities that voted for the PiS. In office, the party froze the so-called fourteenth wage – a bonus of a month's extra payment to miners – and cancelled the right of retirees to receive an annual coal allowance of 8 tonnes a year, perks that were part of the traditional contract between the state and the mining workers. The PiS had promised no further mine closures, yet seven were closed under its leadership. Miners were often glum about their industry, in spite of all the PiS oratory; for them, Poland wobbled rather than stood on coal.[11] Expansion was held back by local opposition – open-cast mines swallowed land; pits made it subside; both exacerbated already

---

8   Kuchler and Bridge, 'Down', 136–7, 142–3; quotation from 143.
9   Ibid., 143–5.
10 Ibid., 135.
11 This is based on fieldwork in Silesia by one of the authors (Allen).

unbearable pollution – and what investors would describe as regulatory uncertainties.[12] But the intentions of the PiS ideologists were not in doubt.

The observation of these intentions was also pursued: the PiS cracked down hard on renewable energy. After joining the EU in 2004, seeking to comply with the directives from Brussels, Poland expanded its wind portfolio at a rapid clip, until in 2015 the installed capacity ranked as the world's twelfth largest. The curve rose steadily up to that year: and then it completely flatlined. Soon after winning the election, the PiS banned construction of turbines close to dwellings, slapped fresh taxes on investors, capped the allowed capacity and imposed a range of other bureaucratic hurdles for anyone who wished to build mills.[13] Overnight, wind farms were pushed into the red. (As for mosques, the PiS came too late to prevent the first such purpose-built edifice in Warsaw from opening in 2015, but a local party branch succeeded in stopping the construction of a proposed Ahmadiyya mosque two years earlier, its petition citing incongruity with the local architecture as well as the potential for 'unwanted and hitherto unknown social problems in the locality'.)[14]

Unsurprisingly, these preferences were coupled with a certain view of the climate problem. In the fall of 2016, acting minister of the environment Jan Szyszko claimed that 'there is no scientific consensus on the topic' and that 'it is not known if the rise in temperatures is related to carbon dioxide emissions', while being perfectly sure that

---

12   Field observations by one of the authors (Allen); *Business Insider Polska*, 'Co z tym węglem? Sytuacja nie jest tak oczywista, jak się wydaje', businessinsider.com.pl, 18 September 2017; *Newsweek Polska*, 'PiS okłamał górników. Partia jednak nie oprze energetyki na (polskim) węglu', newsweek.pl, 18 September 2017; Claudia Ciobanu, 'The Backup Plan: Poles Prepare for Post-Coal Life Despite Government', *Krytyka Polityczna and European Alternatives*, politicalcritique.org, 13 December 2017; Kuchler and Bridge, 'Down', 144.

13   Kacper Szulecki, 'Poland's Renewable Energy Policy Mix: European Influence and Domestic Soap Opera', Working Paper 2017: 01, Oslo: °Cicero (Center for International Climate Research) and University of Oslo, for example 4, 7, 16, 19.

14   Kasia Narkowicz and Konrad Pędziwiatr, 'From Unproblematic to Contentious: Mosques in Poland', *Journal of Ethnic and Migration Studies* 43 (2017): 450; cf. Farid Hafez, 'Street-Level and Government-Level Islamophobia in the Visegrád Four Countries', *Patterns of Prejudice* 52 (2018): 440, 444.

CO$_2$ released from Poland 'is a gas of life for living ecosystems, enabling them to become better and better'.[15] The PiS did not, however, bear any grudges against Paris. Instead it posed as a proud compliant. In the wake of COP21, Prime Minister Beata Szydło proudly declared that 'Poland is working very actively to ensure that the climate on our planet becomes even better', leading *OKO.press*, a source of independent journalism and critical commentary on the regime, to wryly note that either Poland's ambitions were far greater than any other country's – not to save but to *improve* the earth's climate – or the prime minister did not know what she was talking about.[16]

But there was a logic to Szydło's flipping of carbon vitalism into fidelity to Paris. The PiS government pushed the demand that forest management – allowing trees to capture carbon – should be counted as a form of emissions reduction. It greeted Paris as a green light for this tactic: the laissez-faire principles of the accord permitted Poland to claim standing forests as a form of mitigation, while having to do nothing about coal.[17] At COP22 in 2016, Szyszko pleaded for 'forest carbon farms' as the solution, in no disharmony with the mining and burning of the black gold. There was a double irony here, since Szyszko – a professor of ecology, a forester by trade – was the same minister who allowed logging in Białowieża, Europe's last primeval forest of some size, and gave official permission to individuals to cut as many trees as they liked on private property, paving the way for a severe thinning-out of Poland's forests. 'Man has not only the right, but the duty to use natural resources', Szyszko justified the latter reform, citing God's exhortation to 'subdue' the earth in the book of Genesis.[18] On this logic, (1) Poland will burn coal, (2) Poland will cut trees and (3) Poland will make up for its release of carbon by absorbing carbon in its forests – perhaps nourished by the gas of life,

---

15 Quoted in Jula Szulecka, 'Minister Szyszko węglem i wycinaniem lasów chce powstrzymać zmiany klimatu', *OKO.press*, 21 November 2016.
16 Ibid.
17 See for example Krzysztof Tchórzewski, 'Polska odrzuca dekarbonizację. Będziemy szli własnym tempem', *Biznes Alert*, 10 May 2016.
18 Christian Davies, 'Polish Law Change Unleashes "Massacre" of Trees', *Guardian*, 7 April 2017.

in a virtuous cycle that will save the day. The PiS government thereby set a new record in the old game of diversity of tactics, blending denialism and capitalist climate governance in a pro-Paris, pro-coal, pro-forest management, pro-deforestation synthesis. The upshot? 'Zero carbon by 2050? We simply reject it.'[19]

The odd synthesis was on full display when the PiS government hosted its very own COP24 in 2018, symbolically located in Katowice, capital of the coal region in Upper Silesia and the 'clean coal' project.[20] In the months leading up to the summit, the PiS resolved to protect coal interests by heavily curtailing citizen access to the venue. Szyszko stated that Poland wished to use its presidency 'to present climate policy as a tool for sustainable development', by now a euphemism for business-as-usual.[21] When asked if Poland backed the basics of climate science, the Ministry of the Environment was vague. 'It is in the interests of all of us,' a spokesperson responded, 'regardless of the disputes over the causes of this phenomenon, to work towards reducing climate change.' These comments provoked some consternation and were withdrawn by the ministry, then restated and withdrawn again, the spokespersons declining several requests to clarify what the 'disputes' consisted of.[22] So deeply had the far right insinuated itself into planetary developments, so blurred had the lines between denialism and capitalist climate governance become, that a round of UN climate talks could be hosted on these premises.

After visitors to COP24 in Katowice had breathed in air saturated with soot and particulate matter, they entered the venue and were treated to an exhibition of the beauties of coal. The PiS government had decorated its pavilion with sundry types of coal artwork and artisanal products: soap made of coal, earrings and cufflinks and necklaces made of coal. Plaques described the black gold as 'incredibly soft and fragile' and possessing 'a hidden charm'. Below the

---

19 Tchórzewski, 'Polska'.
20 Again, the continuities with the former regime should not be ignored: a similar promotion of coal at a UNFCCC summit was undertaken by Civic Platform during COP19 in Warsaw. See for example Kuchler and Bridge, 'Down', 144.
21 Hanna Schudy, 'Szczyt bez obywateli, czyli PiS broni klimatu (dla węgla)', *Krytyka Polityczna*, 27 July 2018.
22 Arthur Neslen, 'Poland to Put "Common Sense" over Climate Ambition as Host of Critical UN Talks', *Climate Home News*, 27 July 2018.

jewellery, lumps of raw coal were displayed in open boxes, allowing visitors to touch the wonder material with their own hands. Polish president Andrzej Duda, who won the post in 2015 thanks to his collaboration with the PiS, opened the proceedings with a speech in praise of coal, which 'does not contradict the protection of the climate'; Poland, he continued, was not prepared to sever its ancestral ties to the source. Polish companies extracting and burning coal sponsored the conference and filled it with promotion material about their environmental leadership. During the conference, the Polish pavilion organised a 'clean fuels day', which touted the by-products of coal mining as fuels of the future and coal itself as an intrinsic part of Polish biology.[23] Solidarity, the most renowned of Polish trade unions, made a pact with the Heartland Institute: at the summit, the partners issued a joint communiqué in repudiation of the IPCC's claim 'that the world stands at the edge of a climate catastrophe. Solidarity and the Heartland Institute together stress that there is no scientific consensus.'[24] Only a few metres separated the official platform of the presidency from that position. At the twenty-fourth attempt to put the principles of the UNFCCC into practice, it looked as though an unholy alliance of denial and capitalist climate governance had won if not the war, then at least the latest rounds of battle.

On the field of immigration, the borders held. Kaczyński, the de facto leader of the country, announced in a speech uploaded to YouTube in May 2016 that not a single refugee would be accepted 'because there is no mechanism that would ensure safety'; pleas from the EU were met with blanket refusal. The country had long kept its doors shut – a representative figure from 2014 put the number of asylum applicants at 0.2 per thousand inhabitants, to be compared with 2.5 for Germany and 8.4 for Sweden – which did not, however, prevent PiS interior minister Mariusz Błaszczak from declaring that

---

23 Observations by one of the authors (Zoob Carter). See also Brian Kahn, 'Poland Literally Filled an International Climate Conference with Coal', *Earther*, 3 December 2018; Chloe Farand, 'COP24: Climate Science Denial, Disinformation and Fake News at the UN Climate Talks', *DeSmog UK*, 5 December 2018.

24 James Taylor, Jaroslaw Grzesik and Dominik Kolorz, 'Joint Declaration Between Solidarity and The Heartland Institute', Heartland Institute, 5 December 2018.

the previous government 'put a ticking bomb under us' by agreeing to refugees: 'We're defusing that bomb.' Put differently, any reception of asylum-seekers 'is a straight road to a social catastrophe'.[25]

Under PiS, alongside carbon vitalism, such apocalyptic xenophobia – more precisely, Islamophobia – became state ideology. The leaders of Poland continued to chunter on about unassimilable Muslims, no-go zones in Stockholm 'where every few weeks something explodes' (Błaszczak again) and the right of Poles 'to feel masters of their own house' (Kaczyński).[26] Under tightening state control, media disseminated similar stories, while right-wing journalists put out books with titles such as *Caliphate Europe* and *Jihad and the Self-Destruction of the West* and *Freemasonry, Islam, Refugees: Are We Heading for a Great Apocalypse?*[27] Popular online conversations spilled over with tales about Muslims raping their way through Europe and conquering the continent through the incredible fecundity of their uteruses. Arch-conservative bishops – anchors of the PiS regime – dispatched news from Islamised Europe and sermonised on 'these wild people, euphemistically called refugees' who could not be trained to respect anything of value. In the winter of 2016, weekly magazine *wSieci* ('The Network') ran a cover with a blond woman, loosely dressed in the EU flag, violently groped by dark-skinned, hairy arms under the headline 'The Islamic Rape of Europe'.[28]

If early twenty-first-century Poland had plenty of coal left in the ground, however, it had virtually no Muslims. This was the great puzzle of Polish Islamophobia. The country was one of the most ethnically homogenous, monochromatically white in all of Europe – an enduring legacy of twentieth-century history, before which things were, of course, wildly different. Poland could long claim to be one of the most ethnically diverse countries on the continent: just prior to the Second World War, around one-third of Poland's

25 Quoted in Jan Cienski, 'Why Poland Doesn't Want Refugees', *Politico*, 21 May 2017.

26 Quoted in Piotr Cap, '"We Don't Want Any Immigrants or Terrorists Here": The Linguistic Manufacturing of Xenophobia in Post-2015 Poland', *Discourse and Society* 29 (2018): 388, 391.

27 Monika Bobako, 'Semi-Peripheral Islamophobias: The Political Diversity of Anti-Muslim Discourse in Poland', *Patterns of Prejudice* 52 (2018): 457–8.

28 Goździak and Márton, 'Where the Wild', 128–40; Hafez, 'Street-Level', 440.

population was composed of ethnic and religious minorities, including one-fifth of all the world's Jews, the second largest Jewish population in the world. In the 2010s, 97 per cent of the population declared themselves as ethnically solely Polish.[29] The number of Muslims barely reached forty thousand out of 38 million citizens, or less than 0.1 per cent of the population. Some five thousand belonged to the Tatar micro-minority that had inhabited Polish lands for the past seven centuries; others came as students from the Middle East during the post-war era; a few were converts, while the number of Muslim immigrants hovered close to zero.[30] The phantasmagorical nature of contemporary Islamophobia was here pushed to a new level. Muslims were detested like no other ethnic or religious group in Poland *even though they were nowhere to be seen*. Scholars referred to it as 'Islamophobia without Muslims', or 'Platonic Islamophobia' – contactless, without physical referent, much like the anti-Semitism without Jews that continued to thrive in Poland after the Holocaust, but emerging as a political force only in recent years.[31]

One driver of this phobia appears to have been a sense of national peril. Another was the popular wish to retain the white character of the nation: 69 per cent of Poles said they did not want more people of a different skin colour to live in it.[32] Yet another was certainly the global circulation of Islamophobic ideas. Such factors did not, however, fill the gap separating this ideology from reality – how could it be that 47 per cent of Poles believed too many Muslims lived in their country? How could the prospect of a Muslim population haunt them so vividly? One student of the phenomenon, Katarzyna

---

29 Joseph Marcus, *Social and Political History of the Jews in Poland, 1919–1939* (Berlin: De Gruyter Mouton, 1983), 15–17; World Directory of Minorities and Indigenous Peoples, 'Poland', minorityrights.org, July 2018.

30 Katarzyna Górak-Sosnowska, 'Islamophobia without Muslims? The Case of Poland', *Journal of Muslims in Europe* 5 (2016): 190–204; Narkowicz and Pędziwiatr, 'From Unproblematic', 444–5; Goździak and Márton, 'Where the Wild', 131–2.

31 Górak-Sosnowska, 'Islamophobia'; Narkowicz and Pędziwiatr, 'From Unproblematic', 452; Bobako, 'Semi-Peripheral', 449; Goździak and Márton, 'Where the Wild', 125, 131; Krzyżanowski, 'Discursive', 80–1. Yet another term proposed for this syndrome is 'migrational hypochondria'. Mustafa Switat, 'Arabic Community in Poland: Facts and Myths', *Yearbook of Polish European Studies* 19 (2016): 257.

32 Figures from 2013 reported in Górak-Sosnowska, 'Islamophobia', 195. On the sense of peril and Islamophobia as 'hegemonic discourse', see Jaskulowski, *Everyday*, 26.

Górak-Sosnowska, posed these questions to a leading 'critic of Islam', who explained that he was merely being

> forward-looking. Just as ecologists raise awareness against climate change, even though the Earth was still in a relatively good shape, he must raise awareness against Islam. Even if there are now almost no Muslims in Poland, and he cannot attribute to them even the slightest problem, it might easily change in a foreseeable future. While he was unable to provide me with a specific date or even a timeframe of when this might happen,

he clung to his faith.[33] Of a similar hypothetical nature were the civic patrols that entered night clubs and discos in Warsaw, Poznań and Krakow in January 2014 – before the so-called refugee crisis – to chase the mirage of Muslim rapists coming for 'our women'. 'In the space of one evening, in hundreds of places across the country, incidents occur involving the seduction of our female compatriots', the Polish Defence League wrote on its website. 'One of these girls, unwittingly charmed by an exotic prince could, along with her offspring, end up very badly in the Islamic world, which is advancing on us with great strides.'[34] Hence the need for valiant manly interventions.

Of a similar nature too, of course, was the anti-Semitism rearing its head again in Poland. In November 2015, a demonstration against Muslim refugees in the city of Wrocław ended with the crowd burning an effigy of a Hasidic Jew wrapped in an EU flag.[35] The most popular anti-Semitic conspiracy theory in this age of migrations – that George Soros, the Jewish billionaire and 'globalist', planned and funded the flooding of white nations – made the rounds in Poland too, including through the mouths of PiS leaders.[36] Polls suggested a

---

33 Górak-Sosnowska, 'Islamophobia', 203.

34 Quoted in *Radio Poland*, 'Anti-Islamic Group Patrols Clubs to "Protect" Women', thenews.pl, 15 January 2014.

35 JTA, 'Jew Burned in Effigy during Polish Demonstrations against Refugees', *Haaretz*, 19 November 2015.

36 Patrick Strickland, 'What's Behind Hungary's Campaign against George Soros?', Al Jazeera, 22 November 2017; Jason Wilson, ' "Dripping with Poison of Antisemitism": The Demonization of George Soros', *Guardian*, 25 October 2018; Craig Turp, 'Hungary and Poland: Real Friends, Imaginary Enemies', *Emerging Europe*, 3 March 2018.

spike in anti-Jewish attitudes and hate speech after 2014, which seemed to depict a circular movement: classical anti-Semitism serving as inspiration for new Islamophobia, which in turn ushered in renewed hostility to Jews. 'The more people are anti-Muslim, the more they're anti-Semitic', in the words of one specialist at the University of Warsaw.[37] Or, once the fantasies have begun to run amok, there's no reality to hem them in.

Now marry this to coal nationalism, as the PiS did. It interpellated its subjects as members of a nation standing on coal and in need of protection against Muslims and assorted others. This was not, it is important to stress, some spasm of atavistic Polish or Eastern European hysteria, but precisely one branch on a very transnational tree of ideas.[38] (Hence, for example, Lord Lawson of the Vote Leave campaign offered professional commercial advice to the Bełchatów plant.) In Poland, as elsewhere, the patina of nationalist pride masked something no nation could claim as its own. It is also important, however, to recognise the specificities of the case and how they brought out aspects of the global trend with special clarity. Poland sits on the periphery of the EU. After the collapse of the Eastern bloc and absorption into capitalism, the glittering cities of the West came so close yet so far; inequalities at home soared and disappointment deepened.[39] Eventually the EU came to be seen as the latest foreign coloniser – after Prussia, Austria-Hungary, Russia, Germany – bent on devouring the nation. That perception was not necessarily ill-informed: foreign capital did penetrate Poland and seize its means of production to an extent unknown on the western side of the continent.[40]

---

37 Michal Biłewicz quoted in Don Snyder, 'Anti-Semitism Spikes in Poland – Stoked by Populist Surge against Refugees', *Forward*, forward.com, 24 January 2017.

38 As argued forcefully, in the case of Islamophobia, by Ivan Kalmar, '"The Battle Is in Brussels": Islamophobia in the Visegrád Four in Its Global Context', *Patterns of Prejudice* 52 (2018): 406–19.

39 See John Feffer, *Aftershock: A Journey into Eastern Europe's Broken Dreams* (London: Zed Books, 2017).

40 See for example Leonid Bershidsky, 'How Western Capital Colonized Eastern Europe', *Bloomberg*, bloomberg.com, 12 September 2017; Filip Novokmet, Thomas Piketty and Gabriel Zucman, 'From Soviets to Oligarchs: Inequality and Property in Russia, 1905–2016', NBER Working Paper 23712, 2017.

Much of that foreign capital was German, and much of the resentment the PiS voiced was directed against that particular core country. Immigration and renewable energy were both associated with Germany and converted into symbols of its dominance – elements foreign to the once-again-victimised nation of Poland. During the election campaign of 2015, Kaczyński and other PiS leaders accused Germany of rolling out the red carpet for refugees. Fairy tales about the mayhem of Muslim immigration partly transposed Islam from the east to the west in the imaginary political geography: the phantom Muslim came to stand in for the resented West, the quest for autonomy projected against that invisible enemy (but much of this Islamophobia was, of course, imported straight from Western sources).[41] Mixing of the races is what the Westerners who try to rule us do. Now they want us to copy their model. But we do things differently, tend to our own and that which is ours, and nothing is ours more than our black gold – something the West is also bent on stealing from us.[42] PiS coal nationalism blossomed in conjunction with anxieties over the survival of a nation known (since the occupation) as Catholic and white. These anxieties were heightened by the unequal relation to Western Europe, which provoked a felt need to overprove that whiteness. In fact, in an inversion of the logic, and as an attempt to reclaim superiority in the endless seesawing battle with a national inferiority complex, the PiS increasingly presented Poland as the true Europe – the untainted, well-preserved heart of Christian and white civilisation, thanks precisely to the refusal of the corrupted West's new ways. Moreover, this nation was the 'Christ of Europe', an idea consecrated when the PiS formally crowned Jesus Christ king of Poland in November 2016.[43]

---

41 Krzyżanowski, 'Discursive', 86–7; Bobako, 'Semi-Peripheral' 457–8; Kalmar, '"The Battle"', 415. Krzysztof Jaskulowski calls this 'pathological Europeanisaton and transnationalism'. Jaskulowski, *Everyday*, 94. For an analysis of how Eastern European migrants to the West, and specifically to Britain, have developed racist competencies to be viewed more favourably and become integrated in British society, see Jon. E. Fox and Magda Mogilnicka, 'Pathological Integration, Or, How East Europeans Use Racism to Become British', *British Journal of Sociology* 70 (2019): 5–23.

42 Cf. Leszek Koczanowicz, 'The Polish Case', *New Left Review* series 2, no. 102 (2016): 87.

43 Paweł Mączewski, 'Jesus Christ Is Now Officially the King of Poland', *Vice*, 24 November 2016.

Leaders of the party had a penchant for seeing this big picture, in their own way. In 2016, Foreign Minister Witold Waszczykowski summed up the PiS *Weltanschauung* for the German tabloid *Bild*:

> The previous government implemented a left-wing concept, as if the world had to move using a Marxist model in only one direction: towards a mixture of cultures and races, a world of cyclists and vegetarians, who rely solely on renewable energy sources and combat all forms of religion. This has nothing in common with traditional Polish values ... This is against what most Poles have at heart: tradition, historical consciousness, love of the country, faith in God and a normal family life run by a man and a woman.[44]

On the eve of COP24 in Katowice, President Duda retweeted a similarly synoptic article about CO_2 emissions in Europe. He wrote that it was 'very good', 'worth reading', 'I recommend it'. The article spoke about CO_2 as 'a friendly gas'. It condemned the criticism Poland had received for its reliance on coal as nothing but anti-Polish assaults from 'Jewrope'. Immigrants were 'future soldiers of jihad' and wind power an economic cudgel Germany used to beat other nations.[45] Jewrope with its jihadi soldiers knocking around Poland for releasing the gas of life: after a few hours, the president of the country hosting COP24 deleted the tweet. He then posted a new one saying that he had not read the article to the end. A spokesman for the presidential office did, however, stand by the view that CO_2 is a gas of life and offered a lecture on photosynthesis, leading *OKO.press* to sardonically observe that one could just as well tell someone who is about to be drowned in a flood that they need water to live.[46]

These were not the views of some mavericks shouting in the streets. The Poland of the PiS acted as the main drag on climate

---

44 Hans-Jörg Vehlewald, 'Haben die Polen einen Vogel?', *Bild*, 3 January 2016.
45 Violetta Baran, 'Andrzej Duda poleca artykuł w sieci. Przeczytał go do końca?', *WP Wiadomosci*, 6 December 2018; Jakub Szymczak, 'Rzecznik prezydenta Dudy: "Bez CO2 nie ma życia". Aż dziennikarz zaniemówił', *OKO.press*, 7 December 2018.
46 Szymczak, 'Rzecznik'.

policies in the EU, slowing down and hollowing out any ambitions for cutting emissions; when the Union wrangled over 2050 as the target year for 'climate neutrality', it insisted on the utterly non-committal 2070 for itself.[47] Accounting for upwards of 10 per cent of the Union's emissions – more than France, more than Spain, second only to Germany – Poland had the capacity to scuttle any continental decarbonisation.[48] It would be only a slight exaggeration to say that its far right held European climate politics hostage. But then again it was not alone.

Norway Stands on Oil

Apart from a shared history of Nazi occupation, Poland and Norway would not normally be considered mirror images of each other in early twenty-first-century Europe. Yet they also had two other characteristics in common: an economy built around fossil fuels, the extraction of which came to be placed in the hands of the far right. Norway was, by then, the largest European producer of oil and gas. The bulk of the products extracted from its continental shelf were exported, accounting for nearly half of the national exports in value, a fifth of GDP and as much of government revenue: the lifeblood of the social formation.[49] As of 2016, this was the fourteenth producer of oil in the world and the seventh of gas. Luckily for it, CO_2 emissions were officially attributed to the nations where the burning – not the digging up – of fossil fuels took place; if the latter were the accepted metric, Norway, with its 5 million inhabitants, would have

---

47   For example Chloe Farand, 'Poland Expected to Delay EU Carbon Neutrality Deal', *Climate Home News*, 18 July 2019; Jennifer Rankin, 'European Green Deal to Press Ahead despite Polish Targets Opt-out', *Guardian*, 13 December 2019; Jon Birger Skjærseth, 'Implementing EU Climate and Energy Policies in Poland: Policy Feedback and Reform', *Environmental Politics* 27 (2018): 498–518. There were, however, significant continuities in this regard between the Civic Platform and the PiS.

48   Figures for 2018 in Eurostat, 'Early Estimates of CO_2 Emissions from Energy Use', ec.europa.eu, 8 May 2019.

49   Norwegian Petroleum, 'Exports of Oil and Gas', norskpetroleum.no, 2018; *Offshore Energy Today*, 'Norway, the Biggest Oil Producer in Western Europe', offshoreenergytoday.com, 2010.

been the outright leader for per capita emissions, some thirty times above the global average.[50] The fossil treasures were under the control of the Ministry of Petroleum and Energy. Between 2013 and 2020, all four leaders of that ministry came from the FrP, the far-right party led by Siv Jensen, who five years before entering government revealed that she banked on 'climate scepticism' as a voter magnet no less powerful than hostility to immigration.

Starting off in the 1970s as a rambunctious anti-tax party, the FrP picked up the master frame from France in the late 1980s and turned its ire towards the immigrants who hung like leeches on the people's resources. A turning point came in 1987. Timed to the upcoming local and regional elections, the leader and doyen of the FrP, Carl I. Hagen, informed the Norwegian people that he had received a private letter from a certain 'Mustafa'. It read:

> Allah is Allah, and Muhammad is His Prophet! You are fighting in vain, Mr. Hagen! Islam, the only true faith, will conquer Norway too. One day, mosques will be as common in Norway as churches are today, and the children of my grandchildren will live to see this. I know, and all Muslims in Norway know, that one day, the Norwegian population will come to [the Islamic] faith, and that this country will be Muslim! We give birth to more children than you, and many a right-believing Muslim comes to Norway each year, men in fertile age. One day, the heathen cross in the flag will be gone too![51]

---

50   Liv Thoring, 'Førti skitne år med norsk olje og gass', *Arbeidsnotat* (2009); Andreas Ytterstad, *Norwegian Climate Change Policy in the Media: Between Hegemony and Good Sense*, dissertation (University of Oslo, 2012), 4–5; Helge Ryggvik, *Norsk olje og klima: En skisse til nedkjøling* (Oslo: Gyldendal, 2013), 51; Helge Ryggvik and Berit Kristoffersen, 'Heating Up and Cooling Down the Petrostate: The Norwegian Experience', in Thomas Princen, Jack P. Manno and Pamela L. Martin (eds), *Ending the Fossil Fuel Era* (Cambridge, MA: MIT Press, 2015), 259–60; T. A. Boden and R. J. Andres, 'Global, Regional, and National Fossil-Fuel CO_2 Emissions', *Carbon Dioxide Information Analysis Centre*, cdiac.ess-dive.lbl.gov, 2017; Guri Bang and Bård Lahn, 'From Oil as Welfare to Oil as Risk? Norwegian Petroleum Resource Governance and Climate Policy', *Climate Policy* (2019), online first, 1.

51   The letter is included in Sindre Bangstad, 'Norwegian Right-Wing Discourses: Extremism Post-Utøya', in Douglas Pratt and Rachel Woodlock (eds), *Fear of Muslims? International Perspectives on Islamophobia* (n.p.: Springer, 2016), 240. But already the anti-tax predecessor to the FrP had clear racist inclinations: see Bangstad, *Anders*, 114–6.

It is worthy of note that the Front National in the early 1980s had circulated a letter, this one addressed *to* a certain 'Mustafa', describing how the Muslims were on the way to becoming 'lords and masters of Paris'. That document set the stage for the electoral breakthrough of the Front in a municipal election in 1983. Hagen appears to have copied the ruse almost verbatim; like the original Mustafa letter, it was entirely fake, and it did for the FrP exactly what it had done for the Front: handed the party a subnational election victory and thrust it into the centre of political attention.[52] From that moment, immigration was constantly on party leaders' lips.

Putting 'our people first', the FrP now shifted towards an appreciation of some aspects of the welfare state – notably services to the elderly – that it had previously, in its ultra-neoliberal incarnation, sought to tear down. But it still wanted taxes heavily cut or abolished. It also wanted to cover the rugged country with roads and tunnels and bridges, paid for by the state, without tolls. How would that equation be balanced out? By pumping up more oil from the shelf and more oil revenues into the state budget. This had been a siren song ever since Norway struck the black gold, but from the 1970s onwards the government remained bound to the mast of a 'moderate' rate of extraction, so as to protect the economy from overheating and landscapes from overexploitation. During the neoliberal 1990s, the ropes loosened and the national policy changed to extraction at full speed, with the resulting ocean of revenues redirected into a state-controlled Oil Fund. This transformation was not orchestrated by the FrP, but rather jointly by neoliberal economists and social democrats.[53] Even within this framework, however, there were fetters in place, limiting the areas in the high north available for drilling and the amount of money that

---

52 On the original French letter, see Neil MacMaster, 'Islamophobia in France and the "Algerian Problem"', in Emran Qureshi and Michael A. Sells (eds), *The New Crusades: Constructing the Muslim Enemy* (New York: Columbia University Press, 2003), 298–9; on the character and fate of the Norwegian copy, Bangstad, 'Norwegian', 240; Bangstad, *Anders*, 121.

53 For detailed chronicles and analyses, see Helge Ryggvik, *Til siste dråpe: Om oljens politiske økonomi* (Oslo: Aschehoug, 2010); Erik Martiniussen, *Drivhuseffekten: Klimapolitikken som forsvant* (Oslo: Manifest, 2013).

could be transferred from the Oil Fund into the state budget. In the national conversation on how to handle the riches, the FrP occupied a maximalist niche: pump up as much oil as absolutely possible and siphon even more revenues into the state budget – why not enjoy our fruits to the full? In the 1990s, 'petroleum populism' became a mainstay of FrP politics.[54] Two decades later, the FrP had earned the nickname *the oil party*.[55]

Around the turn of the millennium, few countries had such a schizophrenic self-image as Norway. Traditionally identifying as the sons and daughters of the fjords and huts; picturing themselves close to nature by birth, hiking and skiing and being in their element in the open air; idolising Gro Harlem Brundtland, the mother of sustainable development, and Arne Næss, the father of deep ecology, the Norwegians had few equals in the production of oil and gas.[56] How could they hold the two faces together? How could a country proud of having established the world's first separate Ministry of the Environment in 1972 cope with being one its most efficient oil and gas stations? The ideological solution was no less schizoid: we Norwegians are uniquely capable of producing more oil and gas *and* saving the planet at the same time. This ambidexterity was developed over the 1990s by an ideological state apparatus – here truly centred on the state – consisting of the Ministry of Finance, state-owned oil company Statoil, the social-democratic and conservative parties and a cohort of paid journalists, working in concert to inculcate in Norway trust in its fossil fuels.[57]

---

54   Anders Widfeldt, *Extreme Right Parties in Scandinavia* (London: Routledge, 2015), 89; Eirikur Bergmann, *Nordic Nationalism and Right-Wing Populist Politics: Imperial Relationships and National Sentiments* (London: Palgrave Macmillan, 2017), 141–3, 153.

55   Geir Ramnefjell, 'Til siste dråpe', *Dagbladet*, dagbladet.no, 11 May 2017.

56   A magnificent analysis of this schizophrenia is Kari Marie Norgaard, *Living in Denial: Climate Change, Emotions, and Everyday Life* (Cambridge, MA: MIT Press, 2011), particularly ch. 5.

57   See further Ryggvik, *Til siste*; Martiniussen, *Drivhuseffekten*; Eivind Trædal, *Det svarte skiftet* (Oslo: Cappelen Damm, 2018); Anne Karin Sæther, *De beste intensjoner: Oljelandet i klimakampen* (Oslo: Cappelen Damm, 2019); Øyvind Ihlen, 'The Oxymoron of "Sustainable Oil Production": The Case of the Norwegian Oil Industry', *Business Strategy and the Environment* 18 (2009): 53–63; Peder Anker, 'A Pioneer Country? A History of Norwegian Climate Politics', *Climatic Change* 151 (2018): 29–41.

Much the most habitually used argument was that Norwegian oil and gas (just like Polish coal in Poland) are exceptionally clean. They were said to be extracted with less CO_2 than anywhere else. Mimicking the logic of 'I am no racist, but . . .', the CEO of Statoil in 2012 admitted that climate change 'is one of the most serious issues we face. But at the same time, Norwegian oil and gas production is the most climate-friendly oil production in the world – that is, with the least CO_2 emissions per barrel', on which logic reducing Norwegian oil and gas production would be 'a particularly bad climate action, as it will be replaced by less CO_2-friendly [sic] production abroad'.[58] Not only was this statement false on its own terms – Saudi Arabia gave off less CO_2 per barrel – but it blanked out the fact that 95 to 98 per cent of emissions from oil and gas happen *when they burn*, not when they are dug up, and the Norwegian variants burn no less fiercely than others.[59] Equally sanctimonious was the argument that Norway performed an act of altruism through its exports. 'Increased access to Norwegian energy in developing countries will contribute to improved living conditions and pull millions of people out of poverty', claimed a 'fact book' from the Ministry of Petroleum and Energy in 2011; in reality, though, 99.9 per cent of the exported oil and gas ended up in other OECD countries.[60] In line with the same saviour complex, Norway became the largest contributor to REDD+, paying for the preservation of forests in developing countries – the Amazon, not the least – and a veteran champion of other flexible mechanisms.[61] In short, Norway aimed to be top of the heap in capitalist climate governance.

---

58 Helge Lund quoted in Sæther, *De beste*, 47.
59 Ibid., 53; Ryggvik and Kristoffersen, 'Heating', 265–6; Trædal, *Det svarte*, 155; Ihlen, 'The Oxymoron', 59; Andreas Randøy, 'Norsk olje skaper klimaendringer', *Natur og Ungdom*, nu.no, 16 October 2018.
60 Sæther, *De beste*, 32, 66–9, 83–9; quotation from 83.
61 Martiniussen, *Drivhuseffekten*, 145; Trædal, *Det svarte*, 128–9; Sæther, *De beste*, 35, 52; Anker, 'A Pioneer', 37–8. Yet another unsurprising argument was that any reduction in fossil fuel use would have to come from the demand side, and so a supplier like Norway had no responsibility for it. See Taran Fæhn, Cathrine Hagem, Lars Lindholt et al., 'Climate Policies in a Fossil Fuel Producing Country: Demand versus Supply Side Policies', *Energy Journal* 38 (2017): 77–102.

Never far below this veneer of good intentions was another kind of self-congratulation: we Norwegians have become so wonderfully rich thanks to this and no other source. One well-spun story said that old rural Norway was a dreary place before the discovery of oil. The Norwegians rose from the doldrums of fisherfolk and farmers to the pinnacle of global affluence – or, as one Centre Party politician put it in 2018: 'Until we discovered oil and gas in 1969, wealth was nothing we had to any large extent. We are heirs to what was found in the inlets and on the mountain sides. A wilful nation, you can say.'[62] In the same year, NRK, the Norwegian Broadcasting Corporation, produced the most expensive TV drama in its history, called *Lykkeland*, or 'Happy Nation'. It offered a panorama of Norway, and the city of Stavanger in particular, as poor and backward before the windfall. The employers' association Norwegian Oil and Gas seized the opportunity to launch a PR campaign under the hashtag #lykkeland, suggesting that nothing but oil and gas was responsible for everything from the high number of doctors, nurses and midwives to reduced infant mortality, generous maternity leave and kindergartens. When critics pointed out that comparable countries such as Sweden had equivalent standards of welfare without fossil fuels, and that the reforms in question were rather political decisions pushed by the labour movement, Norwegian Oil and Gas responded that its campaign was simply inspired by the TV series, which presumably reflected the state of the nation.[63]

Politicians, TV series, employers: the interpellation had become a national sport: hailing Norwegians as subjects of a nation constituted by fossil fuels. This petro-nationalism was inscribed in the continental shelf, when oil and gas fields opened in the 2010s were named after nineteenth-century national heroes such as Ivar Andreas Aasen (founder of Nynorsk, a Norwegian written language), Aasta Hansteen (artist and advocate of women's rights) and Johan Sverdrup (the prime minister who introduced parliamentarism in Norway). In

---

62 Ole André Myhrvold (Senterpartiet), member of the Standing Committee on Energy and Environment, interviewed by one of the authors (Hesselbjerg).

63 See for example Silje Ask Lundberg and Wenche Skorge, 'Propagandamaskin', *NRK*, nrk.no, 9 December 2018; Tommy Hansen, 'Fakta om lykkelandet', *NRK*, 10 December 2018.

response to the growing awareness of the climatic effects of oil and gas, primitive fossil capital and its friends in the dominant classes worked studiously to fossilise Norwegian national identity, and no one did so with greater ardour than the FrP.

And in the FrP, the hardest petro-nationalism on the spectrum of Norwegian politics merged with the most critical attitudes to immigrants. These people were framed as a drain on the riches, exploiters of 'our' welfare, aberrantly prone to crime, given to the undermining of Norwegian, Western, modern culture.[64] In 1997, then chairman Hagen claimed that 'a society without ethnic minorities is a society in harmony', and the election manifesto of the same year subscribed to the doctrines of ethnopluralism: it is not 'immoral to believe that one should prevent too rapid changes in the unified character of our population. It is incorrect to call this racism when it is not based on ideas about some races being more valuable than others' – only on the idea of one race belonging to *this* country, swimming alone in its sea of oil.[65]

Going with the flow of the right, the FrP sharpened its focus on Islam in the early millennium. Hagen's successor Siv Jensen coined the term 'stealth Islamisation' (*snikislamisering*) for a process, covert yet ubiquitous, by which European countries were transformed step by step into Islamic entities, the Eurabia theory trickling down into this party too.[66] Since its adoption of the master frame, the FrP housed one centre-oriented and one hard-right wing, the latter the most insistent in traducing Islam. To it belonged one Christian Tybring-Gjedde, author of an op-ed in 2010 in which he charged the then ruling Labour Party with 'stabbing the country in the back' and 'replacing' Norwegian culture with the 'un-culture' (*ukultur*) of immigrants.[67] Even so, the FrP

---

64 Bergmann, *Nordic*, 144.

65 Quoted in Lars Waerstad, 'Carl I. Hagen's 15 mest omstridte uttalelser', *Side 3*, side3.no, 18 August 2011; Widfeldt, *Extreme*, 98.

66 For example NTB (Norsk Telegrambyrå), 'Siv Jensen advarer mot snikislamisering', *Dagbladet*, 21 February 2009; Ingelin Haukali, 'Siv Jensen: – Vil fremdeles snakke om snikislamisering', *TV 2*, tv2.no, 18 August 2011. See further Bangstad, *Anders*, 127; Bangstad, 'Norwegian'.

67 Bangstad, *Anders*, 112, 129; on Tybring-Gjedde, see further for example 4, 22, 110, 130–2.

maintained a reputation as an unusually sensible, polished, mild-mannered member of the party family – suddenly shattered when Anders Breivik butchered sixty-nine Labour youth activists held guilty of ceding Norway to the Muslims and it emerged that he had been an FrP activist for ten formative years during his youth. But the Breivik crisis soon passed.[68] Two years after the massacre, the FrP entered a government coalition with the conservatives and laid its hands on the energy department. Its denial had been confirmed all the way up to this threshold, but the party now swallowed the government's intention to pursue 'a proactive climate policy' – the typical assortment of trading and technology – and henceforth officially accepted that climate change might, at least partly, be caused by humans who burn fossil fuels.[69] And dig them up? Certainly not. Everything the FrP did after 2013 aimed at sealing off Norwegian oil and gas production from the remit of climate concern.

Beneath the surface of acceptance, sizeable pockets of open denialism remained.[70] Purveyor of the Mustafa letter Carl I. Hagen staged a comeback in national politics in 2016 with this as his main talking point: 'the climate hysteria is a pure fraud' (note: fraud); it was warm in the time of the Vikings when 'nobody drove a car'; the IPCC 'has closed meetings and deceives the world leaders'.[71] Following these statements, it was revealed that every fourth MP from the party agreed with their old leader (who claimed more would have stepped forth had they not been

---

68 Widfeldt, *Extreme*, 82–3. Some scholars have suggested that the FrP does not belong in the party family: this is refuted by ibid., 83, 114; Bergmann, *Nordic*, 145–6, 154.

69 Regjeringen Solberg I 2013, *Sundvolden-plattformen. Politisk plattform for en regjering utgått av Høyre og Fremskrittspartiet*, regjeringen.no, 7 October 2013, 4. Confirmation of denial: for example *NRK*, 'Jensen kritisk til FNs klimapanel', 31 January 2010; Siv Sandvik, Sigrid Sollund and Astrid Randern, 'FrP-topper tror ikke på klimaforskningen', *NRK*, 24 May 2013.

70 Cf. Johan Falnes, 'Frp snur – stemmer for ny klimaavtale', *Dagsavisen*, 5 December 2013.

71 Fredrik Solvang and Simon Skjelbostad Yset, 'FrP-Hagen avviser sammenheng mellom klimaendringer og CO_2-utslipp', *NRK*, 13 December 2016; Geir Ramnefjell, 'Klimaskeptikere er en stor velgergruppe. Ikke rart Frp plasserer en av dem i regjering', *Dagbladet*, 20 December 2016; Steffen Zachariassen, 'Carl I. Hagen: – Jeg gir full støtte till Trump', *ABC Nyheter*, 2 June 2017.

'mobbed into silence'). So did 45 per cent of party voters. Siv Jensen herself had slipped out denialist lines the year before.[72] Moreover, the party programme for the period 2017–21 retained some instructive ambiguities, admitting that the warnings of the IPCC 'provide a basis for caution' only to, in the next sentence, warn against attempts to connect any one extreme weather event to underlying trends – the kind of late denialism that allows a far-right party to dispute the *actual* impacts of global heating – while throwing a thin gauze of uncertainty over the science as a whole. A red line for the party would be any climate policies that might 'weaken our international competitiveness'.[73]

The key, however, was found in the sections of the programme that sketched a bright future for the oil and gas industry. It behoves the state to ensure continued 'value production, profitability and competitiveness', even more so since Norwegian oil and gas output peaked in 2004. With the rest of the oil located further from the coast, in smaller volumes, deeper into the seabed, the state must guarantee full support and lift taxes. Only thus can continued exploration become 'more profitable, so that we may increase the lifetime on the fields and get the highest possible oil recovery rate'. The programme also took a clear stance on the most intensely disputed question of Norwegian fossil fuel politics since the turn of the millennium: a possible expansion into the relatively pristine waters in the far north, off the islands of Lofoten, Vesterålen and Senja (abbreviated LoVeSe) – which also happen to be the spawning grounds for the largest cod stock in the world – as well as areas in the even more remote Barents Sea and Jan Mayen. The FrP wanted to send in the rigs and drills 'as soon as possible'.[74] Or, as the leader of the parliamentary group put it to the party congress in 2017: 'We will pump up every last remaining drop.'[75]

---

72 *Nettavisen*, 'Syv Frp-ere på Stortinget tror ikke på menneskeskapte klimaendringer', nettavisen.no, 16 December 2016; *NRK*, 'De har landets mest klimaskeptiske velgere', 1 September 2017; Ramnefjell, 'Klimaskeptikere'. See further Jonas Skybakmoen, 'Klimafornekterne i Frp holder stand: – Kulldioksid er selve livets gass!', *Filter Nyheter*, 17 June 2019.

73 FrP, *Prinsipp- og handlingsprogram 2017–2021*, frp.no, 2017, 46–7, 12; cf. Solveig Ruud, 'Frps program er fullt av paradokser', *Aftenposten*, 4 May 2017.

74 FrP, *Prinsipp*, 42; Ryggvik and Kristoffersen, 'Heating', 251. Cf. Kristine Hoff, 'Jensen lokker med 15.000 oljejobber: – Valgkamputspill', *NRK*, 13 August 2017.

75 Ramnefjell, 'Til'.

It bears repeating that this was the party that now exercised direct control over the largest oil and gas reserves in Europe. All the four energy ministers appointed after 2013 – Tord Lien, Terje Søviknes, Kjell-Børge Freiberg, Sylvi Listhaug – strove to reverse the post-peak drop in output and speed up accumulation in the sector. They were flanked by Siv Jensen as finance minister, their own leader heading the second ministry with the power to shape petroleum policy: a placement that betrayed the priorities of 'the oil party'. It couldn't be in a better position to open the subsea coffers. Lien started off by declaring that 'the entire resource base will be exploited, as far as it is profitable'.[76] On 3 November 2016, the government threw a party for four hundred distinguished guests, including the king and the first CEO of Statoil, to celebrate 'the next chapter of the Norwegian oil adventure', and Lien was in a splendid mood: 'For the first time in twenty years, we offer new acreage for exploration', in the south-eastern zone of the Barents Sea.[77] The very next day, the Ministry of Climate and Environment threw its own party to celebrate the ratification of the Paris Agreement; Prime Minister Erna Solberg, manager-in-chief of the schizophrenia, attended both events.[78] She would argue that climate change is a matter of life and death and the greatest challenge of our time *and* that 'profitability will determine whether we will develop oil and gas in the future'.[79] As the undivided representative of the dominant classes, she was happy to leave the actual management of those fuels to the party with the strongest track record of anti-climate politics.

Her second minister, Søviknes, called the idea of closing the oil spigots a 'utopia'.[80] (He got into trouble in 2017 when posting an anti-Muslim video from the organisation Britain First, whose

---

76 Trædal, *Det svarte*, 191.
77 Energidepartementet, '50 år med petroleumsaktivitet i Norge', regjeringen.no, 3 November 2016.
78 Klima- og Miljødepartementet, 'Feirer Paris-avtalen', regjeringen.no, 4 November 2016; Statsministerens kontor, 'Statsministerens program veke 44', regjeringen.no, 28 October 2016.
79 Trædal, *Det svarte*, 189; Brage Lie Jor, Ingrid Kjelland-Mørdre and Sverre Tom Radøy, 'Solberg vil lete etter olje så lenge det er lønnsomt', NRK, 9 October 2019.
80 Mads Fremstad, 'Oljeministern: – Ikke realistisk att vi kun skal satsa på nye næringer', *ABC Nyheter*, 5 January 2017.

name the murderer of Jo Cox shouted; he was better known, however, for having raped a sixteen-year-old FrP activist.)[81] Critics dubbed Søviknes the 'lobby minister' for his cosy relations to Norwegian Oil and Gas, which, upon his resignation, expressed its gratitude for 'great cooperation' and a 'steady course in petroleum policy'.[82] The next relay runner, Freiberg, made a new dash for opening LoVeSe – the question remains unsettled as of this writing – after having dismissed critics of the move as 'climate romantics'.[83] As in Poland, there were obstacles to such expansion, including local opposition, but the aims of the far right were crystal clear.

More reasons to party appeared. Oil and gas output declined between 2004 and 2013, but after the FrP took over the ministry, spreading licences far and wide, it shot up again and stayed on what was indeed a steady upward course.[84] In 2019, a record-high eighty-eight oil and gas fields were operating, and the year was capped off with the official opening of Johan Sverdrup to much fanfare – slated to become by far the largest field in the North Sea, meant to keep pumping crude oil until at least the year 2070. Quite appropriately, the FrP prided itself on a record expansion of Norwegian oil and gas.[85] Primitive fossil capital of all sizes and provenances was invited to the banquet, with investment reported to be pouring in, profits high even when oil prices were low.[86] The output record from the previous peak was expected to be beaten in 2024.[87]

---

81 Olav Eggesvik, 2017, 'Olje- og energiminister Terje Søviknes delte Facebook-video fra britisk høyreradikal antiislamistisk organisasjon', *Aftenposten*, 26 March 2017.

82 Tommy Hansen, director of the association, quoted in Erik Steinsbu Wasberg, Ingvild Sagmoen and Ole Morten Skaug, 'Norsk olje og gass: – En tydelig talsmann', *E24*, 30 August 2018.

83 Toini Thanem, 'Kjell-Børge Freiberg (Frp) blir ny olje- og energiminister', *Verdens Gang*, 30 August 2018; Siv Sandvik and Peter Svaar, 'Advarte mot "klimaromantikere" – nå sliter han med å forklare uttrykket', *NRK*, 31 August 2018.

84 Norwegian Petroluem, 'Exports'; Bang and Lahn, 'From Oil'.

85 FrP, 'Petroleumsvirksomhet', frp.no, 2018.

86 Tsvetana Paraskova, 'Norway's Offshore Boom Is Back On', *Oil Price*, oilprice.com, 16 September 2018; Norwegian Petroluem, 'Investments and Operating Costs', 2018.

87 Stian Nicolajsen, 'Setter ny oljerekord i 2024', *Klassekampen*, 18 December 2018.

Meanwhile the FrP learned to sing the tunes of Norwegian officialdom: insofar as there is a climate problem, our oil and gas help to solve it; emissions can be compensated for elsewhere; the operations on the shelf will be electrified.[88] The cornucopia of that shelf offers some of the greatest potentials for wind and wave power in the world, but the FrP entreated the country to 'hold off large-scale implementation of such technologies until they are demonstrably robust and profitable', and indeed, the extraordinarily intense accumulation of primitive fossil capital crowded out investment in renewable energy, the development of which remained behind both Sweden and Denmark.[89] On immigration and Islam, the tone was for a while more low-key after the peak around Breivik, but matched by hands-on achievements in the tightening-up of migration policies and the deportation of unsuccessful asylum-seekers.[90]

Forces within the party continued to demand that it stay the course and stop flinching. In May 2019, while the school strikes for climate spread through Norway, Christian Tybring-Gjedde began his speech to the FrP congress by declaring himself 'sceptical of Muslim immigration'. The word 'sceptical' here had a somewhat atypical meaning, and the same applied to the central part of the speech, in which Tybring-Gjedde donned a disarming honesty:

> I am a climate sceptic because I don't think we should relinquish a success. Oil and gas are a success in Norway, and it's a blessing for the world that we have this energy that we can sell. The CO_2 hysteria is exaggerated. I just don't believe in it. It's the new idol of the left to believe in this and we should not be a part of it. It's a big lie! We shouldn't speak with a forked tongue! Be clear – be climate sceptics, because that's what the people are! I don't trust the experts![91]

---

88 See for example Helge Rønning Birkelund, 'Folk tar skammelig feil om olje, mener Terje Søviknes', *Fri fagbevegelse*, 31 October 2017; Anders Fjellberg, 'Frp's klimaløsning: – Vi må pumpe opp så mye gass som mulig', *Dagbladet*, 23 October 2018; Trædal, *Det svarte*, 220; Sæther, *De beste*, 74–5.

89 FrP, *Prinsipp*, 41; Ryggvik and Kristoffersen, 'Heating', 264–5.

90 Bangstad, 'Norwegian', 247.

91 Audio recording of the speech included in Steinar Solås Suvatne and Jørgen Gilbrant, 'NÅ: Tybring-Gjedde vant Frp-kamp etter advarsel om muslimer', *Dagbladet*, 5 May 2019.

The party with the Norwegian shelf in its hand interrupted those lines several times for laughter and thunderous applause.

This same unapologetic wing took more direct control of the Ministry of Petroleum and Energy later in 2019, when Sylvi Listhaug became the fourth FrP politician to run it. As the minister of agriculture (2013–15), she had invented a gambit combining anti-Muslim with anti-climate politics: demanding *more pork* on the menus in Norwegian prisons. As the minister of immigration (2015–17), she not only managed to cut the number of asylum-seekers in Norway to a nineteen-year low, but also launched an attack on the 'goodness tyranny' supposedly emanating from the Norwegian church, arguing that Jesus would have assisted refugees 'where they are' (i.e. he would not have let them come to Norway). As minister of justice, in 2018 she claimed that the Labour Party 'cares more about the rights of terrorists than national security', which – Breivik not yet entirely forgotten – caused a furore and sent her stock soaring further with the far right.[92] What of the climate? Eight years before her appointment as oil minister, Listhaug was interviewed while filling up the gas in her Ford Expedition mega-SUV and explaining that CO_2 emissions had no proven negative effects, the bad talk about them merely 'an excuse to introduce more taxes'.[93] At the press conference where her appointment was announced, Prime Minister Solberg vouched for her as a safe non-denier. Minutes later, when Listhaug herself met the media, she contended that climate change is *partly* caused by human action; asked how large that part was, she responded that 'this is not interesting to discuss'.[94] Capitalist climate governance Norwegian-style appeared to have come full circle.

'Climate delayers', Alexandria Ocasio-Cortez tweeted in February 2019, 'aren't much better than climate deniers. With either one if they get their way, we're toast.'[95] She could hardly have found a

---

92 David Vojislav Krekling, 'Her er 11 Listhaug-utspill som har satt fyr på Norge', *NRK*, 13 March 2018.

93 Lars M. Glomnes, 'Ler av klimahysteri', *Verdens Gang*, 16 May 2011.

94 Daniel Røed-Johansen, Solveig Ruud, Kjetil Magne Sørenes and Steinar Dyrnes, 'Klima-kjør mot Solberg da hun presenterte Listhaug som ny oljeminister', *Aftenposten*, 18 December 2019.

95 Matthew Cantor, 'Could "Climate Delayer" Become the Political Epithet of Our Times?', *Guardian*, 1 March 2019. On this distinction, cf. Hoggan, *Climate*, 118–19.

stronger case than Norway. Here the centrally positioned ISA was not denialist by tradition, but rather fine-tuned the message that even in this warming world, Norway is so great that it can delay the one needed policy: capping and stopping fossil fuel production. But the delayers – a category roughly consonant with capitalist climate governance – not only pursued the same policy that denial inspired. They and the deniers became conjoined twins, two halves of a nationalist body. For the former, Norwegian oil and gas were the best, handled by the most virtuous state in a cruel world; for the latter, the whole climate issue was an offence against the same Norwegian interests. Both took for granted that if Norwegians were to cut back their production, someone else would make the money they could have made; both strove for maximum profit from the energy wealth of the nation. Denialism might then seem redundant for primitive fossil capital: its interests were already covered by infinite delay. And yet the whole formation seems to have tilted towards a far right that went for the jugular and put the interests of the nation – identical to those of said class fraction – above reality itself. This apparatus had an existence of its own in the party. It valued denial as a political currency, at one with general campaigning against Islam, the left and other forces foreign to the nation. But it became seamlessly entangled with the relevant bureaucracies and Equinor – the more benevolent-sounding new name of Statoil – because no disunity of interests troubled them, and because something in the structure itself lent weight to the far right.

Seen from another angle, the order had three tiers. On a global scale, capitalist climate governance harboured and revitalised denial; on the national scale, likewise; inside the party, the wing farthest to the right consented to the precepts of such governance, rose within it and took denial close to ministerial power. The diversity within the FrP reflected, monad-like, that of the higher scales. Poland and Norway were here practically spitting images. The unity between the far right and primitive fossil capital was cordial, the slides between denial and governance mostly free of friction, the drive to keep on digging and drilling into the second half of the century unreduced. In Norway as in Poland, the far-right parties took a pre-existing national policy to its logical extreme, embodying the thrust of the social formation and patrolling its boundaries against unwanted others – although in this respect, the PiS, in its special environment, was significantly more extreme than the FrP. But the latter preceded the former both in Mustafa myth-making and fossil fuel nationalism. It should be clear by now that the parties of the far right learned from each other and from global currents of dominant bourgeois ideology *while also* growing out of their own national settings. If the pattern so far looks deadeningly monotonous, it was broken, however, up by some parties that struck out in a slightly different direction.

5
Ecology Is the Border

In 2011, Marine Le Pen succeeded her father as leader of the Front National (FN). In the first years of her reign, she followed in the father's footsteps, scattering around her the usual denialist propositions – 'the world has seen changes in the climate that had nothing to do with human activity', and so on – but before long, she began to remould the party with her distinctive strategic acumen.[1] This included rethinking matters of ecology. In 2014, the FN unveiled 'Nouvelle Écologie', or New Ecology, a front organisation for environmentally inclined sympathisers and members. It spoke of protecting 'family, nature and race', but its exact mission was somewhat unclear.[2] On its behalf, Mireille d'Ornano stated that 'there are pros and cons to the scientific evidence' regarding climate change, condemned the UNFCCC as a 'communist project' and clarified that 'we don't want a global agreement or global rule for the environment'.[3] Yet Nouvelle Écologie did mark the beginning of a greening of the FN, henceforth more attuned to (some) ecological concerns of (some) French voters.[4] The organisation's leader,

---
1 *Terra Eco*, 'Marine'.
2 Schaller and Carius, *Convenient*, 83.
3 Arthur Neslen, 'French National Front Launches Nationalist Environmental Movement', *Guardian*, 18 December 2014.
4 We are here omitting the green elements of Bruno Mégret's challenge to

Philippe Murer, underlined that 'respect for the environment and ecology are very important for the Front National'.[5] Ahead of the 2017 presidential election, he claimed sixteen local chapters; by now the FN had redefined itself as an 'eco-nationalist' party fighting for 'patriotic ecology', which, in Marine Le Pen's vision, meant a decoupling from globalisation and a greater degree of economic self-sufficiency. Early in her election campaign, she maintained that 'by defending protectionism and retaining production within our borders, we combat greenhouse gas emissions'.[6] Gone was the brazen denial. Instead the French nation was extolled as the antidote to environmental degradation, if only its true self were to leave internal exile and take up residence in the Élysée. A properly national French economy would be green.

In fact, seeds of this 'patriotic ecology' were buried already in the father's denialist harangues. At the Nanterre conference on the climate myth in 2010, he had suggested that the best way to reduce CO_2 would be to produce and consume goods locally, within French borders, against the commanding interests of multinational corporations.[7] Such self-contradictions and ambiguities did not disappear with the daughter as FN head. For COP21, the party prepared an 'anti-COP21' event, although it was cancelled due to the terrorist attacks in Paris. The FN members of the National Assembly voted for ratification, except for one, Marion Maréchal-Le Pen – tipped to be next in the line of succession – who abstained. Marine Le Pen later called the Paris Agreement 'wobbly and unworkable', but decried the US withdrawal as 'regrettable, because it sends a negative signal on the

---

Jean-Marie Le Pen in the 1990s. On Marine Le Pen's greening of the party, cf. Salomi Boukala and Eirini Tountasaki, 'From Black to Green: Analysing *Le Front National*'s "Patriotic Ecology"', in Forchtner, *The Far*, 72–87.

5 Anne-Claud Martin, 'Le Front national veut faire de l'écologie', *Euractiv*, 12 December 2014.

6 Rémi Barroux, 'Ecologie: Marine Le Pen repeint le vert en bleu-blanc-rouge', *Le Monde*, 3 December 2016.

7 *Le Parisien*, 'Jean-Marie le Pen dénonce "l'écologisme" et les "bobos gogos"', leparisien.fr, 30 January 2010. Cf. Le Pen's election manifesto for 2007: Jean-Marie Le Pen, 'Programme électoral de Jean-Marie Le Pen, président du Front national et candidat à l'élection présidentielle 2007, mars 2007', *Vie Publique*, 27 March 2007.

necessary action for the environment'.[8] Other party statements upheld the old anti-UNFCCC line – for a 'return to sovereignty' – and suggested that the Paris Agreement was flawed because it disregarded demographic issues.[9] Similar oscillation could, as we have seen, be found in energy policies: no to wind, no to fracking, but yes to 'intelligent protectionism' that would incubate solar, biogas, wood and a hydrogen energy infrastructure, and best of all nuclear power.[10] Clearly the FN had gone some distance from the days of watermelon cutting. 'Brrr brrr, we're freezing our butts off! Those who believe in global warming are beyond help,' Jean-Marie Le Pen tweeted during some cold days in Paris in February 2018. By then, three years had passed since he had been expelled from the party, which no longer even retained his name – it was now Rassemblement National, or National Rally (RN), tricolour plus *vert*.[11]

The name change followed a crisis after the presidential election of 2017. Hopes that decades of painstaking front-building would bear fruit in the ultimate victory were dashed when Emmanuel Macron saw off Le Pen in the second round. During the soul-searching in the aftermath, she fell under the spell of Hervé Juvin, a self-styled essayist in his sixties who sought to outdo all competitors in concern for the environment. He wished to 'defend his biotope against invasive species'.[12] In a rambling opus on the threats to human survival, he threw in the climate crisis and the sixth mass extinction alongside obesity, attention deficit disorder, opioids, sugar addiction, sterility and a couch-potato lifestyle, all conspiring to 'producing by the millions these populations of fat, ugly and stupid individuals' – a veritable biological degeneration of the species. The main driving

---

8   *Le Monde*, 'Les députés français ont approuvé la ratification de l'accord de Paris sur le climat', 17 May 2016; *Libération*, 'Accord de Paris: Un retrait "regrettable" pour Marine Le Pen', liberation.fr, 2 June 2017.

9   Rémi Barroux, 'Présidentielle: Marine Le Pen, l'écologie minimale', *Le Monde*, 1 May 2017.

10   Front National, *Les 144 Engagements Présidentiels: Marine 2017*, rassamblementnational.fr, 2017, 22.

11   Jean-Marie Le Pen's Twitter account, 23 February 2018.

12   Tristian Berteloot, 'Pour Le Pen, le "nomade" se "moque de l'écologie car il n'a pas de terre"', *Libération*, 14 April 2019.

force of this epochal shitstorm was globalisation, defined as 'the end of any links between human life and a territory of its own', human life here synonymous with nation or race. If only the link to territory had remained unbroken, there would have been no alcoholism and no CO_2 emissions. All went wrong the moment human populations were freed to 'settle in the most fertile and diverse ecosystems' – such as France – causing a pressure on natural constraints that is 'totally unsustainable. Just look at the enormous impact of air conditioning on climate change!' Hence the key to survival must be the closing of borders. The human species will save itself and other species when it shuts down entry points for aliens, which translates as 'Europe is the land of Europeans.' Juvin would also slap tariffs on trade with countries that do not commit to zero emissions by 2040. He proposed that nationalist governments set up a nine-member 'Council for Life' to implement these policies and exercise 'final authority upon any issue related to human survival' (presumably including sugar addiction).[13]

By spring 2019, Juvin had become so influential in the Rassemblement National as to draft the party's manifesto for elections to the European Parliament. It called for self-defence against a 'planetary nomadism' in the process of destroying the continent and proposed a 'European ecological civilisation.'[14] During this campaign, Marine Le Pen spoke repeatedly of the enemy as 'nomads', against which the French must protect themselves and their earth: 'He who is rooted is an ecologist. He does not want the land where he raises his children to go to waste. But the nomad does not care, because he has no land!'[15] A fresh new face of the RN, Jordan Bardella, leader of its youth federation, top candidate in the election and subsequently vice president of the party, was as dedicated to the green cause as Le Pen: 'It is by returning to the borders that we will save the planet.'[16] Or, in a soundbite for the campaign and for this current of the far right more generally: 'The best

---

13 Hervé Juvin, *La Grande séparation: Pour une écologie des civilisations* (Paris: Gallimard, 2013); excerpts published as 'Ecology and Human Survival: The Project of a New Alliance for Life', hervejuvin.com, 19 February 2019. Juvin's self-translation has here been grammatically corrected.
14 Rassemblement National, 'Pour une Europe des Nations', 2019, for example 12, 3.
15 Berteloot, 'Pour Le Pen'.
16 Charles Sapin, 'Jordan Bardella: "C'est par le retour aux frontières que nous sauverons la planète"', *Le Figaro*, lefigaro.fr, 23 May 2019.

ally of ecology is the border.'[17] This time, the RN beat Macron, Le Pen enjoying a first sweet revenge, her party winning more seats representing France in Brussels than any other. By now, the French far right had definitively painted its nationalism green.

And the recolouring extended beyond the RN. During the internal crisis in 2017, Nouvelle Écologie joined a splinter group to form Les Patriotes, which sought to outmatch the mother party in environmentalism and insistence on French withdrawal from the EU, or 'Frexit'.[18] On fire with 'patriotic ecology', this party flopped badly in the 2019 election, and Nouvelle Écologie became moribund. But the ideas lived on. Behind the party scene hovered the Nouvelle Droite, the fraternity of far-right activists with intellectual ambitions, led by Alain de Benoist, which had done so much to inspire the original formulation of the master frame. It stuck to the strategy of 'metapolitics' – not stumping for candidates, not winning offices but shifting the premises of the political conversation and pulling the whole culture to the right. It had had some indisputable success in this regard.[19] While the Nouvelle Droite cared more for Europe than France – prefiguring the RN's eventual abandonment of Frexit and turn to a Union of the far right – it was adamant about ethnopluralist separation: each race in its own habitat; no mixing and dilution; conservation of differences as against the homogenising forces of global capitalism – a position that made de Benoist befriend and publish Serge Latouche, maven of the French degrowth movement. The concern for rootedness linked up with a kind of ecology. 'Nature' was here the theatre of permanent combat between species and human races that must defend their own territories and give up on

---

17 Valerie Landrieu, 'Jordan Bardella: "Le meilleur allié de l'écologie, c'est la frontière", *Les Echos*, 7 April 2019.

18 Collectif Nouvelle Ecologie, 'Le Collectif Nouvelle Ecologie s'affranchit du FN', Les Patriotes, 14 November 2017; Les Patriotes: 'Nos Vidéos Thématiques: Ecologie et Patriotisme: Un Marriage Heureux!', n.d.

19 The Nouvelle Droite here claimed insipiration from Antonio Gramsci and referred to its project as 'Gramscianism of the right'. For an excellent overview, see Razmig Keucheyan, 'Alan de Benoist, du néofascisme à l'extrême droite "respectable": Enquête sur une *success story* intellectuelle', *Revue du Crieur* 1 (2017): 128–43; for a detailed analysis, Tamir Bar-On, *Rethinking the French New Right: Alternatives to Modernity* (Abingdon: Routledge, 2013).

artificial, soft-hearted ideals of equality.[20] So strong was the pull of this 'metapolitical' sphere that Marion Maréchal-Le Pen, crown princess of the RN, in 2017 'retired' from party politics, dropped Le Pen from her name and set up her own graduate school in Lyon to spread the ideas. In late 2019, she gave a 'programmatic' speech, omitting anti-capitalist or degrowth rhetoric but combining the Great Replacement theory with love for nature: 'Preserving landscapes, preserving the local – all this is actually an identitarian fight.'[21] A young, fair-haired, more dashing version of Alain de Benoist, Marion Maréchal was tipped to make a comeback and run for the Élysée.

In the ecosystem of the French far right, the Catholic current re-energised by the struggle against same-sex marriage rallied behind something called 'écologie intégrale', or integral ecology: a venture for preserving French men and women in their natural state, where they would respect the limits to how much nature can be bent and exploited. Whereas the Nouvelle Droite remained bitterly anti-Christian and fancied itself true to the pagan origins of the thirty-thousand-year-old Indo-European race, the integral ecologists were out to save Catholic France. In ecumenical spirit, however, they ended up with similar ideas about the borders of nature.[22] The group Terre et Peuple, or Land and People, drove the same ideas towards a violent extreme, while some far-right activists took up survivalism: organising 'survival trainings' and setting up 'autonomous sustainable bases'

---

20 Ibid., 139–40; Stéphane François, 'L'extrême droite française et l'écologie: Retour sur une polémique', *Revue Française d'Histoire des Idées Politiques* 2 (2016): 187–208; Stéphane François, 'L'écologie, un enjeu de l'extrême droite', *Observatoire des radicalités politiques*, Fondation Jean-Jaurès, jean-jaures.org, 23 February 2016. For a list of Latouche's publications in de Benoist's various journals, see Indymedia Nantes, 'Serge Latouche et l'extrême droite', nantes.indymedia.org, 17 August 2019. De Benoist also fraternised with Teddy Goldsmith, founder of *The Ecologist*. Keucheyan, 'Alain', 190.

21 Lucie Soullier, 'Dans le discours de Marion Maréchal, les mots de l'extrémisme', *Le Monde*, 2 October 2019. The graduate school: Institut de Sciences Sociales, Économique and Politiques, issep.fr.

22 The venture was embodied in the magazine *Limite: Revue d'écologie integrale*, revuelimite.fr. It revolved around Eugénie Bastié, a young anti-feminist journalist at *Le Figaro*. On integral ecology, see further for example Jean-Louis Schlegel, 'Les limites de Limite', *Esprit* no 1–2 (2018): 207–12; Simon Blin, 'Les droites dures s'enracinent dans "l'écologie intégrale"', *Libération*, 5 May 2019; Catherine Vincent, '"Écologie intégrale", écofascisme . . . : Une histoire des écologies identitaires', *Le Monde*, 4 October 2019.

in rural areas.[23] A real trouper of these milieus, Alain Soral, head of a publishing house that put out the *Protocols of the Elders of Zion*, also offered courses on permaculture, self-sufficiency and 'how to get back to the land'. 'Healthy and rooted products' were on sale from his online organic food shop. While this could sound like preparation for the coming climate collapse, however, Soral held onto the old dogma of global warming as a racket for eliminating borders.[24] He was one of the exceptions that proved the rule. In the late 2010s, the French far right, inside and outside parliament, stood out for its broad ditching of denialism and acceptance of the reality of an ecological crisis. Had it thereby once again invented a new master frame?

Green Nationalism in Theory

It would not be without antecedents. Much as the US has exported climate denialism to the rest of the world, it has given plenty of nourishment to this second current. In *Border Walls Gone Green: Nature and Anti-immigrant Politics in America*, John Hultgren charts a venerable tradition of environmentalists demanding that immigration should be curtailed so as to heal abused nature; ideals of 'patriotic ecology' are as old as American conservationism itself. The main argument is that more immigrants mean more people who weigh down our ecosystems and resources. They are also said to have a throwaway mentality and scant regard for the value of nature. Here nature is bound up with nation, the one kept clean only if the other is: both risk contamination from foreign sources. 'Nature' serves as a sign for the original order, before the fall into the sump of substances and bodies that do not belong; in this embittered nostalgia, Hultgren

---

23 François, 'L'extrême', 197–8, 201; Terre et Peuple: terreetpeuple.com; Tiphanie Guéret: 'Survivalisme et extreme droite: une romance qui dure', *CQFD*, December 2018.

24 See Soral's outlets 'Egalité et Réconciliation', egaliteetreconciliation.fr; 'Prenons le Maquis', prenonslemaquis.fr; 'Au Bon Sens: Produits Sains & Enracines', aubonsens.fr. On climate change: 'Le "réchauffement climatique": réalité ou phénomène médiatique?', Egalité et Réconciliation, 15 March 2017; 'Soral a (presque toujours) raison – Épisode 11', YouTube, 3 December 2019 (hitting 200,000 views in four weeks).

observes, '"simpler times" is a thinly veiled code for "whiter times"'.[25] Using the term favoured by many of its adherents, we may call this *green nationalism*.

One source of inspiration for green nationalism is, of course, good old-fashioned Malthusianism, which locates the source of ills in a surplus of human numbers. That surplus tends to be produced by others – be they, as in the days of Reverend Malthus, the working poor, or Muslims or some other group of non-whites. In *The Malthusian Moment: Global Population Growth and the Birth of American Environmentalism*, Thomas Robertson has showed how modern environmentalism was tied to Malthusianism by an umbilical cord: when it emerged in the US – most spectacularly on Earth Day in 1970, one of the largest demonstrations the country has ever seen – it came out to fight overpopulation. Spokesmen of the nascent movement would pin all the world's crises on this bane. One of the most prominent was Paul Ehrlich, author of the now discredited bestseller *The Population Bomb* from 1968; ahead of the curve, like the American Petroleum Institute, he had received the news of climate change. He considered it 'even more important' than other environmental problems. All sprang from the same womb. 'Too many cars, too many factories, too much detergent, too much pesticide, multiplying contrails, inadequate sewage treatment plants, too little water, too much carbon dioxide – all can be traced easily to *too many people*.'[26] In the year when the API received its second report on fossil fuels and climate, another explanatory model was thus proposed.

To his right, Ehrlich had Garrett Hardin. In 'The Tragedy of the Commons', probably the most cited text ever written on environmental degradation – nearly forty thousand citations on Google Scholar, compared to three thousand for Crutzen's original Anthropocene paper and sixteen thousand for Carson's *Silent Spring* – this professor of human ecology invented an idyllic

---

25 John Hultgren, *Border Walls Gone Green: Nature and Anti-immigrant Politics in America* (Minneapolis: University of Minnesota Press, 2015), 5.

26 Thomas Robertson, *The Malthusian Moment: Global Population Growth and the Birth of American Environmentalism* (New Brunswick, NJ: Rutgers University Press, 2012), Ehrlich quoted on 266, 143; emphasis in Ehrlich's original.

pasture, onto which herders rush with cattle that devour the grass. Since its publication in *Science* in 1968, the overgrazed common has become an archetypal image. Hardin did not hide the source of his anguish:

> How shall we deal with the family, the religion, the race, or the class (or indeed any distinguishable and cohesive group) that adopts overbreeding as a policy to secure its own aggrandizement? To couple the concept of the freedom to breed with the belief that everyone born has an equal right to the commons is to lock the world into a tragic course of action.[27]

The cause of the tragedy? People of a certain religion, race or class have made the collective decision to bring out enough babies to take over society and trample down its material foundations in the process.

In a second paper, Hardin tackled the issue of immigration head-on. Here he imagined the space of the green nation not as a pasture, but as a lifeboat. It floats on the open sea with a given capacity to carry people. If too many get onboard, the boat sinks. Americans sit on a boat decked with all the amenities one could dream of, but this attracts poor people who have scraped the bottom of their barrels and thrown themselves into the ocean – and so should we extend our hand to them? *Can* we? Writing in the midst of the oil crisis, Hardin accepted that fossil fuels must soon be a thing of the past. With a shrinking energy base, he argued, the carrying capacity of the American lifeboat will be reduced, leaving even less room for more passengers. Meanwhile the waters are filling up with people who have left the overcrowded territories of Morocco and Pakistan and Venezuela and the Philippines, where fertility rates have broken all bounds. (It tends to be someone else who sires too many children: Hardin himself had four, and he opposed efforts to convince white Americans to bear fewer.)[28] Sure, one might apply the ethics of Marxism and Christianity – 'to each according to his needs', a

---

27 Garett Hardin, 'The Tragedy of the Commons', *Science* 162 (1968): 1246.
28 Southern Poverty Law Center, 'Garrett Hardin', n.d., splcenter.org.

principle on which Hardin vented his spleen – and save the poorest souls from drowning, which would guarantee that 'everyone drowns. Complete justice, complete catastrophe.' Food could possibly be produced for everyone, 'but what about clean beaches, unspoiled forests, and solitude?' To treat the lifeboat as a common open for all would be to invite tragedy, much like the owner of a pasture who neglects his duty and permits 'weeds' to take over.[29] Lest we commit idealistic suicide, we – that is, white Americans – must set our priorities straight, consider the lifeboat the possession of *our* children, keep immigrants out and curtail the right to reproduce for those who have already snuck onboard. Garrett Hardin also believed that blacks were by nature less intelligent than whites. He decried that 'black became beautiful' and warned of a 'chaotic Norte Americano Central' and detected a 'passive genocide' in motion, more precisely a genocide against white people, committed by Muslims who – following a plan openly announced in the Iranian parliament – aim to 'outbreed' us.[30]

Neither Ehrlich nor Hardin was an oddball. In American environmentalism, pristine nature has long been held up as a crucifix against the advancing and seething non-white masses, whose proclivity to pullulate brings doom upon the planet. Explaining with steely logic why the environment is best protected through secure property rights and maximum enclosure, Hardin became the guru of bourgeois ecology. Edward Abbey seemed to belong in the opposite corner. Author of *The Monkey Wrench Gang*, the classic novel of eco-sabotage against corporate polluters, Abbey also wanted to call 'a halt to the mass influx of even more millions of hungry, ignorant, unskilled, and culturally-morally-genetically [!] impoverished people', so as to save 'an open, spacious, uncrowded, and beautiful – yes, beautiful! – society.'

---

29 Garrett Hardin, 'Living on a Lifeboat', *BioScience* 24 (1974): 562, 565; 'each according to his need', Garreth Hardin, *The Immigration Dilemma: Avoiding the Tradegy of the Commons* (Washington, DC: Federation for American Immigration Reform, 1995), for example 107, 109. On immigration destroying the environment, see further this book, for example 115–19.

30 Southern Poverty Law Center, 'Garret'; Alex Amend, 'First as Tragedy, Then as Fascism: Ecologist Garrett Hardin's Enduring Gift to the Nativist Right', *Baffler*, 26 September 2019.

The alternative, 'in the squalor, cruelty, and corruption of Latin America, is plain for all to see'.[31]

Abbey penned those words in 1988, the year before his death. Around that time, the environmental movement in the US and the rest of the global North was in the midst of a sharp turn to the left. It was set on that course by African Americans. If the Stonewall Riots in 1969 have become legendary in the LGBT community, a similar status has accrued to the events that took place in 1982 in Warren County, a rural area with a predominantly black and poor population, in the former belt of cotton and tobacco plantations in North Carolina. In 1978, a group of businessmen rid themselves of thousands of gallons of PCBs – a family of highly toxic compounds, the production of which Congress would ban the following year – along North Carolinian highways. The state quickly decided that the waste could not be left on the roadsides but had to be collected and stored somewhere. It settled on Warren County. This was considered a suitable place for a toxic landfill not because the soil was particularly impermeable or the groundwater out of harm's way – scientists declared the contrary – but because of the kind of people who lived there. Those people put up a struggle against the blatant injustice.[32] When the trucks arrived in the fall of 1982, carrying soil mixed up with the PCBs, protestors blocked their path. A film clip shows an African American man leading the crowd, raising his fist and yelling in the faces of the white policemen: 'We will not allow Warren County to become a dump site!'[33] In scenes reminiscent of the heyday of civil rights, the leaders were arrested, scuffles broke out, cops pulled out their batons and demonstrators sat down or jumped in front of the trucks while singing 'Ain't No Stoppin' Us Now' and explaining to journalists their worries about birth defects and diseases, risks the state saw fit to dump on them simply because of who they were. It became known as the moment of birth for the environmental justice movement.[34]

---

31 Quoted in Hultgren, *Border*, 69. Hardin praising enclosure, see for example *The Immigration*, 29.

32 See for example *New York Times*, 'Carolinians Angry over PCB Landfill', 11 August 1982.

33 'PCB Protest in Warren County 1982', YouTube.

34 For a detailed account, see Eileen Maura McGurty, 'From NIMBY to Civil Rights: The Origins of the Environmental Justice Movement', *Environmental History* 2 (1997): 301–23; cf. for example Julian Agyeman, *Sustainable Communities and the Challenge of Environmental Justice* (New York: New York University Press, 2005), 14–15.

'Complete justice, complete catastrophe', was Hardin's axiom; in the 1980s and 1990s, modern environmentalism learned to invert it. Warren County was an all-too-typical case. Some people cause the destruction of ecosystems and others bear the brunt of the impacts. If resources were distributed fairly, the harm done to disadvantaged groups would be stopped and the general drivers of degradation shut down. Businessmen can dump toxins only as long as there are oppressed people on which to offload them; once the latter rise up, the cycles of material throughput will have to be plugged up. Justice, then, is not the negation but the *essence* of sustainability. At the end of the twentieth century, the standard-bearers of American environmentalism – notably the Sierra Club, flanked by Greenpeace – came around to this way of thinking; in campaigns, counter-summits, conversations in plenaries and panels, here was the new master frame of this movement.[35] The green nationalists were expelled and agitation against immigration terminated. Since then, ecology has been twinned with the left. This holds for the US and for most of the rest of the world, where the patterns of degradation and resistance tend to mirror those from Warren County. In places without number, environmental injustice is repeated and multiplied.

But not all environmentalists have been happy with this turn. One self-identified victim of the purge was Dave Foreman, a household name in green circles since the early 1970s. In 1979, he was one of the founders of Earth First!, the most militant wing of the US environmental movement, whose activists lived on platforms in treetops and performed spectacular direct actions under his leadership. When Earth First! caught the wind of justice and turned towards anarchism in the early 1990s, he broke with the organisation in distaste at the 'pressure and infiltration from the class-struggle/social-justice left' and went on to found the Rewilding Institute, associating his name

---

35 For a summary of these developments, see Phaedra C. Pezzullo and Ronald Sandler, 'Revisiting the Environmental Justice Challenge to Environmentalism', in Ronald Sandler and Phaedra C. Pezzullo (eds), *Environmental Justice and Environmentalism: The Social Justice Challenge to the Environmental Movement* (Cambridge, MA: MIT Press, 2007), 1–24.

with that concept.[36] In 2014, Foreman published *Man Swarm: How Overpopulation Is Killing the Wild World*, in which he unloads his resentment against the left that stole his movement and closed its eyes to overpopulation. Climate change, species extinction, invasive species, dammed rivers, tamed wilderness: all derive from reproductive urges running amok. Drought in Syria? Deforestation in the Amazon? CO_2 emissions spiking in China? 'We have come on like a swarm of locusts,' but on closer inspection the swarm does not consist of a universal humanity, for it is *some* people who appear like locusts, first and foremost those who live in Africa, where population growth is '*unbelievably high*'. So it is in Latin America and most of Asia. Europe and the United States have performed an 'awesome deed' by lowering their birth rates, but poor people are now busy sabotaging their success. They are heating up the planet by burning wood in stoves, spreading chemical brews and – the root cause of all evil – emitting ever more babies. 'Like the growth of a metastasizing cancer', the poor non-whites of the global South are ruining the planet in a way prosperous Europeans and Americans would never do: and now they are coming our way.[37]

Immigration, the reader of *Man Swarm* learns, is the number one threat to nature in the northern latitudes. Rich countries have been so stupid as to squander their awesome deeds by opening their borders to mass immigration, currently their sole source of population growth. It is immigrants who call forth new shopping malls, highways, parking lots, pipelines, coal mines in the US and Europe. Even after naturalisation, foreign-born mothers continue to produce a disproportionate share of the babies, which is as much a cultural and political as an environmental calamity. Northern countries are 'being swamped by the ethnic makeup of immigrants', in accordance with a deliberate plan: 'Leaders of ethnic (tribal) immigration outfits'

---

36 Foreman quoted in Ian Angus and Simon Butler, *Too Many People? Population, Immigration, and the Environmental Crisis* (Chicago: Haymarket, 2011), 25. See further Irma Allen, 'The Trouble with Rewilding', *Undisciplined Environments*, undisciplinedenvironments.org, 14 December 2016.

37 Dave Foreman, *Man Swarm: How Overpopulation Is Killing the Wild World* (London: LiveTrue Books, 2014), 9, 21, 97, 47, 53. Emphasis in original. (An earlier edition of the book was published in 2011.)

– that is, Muslim or Latino countries – 'flatly say they want more immigration of their bunch for greater political might that will lead to takeover.'[38] Thankfully, there are solutions – a whole bunch of them, in fact. Detect, detain and deport immigrants. Track visitors. Once they have been deported, ban those people from ever setting their feet on our soil again; force the sending countries to shrink their populations; ensure that skilled workers of foreign descent 'go home to India, Pakistan, Africa, or wherever and make their homelands better'.[39] An efficient border guard is the most effective guardian of life on earth.

If in the early millennium Dave Foreman was the gruffy caretaker of old-school American environmentalism, Paul Kingsnorth was the tortured soul and belletrist of the English kind. The author of acclaimed novels, Kingsnorth was one of the founders of the Dark Mountain Project, a network of writers and artists trying to develop stories that could speak to the realities of ecological collapse.[40] Launched with a manifesto in 2009, Dark Mountain organised annual festivals in Britain, published lavishly designed books and spread to like-minded practitioners of the fine arts – often former activists in the environmental movement, who had withdrawn in disgust or despair – in some other northern European countries. In 2017, Kingsnorth published *Confessions of a Recovering Environmentalist*, a collection of essays in which he laid out the philosophy that birthed Dark Mountain. As a young man, Paul vowed that 'this would be my life's work: saving nature from people'.[41] Thus he joined the green activist milieu burgeoning in England in the 1990s. But after a decade on the front line, he had become disillusioned. Environmentalism, he found, 'was being sucked into the yawning, bottomless chasm of the "progressive" left'. It elbowed out deep ecologists like himself, who 'were told that "(human) social justice and environmental justice go hand in hand" – a suggestion of such bizarre inaccuracy that it could surely only be wishful thinking'.

---

38   Ibid., 44, 105.
39   Ibid., 131, 132, 147. Emphasis in original.
40   Dark Mountain Project, dark-mountain.net.
41   Paul Kingsnorth, *Confessions of a Recovering Environmentalist* (London: Faber & Faber, 2017), 66.

Kingsnorth's essays ooze bitterness against the 'gaggle of washed-up Trots' who took over what used to be his movement.[42]

To cap its victory, the left replaced love for the wild with a 'single-minded obsession with climate change'. The climate issue is about as foreign to true environmentalism – which is all about protecting one's own local turf of nature – as *keffiyehs*; for Kingsnorth, it can be integrated into the storyline of civilisational breakdown, but there's nothing particularly urgent about it, nothing that calls for action or concerted efforts. Indeed, any initiatives of that sort are bound to be a waste of time. Kingsnorth swings between highlighting 'the lack of certainty' around climate change and surrendering to the fact that the climate is 'constantly changing' – and if it brings us down in the end, then that's the way it is, and it is not necessarily a bad thing. 'Every extinction, after all, is also an opportunity. If the dinosaurs had survived, there would have been no humans. Everything changes, and the changes are not always pretty. Who said they had to be pretty?'[43] This is the kind of equanimity Kingsnorth cultivated on the Dark Mountain, whose manifesto likewise rejects the idea that the converging crises of our times have solutions: instead 'we will face this reality honestly and learn how to live with it'.[44] But we will not live with renewables. Part of the crime of the left is to turn the environmental movement into a cheerleader for zero-carbon energy technologies – worst of all wind power. The English moors so revered by Kingsnorth 'are to be staked out like vampires in the sun, their chests pierced with rows of 500-foot wind turbines', proof that environmentalism must be left to its ignominious fate.[45] If Kingsnorth gave up on that cause, however, there was another he learned to champion with gusto: nationalism.

Following the Brexit vote, Paul Kingsnorth elaborated on ideas about 'green nationalism' he had previously sketched. In an essay in the *Guardian* in March 2017, he rehearsed the old-fangled watchword 'small is beautiful' and explained that this had been his reason

---

42   Ibid., 75–8, 43.
43   Ibid., 44, 217, 219.
44   Ibid., 283. Kingsnorth wrote the chapter 'Uncivilisation' together with Dougald Hine.
45   Ibid., 70.

to go green in the first place: he cherished the uniqueness of certain small places.[46] The nation made up the sum of them. 'England', runs the definition in *Confessions*, 'is the still pool under the willows where nobody will find you all day.' The sentimental bond to the willows and the moors is what constitutes an Englishman. A nation 'is about belonging – to a specific place that is not quite like another place'.[47] Now there is a force hell-bent on wrecking this green and pleasant place: Kingsnorth calls it 'globalism'; the Dark Mountain manifesto speaks darkly of 'cosmopolitan citadels'.[48] A nebulous and nefarious alliance of leftists and neoliberals, this force seeks to stamp out national identities, impose a bland global culture, pull up the roots of ordinary people and turn them into 'citizens of nowhere'.[49]

Naturally, this version of green nationalism made for a clear result on the litmus test of immigration. In his post-Brexit essay for the *Guardian*, Kingsnorth affirmed that 'to a globalist, border walls and immigration laws are tantamount to racism or human rights abuse. To a nationalist' – such as himself – 'they are evidence of a community asserting its values', which the English have a right to do as much as anybody else.[50] Mass immigration is the stick that the leftists and the neoliberals force down the throat of resistant Englishmen. *Confessions* contains some figures about the hair-raising levels it had reached in the early millennium: in 2011, 13 per cent of England's population was foreign-born, while in 'the capital, English people had become an ethnic minority'.[51] The latter statement clarifies what skin colour one must have to count as English in the eyes of Paul Kingsnorth. Being English means being white. On this point, the spiritual father of the Dark Mountain Project showed no sign of his usual fatalism: when it came to mass immigration, there was no question of withdrawal or meditation or telling stories to learn to live with it: here was a problem with a

---

46 Paul Kingsnorth, 'The Lie of the Land: Does Environmentalism Have a Future in the Age of Trump?', *Guardian*, 18 March 2017.
47 Kingsnorth, *Confessions*, 203–4.
48 Kingsnorth, 'The Lie'; Kingsnorth, *Confessions*, 284.
49 Essay reprinted in Kingsnorth, *Confessions*, 204.
50 Kingsnorth, 'The Lie'; cf. *Confessions*, 205.
51 Kingsnorth, *Confessions*, 202.

straightforward solution. 'Immigration and emigration should be pretty much balanced. This means a big cut in immigration from the present numbers', as Kingsnorth proclaimed already in 2010.[52] Patrol the borders, in short. Or, in the slightly more poetic words of Kingsnorth's choosing: 'This is a magic island. It knows how to defend itself.'[53]

It speaks to some real achievements of the left that this view of the world was formulated in explicit opposition to the environmental movement. After the millennium, people like Foreman and Kingsnorth had to climb trees that stood closer to the Front National than to Friends of the Earth. Thanks to the bifurcation of the 1980s and 1990s, ecology in northern countries had become more than a preserve for well-to-do white men, with the result that the ideas of green nationalism departed from it and moved over to the far right. There they went into hibernation, but in the rapidly warming 2010s, they came out in the open again. Something in the air beckoned to green nationalism. If Foreman, Kingsnorth and their ilk represented its highbrow version (or perhaps we should say middlebrow), there were also combatants ready to take action.

Green Nationalism in Action – Or, Three Things That Happened on a Friday

Towards the end of the burning Swedish summer of 2018, one young girl could no longer stand the official passivity. Fifteen-year-old Greta Thunberg sat down in front of the Riksdagen building and declared that she would not attend school anymore before the elections. Once they were held, to little effect, she returned every Friday, ignoring her school duties, quietly sitting on the pavement, demanding that rich countries such as Sweden initiate emergency planning and slash

---

52 Paul Kingsnorth, 'Immigration: Truisms vs Cliches', paulkingsnorth.net, 2010, accessed 1 October 2014.

53 Paul Kingsnorth, 'Elysium Found?', *Pastebin*, 22 June 2018. Kingsnorth took down this essay from the website on which it was originally published; for the back story, see London Permaculture, 'Et in Arcadia Ego', *Medium*, 23 June 2018.

emissions by more than 10 per cent per year. Relatively isolated in her native country, Greta Thunberg soon inspired a wave of school strikes across Europe. With a speed hitherto unseen in the struggle against climate change, the movement snowballed towards the date of 15 March 2019, when strikes were planned in places stretching from Mexico City to Delhi. This would be the day when the call for radical action could no longer be ignored.

On the very same day, while children in the town were gathering for the strike, Brenton Tarrant, a twenty-eight-year old Australian gym trainer, loaded his car with semi-automatic rifles and shotguns and drove to the Al Noor mosque in Christchurch, New Zealand. He walked up to the building and was greeted with a 'Hello, brother' from one worshipper at the entrance. The man who uttered these words was among the first to be shot.[54] Tarrant then moved into the mosque, where some five hundred Muslims had assembled for Friday prayers, and walked from room to room with his guns pumping bullets into the mass of humanity; his own film of the attack, livestreamed on Facebook, showed bodies dropping down to his left and right as in a video game. After six minutes, Tarrant returned to his car and drove to another mosque where he resumed his action before being chased away by a worshipper, who wielded only a credit-card reader. When the police arrested Tarrant, he had killed fifty-one Muslims, aged between three and seventy-seven. He had also, naturally, relegated the one and a half million school strikers across the world to a distant second place in the day's newsreels.

Tarrant's lodestar was Anders Breivik, from whom he claimed to have received blessing for his action; just like him, he had composed a manifesto to explain its rationale. It begins in semi-automatic style: 'It's the birthrates. It's the birthrates. It's the birthrates.' Muslims are drowning the world with their children, which is why those children need to be killed. White nations have failed to maintain replacement-level fertility, putting themselves on the chopping block of invaders

---

54   Naomi Klein, *On Fire: The Burning Case for a Green New Deal* (London: Allen Lane, 2019), 40–4; Al Jazeera, '"Hello Brother": Muslim Worshipper's "Last Words" to Gunman', 16 March 2019.

who are bent on 'white genocide'.[55] It follows that white people need to have more children, but they will only be saved if they start reducing the numbers of the enemy. As in all other similar writings, the end is nigh, time running out, apocalypse galore in old Europe. Tarrant describes his travels in France as particularly harrowing. He found villages and cities alike overrun by Muslims, whose conquest had never been opposed. The sole hope he could discern in the country was the election campaign of Marine Le Pen in 2017; when Macron defeated her, Tarrant – according to his own account – broke down in tears and decided to act. The French right appears to have taught him about 'The Great Replacement', the title of his manifesto. In matters of ecology, he likewise took the French line. Where Breivik's mega-manifesto overflowed with quotations from Bat Ye'or and references to Mark Steyn and hundreds of footnotes dripping from the pages, Tarrant – less megalomaniacal, not so fond of books – structured much of his shorter text as a self-interrogation. One question reads: 'Why focus on immigration and birth rates when climate change is such a huge issue?' Answer:

> Because they are the same issue, the environment is being destroyed by over population, we Europeans are one of the groups that are not over populating the world. The invaders are the ones over populating the world. Kill the invaders, kill the overpopulation and by doing so save the environment.[56]

One of the seventy-three pages of 'The Great Replacement' carries the headline 'Green Nationalism Is the Only True Nationalism'. Here Tarrant equates the protection of nature with that of the white ethnic stock and vents a recognisable complaint:

> For too long we have allowed the left to co-opt the environmentalist movement to serve their own needs. The left has controlled all discussion regarding environmental preservation while simultaneously

---

55   Brenton Tarrant, *The Great Replacement*, 3–4. For analysis and contextualisation of the manifesto, see Jeff Sparrow, *Fascists among Us: Online Hate and the Christchurch Massacre* (Melbourne: Scribe, 2019).
56   Tarrant, 'The Great', 21.

presiding over the continued destruction of the natural environment itself through mass immigration and uncontrolled urbanization ... Continued immigration into Europe is environmental warfare and ultimately destructive to nature itself. The Europe of the future is not one of concrete and steel, smog and wires but a place of forests, lakes, mountains and meadows.[57]

With this disquisition, Tarrant parted company with Breivik and the bulk of the European far right. This gave a party like the AfD a chance to take a distance from his ideas, nine-tenths of which it shared. One AfD representative in the state parliament of Berlin and member of that institution's environmental committee, Harald Laatsch, tweeted his condolences to the victims of the massacre. He made sure to add that 'the culprit justified his act with overpopulation and climate protection. The peddlers of climate panic share a responsibility for this development. #GretaThunberg.'[58] With that hashtag, this AfD politician arraigned the strike leader as responsible for the massacre.

It should be rather obvious that two ideas came together in the deed of Brenton Tarrant: the idea that there are too many Muslims/immigrants/non-whites around, and the idea that there are too many people for this planet to stay healthy. It was at that junction he turned towards the Al Noor mosque. Some have suggested that he cared not one iota for any actual lake or mountain or meadow, merely using environmental concerns as a mendacious embellishment of his bloodbath.[59] His passion for the climate might well have been weaker than the AfD's for coal. But it would require interrogation of Brenton Tarrant's soul to settle that question, and he placed himself in a current that exerted a real pull on some people who, to all appearances, loved nature.[60] The actual impact of the massacre is not in doubt. When night fell on 15 March 2019, fifty-one people with no

---

57 Ibid., 37.
58 Harald Laatsch's Twitter account, 15 March 2019. Others on the far right argued that since Tarrant claimed to be an environmentalist, he could not possibly be one of them. See Sparrow, *Fascists*, 81.
59 For example Klein, *On Fire*, 45.
60 Cf. Sparrow, *Fascists*, 81–2.

particular role in stoking the fires of climate breakdown had been murdered in Christchurch. To the school strikes, the action had done nothing but damage. Thus the day seemed to bring home a lesson of much wider applicability: green nationalism is a script for mass atrocity, and the greatest miss any version of environmental politics could possibly come up with.

On that very same day, Cyclone Idai made landfall in Mozambique. The worst ever weather event to strike the southern hemisphere, in the rapid assessment of the UN, it drowned swathes of Mozambique and neighbouring countries in a deluge of water, chewed up beaches, flushed away villages, killed more than one thousand people, made hundreds of thousands homeless, dumped malaria and cholera on the region and obliterated years of hard work by those former European colonies to lift themselves out of poverty. Images from the disaster zone showed black people standing on the roofs of their submerged houses, clinging to treetops, queuing for aid, walking in columns with bundles on their heads, wading in water, stepping around debris, staring into the spaces of washed-out bridges, finding safety in the last moment and crying when burying their dead. Images of black and brown people doing such things had by then become a commonplace of the era, a nearly constant stream in the background of a typical global day. (In the following weeks, it was Iran's turn to stand under water. In late April, yet another cyclone devastated another part of Mozambique. In June, hundreds of Indian villages were deserted after a combined heatwave and drought made them unliveable, and so on.)[61]

Like all contemporary cyclones, Idai had its guns loaded by the surplus of heat in the earth system: raised sea levels send higher storm surges onto land; warmer air holds more water vapour; to make matters worse, the floods rolled over plains racked by drought.[62] Mozambique had done virtually nothing to bring such conditions about. In one calculation, 0.2 per cent of the global

---

61 The latter catastrophe: Sam Relph, 'Indian Villages Lie Empty as Drought Forces Thousands to Flee', *Guardian*, 12 June 2019.

62 Matthew Taylor, 'Climate Change Making Storms Like Idai More Severe, Say Experts', *Guardian*, 19 March 2019; Eric Holthaus, 'Cyclone Idai Lays Bare the Fundamental Injustice of Climate Change', *Grist*, 19 March 2019.

temperature rise caused by emissions up to 2012 could be attributed to this country, compared with, for instance, 20 per cent for the US, 4 for Germany and 3.5 for the UK.[63] In another, the CO_2 emissions of one US citizen equalled that of 270 Mozambicans. Already in the last two decades of the twentieth century, more people were killed in climate-related disasters in Mozambique than in any other country, relatively speaking: 5.5 per one thousand inhabitants, compared with, for instance, 0.03 in the US.[64] When Idai struck, Mozambique ranked 180 out of 189 countries in the human development index, distilled by the UN as a measure of wealth.[65] Awareness of these inequities had, of course, percolated through the country. In the words of Anabela Lemos, a leading activist in Justiça Ambiental, or JA!, meaning 'Now!': 'People in Mozambique know this is climate chaos. They know what's going on. They are going to come and challenge everyone in northern countries and ask – why are you continuing to do this to us?'[66]

Many things can be said about the relation between the events in New Zealand and those in Mozambique on 15 March 2019. One is that there were two ways in which droves of non-white people were killed on that Friday. Another is that green nationalism blames innocents and victims, to the extent that, in its most vicious form, it wants more of them to go out of existence.

A Provisional Definition of Green Nationalism

How shall green nationalism be defined? The core of it seems to be a belief in protection of the white nation as protection of nature. Here the ecological crisis is not denied but enlisted as a reason to

---

63 Marcia Rocha, Mario Krapp, Johannes Guetschow et al., 'Historical Responsibility for Climate Change: From Countries' Emissions to Contribution to Temperature Increase', *Climate Analytics*, November 2015, 39, 41.

64 J. Timmons Roberts and Bradley Parks, *A Climate of Injustice: Global Inequality, North-South Politics, and Climate Policy* (Cambridge, MA: MIT Press, 2007), 78–9, 148.

65 United Nations Development Programme, 'Human Development Reports: Mozambique', hdr.undp.org, 2018.

66 Anabela Lemos, 'People Know This Is Climate Chaos', *Real World Radio*, rwr. fm, 29 March 2019. JA! is the Mozambican chapter of Friends of the Earth.

fortify borders and keep aliens out; applied to the climate, this would imply a break with denial and a rebranding of various nationalist policies as the remedy. Some might be of an economic kind, such as protectionism, now sold for its green virtues. But above all other themes, there still looms, as much as ever, the one motif of stopping and reversing immigration – *ecology is the border*, we can epitomise the central tenet, abbreviating Jordan Bardella. Thus defined, green nationalism possessed a relatively coherent shape in the 2010s, uniting parties such as Rassemblement National, thinkers like Foreman and Kingsnorth and combatants like Tarrant. The latter was not an isolated loner: less than half a year after the Christchurch massacre, on 3 August 2019, a man entered a supermarket in El Paso, Texas, gunned down twenty-two people and left behind a manifesto explaining that he had to kill Latino people to save the environment.[67]

Two main tributaries of green nationalism can be readily identified: Malthusianism and the mysticism of national nature. If the former was paramount for someone like Garrett Hardin, the latter made the heart of Paul Kingsnorth beat fast; less interested in the arithmetics, he sighed for the mysterious, indeed 'magic' communion with the soil – a condition or quality hard to capture in words, because it can only be felt, and only by people with roots in the natural nation.[68] To work up this feeling, green nationalists did not have to look to American sources. There were vintage European bottles to grasp for. But the fear of overpopulation and love for the land flowed easily together, and they tended to come with the idea that the left has established an illegitimate monopoly on environmentalism, which the right must now break up.

On this definition, some far-right posturing in nature would not qualify as green nationalism. SD activists professed a fervid fondness for cats and dogs.[69] Nigel Farage cultivated an image of himself as a

---

67 On this manifesto, see further below.
68 Kingsnorth, 'Elysium'.
69 Martin Hultman, Anna Björk and Tamya Viinikka, 'The Far Right and Climate Change Denial: Denouncing Environmental Challenges via Anti-establishment Rhetoric, Marketing of Doubts, Industrial/Breadwinner Masculinities Enactments and Ethno-nationalism', in Forchtner, *The Far*, 129.

man at home among the hedgerows and hills of rural England, son of a 'fanatical lepidopterist' who would only leave his butterflies for a pint at the village pub or a hunting party.[70] Vox likewise certified its authentic bonds to Spain through hunting. A group of hunters showed their support for the party by posing over some seventy killed rabbits, laid out on the ground so that they formed the letters V – O – X. In Andalusia, the Vox chapter joined the local hunting federation to promote knowledge of this practice in schools, as 'a real vision of the natural world not distorted by animalism [sic] or political ecology'.[71] Santiago Abascal posted pictures and videos of himself riding his horse, drinking from rivers flat on his belly, stroking fields of wheat, looking out over his nation – broad chest, erect stature – from a sharp escarpment, but in all of this, one element was missing: the sense of an ecological crisis, least of all one concerning climate.[72] That would have to be present for green nationalism to come about. As soon as we have proposed this definition, however, we must point out that it is highly provisional, for green nationalism is an exceedingly fluid current. It leaks into its supposed opposite, as we can see in the case of one party that adopted it in the late 2010s: the Finns.

Sever the Extra Hands

In 2015, the PS joined the Finnish government and renounced its denial of climate science. Two years later, the party caused a convulsion in domestic politics, when a leadership stiffer in its hostility to immigration took the reins. The new chairman, Jussi Halla-aho, had frequently expounded on genetic and IQ differences between ethnic groups, suggested that 'loitering and looting' are 'special genetic

---

70 Emily Turner-Graham, ' "Protecting Our Green and Pleasant Land": UKIP, the BNP and a History of Green Ideology on Britain's Far Right', in Forchtner, *The Far*, 65.

71 *Eulixe*, 'Reacciones a los conejos de Vox', eulixe.com, 22 January 2019. *La Razón*, 'PP, Cs y Vox acuerdan promocionar la caza en los colegios de Andalucía', 26 November 2019.

72 Santiago Abascal's Instagram account; Vox, 'Un nuevo comienzo', YouTube, 7 June 2016.

features of Somalis' – a statement earning him a fine in the supreme court – and defended himself as a critic of multiculturalism who simply dared to talk about the problems 'swept under the rug in the name of political correctness'.[73] He was also a friend of animal rights, an eager biker, a supporter of public transport and a PhD holder, having penned a dissertation on Old Church Slavonic. With the party firmly in his hands, he shifted it several steps to the right. The prime minister from the Centre Party could stomach the PS no more and kicked it out of the government; some ministers resigned from the Finns to stay on their posts and formed the more moderate rump 'Blue Reform', while the hard core went back into opposition.

There it retained its acceptance of the fact of anthropogenic climate change. The party of Halla-aho swore to awareness of the problem – which did not, however, imply any revisions regarding peat, Paris or domestic mitigation. To the contrary, Halla-aho argued that any policies making Finland less attractive for business would push industries to other countries, where they would emit more CO_2; best for the climate would be to maintain Finnish competitiveness – or, 'environmental action is move a factory from Poland or China to Finland', in the words of Laura Huhtasaari, second in command.[74] But a different message garnered the highest profile: the catastrophe is driven by overbreeding in and immigration from the global South. 'Enormous population growth in the Middle East, Africa, and Latin America is the cause of climate change and all the other crises', said Juho Eerola, third in command.[75] Muslims in particular produce too

---

73 Jussi Halla-aho, 'Älykkyyden mittaamisen suhteellinen mahdottomuus', *Scripta*, halla-aho.com/scripta, 30 March 2006; Jiri Keronen, 'Jussi Halla-ahon tuomioon johtanut kirjoitus', *Helvetin Puutarha*, keronen.blogspot.com, 8 June 2012; KKO, 'Tiedote 8.6.2012 – Politiikolle tuomio islaminuskoa ja somaleja loukanneista Kirjoituksista', Korkein Oikeus, korkeinoikeus.fi, 8 June 2012; Jussi Halla-aho, 'Lue Jussi Halla-ahon linjapuhe sana sanalta: Paljastaa suunnan', *Uusi Suomi*, 11 June 2017.

74 Halla-aho in *Suomen Uutiset*, 'Halla-aho: Julkisen talouden pitää elää suu säkkiä myöten – "Emme voi paikata jatkuvaa budjettialijäämää velanotolla"', 20 May 2018; Huhtasaari in *Suomen Uutiset*, 'Presidenttiehdokas Huhtasaari teollisuuden ulosliputtamisesta: "Kyllä tulee hintaa päättäjiemme moraaliselle ylemmyydentunnolle"', 17 December 2017.

75 *Suomen Uutiset*, 'Eerola: Maailman väestönkasvu on suurin syy tulevaisuuden ongelmille', 25 June 2018.

many offspring.[76] Environmental action is keeping borders closed; to allow the Muslims and the Africans to come in would be to encourage them to multiply their emissions. Said Olli Immonen, smart young ideologist of the PS: 'We have to prevent big population flows in order to save the earth.'[77]

Immonen was also the leader of Suomen Sisu, the main outfit of Finnish white supremacism, which sprinkled its propaganda with cues to a certain political past. The natural abbreviation for Suomen Sisu, of course, is 'SS' (but the group purported to prefer SuSi). For a symbol, it chose an octagon knot not dissimilar to the swastika or sun cross, while its charter consisted of fourteen principles. As the initiates would know, the number fourteen was shorthand for the fourteen-word slogan-cum-meme 'We must secure the existence of our people and a future for white children', repeated several times in Brenton Tarrant's manifesto; the charter of Suomen Sisu began with the stated aim of 'securing the existence of our people and the preservation of true and natural diversity' (likewise fourteen words). The eighth point talked of 'a responsible relationship between man and nature', the ninth of 'unhealthy consumer society'; but the primary purpose of the SS was to 'fan the flames of nationalism in everyone. We believe that once it bursts into flames, it will create efficacious action.' The website of the group featured pictures of Immonen in a blue tie as well as activists shooting with their guns in a forest, common practice for members during summer camps. The most famous activist of the SS, however, was Halla-aho himself. It was his clique of SS allies that now ran the party machinery of the Finns.[78]

Green shades of brown were present in both branches. One up-and-coming PS politician, Tapio Lämsä, broke some new ground by arguing that 'driving the herds' – i.e. facilitating the movement of refugees – increases emissions from the transport sector. 'Artificial

---

76 Olli Immonen, 'Euroopan väestökatastrofi', olliimmonen.net, 26 September 2017.

77 *Suomen Uutiset*, 'Immonen: Vihreiden politiikka vinksallaan – kannattaa väestönsiirtoa, joka tuhoaa maapalloa', 19 April 2018.

78 *Suomen Sisu*, 'Perusperiaatteet', suomensisu.fi; Antifasistinen Verkosto, 'What Is Suomen Sisu?', *Varis*, 13 December 2018 (the latter is an excellent overview of the SS in English).

displacements' – i.e. receiving asylum-seekers – stimulate immigrants to have even more carbon-hungry babies; furthermore, Finland should respond to foreign mega-emitters of CO_2 by imposing import tariffs from those 'shit-blowing' countries.[79] The shit-blowing countries were, in the official view of the party, the developing ones, guilty of rapidly increasing their fossil fuel consumption and taking away industries from northern countries such as Finland. Finns, by their nature, do not blow shit. 'The Finnish people have a strong and unique connection with nature.' White as innocence, nature has 'always been an inseparable part of Finnish folklore and identity' – an everlasting 'national property' foreigners must not be permitted to sully.[80]

The party that just a few years earlier had portrayed concerns about climate change as a black hoax cooked up to skin the Finns thus flipped the narrative and posed as nature's angels: and the volte-face appeared to cause no inner ructions. It was a smooth passage. It entailed no rethinking of the actual extraction or consumption of fossil fuels in Finland. It did, however, link up with a particular tradition of Finnish ecology, which had its own Garrett Hardin in the figure of Pentti Linkola, an ornithologist, fisherman, forest protector and productive writer, who embraced the grimmest version of lifeboat ethics and urged the lucky riders to 'take the ship's axe and sever the extra hands that clutch at the sides'.[81] A cult figure for hardline green nationalists around the world, Linkola agonised over immigrants maintaining sky-high birth rates long after settling down. 'There is no use in counting the immigrants at the border: one should wait a while and look in their nurseries.' He recommended

---

79   Tapio Lämsä, 'Vaihtoehto vihervasemmistolaiselle ilmastohumpalle', *Perussuomalainen Blogeja Verkossa*, 15 October 2018.
80   PS, 'Ympäristö-ja energiapolitiikka', 15 January, 3.
81   Linkola quoted, and his international influence pointed out, in Jason Wilson, 'Eco-fascism Is Undergoing a Revival in the Fetid Culture of the Extreme Right', *Guardian*, 19 March. A different translation of the original French, 'Grab an axe and sever the hands clinging to the gunwales', appears in Pentti Linkola, *Can Life Prevail?* (n.p.: Wewelsburg Archives, 2009), 92. There were no direct links between the PS and Linkola, however, the latter often referring to the former as 'trash'. On his international influence, cf. Blair Taylor, 'Alt-right Ecology: Ecofascism and Far-right Environmentalism in the United States', in Forchtner, *The Far*, 278.

euthanasia, capital punishment, the abolition of 'overzealous rescue services', forced abortion and sterilisation and possibly also infanticide as the first axes to grab.

By the time of the next national elections, in spring 2019, the moderate Blue Reform had been obliterated, while the Finns sailed ahead of most other parties. In the wake of the IPCC report and the school strikes, climate had become the number one concern of voters. The Finns, of course, argued that their core concern harmonised perfectly with it. In these 'climate elections', as they were generally perceived, parties competed in proposing measures for domestic emissions cuts – with one exception. The PS was still the only party refusing national climate goals. If it no longer denied global heating as such, it vehemently denied any need for Finland to do anything about it, except by shutting out the rest of the world. Variations on the argument became shriller and shriller as election day approached: 'if Finnish people stopped breathing, it would not affect climate change in the slightest' (Halla-aho); 'if every Finn shot himself, it would do nothing to stop climate change' (Eerola); 'Finland has done more than any other country and has the smallest emissions in the world and has already done too much' (an MP and union activist); 'if Finns stopped driving cars, it would delay global Armageddon by one minute' (a candidate from the southeast).[82] When the social democrats broke the taboo on dietary habits and proposed a meat tax, the Finns were up in arms. This would 'take the sausage from the mouth of workers' and make the food of pets so expensive as to cause rivers of tears: 'What do you say to the little girl or boy who cries when Mom and Dad say they can no longer afford their loved little poodle and take it away to be axed? It is a big question.'[83]

No Finns needed mend their ways. Instead the party found faults with Poland, China, India, population growth and development aid.

---

82 Halla-aho in Hannu Tikkala, 'Jussi Halla-aho suomi "ilmastohysteriaa" Ylen tentissä: aiheuttaa ahdistusta ja laskee syntyvyyttä – lue kooste liveanalyysistä täältä', *Yle*, 28 March 2019; Eerola and the candidate from the southeast (Kristian Laakso) in Ellen Barry and Johanna Lemola, 'The Right's New Rallying Cry in Finland: "Climate Hysteria"', *New York Times*, 12 April 2019; Matty Putkonen in the election debate 'Suomesta ilmastotekojen mallimaa? Katso ilmastotentti kello 21', *Yle*, 4 April 2019.

83 Putkonen in 'Suomesta'.

Peat was as dear as ever – 'it's for us what oil is for Norway.'[84] All other parties were guilty of 'climate hysteria', particularly heinous as it discouraged fertile Finns from having children – some reports said young couples had second thoughts, given the state of the planet – and thereby depressed the right sort of birth rates.[85] 'Climate hysteria' became the new far-right keyword. It pulled the Finns back towards default denial, Halla-aho unable to prevent himself from throwing doubt on the science. He tweeted: 'Why is it okay to stir up panic on climate change, even though its negative consequences are still largely theoretical, but at the same time, it is not okay to stir up panic on immigration, even though its negative consequences are very concrete?'[86] Up until the final votes had been counted, it was unclear if the Finns would become the largest or second-largest party. Given its internal split and right-wing turn, many observers had, only months earlier, expected it to perform badly. Halla-aho became 'vote king' – receiving more votes than any other candidate – but the party ended second, 7,666 votes behind the social democrats.[87] Its supporters partied into the night.

The Light of the Danes

In the climate elections of 2019, Finland got its most left-leaning government in two decades. Neoliberal hegemony appeared to be broken, when the newly formed cabinet – a coalition of greens, buoyed by the 'Greta Thunberg effect' here as elsewhere, social democrats and reformist socialists – presented a two hundred-page dossier of policies to make Finland carbon neutral by 2035. No flexible mechanisms were included. The country would see the transition through by its own means. This spelled the end of peat and coal, the

---

84 Ibid.
85 Halla-aho in Anna-Kaisa Brenner, 'Jussi Halla-ahon tytär osallistui nuorten ilmastolakkoon isän vastustuksesta huolimatta – Perussuomalaisten puheenjohtaja ei usko, että pelastumme ympäristötuholta', *Yle*, 1 March 2019.
86 Jussi Halla-aho's Twitter account, 3 March 2019.
87 Blue Reform lost all representation in the parliament. At least five members of the SS now sat in parliament, among them Jenna Simula, whose slogan during the election campaign was 'it is time for home returns', i.e. repatriation.

rolling out of wind and solar, electrification of transport and heating, expansion of the rail network and a series of other infrastructure projects, covered by public investments also slated to raise employment and reverse years of austerity measures. The plan for 'the world's first fossil free welfare state' made waves outside the country.[88] Inside, it appeared to have mass support, some 70 per cent calling for radical climate action; but in the wings waited the Finns. One post-election poll put them on top. Finland thereby instantiated a novel pattern in European politics in the late 2010s: ecology edging itself into the centre; fierce competition between climate and immigration as top concerns; an ever-starker choice of apocalypse; the far right holding firm in its resistance against mitigation *even when it adopted green nationalism*. But sometimes the battle lines were more fluid.

In Denmark, the DF had utterly dominated politics for two decades. The country seemed to be in the grip of a chronic national psychosis about Muslims, expressed in ever more brutal measures against them. In 2016, the Danish state started confiscating jewellery and cash from refugees upon arrival (supposedly to make them pay for the services rendered).[89] Then it designated twenty-five predominantly low-income Muslim neighbourhoods as 'ghettos', whose inhabitants were subjected to a series of special laws, including harsher punishments for crimes and mandatory separation of children from their parents for a period every week (although the DF failed to convince the government to put all 'ghetto children' under curfew from 8 p.m. by fitting them with electronic bracelets).[90] Any parents who took their kids on extended visits to their countries of origin would be sentenced to prison. The Danish state proceeded to isolate asylum-seekers convicted of crimes on a tiny island previously used for researching contagious animal diseases, ferry them to their new abodes in a ship called 'Virus' and proclaim that the time

---

88 For example Greenpeace, 'Finland Strives to Become the First "Fossil Free Welfare State" – Carbon Neutral by 2035', greenpeace.org, 3 June 2019; Jon Henley, 'Finland Pledges to Become Carbon Neutral by 2035', *Guardian*, 4 June 2019.

89 Martin Kingsley, 'Danish Police Seize Valuables from Asylum Seekers for First Time', *Guardian*, 30 June 2016.

90 Ellen Barry and Martin Selsoe Sorensen, 'In Denmark, Harsh Laws for Immigrant "Ghettos"', *New York Times*, 1 July 2018.

had come to give up on integrating refugees in Danish society: they should henceforth be kept apart from the autochthonous population until they could be sent packing. This was referred to as a 'paradigm shift' in immigration policy, as the overarching goal would henceforth be 'repatriation'.[91]

What did the social democrats – the main opposition party, coming out of the Second International and the decades of the Scandinavian welfare state – say about these measures? Not only did they endorse them, but they resolved to beat the far right in its own game. In 2018, the social democrats put forth a document called 'Just and Realist: An Immigration Policy That Unites Denmark'. It began by outlining the enormous burden imposed on the nation: 'Our population has changed rapidly in a short time. In 1980, one percent of the Danish people had non-Western origins. Today, it is eight percent' – a challenge of *origins*, continuing in the offspring of Somali and Vietnamese and Lebanese immigrants and causing no end to the trials and tribulations for the Danes.[92] How would a social-democratic government deal with it? The right to apply for asylum in Denmark would be fully terminated. Any non-Westerner making it to the border should be transferred to facilities in North Africa or the Middle East. But the social democrats also had plans for those already on Danish soil: a cap on people of non-Western origins in urban neighbourhoods, a special labour duty of 37 hours per week, shutdown of Muslim schools and, most importantly, accelerated repatriation. The social democrats proposed that any refugees who still managed to approach Denmark – perhaps via the UN – should be subjected to 'massive counselling' to deter them from staying, that citizens of non-Western origin be stripped of their citizenships if convicted of a crime and that a range of economic incentives be put in place to persuade non-Westerners to 'return to their country of

---

91 Martin Selsoe Sorensen, 'Denmark Plans to Isolate Unwanted Migrants on a Small Island', *New York Times*, 3 December 2018; Christina Jul Nielsen, 'Finanslovsaftale: Udviste asylansøgere skal sendes til nyt udrejsecenter på Lindholm', *TV Øst*, tveast.dk, 30 November 2018; *DR*, 'Udrejsecenter på Lindholm', 27 December 2018; Finansministeriet, 'Aftale om finansloven for 2019 mellem regeringen og Dansk Folkeparti', fm.dk, 30 November 2018.

92 Socialdemokratiet, *Retfærdig og realistisk: En udlændingepolitik der samler Danmark*, socialdemokratiet.dk, 4 (the quote is from chairwoman Mette Frederiksen).

origin.'[93] Waving this plan before the Danish people – the white Danish people, that is – the social democrats placed themselves to the right of the conservative DF-backed government.[94] It was a sign of a country in the throes of *fascisation*, as the process used to be known in the interwar period; we shall return to this concept below.

Perhaps the most remarkable aspect of the document, however, were some other opening lines, just before it recapped the burden of a jump in non-Western origins from 1 to 8 per cent. Here the party, in a statement signed by its chairwoman Mette Frederiksen, explained that the burden will become heavier in the years ahead: 'Climate change will cause even more people to move. Add to that a doubling of Africa's population until 2050.'[95] The Danish social democrats thus came close to another line of reasoning: *in a warming world, it is even more imperative to patrol borders and send people home*.[96] A party that once professed socialism flirted with green nationalism, although in adaptationist form, the climate crisis justifying harsh crackdowns on non-Westerners rather than the latter solving the former.

But the party also had ambitions in the field of mitigation, this being a social-democratic party with no tradition of denial. It went into another Scandinavian 'climate election' in June 2019, in circumstances similar to those in Finland and Sweden: fresh memories of the hot summer of 2018; school strikes and mass demonstrations; right-left polarisation over the issue. Like the AfD, the DF had doubled down on denial.[97] It suffered the most disastrous result in its history, the voter support more than halved. Some observers attributed it to the denialism, for which the DF was apparently punished;

---

93 Ibid., quotations from 12, 31.
94 Cf. European Council on Refugees and Exile, 'Danish "Left" Takes it a Step Further than European Far Right', ecre.org, 8 February 2018.
95 Socialdemokratiet, 'Retfærdig', 4.
96 Cf. Kate Aronoff, 'The European Far Right's Environmental Turn', *Dissent*, dissentmagazine.org, 31 May 2019.
97 Laura Friis Wang, 'DF: Menneskeskabte klimaforandringer er et spørgsmål om tro, og tro hører till kirken', *Information*, 11 August 2018; Nicolas S. Nielsen, 'DF-ordfører tivler på menneskesabte klimaforandringer: Partitop mener mennesker "delvist" har ansvar', *DR*, 11 August 2018; Ritzau, 'DF: Klimaforandringer kræver ikke politisk handling', *Fyens*, 11 August 2018.

Pia Kjærsgaard herself blamed it on *klimatosser* or 'climate fools' (thereby coining a slur many picked up as a badge of honour).[98] Some hailed a fresh era in Danish politics when the far right would no longer set the tone. After all, a social democrat, Mette Frederiksen, would now be prime minister.[99]

While the DF lost out to the social democrats – the denialists to the left-green nationalists, if you will – the far right extruded two new creations. 'Nye Borgerlige', or 'the New Bourgeois', was more neoliberal and even more radically Islamophobic than the DF: 'Things just get worse and worse, the more people come from Muslim countries', declared leader Pernille Vermund, who had ties to alt-right and neo-Nazi groupuscules. Muslims 'must be stopped in every manner we can find.' On climate, the New Bourgeois oscillated between denying the science, attacking 'climate hysteria', trusting that markets and technology would solve the problem and denying that Denmark had any reason to cut its own emissions.[100]

Further to the right was 'Stram Kurs', or 'Hard Line'. This was the one-man show of Rasmus Paludan, a lawyer and YouTuber who had given himself the title 'Soldier of Freedom, Protector of the Weak, Guardian of Society, Light of the Danes and party leader for Hard Line'. The bee in his bonnet was Islam. He made a thing of walking into the officially designated Muslim 'ghettos' and burning the Qur'an or playing football with it; in April 2019, such a happening instigated

---

98 Philip Christoffersen, 'Pia Kjærsgaard: Høj stemmeprocent skyldes klimatosser', *TV2 Nyheder*, 26 May 2019.

99 For an interpretation of the social-democratic victory as 'a massive swing to pro-humanitarian parties', see Michael Strange, 'Europe, Take Note: In Denmark, the Humanitarian Left Is on the Rise', *Guardian*, 6 June 2019.

100 DR, 'Den første partilederrunde', 7 May 2019; Andreas Karker, 'Hård melding fra Nye Borgerlige: Muslimer ska hæmmes på alle måder', *Berlingske*, 2 September 2018; Mathias Bukh Vestergaard and Louise Schou Drivsholm, 'Da Nye Borgerliges spidskandidat afviser, at klimaforandringerne er menneskeskabte, bliver det for meget for Alternativet', *Information*, 8 November 2017; Martin Bahn, 'Nye Borgerliges klimapolitik: Markedet og teknologien er vores bedste våben', *Information*, 25 February 2019; Niels Fastrup, 'Ny, klimaskeptisk forening møder kritik fra klimaforskere: Hvor er dokumentationen?', *DR*, 29 May 2019; Nye Borgerlige, 'Klimapolitik', nyeborgerlige.dk, n.d. On Vermund's ties to groupuscules, see *Redox*, '26 topfolk fra Nye Borgerlige støtter nyfascistisk organisation', redox.dk, 22 May 2019; 'White Pride-veteran deltog i Vermunds bryllup', 30 May 2019.

a riot in Copenhagen, with the youth of the Nørrebro 'ghetto' battling Paludan's police escort and burning barricades. Support for him instantly shot up. The state responded by activating the ghetto laws that doubled sentences for any convicted rioter. Paludan was invited to TV studios to present his manifesto to the public: mass expulsion of people of non-Western origins, a total ban on Islam, a policy of 'ethnonationalist utilitarianism' defined as 'the greatest amount of good for the greatest amount of ethnic Danes'. When asked about the climate crisis, he spoke about immigrants throwing trash on the streets. Hard Line garnered 1.8 per cent of the votes, just below the 2 per cent threshold for entering parliament (but enough to receive hefty state subsidies for political parties). With 2.4 per cent, the New Bourgeois made it through.[101] The story of the Danish far right did not appear to have reached its final chapter. More climate elections were certainly in store. Could the far right stay relevant in them?

Population and Soil

The ideas of green nationalism also sprouted in some central European soils. To the right of Fidesz stood Jobbik, the second largest party in Hungary in the 2010s, which, apart from its even more virulent assaults on Jews and immigrants, distinguished itself by warning – if only parenthetically – about the evils lurking in a warmer world. Party spokesman Kepli Lajos used the occasion of Earth Day in 2018 to remind his fellow Hungarians that they already suffered from droughts and floods. Hence the government should rebuild the environmental protection agency – dissolved by Fidesz – and work to safeguard the natural jewels of the nation (that rare argument on the far right).[102] The party set up its own 'Green Answer', sending

---

[101] Daniel Lingren, 'Islamkritikeren Rasmus Paludan dømt for racisme', *Berlingske*, 9 June 2019; Sandra Meersohn Meinecke, 'Rasmus Paludan angrebet under demonstration på Nørrebro', *DR*, 14 April 2019; Rasmus Paludan, 'Muslimer ska ikke integreres – de skal sendes hjem', *Information*, 29 May 2019; Stram Kurs, 'Politisk grundlag', stramkurs.dk., n.d.; *DR*, 'Den første'.

[102] Kepli Lajos, 'Legyen újra önálló zöld tárca!', keplilajos.jobbik.hu, 21 April 2018; cf. Jobbik, 'Klímavédelem: élharcosok legyünk, ne mártírok', jobbik.hu, 1 April 2015.

activists to hand out flowers to Hungarians, inspect sewage treatment facilities and pick up 'the incredible amount of waste' supposedly thrown around by migrants passing through. 'We cannot put our heads in the sand', Lajos urged in the midst of the so-called refugee crisis in 2015: global warming will cause many more millions to move towards Europe, and for this 'we must be prepared'.[103] Jobbik did not propose any particular action against fossil fuels in Hungary or elsewhere. In fact, it was most steadfast in promoting what remained of Hungarian coal.[104] The climate crisis was rather a reason to look forward to a darker future, in which the defenders of the patria would have to become even tougher.

In Germany, the AfD housed a wing simply known as Der Flügel, headed by Björn Höcke, a high school history teacher who, as he liked to point out, would have loved to continue his job but felt compelled to enter the fray and fight the decay of German culture. He became local chairman of the AfD in the picturesque, forest-clad province of Thuringia. Furthest to the right within the party, by 2017 his Flügel had gathered support from an estimated one-third of its members, possibly reaching half in 2019.[105] Höcke was the main agent of the radicalisation of the party. Straddling street and state, he maintained close ties to the extra-parliamentary right – notably the anti-Muslim PEGIDA, the Nazis rioting in Chemnitz in 2018 and the Institut für Staatspolitik, a private institute striving to educate a new nationalist elite – while dissociating himself from the neoliberal elements within the AfD.[106] Instead he advocated an ethnic version of

---

103 Sarkadi-Illyés Csaba, 'Bevándorlás és klímaváltozás – Európa végleg megváltozik', keplilajos.jobbik.hu, 19 October 2015. On Green Answer, see Anna Kyriazi, 'The Environmental Communication of Jobbik: Between Strategy and Ideology', in Forchtner, *The Far*, 192.

104 Kyriazi, 'The Environmental', 191.

105 Matthias Kamann, 'In der AfD wächst ein zartes Pflänzchen der Mäßigung', *Welt, Die*, 23 July 2017; see for example the website of the current, derfluegel.de; Der Flügel, 'Kyffhäusertreffen 2016 – Rede von Björn Höcke', YouTube, 6 June 2016; *NTV*, 'Höckes "Flügel" kämpft um Einfluss in der AfD', n-tv.de, 22 January 2019; Sebastian Friedrich, 'Sie haben die Zügel in der Hand', *der Freitag*, 2 December 2019.

106 See for example Andreas Maus, Andrea Röpke, Lisa Seemann and WDR, 'AfD-Schulterschluss mit Rechtsextremen', *Tagesschau*, 6 September 2018; Hannes Vogel, 'Alte Kameraden', *Zeit*, 12 September 2018; Institut für Staatspolitik, staatspolitik.de. On Höcke's role in the radicalisation of the AfD, see Rensmann, 'Radical', 46–7.

a post-growth economy. It was a message he had already blazoned: in 2008, he had blamed the financial crisis on a 'monocultured world economy' and 'interest-based capitalism', shibboleths of a certain German political tradition.[107] To break out of the chokehold, the nation must strive towards a *Postwachstumsökonomie*, an economy that no longer has to grow to feed high interest rates and is therefore free to respect the limits of nature.[108] Sustainability would follow from the resuscitation of Germanness:

> I keep stressing that the question of identity is the central one for humanity in the twenty-first century, because it holds the key to ecological and economic homeostasis, or the balancing self-regulation of society. The Germans and the Europeans have the task to rediscover the value of their high culture

and eject their ancient adversaries.[109] Höcke advocated ethnocultural segregation on a continental scale. Once Islam has been removed to its 'own grounds' east of the Bosporus, a pure west and an Islamic east can potentially start cooperating as two separate *Großräume*. But this would require the rewinding of population movements: after seventy years of misrule – commencing after the Second World War, that is – Germany must enforce the 'remigration' of the non-ethnic Germans who have sunk the nation into a morass. That would also be an ecological cleansing. To the 'prohibition of the immigration of foreign people', Höcke added the 'prohibition of investment by foreign capital'; to the purification of the nation, an economy in a steady state.[110]

As a further gloss on this dark green nationalism, Björn Höcke gave a lecture at the Institut für Staatspolitik in 2015 on the theme of

---

107 Text republished by Andreas Kemper, 'Höckes völkische Postwachstumsökonomie', andreaskemper.org, 22 October 2015.

108 For example Björn Höcke, 'Ansprache während des Weihnachtsfests der Jungen Alternative Baden-Württemberg am 22.12. 2014 in Stuttgart', YouTube, 22 December 2014.

109 Björn Höcke, 'AfD als identitäre Kraft', AfD Landesverband Thüringen, 14 August 2014.

110 Quotations from Andreas Kemper, 'Björn Höckes faschistischer Fluss', *Graswurzel*, 1 September 2018.

'population ecology'. Africa has a 'population surplus' of 30 million human beings per year. As long as Europe absorbs that surplus (presumably by letting in those tens of millions of Africans), over-population will keep on galloping. Hence the Africans – here Höcke raised his voice – '*need* the German border, they *need* the European border to find an *ecologically sustainable* population policy. And the countries of Europe need' those same fortified borders against Africa 'and more urgently the Arab world' for 'fully understandable phylogenetic' reasons. More specifically still, Europeans constitute a biological taxon pursuing its own evolutionarily determined reproduction strategy. In Höcke's terminology, the European population is a *Platzhalter-Typ*, roughly 'the type that holds space'. The strategy of this type is 'to optimally exploit the capacity of its *Lebensraum*'. Against it stands the African type, which pursues the antithetical, congenital strategy of 'the highest possible growth rate'. Because of the 'decadent *zeitgeist* that has a firm grip on Europe', the *Platzhalter-Typ* isn't resolute enough in its defence against the Africans (and the Arabs), a disaster from which derives the political mission of Höcke and his friends.[111]

He did not mention global heating. But when he led the AfD to the second place in the Thuringian state election in 2019, Höcke denied attribution and promised to adapt his province to any troubling trend – first and foremost by stopping the construction of wind turbines.[112] He also came up with his own definition of anthropogenic climate change. As an instantiation of the process, Höcke brought up the tragic incident of a mentally ill man of Eritrean origin pushing a mother and her eight-year-old boy onto railway tracks in Frankfurt, instantly killing the child. Since refugees had been allowed to enter Germany, the 'climate' in the country had changed, as evidenced by

---

111   Quoted in Jobst Paul, 'Der Niedergang – der Umsturz – das Nichts. Rassistische Demagogie und suizidale Perspektive in Björn Höckes Schnellrodaer IfS-Rede', *Duisburger Institut für Sprach- und Sozialforschung*, 16 February 2016, 8–9. This report contains Höcke's lecture in full, as well as an extensive analysis of the many sources of the ideas presented in it, including Malthus, E. O. Wilson, John Philippe Rushton and, of course, Thilo Sarrazin.
112   Benjamin Stahl and Jonas Keck, '250 Menschen gegen die AfD – Höcke leugnet Klimawandel', *Main Post*, mainpost.de, 10 August 2018, *MDR*, 'Sommerinterview mit Björn Höcke', mdr.de, 16 August 2019.

the pandemic of sexual assaults and murders and other crimes perpetrated by these foreign populations. The solution to the climate problem, thus redefined by Höcke, spelled deportations.[113] Apart from this sleight of hand, the AfD in Thuringia raged against climate action as much as the rest of the ruck, demanding that funding of renewable energy be cut and speed limits on cars lifted.[114] Leaders of Der Flügel included lignite lovers from Lusatia, who likewise recast environmentalism as the conservation of landscapes free of immigrants and wind turbines. No dissonance between this green nationalism and fossil fuels could be registered. Indeed, the former spilled into the most classical climate denial.

In the spring of 2019, however, rare signs of internal strife over the issue appeared within the AfD. After a poor showing among young voters in the European elections, some of the leaders of the party's youth wing in Berlin rebelled against the official line. In an open letter, they demanded a change of course, lest the AfD 'lose touch' with younger generations; the realities of the climate crisis should be acknowledged and addressed within the existing 'thematic framework' – for example, by making German development aid conditional on 'a one-child policy to confront one of the biggest climate problems, overpopulation'. But it was a non-starter. Denialism easily beat it back and the Berlin youth wing fell apart over the letter.[115] Green nationalism in the AfD only involved the most abstract acknowledgement of an ecological situation, with no pretensions to deal with the climate crisis in the shared sense of the term. Instead, the party drifted further and further away from the hitherto common language.

---

113 Tobias Möllers, 'Wie Thüringens AfD-Chef Höcke sich die Frankfurter ICE-Attacke zu Nutze macht', *Frankfurter Rundschau*, 4 August 2019.

114 Susanne Götze and Sandra Kirchner, *Die Umweltpolitik der Alternative für Deutschland (AfD): Eine politische Analyse*, Heinrich Böll Stiftung, February 2016, 12–14.

115 Matthias Kamann, 'Berliner AfD-Jugend rebelliert gegen Leugnung des Klimawandels', *Die Welt*, welt.de, 28 May 2019; Robert Kiesel, 'Berlin-Vorstand zerbricht an Alleingang bei der Klimapolitik', *Der Tagesspiegel*, 29 May 2019.

A Preliminary Balance Sheet of Green Nationalism

Why would someone on the far right take up green nationalism? It might seem, at first glance, that this was simply the position favoured by the actors farthest to the right, Pentti Linkola standing to the right of Blue Reform, Jobbik to Fidesz, Höcke to the rest of the AfD leadership. But the positioning was slightly more complicated. The Finns were not certifiably more extreme than, say, the SD or the FvD; nor was the RN more so than the AfD, or Tarrant than Breivik. To the right of PiS stood Ruch Narodowy, or the National Movement, even more fanatical about the 'black gold', a guarantee for the coming '200 years of sovereignty'.[116] To the right of Höcke, a milieu of journals and groupuscules honoured and radicalised the notion of climate as conspiracy.[117] One of the most hardcore of the European parties, the Golden Dawn of Greece, gave no hint at recognising the climate crisis, but insisted on full-throttle exploitation of national oil and gas reserves.[118]

The picture becomes yet more complicated when we consider how green nationalism spilled into flat denial. Indeed, even the RN could not hold its tongue: in early 2019, one of its representatives in the European Parliament attributed global warming to 'natural climate cycles' and railed against the idea that humans are 'responsible for everything'.[119] Some denialist parties, conversely, mixed morsels of green nationalism into their rhetoric. UKIP alleged that

---

116 Samuel Bennet and Cezary Kwiatkowski, 'The Environment as an Emerging Discourse in Polish Far-Right Politics', in Forchtner, *The Far*, 247.

117 For those actors, see for example Forchtner and Kølvraa, 'The Nature'; Bernhard Forchtner, Andreas Kroneder and David Wetzel, 'Being Skeptical? Exploring Far-Right Climate Change Communication in Germany', *Environmental Communication* 12 (5): 589–604; Bernhard Forchtner and Özgur Özvatan, 'Beyond the "German Forest": Environmental Communication by the Far Right in Germany', in Forchtner, *The Far*, 224–5, 231.

118 Schaller and Carius, *Convenient*, 11, 14, 31, 86. When this book was being finalised, this party seemed to be down for the count. Helena Smith, 'After Murder, Defections and Poll Defeat: The Sun Sets on Greece's Golden Dawn', *Guardian*, 21 September 2019.

119 Groupe Identité et Démocratie, 'Joëlle Mélin sur le changement climatique', YouTube, 14 March 2019. Cf. Joëlle Mélin, 'Changement climatique: mon message à la jeunesse [Greta Thunberg]', YouTube, 16 April 2019.

'the most significant threat' to English nature is 'unsustainable population growth, which is predominantly fuelled by uncontrolled mass immigration'; the FPÖ avidly championed conservation in the Austrian parliament.[120] At this point, one might question if green nationalism, as we have defined it, had a discrete existence. It is perhaps most accurate to think of it as one tactic *in a diversity of tactics analogous to that previously developed by fossil capital*. Like the latter, the far right departed from denial, but to stay *au courant* with a rapidly warming world, some of its actors resolved to protect the core business – not profit, but hatred of non-whites – by dragging climate through the usual funnel. Much like a chunk of capital presented itself as the solution, so the far right dressed up the white nation. In both cases, the costume swaps were of little consequence and could be reversed every so often. For the far right of the 2010s, denial remained the preponderant position – in Europe, no other party family was so indifferent to questions of climate and environment – and green nationalism a subsidiary. Neither inspired mobilisation against the causes of climate breakdown.[121] If we grant that it can acquire stable form, might green nationalism ever do such a thing?

It is capable of producing two explanatory models for the destabilisation of the climate system. One says that population growth in the global South drives the process – an intelligible hypothesis, which has been disproven over and over again. It was not a baby boom in Africa that ignited large-scale fossil fuel combustion in the early nineteenth century, nor did a multitude of Muslim babies cause the fires to spread in the twentieth, and when Chinese emissions exploded around the turn of the millennium, it was not because families in that country suddenly had three or four children.[122] In the early twenty-first century, fertility rates did remain high in parts of Asia and

---

120 UKIP, *Manifesto for Brexit and Beyond*, ukip.org, 2019, 28. On green nationalism in UKIP, see further Turner-Graham, ' "Protecting" ', 64–6; on FPÖ, Kristian Voss, 'The Ecological Component of the Ideology and Legislative Activity of the Freedom Party of Austria', in Forchtner, *The Far*, 153–83.

121 Cf. Konstantinos Gemenis, Alexia Katsanidou and Sofia Vasilopoulu, 'The Politics of Anti-Environmentalism: Positional Framing by the European Radical Right'. Paper prepared for the MPSA Annual Conference, 2012.

122 On this history, see Malm, *Fossil*.

sub-Saharan Africa – Mozambique, for example (around five births per woman) – in a geographical pattern inverting that of CO_2 emissions. *The people who gave birth to the most children were the ones who consumed the least fossil fuels* and *the people who consumed the most fossil fuels were the ones who had stopped having large families* and where correlation is negative, there can be no causation.[123] Historical responsibility for cumulative CO_2 emissions did not map onto human numbers. Moreover, practically every country in the global South that had moved into the spiral of self-sustaining economic growth predicated on fossil fuels had first experienced a *decline* in fertility rates.[124] The fires rose to the sky where demography had lost importance. They burned through the accumulation of capital, not procreation.

That does not mean that there can be no relation between high numbers of people and high temperatures. But it will pertain to effects rather than causes: if many people inhabit a slowly sinking delta or fetch their water from a river drying out, the hardships can potentially be compounded; if there are only a few scattered families around, they might eke out a decent living.[125] In the former situation, the fires lit and fuelled and refuelled in distant places might push some people to move. Here emerges the possibility of climate breakdown inducing migration – and here, even more conspicuously, we are dealing with effects not causes. The second explanatory model of green nationalism – immigration to the global North drives climate change – does not deserve to be called a hypothesis. It is pure superstition. Someone who thinks that the transportation of refugees is a significant source of CO_2 has furnished their own overheated reality, and so has, indeed, anyone who believes that the trajectory of greenhouse gas emissions in the US and Europe is conditioned by the arrival of migrants. But the

---

123 The seminal paper here remains David Satterthwaite, 'The Implications of Population Growth and Urbanization for Climate Change', *Environment and Urbanization* 21 (2009): 545–67.

124 Judith Stephenson, Karen Newman and Susannah Mayhew, 'Population Dynamics and Climate Change: What Are the Links?', *Journal of Public Health* 32 (2010): 151–2.

125 See ibid., 152–3.

superstition has an evident logic to it, because the far right is only the far right insofar as it filters every problem through its hostility to non-white others.

Green nationalism applied to the climate, then, is one shade of denialism – or perhaps we should say a bastard child of it. The difference between saying 'climate change does not exist' and 'climate change is caused by poor people living in the south and coming towards the north' is rather like that between saying 'the earth is flat' and 'the earth is a golf ball.' Some of the contours of the thing have been hazily recognised, but that's about it. If attribution, trend and impact are no longer denied or ignored, the entirety of the knowledge about the *drivers* is. Concrete proposals that can arise from green nationalism thus seem unable to hit any target other than innocents or victims. In the 2010s, the parties espousing it made no efforts to cut any actual emissions, the stance of the Finns being typical: *we* will continue burning our fossil fuels; it is the others who should not come here. Between 2009 and 2014, the Front National scored highest in consistency by voting against all the thirteen major climate reforms discussed in the European Parliament. It opposed attempts to reduce oil and gas consumption in France, even under the reign of Marine Le Pen's patriotic ecology.[126] No organisation dallying with green nationalism – the RN, the Finns, Jobbik, Der Flügel – made efforts to rid its economy of fossil fuels; instead they seemed drawn to the ideal of undiminished national-energetic strength.

What about borders? Mitigation by border walls would be about as effective as pre-empting another financial crisis by gouging out the eyes of every nine-year-old. If further repression against migrants is all green nationalism can come up with, it would leave fossil capital in peace and, through yet another digression, extend the life of business-as-usual. On its own, it would probably be even more useless than capitalist climate governance, which has at least erected a few wind turbines and lowered the price of solar panels and put vegan options on some restaurant menus. So far, all green nationalism has

---

126 Schaller and Carius, *Convenient*, 30, 83; Boukala and Tountasaki, 'From Black', 79.

to show for it is a body count and a series of contributions to the never-ending denigration of non-white people.

Might this change? Could green nationalism mutate into a force against fossil fuels and shut down real emissions sources? We shall return to this possibility. For now, we must judge green nationalism *another modality of anti-climate politics*, objectively if not subjectively. The fact that it is, in at least one sense of the term, a profoundly *irrational* response to a warming world is clearly not a reason to count on its irrelevance. It might have a future, much as greenwashing had and continues to have for fossil capital. When the 2010s drew to a close, hints of that future were given in one country on the fortified European border against Africa and the Middle East, the only one governed by a party subscribing to green nationalism: Italy.

We Already Have Too Many Here

Lega Nord started off as a separatist Northern party with the typical Islamophobic leanings and default denial. After fiery journalist and orator Matteo Salvini took over in 2013, it flipped into Italian patriotism, abandoned the goal of regional autonomy for the typical master frame and eventually deleted 'Nord' from its name.[127] The Lega opposed the Paris Agreement; as a member of the European Parliament, Salvini voted against it.[128] But he subsequently steered his party towards green nationalism, giving in his election manifesto for 2018 an uncommon amount of attention – eight pages' worth

---

127   On the Islamophobia of Lega Nord, see for example Chantal Saint-Blancat and Ottavia Schmidt di Friedberg, 'Why Are Mosques a Problem? Local Politics and Fear of Islam in Northern Italy', *Journal of Ethnic and Migration Studies* 31 (2005): 1083–104. On the transformation of the party under Salvini, from regionalism to nationalism, see Daniele Albertazzi, Arianna Giovanni and Antonella Seddone, ' "No Regionalism Please, We Are *Leghisti*!" The Transformation of the Italian Lega Nord under the Leadership of Matteo Salvini', *Regional and Federal Studies* 28 (2018): 645–71.

128   Simone Santi, 'Clima, l'Italia ha ratificato l'Accordo di Parigi', *Lifegate*, 19 October 2016; Luca Aterini, 'L'Europarlamento ratifica l'Accordo sul clima di Parigi. I leghisti votano no', *Green Report*, 4 October 2016.

– on environment and energy. It promised a national investment fund for the energy transition, support for enterprises in the renewable sector, a date for banning the sale of cars powered by gasoline and diesel and a range of other measures; it also laid out visions of a 'green' and 'circular' economy. This should not be conflated with ecology of the unpatriotic kind. 'For too many years, environmentalism has spoken only to a limited and biased segment of the population', a euphemism for the left; instead, green technology should be seen as an opportunity to advance homegrown innovation and beef up Italian competitiveness.[129] If the green nationalism of the RN was accented by agrarian lifestyles threatened by free trade, that of Lega had a more high-tech, industrial ambience. It was decidedly friendly to business.

In the election for which that manifesto was prepared, the Lega leapt from 4.1 to 17.4 per cent of the voters – thanks not, of course, to its climate policies, but to its very hard line on immigration.[130] Salvini gained notoriety not for envisioning a circular economy, but for claiming that Italy had lost control over 'entire parts of the country' and needed a 'mass cleansing, street by street, piazza by piazza, neighbourhood by neighbourhood'.[131] Forming a government with the Five Star Movement, he became deputy prime minister and de facto leader of Italy, turning the force of the state against non-white people approaching from the sea or already settled, the Lega growing in the polls by leaps and bounds. Climate policy took a back seat. In May 2018, the government presented an agenda with unspecified intentions to 'accelerate the transition to renewable energy', 'increase energy efficiency' and 'encourage the purchase of hybrid and electric vehicles'.[132] Salvini then spent the summer blocking an NGO-operated ship carrying six hundred African refugees from docking in Italian ports, calling for a census of the Roma population, sending in the

---

129   Lega 2018, 'Elezioni 2018: Programma Di Governo', leganord.org, p. 35.
130   This result was 7.3 percentage points higher than the previous peak in 1996. Albertazzi et al., ' "No Regionalism" ', 650.
131   Hayes Brown, 'Italy's Deputy Prime Minister Once Talked about "Cleansing" the Roma and Migrants', *BuzzFeed*, buzzfeed.news, 22 June 2018.
132   *Qual Energia*, 'Energia e ambiente nel "contratto per il Governo del Cambiamento" M5S-Lega', qualenergia.it, 17 May 2018.

police to clear Roma camps and implementing other high-adrenaline actions.[133]

It was here, in relation to his calling, that climate change came to truly mean something for the Lega leader. He angrily disputed any connection between global heating and migration flows. In tweets and Facebook posts, Salvini sneered at the notion of 'climate refugees' and 'climate migrants'.[134] On Italian television, he questioned the existence of this category of people and pointed to the calamitous consequences if it were to be acknowledged:

> What is a climate migrant? Where do you go in the winter if it's cold and in the summer if it's hot? Enough now, we already have too many here. Should someone who leaves Milan because he doesn't like the fog also be considered a climate migrant?[135]

When the World Bank two weeks after the 2018 election published the report *Groundswell: Preparing for Internal Climate Migration*, anticipating massive population movements within countries in Latin America, South Asia and sub-Saharan Africa in the absence of deep emissions cuts, Salvini, so worked up over the spectre, failed to spot the distinction between movements in the global South and flows to Europe and commented: 'Crazy to exploit a serious issue such as the environment to justify illegal immigration.'[136] He added the hashtag '#stopinvasione'. The logic could be pushed one notch further. In January 2018, Il Populista, a Breitbart-like news site and online channel for the Lega previously co-edited by Salvini, asserted that talk of 'climate justice' and predictions of the displacement of hundreds of millions of people on a warmer planet constitute 'a quick

---

133 See for example Stephanie Kirchgaessner, 'Far-Right Italy Minister Vows "Action" to Expel Thousands of Roma', *Guardian*, 19 June 2018; Angela Giuffrida, 'Italian Police Clear Roma Camp despite EU Ruling Requesting Delay', *Guardian*, 26 July 2018; Alexander Stille, 'How Matteo Salvini Pulled Italy to the Far Right', *Guardian*, 9 August 2018.

134 Matteo Salvini's Twitter account, 17 January 2018; Matteo Salvini's Facebook page, 21 January 2016.

135 Alberto Giannoni, 'L'Europa spalanca le porte: via libera ai migranti climatici', *Il Giornale*, 24 January 2018.

136 Matteo Salvini's Facebook page, 24 March 2018.

and easy way to realise the Soros project of ethnic substitution'.[137] Behind the fake climate refugees, a rich Jew.

Here were the most salient climate-related motifs in the Lega's rhetoric, mortar in the wall into which boats trying to cross the Mediterranean now slammed. By 2019, little of ecology could be seen in the politics of the governing Lega. It was all about the border. Mightier than ever, just before he was elbowed out of government, Salvini posted a report from Sweden, aired on public Italian television, with the war headlines *Svezia Invasa! – Stop Eurabia!*[138] The leader of the fourth most populous EU nation was working hard to stop Eurabia. In this crusade, as one of his last actions as interior minister, he ordered his police force to arrest the captain who had rescued fifty-three migrants on the rough seas off the Libyan coast and brought them to an Italian port.[139] Her name was Carola Rackete; a long-time climate activist, she had earned her skills as a captain on climate research expeditions at the poles. The collision between Salvini and Rackete ended with her going free and him losing his post. The first of its kind, it was unlikely to be the last.

With pretty much absolute certainty, we can expect the far right to refuse climate breakdown as a *reason* to receive migrants. Already at the Nanterre conference, as we have seen, the elder Le Pen had the foresight to reject global warming as a pretext for bringing in 'these so-called climate refugees'. Before the Austrian election of 2017, the FPÖ sent out a press release stating that climate change impacts must never become a recognised reason for asylum in Europe.[140] The AfD labelled the notion of asylum rights for climate refugees 'insane', to be urgently tossed out 'unless we want to become Africa ourselves'. Rasmus Paludan stated that if Bangladesh were to be devastated and migrants from the country reach the borders of Denmark, it would

---

137   Il Populista, 'Ue: grazie a Pd e 5stelle, spalancate le frontiere a milioni di "migranti climatici"', ilpopulista.it, 25 January 2018.

138   Matteo Salvini's Facebook page, 21 May 2018.

139   For example Roland Hughes, 'Carola Rackete: How a Ship Captain Took on Italy's Salvini', BBC News, 6 July 2019.

140   FPÖ, 'FPÖ-Kickl: Klimawandel darf niemals ein anerkannter Asylgrund werden', fpoe.at, 22 September 2017.

be right to mow them down.[141] When figures on the American alt-right pondered the reality of climate breakdown, they concluded that it confirms their mission: as non-white parts of the globe will be wrecked, white nations must 'muster the will to guard their borders and maintain white majorities', in the words of Jared Taylor.[142] The problem itself is not to be addressed; it merely loads the guns of white nationalism. Under Salvini, the Lega did what any far-right party will do. It will use its influence to block the path of victims searching for safer ground.

---

141   Alice Weidel, 'Der Asylanspruch asl "Klimaflüchting" ist wahnwitzig', afd.de, 22 January 2020; Rasmus Paludan's Twitter account, 11 April 2018.
142   Casey Williams, 'What Happens When the Alt-Right Believes in Climate Change', *Jewish Currents*, 13 August 2018.

6
White Presidents of the Americas

It should now be apparent that the policies that Donald Trump pursued from the start were nothing if not ordinary. Their exceptionality pertained to the unique position of the United States in contemporary capitalism. In substance, they were remarkably close to the PiS or the FrP or the AfD, but the United States is a country very different from Poland or Norway or even Germany: it is an empire; it is the superpower of the fossil economy, responsible for more cumulative CO_2 emissions than any other nation, 26 per cent of all that was emitted up to 2016 (compared to less than 22 per cent for the EU's twenty-eight countries, second on the list); it is still – despite predictions to the contrary – the hegemon of world capitalism.[1] For these and many other reasons, it had special significance that primitive fossil capital and the far right blended into the figure of Donald Trump. In his first week in the White House, in late January 2017, he manifested that unity in a cascade of executive orders: build the

---

1 For easily navigated data on historical responsibility, see Hannah Ritchie and Max Roser, 'CO_2 and Other Greenhouse Gas Emissions', *Our World in Data*, ourworldindata.org, University of Oxford, 2018. On the role of the US in world capitalism, see Leo Panitch and Sam Gindin, *The Making of Global Capitalism: The Political Economy of American Empire* (London: Verso, 2012); and for an excellent update, Adam Tooze, 'Is This the End of the American Century?', *London Review of Books* 41, no. 7 (2019): 3–7.

Dakota Access pipeline (Tuesday); build the Keystone XL pipeline (Tuesday); build a wall on the border with Mexico (Wednesday); implement the 'Muslim ban' (Friday), setting the tone for a presidency that, despite all the talk of fitfulness and wild wobbling, followed these two tracks with the tenacity of a locomotive. Another, related source of the singularity was, of course, the role of the United States in the history of climate politics. Homeland of the denialist ISA, here its vindictive victory was the most consummate.

From early 2017 onwards, the most intransigent, rabidly denialist wing of primitive fossil capital governed the US in matters of climate and energy. Trump made arrangements for this transfer of power during his election campaign, when he picked as an informal advisor Robert E. Murray, head of Murray Energy. With thirteen mines in operation and several billions of tonnes of coal in unused reserves, as well as transport terminals, a fleet of barges and factories for making mining equipment, Murray Energy bragged about being 'the largest underground coal mining company in America'.[2] Coal was in decline in the years of Obama, but when the Republican presidential candidate sported a miner's helmet and raised his thumbs to ecstatic supporters who waved signs saying 'Trump digs coal', baron Murray saw a new morning dawning on his line of business.[3] In 2017, this man pronounced that 'the earth has cooled for the last 19 years' and claimed to have four thousand scientists telling him that 'carbon dioxide is not a pollutant'.[4] The hoax of global warming had been perpetrated by 'developing countries of the world to get American dollars'.[5] Murray had a colourful colleague in the second energy advisor picked by Trump on his road to the White House: one Harold Hamm.[6]

Owner of Continental Resources, Harold Hamm was the fortune

---

2 Murray Energy Corporation, murrayenergycorp.com (accessed 12 December 2018).

3 This story is told in detail in Coral Davenport and Eric Lipton, 'How G.O.P. Leaders Came to View Climate Change as Fake', *New York Times*, 3 June 2017.

4 *PBS NewsHour*, 'What Revoking the Clean Power Plan Means, from Both Sides', pbs.org, 10 October 2017.

5 *Economist*, 'Bob Murray, the Coal Baron with the President's Ear', economist. com, 18 January 2018.

6 Matea Gold, 'Here's the Back Story on How Donald Trump Won Over Oil Billionaire Harold Hamm', *Washington Post*, 20 July 2016.

seeker behind the rush to tap the Bakken formation, a vein of oil-rich shale under the states of Montana and North Dakota. His towering rigs impaled land the size of Sri Lanka and outcompeted rivals by pumping at the highest speed, openly flouting environmental regulations, spilling the greatest quantities of oil and toxic sludge, sending injured workers on ambulances shuttling between fields and emergency rooms, all while Hamm steadily climbed up the *Forbes* list of the 400 richest Americans.[7] By 2018, he had reached number twenty-nine. In the business of energy, only the Koch brothers were richer (sharing the number seven spot).[8] For some time, however, Hamm had been irked by one obstacle to even greater profits: high transportation costs, as Continental Resources had to use trucks and trains to carry its crude from Bakken. The Keystone XL and Dakota Access pipelines would solve that problem at a stroke.[9]

Harold Hamm was a pioneer of the technologies of horizontal drilling – a drill descends down a straight line, then turns sharply into the thin bed of the reservoir from the side, like the elongated base of an L – and fracking – a thick liquid is injected at such heavy pressure that it fractures the rocks, shaking loose the hitherto trapped gas or oil and forcing it to escape through the fissures – by which unexpected quantities of fossil fuels suddenly became accessible. Without these keys to the shale chest, Hamm would not have been nearly as rich, nor the US once again the world's top oil and gas producer. After opening Bakken, Hamm became the evangelist of American 'energy independence', the idea that maximum exploitation of domestic fossil fuels with the latest available technologies would make the US self-sufficient in energy, without having to bother with imports from – in particular – Muslim countries.[10] In 2016,

---

7 Josh Harkinson, 'Who Fracked Mitt Romney?', *Mother Jones*, November/December 2012; Russell, *Horsemen*, 80–3.

8 *Forbes*, 'Forbes 400', forbes.com, 3 October 2018.

9 Harkinson, 'Who Fracked'; Steve Horn, 'Company Led by Donald Trump's Energy Aide Says its Oil Will Flow Through Dakota Access Pipeline', *DeSmog UK*, 6 September 2016.

10 For example Stephen Moore, 'How North Dakota Became Saudi Arabia', *Wall Street Journal*, 1 October 2011; Harold Hamm, 'Romney's Billionaire Energy Advisor Lays Out His Vision for Energy Independence', *Forbes*, 21 March 2012; Sharon Udasin, '"Israel Can Follow US Path to Energy Independence"', *Jerusalem Post*, 21 February 2013.

speaking at the Republican National Convention, this man said that 'climate change isn't our biggest problem. It's Islamic terrorism.'[11] He did not like the sight of a wind turbine – 'once they're there, they haunt you' – 'all those things standing out in the distance, we have them all over Oklahoma', his native state. Environmental problems should be dealt with in a different manner. 'Overpopulation – that probably hurts the environment more than anything. Are we going to provide rules to stop overpopulating areas in Africa? Middle Eastern countries? Probably should.' Time to intervene for real in Africa and the Middle East: 'Stop overpopulating areas with people.'[12]

Upon winning the election, Trump recruited Myron Ebell to lead the transitional team at the Environmental Protection Agency (EPA). Ebell was (and as of this writing is) the director of the Competitive Enterprise Institute. In that role, he was the one responsible for the 'Energy' commercial about the gas of life freeing white Americans from a black world of back-breaking labour. Having long served as a chauffeur of denial, Ebell warmed the seat at the EPA for Scott Pruitt, attorney general of the oil state of Oklahoma, in which role he constantly sued the EPA on behalf of oil companies enraged by anything that looked like regulations. Funded up over his ears with fossil fuel money, Pruitt was a close affiliate of, among others, Murray, Hamm and the Koch brothers.[13] When he resigned as head of the EPA, he was succeeded by Andrew Wheeler, a coal lobbyist working mostly on behalf of Murray Energy and a confidante of James Inhofe (the man with the snowball in the Senate).[14]

As his secretary of energy, Trump appointed Rick Perry. Funded in roughly equal measures by ExxonMobil and the Koch brothers,

---

11   Ben Steverman, 'Oklahoma's Future Rests in the Hands of Two Very Different Oil Billionaires', *Bloomberg*, 8 December 2018.

12   Rebecca Leber, 'Oil Billionaire Weighs In on Wind Turbines: "Once They're There, They Haunt You"', *Think Progress*, thinkprogress.org, 20 August 2013.

13   For very extensive documentation, see the entries on Myron Ebell and Scott Pruitt at the ever-invaluable research database of *DeSmog UK*; cf. Steve Eder, Hiroko Tabuchi and Eric Lipton, 'A Courtside View of Scott Pruitt's Cozy Ties with a Billionaire Coal Baron', *New York Times*, 2 June 2018; David Cay Johnston, *It's Even Worse Than You Think: What the Trump Administration Is Doing to America* (New York: Simon & Schuster, 2018), 115–17, 128–9, 141.

14   For Wheeler's career achievements, see the entry on him at *DeSmog UK*.

Texas governor Perry also sat on the board of two companies in Energy Transfer, the pipeline conglomerate behind the Dakota Access project.[15] As his secretary of commerce, Trump selected Wilbur Ross, a billionaire who founded the International Coal Group, Inc. As his first secretary of state, he chose Rex Tillerson, CEO of ExxonMobil; as his second, Mike Pompeo, nicknamed 'the congressman from Koch' for his intimacy with the brothers, who supplied the seed money for his own aerospace company and, on some counts, donated greater sums to this one politician than to any other.[16] Trump's vice president Mike Pence was yet another horse from the brothers' ranch. 40 per cent of Trump's advisors had Koch pedigrees.[17] His first secretary of the interior was Ryan Zinke, who used to sit on the board of a pipeline engineering company; upon his departure, Deputy Secretary David Bernhardt took over the department: he was a former lobbyist for the oil industry.[18] There might have been a high turnover in the early Trump cabinet, but with unfailing consistency it remained populated by stars and intimates of primitive fossil capital.[19]

And that class fraction sent its most obdurate men. All of the names around Trump so far mentioned had years or decades behind them as card-carrying denialists. Indeed, with method in the madness and none of the impetuosity for which he became so famous, Trump let his administration fill up with personalities from the denialist ISA, merging that apparatus with the state apparatus proper so brashly that it would have made Althusser blush. Telling the Oval Office from the Heartland Institute was no longer easy. In 2017 and 2018, the latter organised a new event called the 'America First

---

15 See the entry on Perry at *DeSmog UK*.

16 John Nichols, 'The Koch Brothers Get Their Very Own Secretary of State', *Nation*, 13 March 2018; Adele M. Stan, 'Selection of Pompeo Solidifies Trump's Position with Koch Brothers', *American Prospect*, 14 March 2018; Mayer, *Dark*, xv, 338–9.

17 Mayer, *Dark*, xiii–xiv; Russell, *Horsemen*, xxiv.

18 Julie Dermansky, 'What America Still Stands to Lose as Zinke Leaves Interior and Ex-Oil Lobbyist David Bernardt Stands By', *DeSmog UK*, 16 December 2018; cf. the entry on Zinke at *DeSmog UK*.

19 Cf. Doug Henwood, 'Trump and the New Billionaire Class', in Greg Albo and Leo Panitch, *The World Turned Upside Down? Socialist Register 2019* (London: Merlin Press, 2018), 116–17.

Energy Conference', a combination of cheerleading and planning for the next step, well attended by top officials from the Trump administration, rife with adoration of CO_2, mockery of renewable energy and one-liners such as 'the leftist claims about sea level rise are overblown' (this at the second conference, held in 2018 in New Orleans, Louisiana).[20] The phrase in the conference title was not coined by the Heartland Institute, but by Trump and his advisors: on the campaign trail, at a rally focused on the virtues of oil in North Dakota, the presidential candidate outlined 'An America First Energy Plan', subsequently written up as an official White House document.[21] Then it was soaked up by the Heartland.

The denialist ISA and the federal government could at most be told apart as partners in a call-and-response relationship. Of the first fifteen individuals nominated to Trump's cabinet, seven had ties to ALEC, whose efforts to rewrite state laws in even greater favour of the fossil fuel industry were going at full tilt.[22] 'This is why I love Donald Trump. He has rejuvenated this movement', said arch-denialist Marc Morano.[23] This individual was seated in the front row at a fringe event, held by Trump administration officials at COP24, that waxed at length about how great and clean fossil fuels are and how they ought to be dug up.[24] As dejected as he had been in Paris, Morano was elated in Katowice. In a hoarse voice, barely able to contain himself, he told a reporter from the far-right John Birch Society how proud he was that the US, together with Saudi Arabia, Kuwait and Russia, had refused 'to accept the scientific findings' (his words) of the latest IPCC report – 'the greatest progress we've had since the

---

20   Collin Eaton, 'At "America First Energy Conference", Solar Power Is Dumb, Climate Change Is Fake', Reuters, 9 August 2018.

21   Timothy Cama and Devon Henry, 'Trump Outlines "America First" Energy Plan', *The Hill*, thehill.com, 26 May 2016. The Heartland Institute happily fronted the document on its website: Jim Lakely, 'An America First Energy Plan', Heartland Institute, 11 January 2018.

22   Rob Dennis, 'Trump Sets Up ALEC Administration', *Medium*, medium.com, 13 January 2017; Emily Holden, 'Rightwing Taskforce Secretly Approves Anti-environment Resolutions', *Guardian*, 4 December 2018.

23   Björn af Kleen, 'Klimatskeptikern älskar Trump: "Han har pånyttfött den här rörelsen"', *Dagens Nyheter*, 9 December 2018.

24   Chloe Farand, '"Shame on You": Campaigners Disrupt US Fossil Fuel Event Attended by Climate Science Deniers', *DeSmog UK*, 10 December 2018.

whole nightmare started'.[25] Never before had the men leading the denialist ISA felt so confident about their success.

(And the apparatus eyed a repeat of the American victory elsewhere: the Heartland Institute sponsored the denialist think tank Europäisches Institut für Klima und Energie, or EIKE, which had placed all its eggs in the AfD's basket. After the AfD's entry into the Bundestag, the vice president of EIKE became the party's representative from the Lusatian coal state of Brandenburg. An industrialist in the printing business, he sat on the AfD 'expert group' on energy policies. The Heartland Institute shared podiums with him, as well as its deep pockets with EIKE; for American primitive fossil capital, nothing would be more comforting than Germany falling to the AfD.[26] The same applied to the FvD in the Netherlands, the CLINTEL think tank docking at the Heartland mothership.[27])

The mutual embrace spanned the branches of government. Not only those dealing directly with matters of climate and energy, but pretty much every figure of some importance around Trump had, at one point or another, in passing or as part of a programme, confessed in the faith. One could take Jeff Sessions, the anti-immigration senator from Alabama who served as Trump's attorney general for two years and on his way to the post declared that CO_2 is 'really not a pollutant. It's a plant food, and it doesn't really harm anybody', or Ben Carson, the housing minister who suggested that African slaves headed off to the United States in search of this 'land of dreams and opportunity': he also knew that 'the temperature is either going up or down at any point in time, so it really is not a big deal'.[28] In this roster, Steve Bannon deserves

---

25 'Global Warming Bandwagon Coming Apart, Says Marc Morano', The New American Video, YouTube, 13 December 2018.

26 The EIKE vice president's name was Michael Limburg. See for example AfD, 'Ing. Michael Limburg', afd.de, 2017; Götze and Kirchner, Die Umweltpolitik, 6–7; James Taylor, Dennis Avery and Craig Idso, 'Heartland Institute to Present Latest Climate Science in Poland during COP-24', Heartland Institute, 15 November 2001; James Taylor, 'Skeptic Climate Scientists Gather for Conference in Germany before U.N. Meeting', Heartland Institute, 21 November 2018; Damian Kahya, 'German Far Right Targets Greta Thunberg in Anti-climate Push', Unearthed, 14 May 2019.

27 See for example Collett-White, 'Climate'.

28 Camille von Kanael, 'Trump Picks a Climate Skeptic to Enforce Environmental Laws', Scientific American, 21 November 2015; Jamiles Lartey, 'Ben Carson Incorrectly

special mention for uniting denialism and fossil fuel chauvinism and white supremacy with slightly greater ideological profundity than the president. In his Breitbart world, the threat of the climate hoax corresponded with that of losses for white people.[29] As Trump's chief strategist, Bannon was the architect of both the Muslim ban and the pullout from Paris; after leaving the White House, his permanent world tour as the emissary of the far right confirmed his propinquity with global trends. As for Trump himself, his views on climate, transmitted to the world on a regular basis, are in no need of referencing.[30]

From this platform, the forty-fifth president acted to tear down all barriers to the accumulation of fossil capital – or, to use a favourite verb of his, 'unleash' the energies hitherto blocked. A centrepiece of this agenda was the rollback of the Clean Power Plan. Bequeathed by Trump's detested predecessor, the Clean Power Plan was an attempt to very gently nudge the sector of electricity generation towards lower CO_2 emissions, reaching a level in 2030 that would ideally have been 32 per cent below that in 2005. States were given the freedom to meet this target by any means they saw fit: trading in emissions rights, substituting one fuel for another – for instance, gas for coal – or increasing fuel efficiency in their power plants. Immediately waylaid by a battalion of lawyers working on behalf of people like Robert Murray, the Clean Power Plan never went into effect. Under Ebell and Pruitt and Wheeler, the EPA was assigned the task of finishing it off once and for all, which it did by deleting the 2030 target, leaving it to the states to set whatever aims they liked and handing them only one tool for tackling emissions (if they so wanted):

Suggests African Slaves Were "Immigrants" to the United States', *Guardian*, 6 March 2017; Timothy Egan, 'Exxon Mobil and the G.O.P.: Fossil Fools', *New York Times*, 5 November 2015. See further Emily Holden, 'Climate Change Skeptics Run the Trump Administration', *Politico*, 3 July 2018; Emily Holden and Jeremy C. F. Lin, 'Trump's Climate Science Doubters', *Politico*, 3 June 2018.

29   Cf. for example Erika Bolstad, 'How Steve Bannon Is Shaping Trump's Views on "Climate Change"', *Climatewire/E and E News*, 18 November 2016; *The New Arab*, 'Steve Bannon's Eerie Obsession with the Muslim World', alaraby.co.uk, 19 August 2017.

30   For one collection of early statements, see Dylan Matthews, 'Donald Trump Has Tweeted Climate Change Skepticism 115 Times. Here's All of It', *Vox*, 1 June 2017.

burning less fuel – such as coal – per kilowatt-hour.[31] Capitalist climate governance was again turned into a carte blanche for doing nothing at all.

Making a more immediate difference, however, was a series of measures that opened territory previously out of reach for extractors. The Trump administration lifted the moratorium on leasing federal land to new coal mines. It unblocked nearly the entire American continental shelf for offshore drilling, prepared leases for areas from the Arctic to the Gulf of Mexico and invited investors to plough enormous amounts of fixed capital – the construction of the basic infrastructure might require two decades – into the virgin waters.[32] National monuments such as Bears Ears in Utah were opened, and the Arctic National Wildlife Refuge went the same way; the Trump administration also proposed the privatisation of land in Native American reservations – 2 per cent of US territory, holding an estimated fifth of the oil and gas – so that it could be transferred to wildcatters like Harold Hamm.[33] Another presidential concern was methane. A greenhouse gas several times more potent than CO_2, it leaks into the air from fracking wells – one driver of the massive spike in global methane emissions during the early millennium. Obama's EPA considered requiring companies to reduce the releases, but the rule was never implemented because it was challenged in court. When Trump rescinded it in October 2018, it was the seventy-sixth

---

31 For an excellent analysis, see John H. Cushman, '5 Ways Trump's Clean Power Rollback Strips Away Health, Climate Protections', *Inside Climate News*, 21 August 2018. On the close coordination between EPA officials and leaders of primitive fossil capital in this rollback, see for example Eric Lipton, 'As Trump Dismantles Clean Air Rules, an Industry Lawyer Delivers for Ex-Clients', *New York Times*, 19 August 2018.

32 The White House, 'President Donald J. Trump Unleashes America's Energy Potential', whitehouse.gov, 27 June 2017; Brittany Patterson and Zack Colman, 'Trump Opens Vast Waters to Offshore Drilling', *Scientific American*, 5 January 2018.

33 Emily Holden, 'Trump Opens Protected Alaskan Arctic Refuge to Oil Drillers', *Guardian*, 13 September 2019; Tracey Osborne, 'Native Americans Fighting Fossil Fuels', *Scientific American*, 9 April 2018; Chris D'Angelo, 'Trump Says He "Really Didn't Care" about Drilling Arctic Refuge. Then A Friend Called', *Huffington Post*, 2 January 2018; Valerie Volcovici, 'Trump Advisors Aim to Privatize Oil-Rich Indian Reservations', Reuters, 5 December 2016.

environmental regulation he did away with, the generality related to climate.[34]

Business-as-usual received a shot in the arm. In the late days of Obama, progressive media and movements were mesmerised by the struggle at Standing Rock: but in 2017, the Dakota Access pipeline came into operation, the oil flowing fast under the water sources held sacred by the Lakota and Dakota peoples, making the trains from the Bakken oil fields superfluous.[35] The dark seed of rigs sprouted in places until recently spared. In the fall of 2018, federal lands in the state of Wyoming, rich in rare wildlife, were being auctioned off to euphoric prospectors – 'it's untouched fields' and 'you're making it easy. It's not like there's lots of red tape and documents and paperwork to sign.'[36] The president himself missed no opportunity to take credit for the bonanza. He tweeted self-praise for fresh ExxonMobil investments in Texas and revelled in reports of new coal mines in Pennsylvania.[37] 'Energy independence' à la Hamm was an article of faith for him, but he hardened it by a few degrees into his own doctrine of 'energy *dominance*' – the idea that maximum exploitation of domestic fossil fuels will allow the US not only to detach itself from other countries, but to dominate them.[38] Horizontal drilling and fracking had given the US the means to satisfy the energetic needs of its own economy *and* those of foreigners, who could thereby be kept under the thumb. Or, in Trump's characteristic syntax:

---

34 Russell, *Horsemen*, 83–4; Eliot Weinberger, 'Ten Typical Days in Trump's America', *London Review of Books* 40 (25 October 2018), 3; Andrew Light and Benjamin Hale, 'Year One of Donald Trump's Presidency on Climate and the Environment', *Ethics, Policy and Environment* 21 (2018): 1–3.

35 For example Jarrett Renshaw, 'East Coast Refiner Shuns Bakken Delivery as Dakota Access Pipeline Starts', Reuters, 19 April 2017.

36 Oil and gas consultant Carl Larry and Peter Wold, CEO of World Energy Partners, in NPR, 'Trump Push for "Energy Dominance" Boosts Drilling on Public Land', npr.org, 25 November 2018.

37 Johnston, *It's Even*, 57; Donald Trump, 'Remarks by President Trump at the Unleashing American Energy Event', The White House, 29 June 2017; Stephen Huba, 'Corsa Coal Subsidiary Proposes Another Mine for Somerset County', *Pittsburgh-Tribune Review*, 20 November 2017.

38 Cf. Justin Worland, 'President Trump Says He Wants "Energy Dominance." What Does He Mean?', *Time*, 30 June 2017.

Our country is blessed with extraordinary energy abundance, which we didn't know of, even five years ago and certainly ten years ago. We have nearly 100 years' worth of natural gas and more than 250 years' worth of clean, beautiful coal. We are a top producer of petroleum and the number-one producer of natural gas. We have so much more than we ever thought possible. We are really in the driving seat. And you know what? We don't want to let other countries take away our sovereignty and tell us what to do and how to do it. That's not going to happen. (Applause.) With these incredible resources, my administration will seek not only American energy independence that we've been looking for so long, but American energy dominance. And we're going to be an exporter – exporter. (Applause.) We will be dominant. We will export American energy all over the world, all around the globe

such as to Mexico. A pipeline to Mexico will be built under the border wall. It will 'go right under the wall, right? It's going under, right? [Laughter and applause]', reads the White House transcript of this presidential address.[39] The perfect picture of the idea of energy dominance: a wall to keep non-whites out; a pipeline running under it to keep the same non-whites dependent on *our* oil and gas. In such a world, the power of primitive fossil capital operating in the US would be projected into other countries, whose ability to demand that American fossil fuels be kept in the ground would be proportionately diminished. All should have an interest in keeping the pistons of business-as-usual working.[40]

The doctrine of 'energy dominance' was annotated with some classical American freedom-speak when the Department of Energy in 2019 began to refer to fossil gas as 'freedom gas', to be exported without constraints around the world. Official documents from the Department, still headed by Rick Perry, spoke of fossil fuels as

---

39 Trump, 'Remarks'.
40 Jan Selby, 'The Trump Presidency, Climate Change, and the Prospect of a Disorderly Energy Transition', *Review of International Studies* (2018), online first, 17. For an excellent analysis of 'energy dominance', see also Katy Lederer, 'What Is Energy Dominance?', *n+1*, 23 July 2018.

'molecules of US freedom'.[41] The president concurrently stepped up his stout campaign against wind power. 'They say the noise causes cancer', he declared at a Republican fundraiser in the capital – the turbines causing cancer by their very *sound*, apart from, as Trump had previously claimed, inducing sleep disorders, killing bald eagles and making America ugly.[42] Dominance and freedom in the realm of energy could only come from under the ground. ALEC was enamoured with the doctrine, and so was the Heartland Institute. From the latter, Senior Fellow Peter Ferrara exulted over the statistics showing the US to be the number one producer of oil and gas in the world and couldn't wait for it to also overtake China and India in coal: 'America has enough coal, the original energy resource fueling the industrial revolution, which built the modern world and modern prosperity, to last another 500 years.'[43] Coal to the boilers for another 500 years. The White House used only slightly less exuberant figures: it gloated over data placing the world's second largest reserves of coal in the US and basked in a report saying that at least 60 per cent of all the world's oil pumped between 2018 and 2023 would come from its territory.[44]

The backers ticked off the items on their wish lists. In January 2018, the *New York Times* revealed that Robert Murray had served the newly elected president with an 'action plan' detailing sixteen requests, most of which were either fulfilled or on their way to being so.[45] The Heartland Institute handed him thirteen recommendations. In August 2018, it was still waiting for him to dismantle the EPA as

---

41 Tara Law, '"Freedom Gas?" "Molecules of U.S. Freedom?" Trump's Energy Department Unveils New Nicknames for Natural Gas', *Time*, 29 May 2019.

42 Brad Plumer, 'We Fact-Checked President Trump's Dubious Claims on the Perils of Wind Power', *New York Times*, 3 April 2019; Bess Levin, 'Certified Moron Donald Trump Thinks Wind Turbines "Cause Cancer"', *Vanity Fair*, 3 April 2019.

43 Peter J. Ferrara, 'Trump's Policies Bring Not Just Energy Independence, but Energy Dominance', *Investor's Business Daily*, 12 March 2018. At its forty-fifth annual meeting in August 2018, ALEC held a workshop on the theme 'Energy Dominance: Exploring the Trump Energy Agenda'. ALEC, August 2018.

44 The White House, 'President Donald J. Trump Unleashes America's Energy Potential', 27 June 2017; The White House, '"US Will Be the World's Largest Oil Producer by 2023, Says IEA"', 5 March 2018.

45 Lisa Friedman, 'How a Coal Baron's Wish List Became President Trump's To-Do List', *New York Times*, 9 January 2018.

such, end 'climate profiteering' in the form of every manner of direct and indirect subsidy to wind and solar and, crucially, repeal the so-called Endangerment Finding from 2009, in which the EPA classified CO_2 as a danger for public health, anathema for carbon vitalists. The Institute also clamoured for Trump to nullify fuel-efficiency standards for cars, but that wish was fulfilled the very same month.[46] One of the last decisions Obama made in office was to review and tighten those standards. Not only the Heartland Institute, but the giants of the auto industry – GM, Ford, Fiat Chrysler – lobbied hard against it.[47] When Trump made clear that the standards would be frozen, meaning that new cars would not have to become any more efficient in fuel consumption over the next half-decade, it was one of his most consequential counter-reforms. It would spur another gasoline binge on US roads. One team of researchers estimated that the ensuing growth in annual emissions by the year 2035, relative to a scenario where the standards would indeed have been tightened, will equal the *combined* annual emissions of the world's seventy smallest countries in 2018. It will also exceed the aggregate from some rather populous nations, such as Nigeria or Bangladesh.[48] And the Heartland and the auto giants cheered.

The other recommendations had already been implemented, including the discontinuation of climate research and, of course, withdrawal from Paris. Just one more remained: creating a 'President's Council on Climate Change' charged with spreading the faith of the denialist ISA to the world. This wish appeared to be nearly fulfilled in the spring of 2019, when the Trump administration did indeed prepare to set up its own climate panel to rival that of the UN. It would be headed by William Happer. A Princeton physicist with a good claim to being the most over-the-top carbon vitalist, funded by Peabody Energy and other such companies, he had argued that the abuse suffered by CO_2 equals that endured by Jews under Hitler: 'Let

---

46 Coral Davenport, 'Trump Administration Unveils Its Plan to Relax Car Pollution Rules', *New York Times*, 2 August 2018.
47 Sandra Leville, 'Exclusive: Car Makers among Key Opponents of Climate Action', *Guardian*, 10 October 2019.
48 Trevor Houser, Kate Larsen, John Larsen et al., 'The Biggest Climate Rollback Yet?', Rhodium Group, 2 August 2018.

me point out that if you have a well-designed coal plant, what comes out of the stack of the plant is almost the same thing that comes out of a person's breath.' It was John Bolton who recruited him to the task of finishing off climate science; both men had received money from the hedge-fund capitalist Robert Mercer, whose carbon vitalism was accompanied by the belief that African Americans were better off before they were granted civil rights.[49]

With carbon vitalism filling the halls of the White House, the US now came closer to the position that global heating was a good thing. In May 2019, Mike Pompeo from Koch praised the disappearance of ice from the Arctic. That region, he said,

> is at the forefront of opportunity and abundance. It houses 13 percent of the world's undiscovered oil, 30 percent of its undiscovered gas, an abundance of uranium, rare earth minerals, gold, diamonds, and millions of square miles of untapped resources, fisheries galore. Steady reductions in sea ice are opening new passageways and new opportunities for trade.[50]

When Interior Secretary Bernhardt was asked about the news that the atmospheric CO_2 concentration had passed 415 ppm, he said he hadn't 'lost sleep' over it.[51] Five hundred years of coal burning; joy over the melting Arctic; no sleep lost: one would be forgiven for the impression that these men on some level of their psyche actually *desired the destruction of the planet*. In the second part, we shall return to the question of what kind of drive this might represent.

---

49 Lawrence Carter and Maeve McClenaghan, 'Exposed: Academics-for-Hire Agree Not to Disclose Fossil Fuel Funding', *Greenpeace Energy Desk*, 8 December 2015; Gahame Redfearn, 'Physicist William Happer, the "Unmoored" Climate Science Denier Heading a White House Climate Probe', *DeSmog UK*, 20 February 2019; Bess Levin, 'Trump's New Climate Czar: Carbon Dioxide Has Been Treated Just like "Jews under Hitler"', *Vanity Fair*, 20 February 2019; Coral Davenport and Mark Landler, 'Trump Administration Hardens Its Attack on Climate Science', *New York Times*, 27 May 2019; Jane Mayer, 'The Reclusive Hedge-Fund Tycoon behind the Trump Presidency', *New Yorker*, 17 March 2017.

50 Jennifer Hansler, 'Pompeo: Melting Sea Ice Presents "New Opportunities for Trade"', CNN, 7 May 2019.

51 Emily Holden, 'Trump's Interior Secretary: I Haven't "Lost Sleep" over Record CO_2 Levels', *Guardian*, 15 May 2019.

All of that desire would probably not be satisfied. As in Poland or Norway, there were factors preventing some of the dreams from coming true; five hundred years of unchecked coal burning, for instance, was impossible by definition (the people burning the coal would themselves be burnt before halfway to the bottom). Coal was unlikely to undergo the same renaissance as oil and gas, as even Robert Murray admitted.[52] Its decline during the years of Obama was induced by the falling prices of gas – ironically, thanks to Murray's fellow advisor Hamm and his band of frackers – as well as of renewables, both undercutting the black rock in power generation. Few believed that Trump could do anything to reverse that trend, especially since he would not curtail fracking.[53] But the gutting of the Clean Power Plan did give owners further incentive to keep their existing coal-fired power plants running and squeeze the last drops of profit from all that fixed capital.[54] And five hundred years are not needed to damage the climate system beyond repair: a few decades of extended lifetime for American business-as-usual would be enough.

Regardless of the exact fallout of any of Trump's specific reforms, he ensured that the federal state apparatus became in full thrall to primitive fossil capital – a mighty blockade against any meaningful measures to abate climate breakdown. The territory was secured by the praetorian guard that was never much tempted by greenwashing. In a penetrating analysis, Doug Henwood has explained why the most eminent men – the Koch brothers, Murray, Hamm – all happened to run private companies, rather than publicly traded corporations: inside their own fiefdoms, they did not need to bother with shareholders who might want to put a green tinge on their businesses. They could stick it out to the end, impervious to negative stimuli from science and society.[55] The old company form made for

---

52 Dominic Rushe, 'Top US Coal Boss Robert Murray: Trump "Can't Bring Mining Jobs Back"', *Guardian*, 27 March 2017.

53 Cf. for example Jeff Brady, '"America First" Energy Plan Challenges Free Market Realities', NPR, 7 February 2017.

54 Lisa Friedman and Coral Davenport, 'New E.P.A. Rollback of Coal Pollution Regulations Takes a Major Step Forward', *New York Times*, 20 August 2018.

55 And it allowed them to plough profits back into unceasing expansion, not having to share any dividends with stockholders. Henwood, 'Trump', 119–20; Mayer, *Dark*, 171.

some of the crudely personalised influence in the web surrounding Trump – a narrow phalanx of capitalists who did not, however, operate in antagonism to the rest of the class. US capital might originally have preferred Hillary Clinton over Trump, but after less than two years in power, it had warmed to him. In May 2018, the *Economist* published a summary of 'Why Corporate America Loves Donald Trump': taxes cut, regulations removed, profits soaring, fossil fuels pouring out in ever greater quantities.[56] Primitive fossil capital had, of course, always existed in symbiosis with fossil capital in general, every Ford suckling a Rockefeller, every Bezos a Hamm. Politically as well, the former fraction did the work for the entire capitalist class, in that it insulated it from any potentially traumatic weaning from fossil fuels.

Some continuities with the previous Republican administration could be noted. Leaving Kyoto, voicing denial, axing regulations, cultivating ties to mining and dirty industry: Bush the younger prefigured several of Trump's moves.[57] But his administration was not nearly as retrogressive and predatory and – yes – consistent. During his eight years in power, the rest of the world, led by the EU, experienced the great climate awakening that eventually left the Republicans looking like misfits. Trump was the mightiest bull in a herd that charged ahead even in northern Europe. On the other hand, there were also continuities with the previous Democratic administration that shouldn't be overlooked. It was Obama who oversaw the renaissance of American fossil fuel production, sought to turn the bottomless well into a source of export commodities, lifted a forty-year ban on exporting crude oil – incidentally, in the week after he had returned home from COP21 – approved every single export licence for liquefied fossil gas put before him, expedited the licencing of pipelines, opened waters off the Alaskan coast for drilling and expressed his pride in all of this in terms not dissimilar

---

56 *Economist*, 'Why Corporate America Loves Donald Trump', 24 May 2018; cf. Henwood, 'Trump', 117–20. (But the *Economist* upbraided 'America Inc' for short-sightedly succumbing to the siren songs of Trump.)

57 For example McCright and Dunlap, 'Anti-Reflexivity', 115–21; Dunlap and McCright, 'Climate', 249; Oreskes and Conway, *Merchants*, 213; Powell, *The Inquisition*, 43, 105–6.

from Trump's.[58] 'Over the last three years,' Obama said, vaunting these achievements in his 2012 State of the Union address, 'we've opened millions of new acres for oil and gas exploration' and 'right now – right now – American oil production is the highest it's been in eight years . . . we have a supply of natural gas that can last America nearly 100 years'. (Trump: 'We have nearly 100 years' worth of natural gas'). In Obama's eyes, none of this came into conflict with a simultaneous expansion of renewables. 'This country needs an all-out, all-of-the-above strategy that develops every available source of American energy.'[59] (Obama found it politic to repeat the boast in 2018: 'That was me, people,' he told a cheering audience of oilmen in Houston, pointing to the renaissance.)[60]

Why, then, did primitive fossil capital whine so loudly? Why did Harold Hamm, in the same year as Obama gave that address, write an op-ed in *Forbes* in which he reproached this president and his administration for having 'done everything in their power to stop fossil fuel usage'?[61] When Trump took over, Hamm told Fox Business that he was 'very excited about the fact that we can get back from the brink of extinction and have a future here in America as oil and gas producers'.[62] Trump later reciprocated by telling a gathering of such producers that 'you've gone through eight years of hell, and actually I could say even a little bit more than that'.[63] Extinction, hell, even worse: clearly not what transpired. So why the burning desire to annex the state apparatus? The answer seems to lie in the contradictions of capitalist climate governance: based on adherence to the science, it has regularly aired the pretension to mitigate climate change. In his speech accepting the nomination as the Democratic

---

58 Selby, 'The Trump', 16; Lederer, 'What Is'; John M. Broder, 'US Approves an Initial Step in Oil Drilling near Alaska', *New York Times*, 30 August 2012; BBC News, 'US Spending Bill Lifts 40-Year Ban on Crude Oil Exports', 18 December 2015.
59 Barack Obama, 'President Obama's State of the Union Address', *New York Times*, 25 January 2012.
60 Edward Klump, 'Obama on US Energy Boom: "That Was Me, People"', *E and E News*, 28 November 2018.
61 Hamm, 'Romney's'.
62 Fox Business, 'This Oil Tycoon Is Not Considering a Job in Trump's Cabinet', foxbusiness.com, 1 December 2016.
63 Trump, 'Remarks'.

presidential candidate in 2008, Obama predicted that people 'generations from now' would look back upon his win as 'the moment when the rise of the oceans began to slow and our planet began to heal'.[64] Just like the testimony of James Hansen back in 1988, but two decades deeper into a warming world, this carried the implication and risk of a sharp reduction in the burning of fossil fuels. Or, if Obama had been serious about slowing the oceans' rise and his administration willing and capable of initiating the showdown with fossil capital, extinction *would* indeed have been pending. The capitalists first in the line have always been most acutely aware of this logic, always most jittery about it and always, from the early days of the Global Climate Coalition, most adept at pre-emptive strikes.

But perhaps more importantly, capitalist climate governance of the enlightened kind Obama represented was susceptible to pressure from popular movements. Unlike institutionalised denial, it could be requested to adhere to science in practice as well in the effusion of words. Hence the suspension of Keystone XL in 2015, after the broadest and most persistent campaign ever undertaken by the US climate movement and its allies, and of Dakota Access in 2016 after Standing Rock. These concessions augured ill for primitive fossil capital. They demonstrated the instability of capitalist climate governance and the danger that, under the pressure of mass mobilisation, it might go beyond itself into actual limits on extraction. If Obama effused high-flown rhetoric during his 2008 campaign, only to superintend the oil and gas boom in his first term, he gave in to some of that pressure at the end of his second and veered back towards mitigation, at least in the form of shelving the most egregious expansionary projects.[65] The importance of his late decision on Keystone XL in particular can hardly be overestimated: like the Hansen testimony once did, it sent shock waves through primitive fossil capital, reinforcing a narrative of near-death experience at the hands of this president. That story had been spun since 2008 by the API, Peabody, the Koch brothers

---

64 Barack Obama, 'Final Primary Night: Presumptive Democratic Nominee Speech, St. Paul, Minnesota, June 3, 2008', obamaspeeches.com.

65 See Andrew Cheon and Johannes Urpelainen, *Activism and the Fossil Fuel Industry* (Abingdon: Routledge, 2018), for example 8–9, 13, 68, 74, 83–9.

and their sort, but this was a rebuff with a qualitative difference.[66] In its wake, Andrew Wheeler vented alarm on behalf of his clients:

> The environmental organizations are going to be emboldened by this and going to go after more projects ... On the industry side, they're going to have to step it up. People are going to take [a] step back and reassess what went wrong and what they could have done differently. Anybody representing any energy or infrastructure project is going to have to take a look at the tactics. Do you politicize it more or try to take politics out of it?[67]

Evidently 'the industry side' did step it up, reassessed the tactics and chose to politicise energy *by jumping on the bandwagon of ethnonationalism*. The nature of the latter was never in doubt.

Spillover, Smokescreen, Misfit?

Donald Trump's career in racism is by now at least as proverbial as that in climate denial. 'Laziness is a trait in blacks', he said some three decades before his presidential campaign.[68] When he launched it, his website contained opinions on a single issue: immigration.[69] From the moment he hurled himself into orbit with his speech on Mexican immigrants as 'rapists', he hailed Americans as members of a white nation. 'Make America Great Again' did not, of course, refer to African or Latin or Muslim or Native American, but to a specifically white America that had lost its greatness – or, 'that space that we once occupied before Mexicans, immigrants, gays, and black presidents became an occupying force', in one decoding of the

---

66   See Lukas Hermwille and Lisa Sanderink, 'Make Fossil Fuels Great Again? The Paris Agreement, Trump, and the US Fossil Fuel Industry', *Global Environmental Politics* (2019), online first; Mayer, *Dark*, 245–6, 334–8; Cheon and Urpelainen, *Activism*, 148–50.

67   Quoted in Emily Kopp and Jesse Rifkin, 'Chevron Hires Sidley Austin', *Politico*, 9 November 2015.

68   Ibram X. Kendi, *How to Be an Anti-Racist* (London: Vintage, 2019), 8.

69   Ashley Jardina, *White Identity Politics* (Cambridge: Cambridge University Press, 2019), 233.

interpellation.[70] A wealth of research has demonstrated that a significant enough segment of the white population felt summoned by this call. For a number of white Americans, their country was a rug being 'pulled out from under them', as Ashley Jardina sums up the mood in *White Identity Politics*, amassing an abundance of evidence from polls, media, surveys: anxious about their privileges, fearful that non-whites were usurping them – including becoming more numerous – they were, by the time Trump stepped onto the stage, ready to defend themselves.[71] In another sign of transatlantic convergence, much of the enmity was directed against immigration. Whites were far more opposed to it than blacks or Latinos, although not if the immigrants were reliably white – 'We should have more people from places like Norway', as Trump would say.[72] He was the right president for racially conscious whites, the man to redress the affronts they had put up with for so long.

Having a black president was not the least of these affronts. For the first time in history, the highest office had been lost to the white man. It concentrated the perceived loss of dominance at one point, as evident in the Tea Party movement – replete with Obama effigies and posters of Obama as witch doctor with a bone in his nose – and birtherism, the claim that this black man must be an illegitimate foreigner in the White House. It was in this moment the political phenomenon of Donald Trump arose: first as birtherist, then as

---

70 William Major, 'Donald Trump's Wall of Whiteness', in Jeremy Kowalski (ed.), *Reading Donald Trump: A Parallax View of the Campaign and Early Presidency* (Basingstoke: Palgrave Macmillan, 2018), 61. Major should have also mentioned Muslims, as Islamophobia was a most prominent theme in Trump's election campaign: see Khaled A. Beydoun, *American Islamophobia: Understanding the Roots and Rise of Fear* (Oakland: University of California Press, 2018), 9, 12, 45, 190–2; Lean, *The Islamophobia*, 7. See further for example Lawrence D. Bobo, 'Racism in Trump's America: Reflections on Culture, Sociology, and the 2016 US Presidential Election', *British Journal of Sociology* 68 (2017): 97; Joshua Inwood, 'White Supremacy, White Counter-revolutionary Politics, and the Rise of Donald Trump', *Environment and Planning C: Politics and Space* 37 (2019): 580.

71 Jardina, *White*, 267. Cf. for example Brian F. Schaffner, Matthew MacWilliams and Tatishe Nteta, 'Understanding White Polarization in the 2016 Vote for President: The Sobering Role of Racism and Sexism', *Political Science Quarterly* 133 (2018): 9–34; Diana C. Mutz, 'Status Threat, not Economic Hardship, Explains the 2016 Presidential Vote', *PNAS* 115 (2018): E4330–9.

72 Jardina, *White*, 161–3, 184, 233–4; Kendi, *How To*, 62.

candidate for obliterating all traces of Obama. 'For Trump', Ta-Nehisi Coates argued in a much-read essay, 'it almost seems that the fact of Obama, the fact of a black president, insulted him personally.' Hence 'replacing Obama is not enough – Trump has made the negation of Obama's legacy the foundation of his own.' The idea of being white 'is the idea of not being a nigger' and so it has always been, but Trump was the first occupant of the White House to arrive

> in the wake of something more potent—an entire nigger presidency with nigger health care, nigger climate accords, and nigger justice reform, all of which could be targeted for destruction or redemption, thus reifying the idea of being white.[73]

With regards to the 'nigger climate accords', Coates's essayist intuitions have been substantiated with rigorous analysis of attitudinal data by Salil D. Benegal: in 2008, the recognition of climate change as a reality and cause of concern suddenly plunged, but only among white Americans. Blacks retained it in full. Why? Benegal attributes it to a 'spillover of racialisation', a process by which the primary perception of Obama as a black and thus anti-white president sloshed onto secondary fields of policy, such as climate.[74] If global warming was something Obama believed in, it must be a hoax to disinherit the whites.

Of a piece with this edgy atmosphere – or paranoia, if you will – was the rhetoric coming from primitive fossil capital and the denialist ISA about Obama killing coal, extinguishing oil and causing general hell.[75] Needless to say, neither whites nor fossil fuels *actually lost* their dominance under his eight years; had they done so, had they undergone the feared replacement and disappeared into the margins, no vengeance on the ensuing scale would have been possible. But the privileges of both were – or were at least felt to be, in some minimal sense of the term – in question. Hence the fusional reaction: under Trump, the combustion of fossil fuels came to connote propping up

---

73   Ta-Nehisi Coates, 'The First White President', *Atlantic*, theatlantic.com, October 2017. Cf. Jardina, *White*, 257–8.

74   Salil D. Benegal, 'The Spillover of Race and Racial Attitudes into Public Opinion about Climate Change', *Environmental Politics* 27 (2018): 733–56.

75   Ibid., 738, 750–1.

the white nation. Coates argued that Trump was 'the first white president', since no one had previously come to the highest office to negate an African American predecessor. On the same logic, he was the first president to fill up white skin with black fuel.

How were the two moments related during the presidency? Laura Pulido and her colleagues sieved the output from Trump's first year in office – tweets, speeches, appointments, executive orders, memoranda, budget decisions – and ordered them into 195 environmental and 354 racial 'events'. While the latter category was larger by number, it had less material substance. The bulk were rhetorical excrescences – in the nature of Trump talking about 'shithole countries', tweeting that congresswomen of colour should 'go back' to such countries, calling neo-Nazis 'very fine people' and immigrants 'animals': the whole unceasing discharge – whereas the environmental events tended to be earthy actions, like opening federal land to coal mines. The transfer of resources to private hands came with little presidential flourish. It proceeded quietly, methodically, indicating 'some level of preplanning' – the ground prepared by ALEC, the Heartland, Murray et al. – while the racism was enacted as spectacle on a whim.[76] The more outrageous the Trump show, the more spellbound the media and the less attention paid to the environment. In this analysis, the performance of whiteness worked as a smokescreen.

But it could be argued that the two moments rather interpenetrated all the way down the line, up the sphere of virtual palaver. In Trump's world, the main global threat to the supremacy of white America was China, and it was in relation to China that he hammered out his view on climate, as in his single best-known pronouncement on the matter: 'The concept of global warming was created by and for the Chinese in order to make US manufacturing non-competitive', a tweet repeated frequently with tiny variations.[77] Just as in the True

---

[76] Laura Pulido, Tianna Bruno, Cristina Favier-Serna and Cassandra Galentine, 'Environmental Deregulation, Spectacular Racism, and White Nationalism in the Trump Era', *Annals of the American Association of Geographers* 109 (2019): 520–32; quotation from 526.

[77] Donald Trump's Twitter account, 6 November 2012; see further Matthews, 'Donald'. On China as the global threat to white America in Trump's and his supporters' worldview, see for example Mutz, 'Status'.

Finns cartoon and innumerable other statements on the far right, the signal from the climate was here attributed to a racial other trying to rob whites of their wealth. The function of such communication was not merely to maintain the pandemonium and 'deflect attention' from environmental destruction, but rather to build the case for the latter and lend it ideological content.[78]

Conversely, some of the private proprietors that benefitted most from Trump's silent deregulations had their own legacy of racist swagger. In her study of the Koch brothers, *Dark Money: The Hidden History of the Billionaires behind the Rise of the Radical Right*, Jane Mayer reconstructs their education from father Fred, who in 1960 published a pamphlet claiming that black commotion was part of the Communist plan to take over America. One of the founders of the John Birch Society, Fred Koch sponsored a colleague to run for president on a platform calling for racial segregation (and total abolition of income taxes) in the year 1968. The young brothers further learned from the Freedom School, which taught that it was a mistake to forcibly terminate slavery – people should have the right to sell their bodies if they wish – and introduced them to neoliberal theory. David Koch went on to explain Obama's vile radicalism as a product of his foreignness and more specifically his Africanness. 'His father was a hardcore economic socialist in Kenya.'[79] These were not errant cases. One of the Koch brothers' closest cronies was Corbin Robertson Jr., owner of the largest private coal cache in the nation: he had inherited his wealth from an oil company set up by his grandfather, a member of the fallen gentry of the Confederacy, who hated Roosevelt's 'Jew Deal' and sought to form a third party for 'the restoration of the supremacy of the white race'.[80] How far back do these entanglements go? What is their significance for events in the early twenty-first century? We shall argue that white skin and black fuel are more deeply interfused than either the spillover or the smokescreen model – as much as they capture parts of the dynamics – would seem to allow.

---

78 Pulido et al., 'Environmental', 522.
79 Mayer, *Dark*, 226; further 50–7.
80 Ibid., 249–50.

There is another reason to suspect greater depths beneath Trump: in his laundry list of Obama achievements to reverse, as written up by Ta-Nehisi Coates, only one item resonated with ongoing attempts to shore up whiteness on its continent of origin. 'Nigger health care' and 'nigger justice reform' had no prominent places on the European far right. But 'nigger climate accords' and their equivalents had: Trump emerged in a conjuncture when the themes of race and climate traversed the Atlantic. This has implications for our assessment of his historical import. Even as perceptive an observer as Doug Henwood maintains that 'no candidate had ever been as ideologically un-anchored as Trump'; but it would be more correct to say that he was the first candidate to become firmly anchored in a new ideological constellation sweeping through the global North.[81] Dylan Riley has built a theory of Trump as a 'patrimonial misfit', meaning that he entered the White House as the father of a household and counted on personal loyalty as the adhesive of the apparatus. Since the US had the most advanced capitalist state in history, the project was a contradiction in terms: this state simply could not be governed as a private domicile. Hence the erraticness and weakness of Trump's rule, the dismissals from and backstabbing within the administration, the jerks of a bureaucratic machinery suddenly subjected to a personage who treated it as a family revolving around himself. 'Flukey in origin', this arrangement had no 'staying power', no ideology or cause, no moorings in any capitalist base layer. A sorry, lonely real-estate mogul, Trump suffered the absence of 'any organic connection to the class of which he is part' and tried in vain to make up for the lack with his all-too-few buddies.[82]

In light of the above, this must be deemed an astonishing misjudgement. Only the conviction that matters of energy are utterly immaterial to a capitalist economy could buffer it against basic data on the campaign and administration of Donald Trump. In the *longue durée* of global warming, none of it was a fluke. If that is recognised

---

81 Henwood, 'Trump', 104.
82 Dylan Riley, 'What Is Trump?', *New Left Review* series 2, no. 114 (2018), quotations from 28, 22.

as a relevant context, it would be more accurate to say that Trump properly shouldered the burden of the American empire by elevating trends in world capitalism to the highest office and giving vent to the associated ideas. It would also be more accurate to follow Nicos Poulantzas and say that in the homeland, the state of Trump reorganised the dominant classes under the leadership of one fraction – namely, primitive fossil capital – and drastically shrank whatever relative autonomy the state had possessed vis-à-vis the latter. Poulantzas intuited that such a process would entail the decaying of parliament and other institutions of representative democracy and the rise of an ever more self-willed and capricious executive. But he also thought it would stir up tensions with other class fractions that would not – his own example – indulge big oil.[83] This did not transpire.

The leadership of the Murrays and the Hamms induced no bourgeois infighting. One reason might be that renewables, or the flow of energy, had yet to support a well-endowed, well-organised counterweight to primitive fossil capital – indeed, such a fraction was nowhere in sight. Panels and turbines had no Koch brothers. Given that, *qua* fuels, sun and wind cannot be extracted for profit, they might not ever sustain anything like the fraction that led US capitalism to class-wide acclaim. In fact, the trifling political donations from firms in the renewable energy sector went increasingly to the Republicans under Trump, demonstrating that even they enjoyed him.[84] Clearly, they lacked any counter-hegemonic ambition to lead an actual shift from one type of energy source to another.

Below the froth and show of Washington politics in the first three years of Trump's reign, consonance, not discord, characterised the inner relations of the US capitalist class. 'The capital accumulation process now directly dictates the rhythm of state activity', writes

---

83   Poulantzas, *State*, 127–8, 137, 172, 181–2, 224. In terms of the classical Poulantzas-Miliband debate, a friend of the latter would perhaps point out that the formation of pro-fossil cabinets in the US, Norway, Poland and elsewhere rather demonstrantes the importance of the personnel of the state apparatus.

84   Nichola Groom, 'Clean Energy Sector Swings Republican with US Campaign Donations', Reuters, 2 May 2018; Julia Pyper, 'Solar and Wind Companies Spend More on Republican Candidates Than Democrats', *Green Tech Media*, 3 May 2018.

Poulantzas: and the accumulation was one of ever-so-ascendant fossil capital.[85] But this configuration was emphatically not unique to the US. We have seen similar things in Poland and Norway, and it is not unlikely that they would evolve in a Germany under the AfD or a Netherlands under the FvD or PVV. Yet the US is still observed with a degree of myopia. 'You have this enormous discrepancy between the White House and, essentially, everyone else', claimed top climate scientist Johan Rockström in 2018.[86] Methodological nationalism cannot register the phenomenon, let alone plumb its depths. Instead our many cases suggest that *in the climate emergency, the far right is a vehicle in which primitive fossil capital takes a seat, driving the state away from any limitations on fossil fuel use*, and the rest of fossil capital (in the US, even non-fossil capital) comes along rather blithely. This clearly has not happened everywhere and might well be prevented from repetition. But Donald Trump had not even completed the first half of his first term before something similar unfolded in another giant country in the Americas.

Seize the Wealth While There Is Time

In the presidential election in Brazil in 2018, Jair Bolsonaro was the favourite candidate of domestic and global capital. Every single business association backed him in the second round; the *Wall Street Journal* endorsed him as the 'Brazilian Swamp Drainer', a southern duplication of Trump; after his victory, stock markets bounced with enthusiasm.[87] Bolsonaro had by then become a household name for his statements in support of torture and military dictatorship and mass killings of dissidents, threats to close down NGOs, advocacy of lax gun laws, loathing of LGBT people, blatant misogyny and

---

85 Poulantzas, *State*, 169.
86 Mark Landler and Coral Davenport, 'Dire Climate Warning Lands with a Thud on Trump's Desk', *New York Times*, 8 October 2018.
87 Jeffery R. Webber, 'A Great Little Man: The Shadow of Jair Bolsonaro', *Historical Materialism*, historicalmaterialism.org, 28 August 2019; *Wall Street Journal*, 'Brazilian Swamp Drainer', 8 October 2018; Martin Baccardax, 'Brazil Stocks Hit Record High Bolsonaro Sweeps to Power amid Pro-Business Agenda', *The Street*, 29 October 2018.

contempt for marginalised communities of Afro-Brazilians.[88] He also had a PowerPoint government plan waxing lyrical about entrepreneurship, Brazil's 'true values' (the conservative ones) and slim government. It contained not a word on the protection of climate or forests or anything else related to the environment: not a minus for the *Wall Street Journal* and its ilk.[89] But he did have things to say with bearing on such matters.

When Bolsonaro prepared for the installation of his government, he fulfilled several of the expectations for a tropical Trump: for instance, by appointing Roberto Castello Branco as the CEO of state-owned oil company Petrobras. Until 2014, this man was the CEO of Vale, a multinational corporation producing more nickel and iron ore than any other in the world, as well as plenty of coal from mines in Brazil, Australia and Mozambique. Bolsonaro's PowerPoint plan included the assessment that the energy sector of the country 'needs a liberal shock'.[90] After the election, he selected a clique of ultra-neoliberals trained in the Chicago school of economics to take the rein of the government, Castello Branco being one of them. As the director of Petrobras, he would not, he said upon his appointment, go for all-out privatisation. Instead the company should sell off some assets and focus on the core activities of oil exploration and production and speed those up.

Petrobras had been at the forefront of deep-water drilling over the past decades. Castello Branco aimed to go further and maximise the potential of the so-called pre-salt reservoirs, or oil embedded under salt and rock layers deep in the sea off the Brazilian coast, expensive to reach but in enormous supply. In a declaration of intent noteworthy for its candid expression of the outlook of his class fraction, Castello Branco said: 'We need to seize this wealth while there is

---

88   The best compilation is Sean Purdy, 'Here's What Jair Bolsonaro Thinks', *Jacobin*, 28 October 2018; see also for example Eliane Brum, 'How a Homophobic, Misogynist, Racist "Thing" Could Be Brazil's Next President', *Guardian*, 6 October 2018; Sam Meredith, 'Who Is the "Trump of the Tropics?": Brazil's Divisive New President, Jair Bolsonaro – in His Own Words', CNBC, 29 October 2018.

89   Bolsonaro 2018, 'O Caminho da Prosperidade: Proposta de Plano de Governo', docs.google.com.

90   Ibid., 71.

time. In some decades, oil will lose the relevance it has today.'[91] He reiterated this position shortly after the cabinet was sworn in: Petrobras would not be fully privatised, but its monopoly – something Castello Branco considered 'inadmissible in a free society' – should be broken up. Above all, Petrobras was now assigned the task of cementing Brazil's position as the world leader in the pre-salt field and seize the wealth at maximum speed.[92] (Less than three weeks later, Vale, Castello Branco's old company, had a brief moment of fame when some three hundred people lost their lives after a tailing dam at an iron mine collapsed and sent out a torrent of mudflow.)

Brazil does not carry a heavy climatic weight only in its very considerable holdings of fossil fuels. It also has sovereignty over 67 per cent of the world's tropical rainforests. Left intact, such forests store immense amounts of carbon. But from the late 1940s to the end of the century, Brazil underwent agricultural 'modernisation' in the form of Green Revolution technologies, a mass exodus of small farmers, the spread of ranching and, finally, an opening of the soil to foreign investment, and intrinsic to these waves of capitalist incursion into the interior was the clear-cutting of vast tracts of Atlantic and Amazonian rainforest. Deforestation in the Amazon became particularly devastating from the mid-1960s onwards, when the military dictatorship – so admired by ex-paratrooper Bolsonaro – forcibly accelerated 'development'. In later decades, the chainsaws and bulldozers were driven by the expansion of the soy and cattle frontiers into the hinterlands, fuelled, in turn, by speculation in commodities and investments flowing into agrobusiness.[93]

---

91 Quoted in Gram Slattery, Rodrigo Viga Gaier and Marta Nogueira, 'New Petrobras CEO Rules Out Privatization, Wants Non-Core Asset Sales', Reuters, 19 November 2018; see further for example Mariana Schreiber, 'Quem serão os quatro homens fortes no governo Bolsonaro', BBC Brasil, 28 October.

92 Marta Nogueira and Rodrigo Viga Gaier, 'Novo CEO da Petrobras quer acelerar produção e diz que parcerias são bem-vindas', Reuters, 3 January 2019.

93 For these processes, see for example Claiton Marcio da Silva, *De agricultor a farmer: Nelson Rockefeller e a modernização do agricultura no Brasil* (Guarapuava: Editora Unicentro, 2015); Phillip Fearnside, 'Soybean Cultivation as Threat to the Environment in Brazil', *Environmental Conservation* 28 (2001): 23–38; Phillip Fearnside, 'Deforestation of the Brazilian Amazon', in Hank H. Shugart, Jr. (ed.), *Oxford Research Encyclopedia of Environmental Science* (New York: Oxford University Press, 2017), oxfordre.com/environmentalscience; Liz-Rejane Issberner and Philippe Léna (eds),

The Amazon received a respite from 2005 to 2012, the heydays of the Lula and Dilma regimes. Deforestation decelerated markedly in those years, thanks to fetters on the soy and cattle frontiersmen, tougher monitoring, enforcement of laws against illegal cutting, the designation of new protected areas and a host of other measures directed from the central state. Some forests even rebounded. In 2012, the rate of deforestation was 84 per cent lower than the 2004 peak; as a result, total emissions had gone down by some 40 per cent.[94] Few countries had ever achieved a similar feat. But in 2012, the capitalist crisis from four years prior washed ashore in Brazil in the shape of a collapse in commodity prices, ending the social peace and sending investors back into the Amazon to recuperate their profits. Deforestation picked up further speed after the institutional coup against Dilma in May 2016. The right-wing interim government immediately set about lifting fetters and hastening the shift of the tropical rainforests of Brazil – and thereby of Latin America, and the world as a whole – from sinks to net sources of carbon, losing more into the atmosphere than they draw down.[95] That was the overture to the rise of Jair Bolsonaro.

'There is still space for deforestation in the Amazon', said the swamp drainer.[96] During his campaign, he vowed to relax law

---

Brazil in the Anthropocene: Conflicts Between Predatory Development and Environmental Politics (Abingdon: Routledge, 2017); Maria Luisa Mendonça and Fábio Teixeira Pitta, 'International Financial Capital and the Brazilian Land Market', *Latin American Perspectives* 45 (2018): 88–101.

94 Doug Boucher, Sarah Roquemore and Estrellita Fitzhugh, 'Brazil's Success in Reducing Deforestation', *Tropical Conservation Science* 6 (2013): 426–45; Joana Castro Pereira and Eduardo Viola, 'Catastrophic Climate Risk and Brazilian Amazonian Politics and Policies: A New Research Agenda', *Global Environmental Politics* 19 (2019): 95.

95 A. Baccini, W. Walker, L. Carvalho et al., 'Tropical Forests Are a Net Carbon Source Based on Aboveground Measurements of Gain and Loss', *Science* 358 (2017): 230–4; Jeff Tollefson, 'Brazil's Lawmakers Renew Push to Weaken Environmental Rules', *Nature* 557 (2018): 17; Otavio Cavalett, 'From Political to Climate Crisis', *Nature Climate Change* 8 (2018): 663–4; Pedro R. R. Rochedo, Britaldo Soares-Filho, Roberto Schaeffer et al., 'The Threat of Political Bargaining to Climate Mitigation in Brazil', *Nature Climate Change* 8 (2018): 695–8. For a characteristically majestic account of the coup against Dilma and the rise of Bolsonaro, see Perry Anderson, 'Bolsonaro's Brazil', *London Review of Books* 41, no. 3 (2019): 11–22; also in *Brazil Apart: 1964–2019* (London: Verso, 2019), ch. 5.

96 Quoted in Felipe Milanez, 'Bolsonaro Calls for Carnage and Environmental Holocaust in Brazil', *Entitle Blog*, 23 October 2018.

enforcement in the Amazon, disburse licences to entrepreneurs, open indigenous territories to mining, expel environmental NGOs such as Greenpeace and WWF from the country and essentially hand over management of the rainforests to the *bancada ruralista*, the staunchly anti-environmental political bloc representing large landowners and capitalists active in agriculture and livestock. This agribusiness fraction wholeheartedly backed Bolsonaro – not for his famed promise to restore 'law and order', but for his pledge to liberate agribusiness and mining from 'an industry of fines' and ensure a free-for-all on the commodity frontiers.[97] As his minister of agriculture, Bolsonaro chose Tereza Cristina Dias, known colloquially as 'the Muse of Poison' for having sponsored a bill facilitating the approval of harmful chemical pesticides. She also had a history of commercial partnerships with JBS, the largest meat-processing company in the world.[98] With her in charge of Brazilian soil, Bolsonaro clinched his alliance with the *bancada ruralista* – not to forget 'the mining community', which, according to a *Bloomberg* dispatch, was 'salivating' over the new president.[99] In a separate but related move, he promised to criminalise the Landless Workers' Movement, or Movimento dos Trabalhadores Rurais Sem Terra (MST), so as to uphold 'the sanctity of private property'.[100] Even before Bolsonaro took power, in the first three months of the election campaign the rate of deforestation in the Amazon spiked by

---

97 Fabiano Maisonnave, 'Amazon at Risk from Bolsonaro's Grim Attack on the Environment', *Guardian*, 9 October 2018; Sarah Mota Resende, '"No que depender de mim, não tem mais demarcação de terra indígena", diz Bolsonaro a TV', *Folha de S. Paulo*, 5 November 2018.

98 Bruna de Lara, 'The Corruption Cabinet', *The Intercept*, theintercept.com, 9 December 2018; Tatiana Ramil, 'Tereza Cristina diz que "cumpriu legislação" ao conceder incentivos fiscais à JBS', Reuters, 19 November 2018; Filipe Matoso, Lucas Salomão and Yvna Sousa, 'Ruralistas indicam, e Bolsonaro anuncia Tereza Cristina como ministra da Agricultura', *Globo*, 7 November 2018; Cristina Boeckel, 'Bolsonaro recebe representantes do agronegócio e defende agricultura familiar', *Globo*, 24 October 2018.

99 David Biller and R. T. Watson, 'Bolsonaro's Love Affair with Mining Aims at the Amazon's Treasures', *Bloomberg*, 24 October 2018.

100 On MST, see for example Sue Branford and Jan Rocha, *Cutting the Wire: The Story of the Landless Movement in Brazil* (London: Latin American Bureau, 2002); Miguel Carter, *Challenging Social Inequality: The Landless Rural Workers Movement and Agrarian Reform in Brazil* (Durham: Duke University Press, 2015).

50 per cent, as fortune-hunters jumped the gun and converted forest to pasture.[101]

There was a racial dimension to this whole enterprise that Bolsonaro did nothing to conceal. 'Minorities have to bend down to the majority', he declared in a speech in 2017, referring to non-white peoples inhabiting the Amazon. They faced the choice to 'either adapt or simply vanish'.[102] Slated for such disappearance were indigenous populations and *quilombolas*, or descendants of escaped slaves, also known as 'maroons'. Both groups inhabited lands in the Amazon set aside for them by previous governments, much to the chagrin of Bolsonaro and his followers. Here was a cardinal principle of the Brazilian far right: backward, savage communities have to be subordinated to the modern, developing nation, their resources subjected to maximum extraction.

To the fossil fuels and the forests must be added Brazil's role in international climate politics. It is the country where it all started, back in 1992, and during the very first COP summits, Brazilian delegations advocated a revolutionary principle for mitigation: industrialised countries should reduce their emissions in proportion to their *historical* responsibility for rising temperatures. The burden for phasing out fossil fuels should fall on those that have burnt the most – meaning the US and Western Europe (with a special place in hell and extra-large quota for the country that invented the fossil economy: the United Kingdom). Failure to honour the binding national commitments must result in financial penalties. The fines would then be redistributed to developing countries, free for the time being from any obligations to cut emissions. This became known as 'the Brazilian proposal'. It was thrown out at Kyoto in 1997 because the industrialised countries considered it unfair to them; by any defensible ethical standard, it is one of the soundest ideas ever put forward in the UN

---

101 Fabiano Maisonnave, 'Bolsonaro's Deforestation of the Amazon Has Already Begun', *Climate Home News*, 14 November 2018.

102 'Jair Bolsonaro Diz Que a Minoria Tem Que Se Adequar a Maioria 10/02/17', YouTube, 15 February 2017. On race in Brazil, see for example Mônica Grin, *"Raça": Debate Público no Brasil* (Rio de Janeiro: Mauad Editora Ltda, 2010); Lilia Moritz Schwarcz, *Nem preto nem branco, muito pelo contrário: Cor e raça na sociabilidade brasileira* (São Paulo: Claro Enigma, 2012).

climate negotiations and stands, to this day, as a monument to what could have been achieved.[103]

Ironically, the Brazilian proposal was tabled by negotiators from the liberal government of Fernando Henrique Cardoso.[104] In 2009, at the height of his powers, Lula played a rather different role. At COP15 in Copenhagen, Barack Obama provoked nearly universal shock when, at the last minute of the drawn-out summit, he proposed that the principle of binding commitments be ditched in favour of voluntary pledges of no legal consequence. The COP ended in disgraceful acrimony, as all developing countries rejected the flagrant attempt to relieve mega-emitters of any duties to cut back – all but China, India, South Africa and Lula's Brazil. In taking that position, the Brazilian government made a decisive contribution to the splintering of the global South as a front in climate diplomacy.[105] By the time of COP21 in Paris, the proposal of 'do what you want if you want to' was accepted as the axiom of capitalist climate governance on a worldwide scale.

And with Bolsonaro, Brazil, as the first major country of the global South, took the next step: into barefaced denial. When Trump quit Paris, Bolsonaro responded with a supportive tweet about 'the greenhouse fables'.[106] He promised to do for Brazil what Trump had done for the US; his sons and political companions expatiated more. Carlos Bolsonaro, a city councillor in Rio de Janeiro, claimed that the earth is currently in the midst of the most extreme period of cooling ever

---

103   For an overview of the proposal and its fate, see Emilio L. La Rovere, Laura Valente de Macedo and Kevin A. Baumert, 'The Brazilian Proposal on Relative Responsibility for Global Warming', in K. A. Baumert, O. Blanchard, S. Llosa and J. F. Perkaus (eds), *Building on the Kyoto Protocol: Options for Protecting the Climate* (Washington, DC: World Resources Institute, 2002), 157–73.

104   This might have had something to do with Cardoso's early career as a Marxist theorist of underdevelopment: see Fernando Henrique Cardoso and Enzo Faletto, *Dependency and Development in Latin America* (Berkeley: University of California Press, 1979 [1971]); Anderson, *Brazil Apart*, 5–9.

105   Ciplet et al., *Power*, 65–89. On Brazil's role in the various stages of the UN negotiations, see further Marieke Riethof, 'Brazil and the International Politics of Climate Change: Leading by Example?', in Michela Coletta and Malayna Raftopoulos (eds), *Provincialising Nature: Multidisciplinary Approaches to the Politics of the Environment in Latin America* (London: Institute of Latin American Studies, University of London, 2016), 89–114.

106   Jair Bolsonaro's Twitter account, 2 June 2017.

and blamed leftists for spreading the contrary idea.[107] Senator Flavio Bolsonaro considered global warming 'a fraud'.[108] Senator Eduardo Bolsonaro, who received more votes than any other lawmaker in Brazilian history, produced a not-so-artful video in which he stands in a snow-covered landscape somewhere in the US, ski hat down to the eyebrows, launching into a diatribe against the 'globalism' of the Paris Agreement and its attempts to force nations such as this cold snowy America down on their knees for no good reason.[109] 'They don't want to let developed countries keep polluting', in his own English translation, but still they give 'the right to pollute [to] undeveloped countries. It searches on a world level [for] a bigger equality between the countries. So the American society would be punished... This world equality search doesn't make any sense', he says, before claiming that all laws introduced in Brazil nowadays come from foreign agents such as George Soros.[110] The oracle conveying that latter truth in the video is Olavo de Carvalho, a thinker with a very special status in the fourth most populous nation of the world in the late 2010s, to whom we shall return.

As his minister of the environment, father Bolsonaro picked Ricardo Salles. Amigo of the *bancada ruralista* and the salivating miners, this man had never even visited the Amazon region. He contended that 'the discussion over whether there is or isn't global warming is secondary', an unusually self-effacing twist on denial.[111] When he ran for a seat in Congress in 2018, he designed a poster with himself smiling in the upper left corner, positioned above a sort of diagram or mind map centred on a cluster of bullets. Arrows connected a wild boar and a group of MST activists to those bullets. Both stood as enemies of wealthy landowners and should be dealt

---

107 Carlos Bolsonaro's Twitter account, 6 August 2016.
108 Flavio Bolsonaro's Twitter account, 22 June 2017.
109 Sam Cowie and David Child, 'Brazil: Bolsonaro and Haddad Face Off in Second Round of Election', Al Jazeera, 8 October 2018.
110 'Eduardo Bolsonaro e a farsa do aquecimento global', YouTube, 22 January 2018.
111 Dom Phillips, 'Climate Change a "Secondary" Issue, Says Brazil's Environment Minister', *Guardian*, 10 December 2018. On the minister not visiting the Amazon: Jonathan Watts, 'Deforestation of Brazilian Amazon Surges to Record High', *Guardian*, 4 June 2019.

with accordingly – 'zero tolerance' – a solution that earned Salles the responsibility of overseeing the environment in the single most biodiverse country in the world.[112]

As his foreign minister, Bolsonaro chose Ernesto Araújo, an eccentric career diplomat who had attracted a large, youthful audience through his blog *Metapolítica*, or 'Metapolitics', the term of the Nouvelle Droite. There he wrote, five days after Bolsonaro won the first round:

> The left has appropriated the environmental cause and perverted it to the point of paroxysm over the last 20 years with the ideology of climate change, the climatism. The climatism gathered some data suggesting a correlation between rising temperatures and an increasing concentration of CO_2 in the atmosphere, ignored data suggesting the opposite, and created a 'scientific' dogma that no one can contest ... Climatism is a globalist tactic to scare people and gain more power,

but this seemed among the minor misdemeanours.[113] Eight days before Bolsonaro won the second round, Araújo was back with a new post on how the left, and in particular the Workers' Party of Lula and Dilma, hates freedom and hates the human being. It therefore enforces the criminalisation of

> everything that is good, spontaneous, natural and pure. Criminalisation of the family on charges of patriarchal violence. Criminalisation of private property. Criminalisation of sex and reproduction, justified by the claim that any heterosexual act is rape and every baby is a risk to the planet because it will increase carbon emissions ... Criminalisation of red meat. Criminalisation of air conditioning. Criminalisation of beauty. Criminalisation of all Western thinkers since Anaximander. Criminalisation of oil and

---

112 Olivia Rosane, 'Brazil's New Environment Minister Is Bad News for the Amazon and the Climate', *EcoWatch*, 10 December 2018. On biodiversity in Brazil, see for example UN Environment, 'Megadiverse Brazil: Giving Biodiversity an Online Boost', 28 February 2019.

113 Ernesto Araújo, 'Sequestrar e perverter', *Metapolítica*, 12 October 2018.

any cheap and efficient energy. Criminalisation of the existence of the human being on earth,

and the list went on, enumerating everything Araújo would set free and protect as foreign minister.[114] His words resounded through some camps. At COP24, Brazil retracted its earlier promise to host the next COP, in a first step of disengagement. Marc Morano thanked the heavens again: 'There are countries now joining in, questioning the UN, Brazil announcing that they're not going to host a summit and that they think global warming is a Marxist plot – these are huge developments.'[115] We shall revisit the origins of these ideas.

As his vice president, Bolsonaro recruited General Hamilton Mourão. Himself claiming indigenous descent, the general on one occasion attributed Brazil's problems to the inheritance of 'a culture of privileges from Iberians, indolence from indigenous people, and *malandragem*' – a term denoting a life of idleness, petty crime and dishonesty – 'from Africans', refreshing centuries-old racial classifications.[116] On another, he stated that his grandson was 'a beautiful guy, exemplifying racial whitening', thus making clear what phenotypes he saw as ideal.[117] When the new president was inaugurated on 1 January 2019, one journalist reported an encounter with a black street cleaner, 'one of few black persons among the hundreds of thousands of white supporters who had congregated to praise Bolsonaro.'[118]

---

114 Ernesto Araújo, 'Eu vim de graça', *Metapolítica*, 20 October 2018. As for the making of babies, we could point out here that Jair Bolsonaro accepted the reality of environmental degradation on one count: 'This explosive population growth leads to deforestation. Because you will not grow soy on the terrace of your building or raise cattle in the yard. So we have to have a family planning policy. Then you begin to reduce the pressure on those issues that lead, yes, in my opinion, to global warming, which could be the end of the human species.' Quoted in Maisonnave, 'Amazon at Risk'.

115 'Global Warming Bandwagon Coming Apart, Says Marc Morano', New American Video, YouTube, 13 December 2018.

116 Luiz Raatz and Filipe Strazzer, 'Mourão liga índio à "indolência" e negro à "malandragem"', *Estadão de S. Paulo*, 6 August 2018.

117 Mariana Carneiro , '"Meu neto é um cara bonito, branqueamento da raça", diz general Mourão', *Folha de S. Paulo*, 6 October 2018.

118 Henrik Brandão Jönsson, 'Bolsonaro svors in inför sina trogna kärnväljare', *Dagens Nyheter*, 2 January 2019.

Brazil being a social formation very different from, say, Norway or Italy, immigration was not the funnel of its far right (although Bolsonaro did make clear that any immigrants reaching the country were 'the scum of humanity' and should be dealt with 'by the army').[119] In the landscape of Brazilian racism, Bolsonaro interpellated white people as against *quilombolas* and indigenous populations in particular. But he also drew upon resentment against the modest redistributive gains from rich to poor, white to non-white under Lula and Dilma. 'That these measures granted a novel quotidian presence of Afro-Brazilians and working-class citizens in the heretofore exclusive spatial domains of the rich and the white – shopping malls, universities, and aeroplanes – was an affront to an elite way of life', Jeffery Webber has observed; singular yet similar to trends in the US and Europe, such grievances made up a 'powerful psychosocial component of upper middle class support for Bolsonaro'.[120] His victory was most resounding in the relatively affluent south, where identities and origin myths are often derived from the white European migrants who settled there. He performed worse in the poorer, blacker north. The results of the second round triggered an outpouring of online abuse against black people. One jubilant Bolsonaro supporter dressed up his son as a blackface in chains.[121]

Consequently, Bolsonaro's very first act as president consisted in seizing non-white land. Hours after swearing the oath, he signed an executive order transferring the regulation of two types of Amazonian territory to the Ministry of Agriculture: indigenous land and *quilombos*. Their inhabitants were 'isolated from true Brazil', but 'we will integrate these citizens', Bolsonaro offered as rationale for the decision on Twitter.[122] The year before the election, he claimed to have visited a *quilombo* and found the people there to be 'parasites' not

---

119 Quoted in Pedro Henrique Leal, 'Bolsonaro and the Brazilian Far Right', *openDemocracy*, 24 April 2017.

120 Webber, 'Great'. For an excellent analysis of the determinants of support for Bolsonaro in other strata, see further this text.

121 Kiratiana Freelon, 'Many Afro-Brazilians Are Bracing for the Worst after the Election of a Far-Right, Racist Candidate for President', *The Root*, 30 October 2018.

122 Rubens Valente, 'Governo Bolsonaro paralisa reforma agrária e demarcação de territórios quilombolas', *Folha de S. Paulo*, 8 January 2019; Phillips, 'Climate'.

even 'fit for breeding'.[123] Around the same time, hundreds of *quilombolas* had celebrated newly issued titles to their land, on which they had lived since the days of slavery.[124] On 1 January 2019, these territorial possessions were handed over to the Ministry of Agriculture run by the white 'Muse of Poison'.

Of greater geographical extent, however, were the indigenous lands likewise allocated to the ministry – to all intents and purposes, now an executive committee of the *bancada ruralista*. No more land would be added to reserves; instead, the state began the process of opening parts of the Amazon hitherto off-limits to entrepreneurs. A spate of invasions of indigenous reserves followed. Miners streamed into the vast Yanomami territory in the northern Amazon. Bolsonaro egged the land-grabbers on by proposing to arm ranchers, cut up reserves and allow indigenous people to privately sell their lands.[125] Entirely predictably, the result was another surge in the rate of deforestation: in 2018 and early 2019, loggers and cattle ranchers acting on the mandate of Bolsonaro were responsible for the largest losses of rainforest on earth. From the Ituna Itata reserve, home to some of the last uncontacted peoples, satellite images revealed illegal clearing more than twice as extensive as all that had occurred in the previous sixteen years. By May 2019, the rate of deforestation in the Brazilian Amazon had reached two football pitches every minute, the highest record since monitoring began.[126] By late July, the total area chopped out since Bolsonaro swore the oath was more than double that cleared in the same period in 2017. And now the rainforest was burning.[127]

On 20 August 2019, around two o'clock in the afternoon, the coastal city of São Paolo went dark. Smoke from fires deep in the Amazon

---

123 Leal, 'Bolsonaro'.
124 Phillips, 'Climate'.
125 José Ricardo Wendling, 'Resistência Indígena', *Amazonas Atual*, amazonasatual.com, 24 January 2019; Rubens Valente, 'Invasão Em Terra Indígena Chega a 20 Mil Garimpeiros, Diz Líder Ianomâmi', *Folha de S. Paolo*, 16 May 2019; Cristiane Agostine and Carolina Freitas, 'Bolsonaro promete rever demarcações e quer explorar Amazônia com EUA', *Valor*, 8 April 2019.
126 Damian Carrington, '"Death by a Thousand Cuts": Vast Expanse of Rainforest Lost in 2018', *Guardian*, 25 April 2019; Watts, 'Deforestation'.
127 Herton Escobar, 'Brazilian President Attacks Deforestation Data', *Science* 365 (2019): 419.

had travelled some 3,000 kilometres to snuff out the lights in the metropolis: 'It was as if the day had turned into night', attested one amazed resident.[128] Unlike other wildfires sweeping through forests across the globe in recent years, these were not self-ignited by higher temperatures; hotter though it was, the rainforest would not catch fire on its own. Instead the fires were the work of loggers. To cut down a patch of the Amazon, they would move in with bulldozers and tractors during the wet season, leave the tree trunks on the floor and then torch them during the dry season, starting in July.[129] Early reports in the Brazilian press spoke of an 'explosion' of fires in July 2019.[130] The initial estimates of an increase by 278 per cent over the same period the previous year were later confirmed; by August, the number of fires – some eighty thousand – was indeed three times higher. Although hot and dry conditions can exacerbate blazes and spin them out of control, this year the Amazon had no drought. But it had a deforestation orgy inaugurated by the highest national office.[131] On 6 August, in the midst of the inferno, Bolsonaro complained that '60 per cent of our territory is rendered unusable by indigenous reserves and other environmental questions': this to a gathering of car dealers in São Paolo.[132] Others saw another scenario building up. Before the end of the year, scientists were discussing whether it would take two more years or two decades for the maimed Amazon to reach the tipping point where the rainforest no longer generates enough rain to sustain itself, switches into a savannah and, in the process, regurgitates its colossal stocks of carbon into the atmosphere.[133]

---

128 Danielle Garrand, 'Parts of the Amazon Rainforest Are on Fire – and Smoke Can Be Spotted from Space', CBS News, 22 August 2019.

129 Jos Barlow, Erika Berenguer, Rachel Carmenta and Filipe França, 'Clarifying Amazonia's Burning Crisis', *Global Change Biology* (2019), online first; Herton Escobar, 'Amazon Fires Clearly Linked to Deforestation, Scientists Say', *Science* 365 (2019): 853; Jessie Yeung, 'Blame Humans for Starting the Amazon Fires', CNN, 23 August 2019.

130 André Borges, 'Destamento explode em julho e chega a 2.254 km², um terço dos últimos 12 meses', *Estadão de S. Paolo*, 6 August 2019.

131 Barlow et al., 'Clarifying'; Tom Phillips, '"Worst of Wildfires Still to Come" despite Brazil Claiming Crisis Is under Control', *Guardian*, 28 August 2019.

132 Tom Phillips, 'Bolsonaro Rejects "Captain Chainsaw" Label as Data Shows Deforestation "Exploded"', *Guardian*, 7 August 2019.

133 Dom Phillips, 'Amazon Rainforest "Close to Irreversible Tipping Point"', *Guardian*, 23 October 2019.

Meanwhile on the petroleum frontier, Petrobras prepared for the world's largest expansion of offshore oil and gas production by putting out tenders for seven huge vessels in that business. The state held an auction where oil companies bid for drilling rights. Production climbed steeply already in Bolsonaro's first year. *The New York Times* reported that a 'flood of oil is coming, complicating efforts to fight global warming' and named the sources of additions from fresh fields in the years ahead: in the top, Norway and Brazil.[134] On his way to COP25, Environmental Minister Salles met up with the Competitive Enterprise Institute.[135] Foreign minister Araújo dispatched his diplomats to represent the nation at the annual conference of the Heartland Institute (Myron Ebell was there too).[136] President Bolsonaro cut 96 per cent of the funding to domestic mitigation efforts – why he left the remaining 4 per cent was unclear.[137] Furiously ripping up every measure and institution, every limit and fetter put in place since the Rio summit of 1992, with the Amazon falling and burning, he served the world the most vivid memento mori from the far right so far. But this was only what happened during the first year of his rule.

---

134 Clifford Krauss, 'Flood of Oil Is Coming, Complicating Efforts to Fight Global Warming', *New York Times*, 3 November 2019; *Guardian*, 'What Do We Know about the Top 20 Global Polluters?', 9 October 2019.
135 Dom Phillips, 'Brazil Environment Minister to Meet US Climate Denier Group before UN Summit', *Guardian*, 13 September 2019.
136 Patrícia Campos Mello, 'Governo brasileiro participa de reunião com negacionistas do climate', *Folha de S. Paolo*, 30 July 2019.
137 André Borges, 'Ministério do Meio Ambiente quase zera verba de combate á mudança climática', *Estadão de S. Paolo*, 7 May 2019.

PART II

7
Towards Fossil Fascism?

Sometime after the election of Donald Trump, comparisons with Europe between the two world wars became a cliché of the political conjuncture. Premonitions of fascism filled the air like some heavy smoke, setting off alarms for liberal and leftist observers alike; the prospect of a relapse into the interwar conflagration had haunted the West ever since 1945, but rarely before with such intensity as in the second decade of the twenty-first century. Leaders of the far right dutifully fanned the flames. In one instance, in October 2018, Alexander Gauland wrote a piece for the *Frankfurter Allgemeine Zeitung* on how the 'globalist elite' rules the world. It joins together transnational corporations, the UN, media, universities, NGOs and established parties; controls information; steals billions of dollars from taxpayers and showers them on banks and immigrants in equal measure and makes sure to stay 'culturally "colourful" '. The AfD leader also used the following words:

> The members [of this elite] live almost exclusively in big cities, speak fluent English and when they move from Berlin to London or Singapore for jobs, they find similar flats, houses, restaurants, shops and private schools everywhere ... Their bond to any homeland is weak. In a detached parallel society, they feel that they are world

citizens. The rain that falls in their homelands does not make them wet.[1]

German historians were quick to spot a resemblance. In November 1933, Adolf Hitler gave a speech in the Berlin district of Siemensstadt, or the 'City of Siemens', on how a 'small, rootless, international clique' ruled the world. He used the following words:

> [These are] people who are at home both nowhere and everywhere, who do not have anywhere a soil on which they have grown up, but who live in Berlin today, in Brussels tomorrow, Paris the day after that, and then again in Prague or Vienna or London, and who feel at home everywhere,

to which the audience responded by howling: 'The Jews!' The historians accused Gauland of having had the Führer's speech in front of him on his desk when writing the column. He denied the charge and any knowledge of the passage in question.[2] Let us assume he spoke the truth. In that case, Hitler's words structured the text of the leader of the most successful far-right party in Germany after 1945 like an invisible ink – hardly a reassuring disavowal. The resemblance would not be any less uncanny for being unintentional. The same went for keywords appearing in certain AfD statements, like *Lügenpresse* or *Volkstod*, not to mention *Volksgemeinschaft*, which appeared to have been plucked from the NSDAP lexicon. If AfD representatives just happened to use those same words, they did not thereby assuage concerns – but then again it was unlikely that the leaders of the German far right did not know what they were doing.[3]

When Bolsonaro's secretary of culture Robery Alvim announced a new prize for artists, he laid out the criteria for good art under the new regime: 'The Brazilian art of the next decade will be heroic and

---

1 Alexander Gauland, 'Warum muss es Populismus sein?', *Frankfurter Allgemeine Zeitung*, 6 October 2018.
2 Kate Connolly, 'AfD Leader Accused of Echoing Hitler in Article for German Newspaper', *Guardian*, 10 October 2018.
3 Witte and Beck, 'The Far'; Patrick Gensing, 'Poggenburg und die "Volksgemeinschaft"', *Tagesschau*, 2 January 2019.

it will be national, it'll be endowed with great capacity for emotional involvement and deeply committed to the urgent aspirations of our people, or it will be nothing.' Six years previously, a biography of Joseph Goebbels had been published in Brazil, containing the following line from one of his speeches: 'The German art of the next decade will be heroic, it will be steely-romantic, it will be factual and completely free of sentimentality, it will be national with great pathos and committed, or it will be nothing.' A 'rhetorical coincidence', said the secretary.[4] When Marine Le Pen rechristened her party 'Rassemblement National', she picked a name used by Nazi collaborationists in the early 1940s.[5] During his time at the interior ministry, Matteo Salvini spilled phrases from Benito Mussolini and chose to address a crowd from the balcony where he used to watch executions.[6] Santiago Abascal of Vox was the grandson of a Francoist mayor and remained close to the loyal parts of the Franco family. Other party representatives attended mass for the old dictator in which participants made the fascist salute, or published manifestos in defence of his honour; uniting Francoist colonels and veterans of the Falange, Vox called for his name to be respected.[7]

All of this and much more had the feel of an encore. It might as yet be less severe than the original event, a simulacrum or second-time-as-farce kind of recurrence, maybe an obsessive-compulsive Nazi ventriloquism. It seemed more banal and insipid than classical fascism – but

---

[4] Journalistas Livres, 'Brazilian Secretary of Culture Quotes Goebbels in Official Statement', journalistaslivres.org, 17 January 2020. The night when Roberto Alvim recorded his video, Bolsonaro said of him: 'Now we have a real secretary of culture.' After the outrage, Bolsonaro sacked him. Sam Cowie, 'Brazil Culture Secretary Fired after Echoing Words of Nazi Goebbels', *Guardian*, 17 January 2020.

[5] Kim Willsher, 'Marine Le Pen Sparks Row over New Name for Front National', *Guardian*, 13 March 2018.

[6] Nick Squires, 'Italy's Anti-immigrant Deputy PM Matteo Salvini Under Fire for Citing Mussolini', *Telegraph*, 30 July 2018; Josephine McKenna, 'Salvini Under Fire for Address from Notorious Balcony Used by Mussolini to Watch Executions', *Telegraph*, 5 May 2019.

[7] *Público*, 'Vox tira de generales retirados para sus listas, algunos abiertamente franquistas', publico.es, 18 March 2019; Diego Rodríguez Veiga, 'La conexión de dirigentes de Vox con Franco más allá de la amistad de Abascal y el bisnieto Luis Alfonso', *El Español*, 14 November 2019.

did it not come close to deserving that epithet?[8] Just how to understand the surge of the far right in Europe and indeed the world was far from clear: and here too the conjuncture mirrored the interwar period. When fascism burst into history, its enemies did not know what to make of it. 'There has been great confusion regarding fascism', Clara Zetkin opened her 1923 essay.[9] She and her peers searched for clues in Marx's writings on counter-revolutionary politics, primarily in France after 1848, where the prototype of 'Bonapartism' emerged, as a despotic state that appeared to stand above the fray of the struggling classes while beating back all democratic aspirations of working people, but the parallel did not fully capture what was going on. Fascism had no real precursor. Marxists and other anti-fascists continued to testify to theoretical confusion deep into the interwar period, and to feelings of being caught up in some strange social geology or of 'sliding down a steep slope, and being swept along by the laws of gravity'.[10] A difference in the early twenty-first century was, of course, the very widespread knowledge precisely of fascism. This time, the analogy of that experience was fully available, the attempts at classification and comprehension pulled to it as to an irresistible magnet. But confusion prevailed. For some, the contemporary far right represented a 'neofascism'; for others a 'postfascism', a 'quasifascism', a 'prefascism', a 'protofascism' or a 'late fascism', not exhausting the flurry of suggested labels.[11]

---

8   Poulantzas ended his study of fascism by predicting that if it returned, it would do so as a farce: Nicos Poulantzas, *Fascism and Dictatorship: The Third International and the Problem of Fascism* (London: Verso, 1979), 358.

9   Zetkin, *Fighting*, 23.

10   Léon Blum, in the context of Popular Front policy: 'The Socialist-Communist Pact', in David Beetham (ed.), *Marxists in Face of Fascism* (Manchester: Manchester University Press, 1983 [1934]), 175.

11   'Neofascism': John Bellamy Foster, *Trump in the White House: Tragedy and Farce* (New York: Monthly Review Press, 2017). 'Postfascism': for example Chamsy el-Ojeili, 'Reflecting on Post-Fascism: Utopia and Fear', *Critical Sociology* (2018), online first; Mikkel Bolt Rasmussen, 'Postfascism, or the Cultural Logic of Late Capitalism', *Third Text* 32 (2018): 682–8; Enzo Traverso, *The New Faces of Fascism: Populism and the Far Right* (London: Verso, 2019). 'Quasifascism': Robert Paxton quoted in Federico Finchelstein, *From Fascism to Populism in History* (Oakland: University of California Press, 2017), 11. 'Prefascism': Liz Fekete, *Europe's Fault Lines: Racism and the Rise of the Right* (London: Verso, 2018), 108. 'Protofascism': Carl Boggs, *Fascism Old and New: American Politics at the Crossroads* (New York: Routledge, 2018), 2. 'Late fascism': Alberto Toscano, 'Notes on Late Fascism', *Historical Materialism*, 2 April 2017.

All such proposals came with a familiar risk: sloppy and hasty and politically misleading usage of 'fascism', whatever the preferred prefix. This was the cardinal error of the Stalinist Comintern, when it branded German social democracy a carrier of 'social-fascism', twin of the Nazi variety. By foreclosing a united front between social democrats and communists against the NSDAP, such conceptual inflation merely facilitated the ascent of the real thing.[12] In the early twenty-first century, on the other hand – so different and yet so similar – the notion of 'Islamofascism' gained purchase for the sole benefit of Islamophobia: 'I want the fascist Qur'an banned', said Geert Wilders.[13] In between 'social-' and 'Islamo-', inaccurate overextensions and cries of wolf were legion, providing another lesson for the present conjuncture. Some working definition – analytically precise, politically on target – is clearly of the essence.

Here it is impossible to bypass the work of Roger Griffin, the dean of 'comparative fascist studies', who has spent an academic career distilling the essence of fascism. He treats it squarely as a set of ideas. It can be reconstructed from the texts that fascists wrote. The common belief setting their hearts aflame was fairly straightforward: the nation must, above all else, be reborn. The nation in question was not a constitutional entity defined by the rights and obligations of citizens, but instead conceived as an organic community, the patrimony of a race or culture or civilisation united by heritage and blood. As such it went beyond the liberal conception of the nation, and hence Griffin calls the faith in it 'ultranationalism', roughly corresponding to what we have previously termed 'ethnonationalism'. In the eyes of fascists, the noble nation had to be snatched from its deathbed. Over it hovered an existential threat, in the form of geopolitical demotion, military humiliation, racial miscegenation, economic reduction, loss of identity, cultural degeneracy, ideological deviancy or some other grim reaper: the nation had to be rescued and given a new birth. All fascists posed as deliverers of their

---

[12] The most acute critique of the Comintern lines on fascism remains Leon Trotsky, *The Struggle against Fascism in Germany* (New York: Pathfinder, 1971).

[13] Ian Traynor, '"I Don't Hate Muslims. I Hate Islam," Says Holland's Rising Political Star', *Observer*, 17 February 2008. For an excellent, nuanced critique of the notion, see Traverso, *The New*, 82–93.

invaluable endowment. In Italy, they wanted to lead the inheritors of the Romans to take wing anew; in Germany, the descendants of the Nordic and Germanic race to fly *über alles*; in other countries, the heirs of some other glorious caste that risked being lost forever but would, by resolute action, soar to majestic heights. For this myth, Griffin has adopted the term 'palingenesis', combining the Greek word for birth (*génesis*) with that for again (*palin*). Fascism can thus be defined as *palingenetic ultranationalism*. Permutations were and remain possible. But whenever and wherever fascism appears, it will posit the sequence of past grandeur to present crisis to coming rebirth of an exalted and exclusive nation.[14]

This definition has its undeniable merits. As Zetkin already appreciated, fascism cannot be understood without due attention to its stirring power, its appeal as a worldview *sui generis* and ability to kindle souls:

> Masses in their thousands streamed to fascism. It became an asylum for the politically homeless, the socially uprooted, the destitute and disillusioned. And what they no longer hoped for from the revolutionary proletarian class and from socialism, they now hoped would be achieved by the most able, strong, determined, and bold elements of every social class. All these forces must come together in a community. And this community, for the fascists, is the nation.[15]

Or, in the words of Arthur Rosenberg, one of the most astute Marxist witnesses to fascism, writing in 1934: 'The leading idea of Nazi propaganda is national renewal.'[16] Although Griffin would be loath to

---

14 The most recent statement of this theory of fascism is Roger Griffin, *Fascism: An Introduction to Comparative Fascist Studies* (Cambridge: Polity, 2018), for example 40–5, 68–72, 129–30. It was first elaborated in *The Nature of Fascism* (London: Routledge, 1993). See also for example 'The Primacy of Culture: The Current Growth (or Manufacture) of Consensus within Fascist Studies', *Journal of Contemporary History* 37 (2002): 21–43; *A Fascist Century* (Basingstoke: Palgrave Macmillan, 2008); 'Studying Fascism in a Postfascist Age: From New Consensus to New Wave?', *Fascism* 1 (2012): 1–17.

15 Zetkin, *Fighting*, 31.

16 Arthur Rosenberg, 'Fascism as a Mass Movement', *Historical Materialism* 20 (2012 [1934]), 188. Cf. also Ernst Bloch, *The Heritage of Our Times* (Cambridge: Polity, 1991 [1935, 1962]), 90–1, 97.

admit it, since he fancies himself the first discoverer of this kernel of fascism, perspicacious militants like Zetkin and Rosenberg again and again stressed that the enemy could never be defeated unless its ideological firepower be taken seriously and neutralised.

But it would be silly to argue that the Marxists who wrestled with the behemoth in real time also laid down an unimprovable scientific understanding of it. In hindsight, its contours can be more precisely delineated. The widely influential formula of 'palingenetic ultranationalism' does pin down an essential aspect of fascism, something like the equivalent of class struggle and equality in socialism or property rights and freedom in liberalism. It highlights something that has indeed often been left in the shadows: the *ideas* of the farthest right, strange to the non-initiated but of the greatest cogency to its adherents. It urges close engagement with these ideas. It allows Griffin to identify a whole plethora of movements and groupuscules in interwar Europe, post-war Europe, modern Latin America and South Africa and elsewhere as specimens of fascism, insofar as they shared the 'mythical core' or 'fascist minimum' he has specified. For this, they needn't be of any relevance to anyone. A lonely, pimpled, seventeen-year old skinhead could sit in southern Sweden in the 1980s and put swastikas on his wall and dream of a new millennium for the Aryan race and clearly be a fascist, and maybe also member of a fascist sect, without him and his friends being able to influence the course of events on more than a purely local scale.

A distinction is here called for: between fascism as a set of ideas and fascism as *a real historical force*.[17] A definition of the latter must go beyond doctrines and diction. In fact, programmatic content and philosophical coherence of the kind Marx supplied to socialism or Mill to liberalism were rather tangential to fascism in the two countries that produced the paradigmatic cases of its power: Italy and Germany.[18] Such accoutrements mattered less to fascism than to any

---

17 Griffin implicitly makes such a distinction – see for example Griffin, *The Nature*, 45 – but invests most if not all of his analytical work in the former category.

18 As argued by Robert O. Paxton, *The Anatomy of Fascism* (London: Penguin, 2004), 16–18, 214–5, 221. Cf. Dylan Riley, 'Enigmas of Fascism', *New Left Review* series 2, no. 30 (2004): 140. In response to Paxton's critique, Griffin offers a poor retort in 'The Primacy', 34–5. Others who question the search for the exactly defined nominal essence

of its contenders. Griffin himself quotes Mussolini spurning carefully crafted belief systems for the elemental force of the 'living act' and 'will to power'.[19] It then seems a bit inapposite to boil down fascism – of all historical forces – to what it thought and said, rather than concentrating on what it did and how it could do it. This is the argument of Robert Paxton, whose synthesis of the actual history in *The Anatomy of Fascism* competes with Griffin for authoritative influence. The perfect lexical definition frames 'a static picture of something that is better perceived in movement' – 'something better understood *as a process*'.[20] More particularly, fascism would never have become a historical force had it not come to power. Grasping it by pinpointing its ideational essence is rather like trying to taste a bread by looking at its recipe. There have to be ingredients, fermentation, hands kneading the dough, an oven set at just the right temperature: a series of historical events and conditions, without which the myth of the reborn nation would have been a passing fancy on a piece of paper. Or, from the heat of the combat, with Leon Trotsky:

> To our sage schoolteacher only the 'essence' of fascism is important . . . But the whole matter lies in the fact that the pogrom 'essence' of fascism can become palpable only after it comes to power. And the task consists precisely in not permitting it to attain power . . . It is not enough to understand only the 'essence' of fascism. One must be capable of appraising it as a living political phenomenon, as a conscious and wily foe.[21]

On the other hand, retaining a definition of fascism as a set of ideas is necessary for pointing it out in circumstances where it is *not*

---

of fascism on similar grounds are Richard Saull, 'Capitalism and the Politics of the Far Right', in Leo Panitch and Greg Albo (eds), *The Politics of the Right: Socialist Register 2016* (London: Merlin Press, 2015), 29; Finchelstein, *From Fascism*, 38–40, 54; Jean-Yves Camus and Nicolas Lebourg, *Far-Right Politics in Europe* (Cambridge, MA: Harvard University Press, 2017), 20; for an effective critique, see in particular David Renton, *Fascism: Theory and Practice* (London: Pluto, 1999), particularly ch. 2.

19 Griffin, *Fascism*, 61.
20 Paxton, *The Anatomy*, 14. Cf. David Renton, 'On Benjamin's *Theses*, or the Utility of the Concept of Historical Time', *European Journal of Political Theory* 11 (2012): 389–90.
21 Trotsky, *The Struggle*, 229–30.

anywhere near attaining power: it can be a moribund phenomenon, a boy's dream or a nostalgic reverie and still be a case of fascism, by dint of its ideational content. It can also travel from one context to another and keep its 'pogrom essence' intact, even if unable to live it out. For fascism in this sense, we shall accept – but only for the time being – Griffin's definition.

For fascism to exist as a *real historical force*, however, more was required, the first condition being a *crisis*. The ideas of a virulent nationalism – the nation as the supreme value, the conquest of foreign lands as patriotic duty, the demonisation of Jews, a fad for uniforms – had achieved currency already in late nineteenth-century Europe, but they could not seize the day as long as relative stability prevailed. Only the First World War opened the space for them. The mass slaughter in the trenches traumatised European societies, normalised violence and discharged roving bands of veterans who considered themselves entitled to the countries for which they had fought, forming the nuclei of both the *squadristi* and the *Freikorps*. From Russia blew the winds of revolution, very nearly sending capitalist state apparatuses crashing down across the continent. Italy and Germany were especially close to falling. Those two nations also had their territorial ambitions denied by the peace treaties, their throbbing impulses to expansion held back and, in the case of Germany, reversed into outright loss of national and colonial land. On top of all this came, after 1929, the global depression.[22]

Beset by enemies and corrosive forces external and internal, the states of Italy and Germany entered a post-war crisis so deep that their basic functions – to maintain the cohesion of their social formations and ensure continued capital accumulation – could no longer be exercised with anything like normal routine. This crisis was *constitutive* of fascism as a real historical force and not a spit in the wind, as Zetkin, Rosenberg, Trotsky and most other Marxists immediately saw, and as Griffin too acknowledges, although he does not make

---

22 This draws particularly on Geoff Eley, 'What Produces Fascism: Preindustrial Traditions or a Crisis of a Capitalist State', *Politics and Society* 12 (1983): 69–71, 77–81; Geoff Eley, *Nazism as Fascism: Violence, Ideology, and the Ground of Consent in Germany 1930–1945* (London: Routledge, 2013), 205–7, 214; Geoff Eley, 'Fascism Then and Now' in Panitch and Albo, *The Politics*, 96, 100–5.

much of it.[23] For Paxton, on the other hand, this is the first item on a long list of passions that mobilised fascist action: 'a sense of overwhelming crisis beyond the reach of any traditional solutions'.[24] But the modern historian who has done most to foreground this element is Geoff Eley, who speaks of 'fascism-inducing' or 'fascism-producing' crises. For anyone concerned with the possible reappearance of fascism, it follows that 'the key question becomes: what kind of crisis calls this politics to the agenda?'[25]

But no crisis has ever induced fascism through automatic causation, just as no ingredients bake themselves into a bread. Someone is always running the bakery. Paxton highlights the fact that both Mussolini and Hitler came into office by order of traditional powerholders. Both men were invited to rule by the legitimate representatives of their respective states – King Victor Emmanuel III in October 1922, President Paul von Hindenburg in January 1933 – who acted out a shared resolve among their dominant classes to bank on fascist forces as the best way out of the impasse. Both Il Duce and the Führer had taken a previous stab at seizing power on their own – the former in the election campaign of 1919, the latter in the Beer Hall Putsch in 1923 – and failed abysmally. Their route to government passed through an alliance with the existing establishment. Neither wished to break into palaces on the back of a general strike or armed insurrection or some other social upheaval that might unchain the forces of the left they were out to eradicate. Both craved the embrace of the army and police. An ordered transfer of authorities was the only conceivable scenario, and none other came to pass: no fascists ever installed themselves on the throne against the will of incumbent rulers. As long as they were spurned – as in France, Belgium, the United Kingdom – they stood not a chance, no matter how feverish the propaganda they churned out or how much muscle they mobilised on the streets.[26]

---

23 For example Zetkin, *Figthing*, 27, 67; Rosenberg, 'Fascism', 156, 181–4; Trotsky, *The Struggle*, 17, 26; Griffin, *The Nature*, 40, 213; Griffin, *Fascism*, 88.
24 Paxton, *The Anatomy*, 291.
25 Eley, 'Fascism', 93.
26 Paxton, *The Anatomy*, 68–115; cf. for example Poulantzas, *Fascism*, 333; Riley, 'Enigmas' 144; Griffin, *The Nature*, 78–9.

Such an analysis should not be mistaken for a facile view of fascism as the string doll of big capital, designed by it and moving as it did. Keen observers from Zetkin onwards recognised fascism as a mass movement in its own right, with an authentic following – even among some proletarian strata – and a winning nationalist zeal irreducible to the needs of any dominant class fraction.[27] It was never the first choice of a king or a president. Rather it served as a last resort, to which official powerbrokers and bourgeois layers turned in an hour of desperate need. For a long time, they often found fascists repugnant, vulgar, eccentric and potentially destabilising and would have much preferred the calmer waters of London or Stockholm.[28] In a study written under the shadow of Trump, Benjamin Carter Hett offers a careful reconstruction of *The Death of Democracy: Hitler's Rise to Power* and shows how the uppermost echelons of the Weimar Republic – businessmen, top brass, landholders, conservative politicians – step by step moved towards a compact with Hitler. After his appointment as chancellor, his predecessor Franz von Papen described the character of their choice: 'We have hired him.'[29] The second condition of the existence of fascism was *the willingness of sections of the dominant class to call upon the fascists* to relieve the crisis.[30]

Fascism, then, was not for ordinary times. No masses would have found it attractive had they not felt the ground disappearing beneath

---

27 For example David Beetham, 'Introduction', in Beetham, *Marxists*, 8, 11–12; Karl Radek, 'Capital's Offensive', in ibid. [1922], 101; August Thalheimer, 'The Development of Fascism in Germany', in ibid. [1928], 193; Zetkin, *Fighting*, 24, 27, 32–3, 44, 59–60; Rosenberg, 'Fascism', 149; Trotsky, *The Struggle*, 59; Ernest Mandel, 'Introduction', in ibid., 24; Poulantzas, *Fascism*, 85, 112–13.

28 See for example Trotsky, *The Struggle*, 278, 283, 340; Beetham, 'Introduction', 8, 11–12; Thalheimer, 'The Development', 200; Zetkin, *Fighting*, 32–3; Mandel, 'Introduction', 24; Poulantzas, *Fascism*, 85, 112–13; Ernesto Laclau, *Politics and Ideology in Marxist Theory: Capitalism, Fascism, Populism* (London: Verso, 2012 [1977]), 120.

29 Quoted in Benjamin Carter Hett, *The Death of Democracy: Hitler's Rise to Power* (London: William Heinemann, 2018), 182.

30 Cf. Paxton, *The Anatomy*, 68, 72–3, 115, 220; Richard Saull, 'Capitalism, Crisis and the Far-Right in the Neoliberal Era', *Journal of International Relations and Development* 18 (2015): 34–6; Saull 'Capitalism and the Politics', 142; Eley, *Nazism*, 207; Eley, 'Fascism', 99–101; Rosenberg, 'Fascism', 153, 156. Griffin himself nearly accepts this: 'Thanks to the collusion of political and social elites who feared chaos and the threat of the left more than they feared totalitarianism, the Nazis were allowed to enter the citadel of state power in 1933, with catastrophic consequences.' Griffin, *Fascism*, 66.

their feet. No dominant classes would have called on its services had their dominance remained secure. As long as lives in advanced capitalist states proceeded in well-worn furrows, with only average levels of dislocation and disorder, as was the case for a long time before the First World War and again after the Second, politics could be managed with the normal procedures of a bourgeois democracy: peaceful party competition, regular elections, alternating governments, civil conversations, rule of law, at least some basic freedom of expression and other democratic rights. Countries barely ruffled by the interwar crises never developed fascism as a real historical force. Sweden, for instance, had its fair share of fascist dreamers and demagogues, but its territory was untouched by the First World War, its capitalist class free-riding on larger imperial powers, its revolutionary working-class movement defanged by reformism, the leaders of its contending classes partnering in a lasting peace during the Depression.

But in the countries where the crises reached the greatest depths, the normal state gave way to an *exceptional state* in brown or black uniform. Its behaviour massively overstepped the boundaries of normality. Poulantzas made this modality of a capitalist state central to his theory in *Fascism and Dictatorship: The Third International and the Problem of Fascism*; under Mussolini and Hitler, the exercise of power was characterised by a general recourse to violence. The law was suspended to make way for the arbitrary will of the leader. Elections were discontinued and parliaments maintained as facades of one-party rule, the leaders of rivalling parties placed behind bars or worse; ideological state apparatuses were fully merged with other branches, so that judges and police chiefs doubled as propagandists.[31] More so than with normal states, the fascist exceptions entered the economy with boots on the ground to buttress individual capitals. This aspect of exceptionality was later emphasised by Jane Caplan, who argued that what really set the fascist state apart was its far more direct role in managing the day-to-day accumulation of capital – sourcing raw materials, opening markets, supplying

---

31 Poulantzas, *Fascism*, 11–16, 310–13. Cf. Bob Jessop, *The State: Past, Present, Future* (Cambridge: Polity, 2016), 211–20.

and controlling workers – like a guardian caring for a minor in trouble.[32]

The *differentia specifica* of fascism-as-force appears, however, to have been the level and extent precisely of physical violence in the internal affairs of the nation (as distinct from in the colonies, where it was nothing new). 'What is fascism?' asked Antonio Gramsci in 1921: 'It is the attempt to resolve the problems of production and exchange with machine-guns and pistol-shots.'[33] Eley points out that socialists might have been harassed and discriminated against in imperial Germany, but killing them amounted to a break with the relatively respectful manners of constitutionally constrained politics.[34] But socialists were not, of course, the sole target of such violence. 'A demagogic nationalism spontaneously seeks an object through which it can daily demonstrate its own superiority and onto which it can release the delirium of its racial frenzy', writes Rosenberg.[35] Fascism let loose a frantic mass terror on the vilified *Volksfeinde*, with a view to the elimination of their presence. As Griffin correctly notes, fascism was and always will be 'essentially racist', although its conceptions of race and choice of victims might vary; as pipedream or practise, it seeks the dominance of the insiders of the ultra-nation over those outside it.[36]

On the basis of this rather well-known history, we might venture a simple, provisional definition of fascism as a real historical force. *Fascism is a politics of palingenetic ultranationalism that comes to the fore in a conjuncture of deep crisis, and if leading sections of the dominant class throw their weight behind it and hand it power, there ensues an exceptional regime of systematic violence against those identified as enemies of the nation.* Such a definition takes on Griffin's common denominator of ideas but couples it to the temporal process and social relations pressed by Paxton and others. Rather than freezing the object in a still image, the motion can be

---

32 Jane Caplan, 'Theories of Fascism: Nicos Poulantzas as Historian', *History Workshop* no. 3 (1977), particularly 95–6.
33 Antonio Gramsci, 'On Fascism', in Beetham, *Marxists*, 82.
34 Eley, *Nazism*, 208–9; Eley, 'Fascism', 95–8.
35 Rosenberg, 'Fascism', 151.
36 Griffin, *The Nature*, 48.

followed from platforms and rallies all the way into an ironclad state of affairs.

But, while Europe between the wars must serve as the template, we should reject an understanding of fascism-as-force as indissolubly 'tied to the unique conditions of the interwar period', since this a priori excludes a recurrence.[37] Those conditions will obviously never come back. There will be no second round of trench warfare in northern France, no toppling of a Russian tsar, no pogroms smashing through Jewish villages in Ukraine, no black-shirted columns getting on the trains to Rome: but as has been frequently pointed out, anyone who looks for an exact reincarnation is guaranteed to miss the return of precisely what they fear.[38] 'Historical phenomena never repeat themselves completely', Trotsky admonished those who searched for parallels in Bonapartism, but this does not prevent us from dividing human affairs into taxa that might come and go (capitalism, famine, democracy, collapse) rather than dissolving history into an infinite number of unrecognisable occurrences.[39] What cannot be ruled out, in our case, is the formation of a political pole with reminiscent ideas, a crisis of a similar magnitude and rulers and vested interests that arrange for the same sort of coalition. If something along those lines were to happen, 'then we can surely deploy the same term', to quote Eley.[40]

And something along those lines might be foreshadowed in the present. 'We should', Eley urges, 'concern ourselves with the production of fascist *potentials*'; be alert to 'the kind of crisis where a politics *that begins to look like fascism* can coalesce'.[41] A timebound image makes for closed eyes, but 'a portable concept of fascism helps make these dangers legible'.[42] And since we are dealing with a wily foe, the concept may have to be further adapted; indeed, we shall revise our definition shortly. For now, it will do for heuristic and vigilant

---

37 The position taken by Golder, 'Far Right', 481.
38 Poulantzas, *Fascism*, 53, 358; Griffin, 'Studying', 16; Paxton, *The Anatomy*, 174–5; Nigel Copsey, 'The Radical Right and Fascism', in Rydgren, *The Oxford*, 109.
39 Trotsky, *The Struggle*, 330.
40 Eley, *Nazism*, 218.
41 Eley, 'Fascism', 104, 93. Emphases in original.
42 Eley, *Nazism*, vi.

purposes. It maintains that any rise of fascism is resistible at every stage: the politics of palingenetic ultranationalism can be disarmed through ideological and other modes of attack. The deep crisis can be resolved in a different direction. If the dominant class leans towards a fascist solution, it can be thwarted, and if it succeeds and an exceptional regime is instituted, there is still scope – so the historical record suggests – for some forms of late resistance.

Now, up until very recently, the prospects for fascism-as-force seemed non-existent. The consensus in the field held that its two chief conditions were absent. First, there was no deep crisis in sight. There might have been ruffles on the surface of progress, but at least in Europe, 'peace, prosperity, functioning democracy, and domestic order' had been safely in place since 1945 and showed no sign of profound reversal, Paxton concluded his *Anatomy of Fascism* in 2004.[43] In his magnum opus *The Nature of Fascism* from 1991, Griffin accepted that palingenetic ultranationalism will always appeal to some marginal existences; 'as an active factor in the transformation of history, however, fascism is a spent force.' Tranquillity ahead, 'there is every reason to assume that fascism will continue to be denied the political space' it requires.[44] When Griffin updated his assessment in 2018, he took the deed of Anders Breivik as proof of the status of palingenetic ultranationalism – the mass murderer's ideas reflect the essence perfectly – as 'curiously *private* and *socially isolated*', 'no longer synchronized with the objective state of the world'.[45] Breivik's was a shot in the dark. The misfires of fascism after 1945 could be attributed to the absence of 'a generalized subjective sense of an existential crisis of the nation and of modern liberal civilization'.[46] No opening existed for the madmen. No convulsions that could shake bourgeois society like the First World War and its aftermath were in view; no sense of an overwhelming crisis beyond the reach of traditional solutions; no anxieties about the future intense enough to twist people's heads.

Second, it followed, capitalist classes seemed to have very little appetite for actually existing far-right parties. The scholar with the

---

43 Paxton, *The Anatomy*, 186.
44 Griffin, *The Nature*, 45, 220–1.
45 Griffin, *Fascism*, 123–4. Emphases in original.
46 Ibid., 100–1.

most comprehensive such assessment was Richard Saull, who argued that capital had undergone a metamorphosis since the interwar era, no longer formatted in national compartments but thoroughly globalised. In the early twentieth century, capitalists latched on to their respective states and expected them to conquer territories and erect tariffs and fight it out with their rivals: in the early twenty-first, they merge across borders and send off stocks and factories without paying heed to nations. They have no desire for imperial adventures or walls. Instead they tend to look down on nationalists as uncouth bigots who imperil the free circulation of commodities, including labour-power. Small firms cornered on national markets and other petty enterprises might be responsive to a nationalist agenda, but the commanding heights of capital are 'committed to a political economy directly at odds' with it; moreover, the far right has become 'increasingly *decoupled*' from the leading fractions of capital, in the starkest possible contrast to interwar Italy or Germany.[47] On this view, the far right in the era of late globalised capitalism is an essentially plebeian force, doomed to impotent screams of rage against immigration. It will not morph into fascism, since the premise of dominant class support is bound to go unfulfilled. As of the early millennium, researchers from diverging schools were thus 'in complete agreement that anything like historic fascism is impossible in the advanced capitalist world today, because of the basic solidity of liberal democracy in this region'.[48] These happy appraisals were logically consistent with our definition. Were they also empirically perceptive?

The consensus was not without dissenters. In 2013, Geoff Eley ended his *Nazism as Fascism: Violence, Ideology, and the Ground of Consent in Germany 1930–1945* on a bleaker note, for he had picked up the signal of a deep crisis in the making. 'The global environmental catastrophe, and climate change in particular, now challenge the

---

47 Saull, 'Capitalism, Crisis', quotations from 44–5. Emphasis added. See also Saull, 'Capitalism and the Politics', 145–9, and cf.: 'In the contemporary situation, all that may remain are those aspects of the far-right programme *which are irrational for capital*.' Neil Davidson, 'The Far-Right and "the Needs of Capital"', in Saull et al., *The Longue*, 137. Emphasis added. A similar theory of decoupling appears to be espoused by Traverso, *The New*, 13.

48 Riley, 'Enigmas', 146.

possibilities for effective and accountable governance' and 'may well enable a politics that resembles fascism to coalesce.'[49] Two years later, another prominent historian of the original cataclysm, this one of a liberal persuasion, Timothy Snyder, devoted the concluding chapter of *Black Earth: The Holocaust as History and Warning* to global warming as the sort of crisis for which fascism waits in the wings. 'When an apocalypse is on the horizon', he ventured, 'demagogues of blood and soil come to the fore.'[50] We shall return presently to the content of these scenarios. As for the dominant class component, we have acquainted ourselves with some fractions of rather considerable weight that seem tolerant, to put it mildly, of parties and presidents on the far right. It does not necessarily have to be the nationalism per se that beguiles them. Other attributes and corollary preferences could matter more. We shall return to this question too, but first we must ask what the term *fossil fascism* could possibly mean.[51] We can approach the phenomenon – if only *in potentia* – from two angles. We can see it as a compound of notions of energy and nation, or as a conjunction of a deep crisis and dominant class interests rooted in fossil fuels; a set of ideas or a real historical force. Let us begin with the latter.

Some Scenarios of Fossil Fascism

One could envision several kinds of crises inducing fossil fascism. Imagine that, at some point in the future – and considering the state of things, it cannot be too distant – a campaign for radical emissions cuts gains momentum. It puts on the agenda of some advanced capitalist country reductions of 5 or 10 per cent per year. Oil wells and coal mines are marked out for immediate closure. A motion banning the construction of any additional infrastructure – not one more pipeline, not one more digger – is endorsed by a group of parliamentarians swiftly growing towards a majority. There is talk of prohibiting the still-lucrative export of liquified fossil gas already at the start

---

49   Eley, *Nazism*, 217.
50   Timothy Snyder, *Black Earth: The Holocaust as History and Warning* (London: Penguin, 2015), 326–7.
51   Although Cara Daggett uses this term, she supplies no definition.

of next year. A committee has drawn up a road map for the complete cessation of fossil fuel burning in electricity generation as well as transport; all sectors of the economy are to be subjected to the annual quotas, CEOs held legally responsible for compliance, overconsumers sanctioned with penalties and leakages plugged, so as to reach – no missed targets this time – zero emissions within less than two decades. Commensurate preparations are underway for scaling up renewable energy production and zero-carbon solutions as all-encompassing and, among some layers, wildly popular substitutes. The realities of global warming have caught up with the polity, its material foundations slated for extreme makeover: here is a *mitigation crisis*.

For some fractions of the capitalist class, this is a life-threatening situation. Primitive fossil capital is about to be liquidated, with no ability to reinvent itself for another life in the fossil-free economy: coal mines cannot produce wind. The assets buried in the seams will be irretrievable. Mouth-watering business opportunities, fixed capital of mammoth size, an entire department of accumulation will be forever gone. Now dawns the existential crisis this class fraction has dreaded since at least the 1980s, mixed up with a structural crisis for other capital, all the more convulsive for having been postponed for so long: there will be little more than a dozen years to complete the transition, under the rigorous guidance of the state. The prime capitalist prerogative – control over the privately owned means of production – can no longer be considered untouchable. The structural crisis could grow into threats to the life of wider segments of capital too.

But imagine there is also, in this moment of truth, a force on the far right picking up its own momentum. It sees the world in other colours. It recommends no action; it refuses to believe the hype; it rather wishes to attend to the health of the ultra-nation. It is fully available for a quid pro quo with the establishment and points its finger at some other group that ought to be removed from the body politic. In this scenario, we can imagine primitive fossil capital operating as the elite force of the dominant class, stepping forth to strike a deal with the far right – through funding, negotiating a coalition, siding with sympathetic elements within the repressive state apparatus or some other act of hiring – so as to protect itself from the existential and fossil capital in general from the structural crisis. The rest can be left to the imagination.

Such a scenario would seem to be most likely in a country with extensive fossil fuel extraction *and* some tradition of ethnonationalism. The US comes to mind, as does Germany, Poland and Norway, to which we might add Australia, Canada, Russia and perhaps other producing countries too. But we might also think of social formations where primitive fossil capital is less ponderous than the fraction presiding over agribusiness or the meat industry, whose very existence would likewise be in peril on the day real mitigation begins. Here it would also be a matter of ending entire lines of business that collide with the imperatives of climate stabilisation: just as gas reserves would have to be left untouched, trees would have to be left standing, deforestation turned into its opposite for the long haul.[52] Nor should we discount fossil capital in general. Owners of the car industry in Hungary or Austria or Germany could face huge losses, even if they have a better chance of surviving a mitigation crisis. So far, there is little to suggest that these dominant classes would be prepared to give up their existence or undergo the required transmutation at the drop of a hat; they look rather more inclined to defend their crassest interests by any means necessary.

In the absence of multiple mitigation crises with a non-fascist outcome, in which all of the above and more is implemented, we are, on the other hand, certain to experience a proliferation of *adaptation crises*. Imagine that repeated climate shocks chip away at the material foundations of societies at a level far deeper than in the first two decades of this century: heatwaves five or ten degrees hotter; wildfires roaring through regions for months on end; food provisioning systems at breaking point; storms pushing the sea dozens of kilometres inland – here there is little need to exercise the faculty of the imagination. It is portended in the science. Adaptation crises would disrupt established stores and circuits of biophysical resources. Emergencies that put 'peace, prosperity, functioning democracy, and

---

52   See for example Troy Vettesse, 'To Freeze the Thames: Natural Geoengineering and Biodiversity', *New Left Review* series 2, no. 111 (2018): 63–86. Perhaps agri-fascism would be a more accurate moniker for this scenario in Brazil.

domestic order' to the test are not, after all, quite so improbable – they are literally in the pipeline, and it is with them scholars of fascism such as Eley and Snyder are concerned. The former worries what will happen when serious shortages set in. If things like inhabitable land and edible food become scarce, those still blessed with plenty are likely to guard it more jealously than ever and keep outsiders at bay: we can look forward to 'fortress mentalities, idioms of politics organized by anxiety, *gatedness* as the emerging social paradigm'.[53] The gates will be locked to keep ever more precious resources secure – a prognosis that has trailed climate projections for some time now.

On a planet with shrinking habitable areas, the reasoning continues, fortunate strata might not only want to hold on to what they've got, but also create buffers. Cara Daggett alerts us 'to the possibility that climate change can catalyse fascist desires to secure a *lebensraum*, a living space, a household that is barricaded from the spectre of threatening others'.[54] Snyder follows this train of thought all the way back into the concentration camps:

> In a scenario of mass killing that resembled the Holocaust, leaders of a developed country might follow or induce panic about future shortages and act pre-emptively, specifying a human group as the source of an ecological problem, destroying other states by design or by accident. There need not be any compelling reason for concern about life and death, as the Nazi example shows, only a momentary conviction that dramatic action is needed to preserve a way of life.[55]

The people to be done away with could be Muslims, Jews, gays or any other group to which anxieties attach. Even if they have nothing to do with rising temperatures, an extreme crisis – or just the subjective sense of one – might make some blood boil: 'No green politics will ever be as exciting as red blood on black earth.' Snyder also thinks the Muslims of the Middle East could start blaming the Jews for cooking

---

53 Eley, *Nazism*, 217. Emphasis in original. Cf. Eley, 'Fascism', 92–3.
54 Daggett, 'Petro-masculinity', 20.
55 Snyder, *Black*, 326.

them and take it out on the state of Israel. Or, China might do in the 2030s what Germany did in the 1930s and strike out towards its own *Lebensraum*, colonise Africa or Russia and slay every *Untermensch* in its way.[56]

Here is a risk of being carried away. Snyder does let his imagination run freely, to the point of projecting his future climate Holocaust scenario back onto Nazism itself: now he portrays Hitler as an ecological warrior. The Führer followed 'an urgent summon from the future (ecological panic)'. 'In Hitler's ecology, the planet was despoiled by the presence of Jews', whose disappearance was 'part of an ecological restoration'; contrary to received wisdom, 'the struggle against the Jews was ecological, since it concerned not a specific racial enemy or territory but the conditions of life on earth.'[57] These are some rather bizarre anachronisms that come with no substantiation. The Jews were nothing if not a specific racial enemy in the eyes of the Nazis, who did not speak in terms of 'ecological restoration' and never bothered about life on earth in general.[58] When Snyder concludes his *Black Earth* with the assertion that 'we share Hitler's planet and several of his preoccupations', he violates what must be the first rule for any such comparison: *this time will be different*, not the least because a climate crisis *did not* figure in the concatenation of disasters that linked the two World Wars. Even speculations about fascism rearing its head must be subject to analytical restraint.

Sticking to our ideal types, the accumulation of fossil capital would hang in the balance in a mitigation crisis. In an adaptation crisis, all categories of property might be thrown into disarray. Dominant classes would have interests to guard in both scenarios, and a politics of ultranationalism – palingenetic or with other banners, as we shall soon see – might come in handy. It could be used to beat back the challenge of a mass movement or other actor pushing for an across-the-board transition, or for redistribution of resources from those who have more than enough to those teetering on the edge; indeed,

---

56 Ibid., 328–35; quotation from 344.
57 Ibid., 323, 28, 8.
58 As pointed out by Jeffrey Herf in Michael Berenbaum and Jeffrey Herf, 'Conflicting Perspectives on Snyder's *Black Earth*', *Journal of Cold War Studies* 19 (2017): 231–2.

any adaptation crisis is likely to provoke cries for the latter. Luxury enclaves should not expect peace in the heat. Both crisis types would then seem to hold potentials for a reappearance of several of the original components: states struggling to maintain the cohesion of their formations and ensure continued accumulation; traditional powerholders scrambling for solutions; a readiness to experiment with more direct management of the economy to keep up business-as-usual; the hand reaching for pistols and machine guns. Race could be a vector of diversion and deflection. It might channel energies towards targets that can be safely destroyed with the status quo in one piece, or at least a semblance of it in place: an opportunity for the far right to realise its programme of elimination, by means of an exceptional state.[59] We are here talking of strictly hypothetical scenarios.

After a mitigation crisis with a fascist outcome, there would be fossil fascism as historical fact and force. What of adaptation? Here fossil energy would also be the firebox – not as object of contention, but as the material that has generated the immediate climatic stress. Furthermore, defensible concentrations of resources are likely to correlate with histories of such combustion. We know that the most vulnerable populations will be those that have done the least to cause global heating; conversely, those who might garrison themselves on islands of riches will have built their stations on a century or two of burning. Fossil fascism would seem an accurate term for both scenarios.[60]

The presence of racial others could be perceived as related to adaptation: imagine the northern Nile Delta goes under water. Imagine some of the Egyptians unable to squeeze into Cairo and unwilling to travel into overheated Upper Egypt try to cross the Mediterranean. Imagine this coincides with rising food prices in Europe, due to strains on the supply chains: then some might want to bring on the gunships and burn the boats and, why not, proceed with the Muslims

---

59 Cf. Alex Alvarez, *Unstable Ground: Climate Change, Conflict, and Genocide* (Lanham: Rowman & Littlefield, 2017), 47, 52, 70

60 And so we need not take up the term 'climate fascism', suggested for the adaptation crisis scenario by Christian Parenti, *Tropic of Chaos: Climate Change and the New Geography of Violence* (New York: Nation Books, 2011), 11, and relayed by Daggett, 'Petro-masculinity', 3.

already on European soil – time to get them back to their shores. Entertaining such a scenario does not require subscription to a view of the world as a *bellum omnium contra omnes*. It merely presupposes that *some* regard it that way. In *The Malthusian Moment*, Thomas Robertson shows how American neurosis over population growth around the year 1968 was a mechanism for processing the turbulence in the Third World – Vietnam, the national liberation movements, the gangrenous poverty that carried 'the rising tide of communism', all conveniently projected onto a riotous urge to breed.[61] Now this does not mean that these developments *were* caused by overpopulation, only that some interpreted them through a Malthusian lens, and it is fairly likely that the far right will step forward to do the same whenever normal routines of resource consumption are stretched. There might well be enough food and land for everyone already in Europe plus the needy from the Nile Delta to survive, on condition that they are shared equitably. But ultranationalists will not accept that.

For an adaptation crisis to break out and spiral into attacks on racial others, the scarcities, then, could be more imaginary than real. But they could also be more real than in the twentieth century: one cannot rule out that this breakdown drastically shrinks resource bases in the aggregate. In that case, too, ultranationalists would leap forth with the demand to sever the extra hands, using racial classifications to sort those deserving of survival from those to cut down.[62] They can be counted on to regard non-whites as having no legitimate claims on the nation. Indeed, this would be prime time for a far right that has for decades obsessed over demographic apparitions – 'It's the birthrates. It's the birthrates. It's the birthrates' – a fossil fascism of this type easily subsuming green nationalism. The Malthusian element of that current would have special traction. But the same muddle might develop in a mitigation crisis as well, as indicated by one text that begins to dissolve these distinctions.

In August 2019, in response to the mass actions of Extinction Rebellion – a very gentle foreshadowing of a mitigation crisis – the

---

61 Robertson, *The Malthusian*, 89; see further for example 62, 139–44, 165, 223–6.
62 Cf. Klein, *On Fire*, 50; Alvarez, *Unstable*, 60, 83–7, 129.

Spectator published a piece on the real problem. The writer, Lionel Shriver, claimed to be 'agnostic' about the reality of climate change, but accepted that it is caused by overpopulation. Contrary to what the 'rabble-rousers' alleged, this crisis (insofar as it exists) is not at all 'the West's fault': it is the doing of the people of Africa and the Middle East, who insist on having too many children. Soon those children will move towards Europe, and they 'will all be branded climate refugees. But in the main, on a continent [Africa] that has never been especially hospitable to human life, they will really be population refugees.'[63] Some sort of inherent deficiency in the African continent makes it extrude a surfeit of human beings who have no moral right to come here. The *Spectator* thus seemed to align with Der Flügel. Shriver pretended that the solution to any climate crisis is the distribution of contraception in Africa and the Middle East, but her argument cast another shadow onto the future: when conflicts over this crisis heat up, some will – they already do – point their fingers at non-white people simply because they exist. It is a line pregnant with violence.

Whether it emerges out of a mitigation or an adaptation crisis, fossil fascism would have similar dispositions. And then one can, of course, envision climate breakdown interacting with any number of other crises, pertaining to finance, unemployment, inequalities, geopolitics, armament, soil quality, biodiversity, insect collapse, the state of the oceans; indeed, a *pure* climate crisis is hard to conceive. Candidates for entry into an overdetermined 'existential crisis of the nation and of modern liberal civilization' – what Breivik supposedly lacked – do not seem in short supply. Each ecological component has an inbuilt logic similar to global heating. Allowed to fester, it may, sooner or later, spark a mitigation crisis in which some stratum of the material base of the capitalist mode of production – intense industrial agriculture being perhaps first in line – has to be brought under public control and restructured root and branch, or else – or at the same time – a series of adaptation crises will kick in. The far right could respond similarly to them. (Indeed, for Shriver of the *Spectator*,

---

63 Lionel Shriver, 'Contraception is the Answer to Climate Change', *Spectator*, 17 August 2019.

not only the climate crisis but 'every other global headache you care to name' – from species extinction to desertification – is driven by overbreeding in Africa and the Middle East.)

But one could also imagine a scenario where the far right rapidly sinks into irrelevance in a kairotic moment for climate action, or in a mega-firestorm or desiccation event that might not be conducive to any kind of politics at all. The climate crisis could perhaps run its course in an entropic fashion, the human enterprise unravelling and its internal quarrels coming to an end not with a bang but a whimper. Or solar geoengineering could work so well that the whole dynamic is defused. Or put on hold. Or switched onto some other track, perhaps a kind of biogeochemical Bonapartism where the executive power of the aerosol injectors becomes so autonomous as to approach a dictatorship, with or without mass support – in short, anything can happen.[64] All depends on political battles and how they work their way through nature. Its just that some scenarios do a better job than others in extrapolating tendencies present in the early twenty-first century. Novelists working in the genre of climate fiction, or cli-fi, have recently showed how.

Imagining the Change

One can, of course, envision mitigation and adaptation crises fusing into one. Given the resilience of business-as-usual over the past decades, a mitigation crisis might be unlikely without a very deep attendant adaptation crisis: radical emissions cuts enacted during a summer when, say, a fifth of European territory is on fire. But can we count on reason to prevail even then? In his novel *American War* from 2017, Egyptian-American writer Omar El Akkad imagines the fate of a United States battered by global heating in the second half of the twenty-first century. Whole chunks of the Southern states have been swallowed by the sea. The levees have been washed away for

---

64 For considerations on geoengineering along these and other lines, see J. P. Sapinski, Holly Jean Buck and Andreas Malm (eds), *Has It Come to This? The Promises and Perils of Geoengineering on the Brink* (New Brunswick: Rutgers University Press, 2020).

good; storm surges move back and forth across the remains of Mississippi and Louisiana; waves of refugees flow northwards; the capital has been relocated inland from Washington, DC. In the early 2070s, the president finally introduces 'the Sustainable Future Act, a bill prohibiting the use of fossil fuels anywhere in the United States'.[65] He is promptly assassinated. In 2074, the string of Southern states most devastated by the rising sea and the scorching heat secede from the Union in protest against the Act and plunge the country into civil war.

The analogy with slavery is, of course, no accident. A major difference is the circumstance that the Southern states would now seem to have the most to gain from the prohibition – and yet 'here, in the South, an entire region decided to wage war again, to sever itself from the Union rather than stop using that illicit fuel responsible for so much of the country's misfortune'. The exceptional regime in *American War* is a war machine, a Sparta of fossil fuels, geared to defend the rights of combustion against the infringements of the North. Despite the analogy with slavery, we hear little of racist violence; instead the violence, which includes mass terror against the civilian population, is directed straight at the Northern enemy. Here ultranationalism does not deflect from the threat of mitigation, but attacks it frontally. The ideology of the Southern cause is a generalised reverence for all things fossil. Anyone caught burning the fuel in territories controlled by the North is arrested, but in the South it is the object of a cult. Owning a vehicle powered by oil speaks 'not only of accumulated wealth, but of connections, of status'.[66] Drawn to the memories of oil-fuelled grandeur, clamouring for a new nation and 'excited for a fight', masses in the South rally to the flag; the secessionist armies have an authentic following far beyond the magnates of the oil industry. Young Sarat, the protagonist of *American War*, eagerly volunteers to join. On her way to the front, 'an old, fossil-powered muscle car scream[s] past the bus, its hood decorated with a stylized rattlesnake'. The war claims millions of lives. As a veteran, Sarat

---

65 Omar El Akkad, *American War* (New York: Alfred A. Knopf, 2017), 22.
66 Ibid., 18, 20. 'A small group of children ran aside the old fossil car, knowing through instinct that the driver of a such a thing must reside above the masses.' Ibid., 232.

remembers 'the old wartime footage of hollering Southerners on the back of huge fossil trucks, revving their engines in defiance.'[67]

If an adaptation crisis touches off a mitigation crisis in *American War*, it could also go the other way. Fascist outcomes of mitigation crises would close the window for preventing ever-deepening adaptation crises and increase the likelihood that these in turn have fascist outcomes. A rightwards management of a crisis at one pole presumably boosts a further shift to the right at the other. Something along these lines appears to have happened in the United Kingdom described at an unspecified future date in *The Wall*, British writer John Lanchester's novel from 2019. A wall has been built around the entire shoreline of the shrunken island country, ten thousand kilometres long, on average five metres high, with a watchhouse every three kilometres and ramparts and barracks and helipads made of concrete. The Wall protects against 'the Others'. A fittingly floating residual category, 'the Others' are those people whose lives elsewhere have been ravaged by 'the Change', as global heating is officially referred to. Some of them try to make their way to the relative safety of Britain, whose people still enjoy decent living standards and occasionally receive 'some big-picture news about crops failing or countries breaking apart or coordination between rich countries, or some other emerging detail of the new world we were occupying since the Change'.[68]

Lanchester drops enough hints about the ethnicity of the Others: one individual is black; one family speaks Swahili; a group of infiltrators has embarked from sub-Saharan Africa. They come in boats. 'They come in rowing boats and rubber dinghies, on inflatable tubes, in groups and in swarms and in couples, in threes, in singles; the smaller the number, often, the harder to detect.' If any of them manage to overrun the Defenders and breach the Wall, they will be caught, because all members of the nation have a digital chip necessary for residence. Captives then face three choices: euthanasia, deportation back into the sea or enlistment as slave servants for the rich. As in *American War*, possession of a fossil-fuelled vehicle is a

---

67   Ibid., 65, 58, 278.
68   John Lanchester, *The Wall* (London: Faber & Faber, 2019), 11.

token of might; the sound of a motorboat is reassuring for the insiders. The elite continues to take to the sky in private jets. Planes and drones are used for shooting down approaching Others, who, in the middle of the novel, are projected to arrive in new groups of 'drowning, dying and desperate' people fleeing temporary refuges serially devastated by the advancing Change. A member of the elite prepares the Defenders for the coming killing spree by praising the Wall as 'the best in the world. You are the best in the world. This country is the best in the world.'[69] The sea around the nation that first developed a fossil economy is a single moat of blood.

A lesser cli-fi novel from 2019, which incidentally also refers to climate collapse as 'the Change', paints an even grislier picture: the heat drives refugees towards Scandinavia, where the governments set up extermination camps as 'the only way to control the influx. At least, that's what they hoped. They ran liquidation shifts day and night and the skies turned black with incinerated flesh.'[70] But the surge of the far right in the 2010s was not a precondition for such dystopias; already in 1993, from her black feminist vantage point, Octavia Butler imagined similar devolutions in *Parable of the Sower*. Here it is Southern California suffering a withering drought in the late 2020s. A hurricane kills seven hundred in one day, tornadoes are 'smashing the hell out of' Southern states, 'sea levels keep rising with the warming climate' and Americans still debate whether it's all real. 'People have changed the climate of the world', says the protagonist, whose friend retorts: 'Your father says he doesn't believe people changed the climate in spite of what scientists say.'[71] A newly elected president promises to take the country back to the good old days – the days when gasoline still fuelled cars and trucks – and sets about dismantling environmental protection laws and breaking up scientific outfits. Refugees trek towards northern latitudes, trying to make their way through white enclaves, work conditions bordering on slavery and hostility to newcomers of colour: elements of *American War* and *The Wall* wrapped into one narrative.

---

69 Ibid., 35–6, 112.
70 Grant Price, *By the Feet of Men* (Winchester: Cosmic Egg Books, 2019), 182.
71 Octavia E. Butler, *Parable of the Sower* (New York: Grand Central, 2007 [1993]), 54, 118, 57.

Obviously, future horror is the staple of this genre, which cannot aspire to the accuracy of something like integrated assessment models – but then again political crises in a warming world might not be amenable to computer simulations. Cli-fi authors can take the pulse of the present. 'The logic of the situation' is now such that the plots of *American War* and *The Wall* do not come across as crazily implausible. Marxist anti-fascists used 'the logic of the situation' as a barometer of the pressure in the interwar period and found that it forecast danger; *mutatis mutandis*, we can do the same and speak of *fascist tendencies* in the present.[72] Full-blown fascism as a real historical force – fossil or otherwise – is not yet found in any country we have examined, but we can make out tendencies operating in that direction: nationalist politics on the rise; deep crises afoot; dominant class interests in realignment. Another term for such tendencies is *fascisation*, used by anti-fascists in the interwar period who saw their societies sliding down the slope.[73] Perhaps we can now talk of fossil fascisation.

The logic of a mitigation crisis with a fascist outcome appears to have been mildly prefigured in recent history: in primitive fossil capital affiliating with Trump after the victories of the climate movement in the late Obama era; in the limits on deforestation spurring Brazilian agribusiness to cast its lot with Bolsonaro; in the AfD's response to the German mobilisations; in the spread of the syndrome of selection. The defiant revving up of engines can be faintly discerned in the behaviour and result of the SD in the Swedish elections of 2018. Indeed, one way of understanding the rise of the far right in the early twenty-first century is to see it as a reaction – first pre-emptive, later direct – to the quickly approaching crunch time of the climate crisis. Some of the deep structural forces in society that resist any transition appear to have gravitated towards this political pole. The worse the crisis, on this logic, the stronger the attraction, and these tendencies pertain not only to mitigation.

---

72   See for example Beetham, 'Introduction', 30; cf. Poulantzas, *State*, 209–10.

73   The term is primarily associated with the Comintern and its disastrous theory of social-fascism, but it was also used by its sharpest critic, Trotsky, who likewise identified a process of 'fascisation' in Germany, the reactionary governments of late Weimar period paving the way for Hitler. See Trotsky, *The Struggle*, for example 95, 263.

If we accept the indications that an unprecedented anthropogenic drought in the Fertile Crescent was one of the many sparks that set Syria on fire, then we can see the European reaction to the so-called refugee crisis of 2015 as a portent of adaptation crises to come.[74] The treatment of migrants along the US border is another omen. The causal link between climate shocks and migration from Central America might have been more direct than in the Syrian case, the poverty exacerbated by dislocated weather patterns, floods alternating with droughts, inundated towns, lost harvests.[75] When Hurricane Dorian destroyed the Bahamas in September 2019, 199 survivors landing in Florida were turned around, Trump explaining that he wouldn't let in a group with no right to be in the US, as it included 'some very bad people and some very bad gang members and some very, very bad drug dealers'.[76] It did not require a vivid imagination to believe that this designation had something to do with the blackness of the flotsam.

We do not seem to be on a trajectory that takes us safely away from scenarios of fossil fascism. To the contrary, some major strands of recent Western history, from the responses to the Hansen testimony in the unseasonally hot summer of 1988 to the walls going up in the

---

74 See for example Colin P. Kelley, Shahrzad Mohtadi, Mark A. Cane et al., 'Climate Change in the Fertile Crescent and Implications of the Recent Syrian Drought', *PNAS* 112 (2015): 3241–6; Colin P. Kelley, Shahrzad Mohtadi, Mark A. Cane et al., 'Commentary on the Syria Case: Climate as a Contributing Factor', *Political Geography* 60 (2017): 245–7; Guy J. Abel, Michael Brottrager, Jesus Crespo Cuaresma and Raya Muttarak, 'Climate, Conflict and Forced Migration', *Global Environmental Change* 54 (2019): 239–49; and for more references and an attempt at historical-materialist analysis, Andreas Malm, 'Revolution in a Warming World: Lessons from the Russian to the Syrian Revolution', in Leo Panitch and Greg Albo (eds), *Rethinking Revolution: Socialist Register 2017* (London: Merlin Press, 2016), 120–42. The nexus of climate and migration is highly complex and the literature voluminous; we cannot review it here. For a typical survey of the research, see Cristina Cattaneo, Michel Beine, Christiane J. Frölich et al., 'Human Migration in the Era of Climate Change', *Review of Environmental Economics and Policy* 13 (2019): 189–206.

75 See for example the dispatches in the *Guardian* by Nina Lakhani: ' "People Are Dying": How the Climate Crisis Has Sparked an Exodus to the US', 29 July 2019; 'Living without Water: The Crisis Pushing People Out of El Salvador', 30 July 2019; ' "It Won't Be Long": Why a Honduran Community Will Soon Be Under Water', 31 July 2019.

76 Arwa Mahdawi, 'The Plight of Hurricane Dorian Evacuees Offers a Frightening Vision of "Climate Apartheid" ', *Guardian*, 20 September 2019.

2010s, can be read as adumbrations of this sort of future. It seems hard to demarcate sharply from the recent past. Primitive fossil capital already has operated as the de facto elite force of the totality of fossil capital and partnered with the far right, hiding behind and benefitting from its interpellation of the white nation. Its bankrolling of denial and obstructionist lobbying have protected fossil capital in general from a structural crisis *and the partnering with the far right has had the same effect.* All of this has occurred because fossil capital – the totality of it – has faced a gestating mitigation crisis and spared no efforts in preventing it from breaking out. And the far right already has a radiative forcing in the climate system. The atmosphere of the future will have a political composition: some of the CO_2 will have been conducted there by forces like the AfD, the FrP, the PiS, Fidesz, the Trump and Bolsonaro cabinets. They have fanned the actual, material flames, which in turn could send adaptation crises their way; the feedback loop might already be in motion.

'Postfascism' is an infelicitous term in this context. It implies that the contemporary far right is that which comes *after* fascism has happened, much like postmodernism succeeded modernism.[77] The question to ask rather seems to be if and how it *anticipates* fascism-as-force: in what ways it represents a *pre*-fascism or *proto*-fascism that may develop towards the end station *unless actively and effectively resisted.*[78] ('*Late* fascism' is also a preferable term, since lateness is a feature of the whole ordeal.) But we should not run ahead of events. Instead of staying with the task of trying to foresee the political future of a rapidly warming world, we can return to the ideas about nation and energy on the actually existing far right. More lurid speculations about how bad things can become are best left to the novelists.

---

77   Cf. Traverso, *The New*, 4.
78   This follows Richard Seymour, 'What Trump Isn't', *Patreon*, patreon.com, 28 December 2018.

8
Mythical Energies of the Far Right

If someone were to ask for a pedagogic example of the myth of palingenesis, 'Make America Great Again' would be an easy choice. Once upon a time, (white) America was great, probably greatest in the world, a nation with an innate greatness living out its destiny. Then a period of decay set in, thanks to the evil ones who captured the nation. Now America will become great again, almost born anew – not by regressing to the 1950s or 1910s, but by resuming the journey the evildoers cut short.[1] The myth of palingenesis never recommended a return of the same-old, same-old, a literal restoration of the Roman or Germanic order from one or two millennia ago; rather it pictured that past as a launching pad for a sparkling future. Looking back at the virtues and achievements of the Romans and Aryans helped reignite the nation for new great deeds, with the emphasis being precisely on the *new*, as in a new age, a New Man, a total renewal shaking off the slow death and taking to the skies: fascism had its eyes set on the future.[2] For all the crises that afflicted interwar Europe, fascism was thoroughly permeated by optimism about the enhancing forces of technology and a confident faith in industry (as we shall see in more detail below). Palingenesis was a myth in step

---
1 Cf. Finchelstein, *From Fascism*, 235.
2 Griffin, *The Nature*, for example 33–9, 47.

with its times. It worked as the 'mythical core' of fascism, in the sense that it spoke to the gut feelings of broad layers of people and inspired them to believe in the project of lifting the nation by extirpating its enemies.[3]

It may be that the myth of palingenesis will never go fully out of fashion. But it is now also joined by another myth, with a different conception of time and political tonality, of particular prominence in Europe. It encapsulates the ideas with which the far right moves deeper into a warming world. As related above, on the day after the Swedish election in 2018, Mattias Karlsson of the SD posted a picture of soldiers solemnly bowing to their great eighteenth-century king. He claimed that Sweden faced a time of tribulation, just as it had done

> before in history. It has seemed hopeless many times. We have been occupied by alien states, foreign lords have oppressed the people, we have been under attack and stood alone against a united force of European superpowers and experienced overwhelming numerical inferiority. Yet we have always resisted

and defended ourselves and pushed the enemy back to where he came from, thanks to the supreme heroism of a patriotic guard, whose clothes were donned by the Sweden Democrats.[4] Then as now, it is 'victory or death'.

Clearly meant to galvanise disciples in their fight against today's invaders, the message was mythical in another sense too: it had no relation to actual history. On any count, Sweden must be one of the countries in the world that has suffered *least* from occupation or incursion or the meddling of foreign states (unless one counts the colonisation of Sami territory in the north of present-day Sweden). The trials Karlsson described never happened. The fact that he invented a history of victimisation out of thin air, however, did not necessarily impair the efficiency of his communiqué. The myth in

---

3 For example ibid., 29–34, 73–5, 105–6; Griffin, 'The Primacy', 27; Griffin, *A Fascist*, 4–5, 7–8, 17; cf. Eley, *Nazism*, 210–12; Eley, 'Fascism', 105.
4 Mattias Karlsson's Facebook page, 10 September 2018.

question has both an exceedingly tenuous link to past (and present) realities *and* an ability to speak to the gut feelings of people. It had been popular among leaders of the Sweden Democrats long before Karlsson's post, such as when Ted Ekeroth explained that 'Islam has been in permanent war with the rest of the world since the days of Muhammed'. A first invasion occurred in the 630s. A second was halted at Poitiers in 732. A third was repulsed in the Battle of Vienna in 1683 *and here we are again, fighting the same enemy.* Or, as an SD intellectual using the pseudonym Karl Martell, after the hero who struck down the Muslim armies at Poitiers, wrote in the party journal: this same old foe is now 'invading our territories' by means of 'demographic warfare agents'.[5] Only Martell has yet to strike this time.

This is *the myth of palindefence*. It says: we defended ourselves and our inestimable estate in the past; we were under siege but eventually rebuffed the enemy; we fought hard and gallantly for what will always be ours and *now we have to do it again.* Much like the myth of palingenesis, it deploys a certain vision of the past to stake out the present duties in the service of the nation. But where the former works in a *generative* mode and aims at a resplendent future, the latter operates in a *defensive* register and seeks to foment aggressive protection of existing traits and property. In palingenesis, a period of decadence has plagued the nation and must now come to an end. A new greatness will ensue, the inner energy of the nation reactivated. In palindefence, the existential threat has been encroaching on the nation since time immemorial, sometimes raiding the homeland, sometimes sulking around its borders, never entirely absent and now standing at our gates again. Only if we act with the energy of our forefathers will our way of life be saved for present and future generations.

The myth of palindefence became immensely popular in the early twenty-first-century European far right and the body of ideas that nourished it, such as Eurabia and the Great Replacement. Bat Ye'or did more than anyone to embellish the notion of Islam as a

---

5 Ted Ekeroth, 'Åsikter är åsikter, Malm!', sdblogg.se, 15 February 2008; Karl Martell, 'D-stridsmedel', *SD-Kuriren*, no. 54, October 2003.

permanent war against the West. From the seventh until the late twentieth century, the leaders of Europe were wise and valorous enough to fight back, but then the fateful treachery of the Euro-Arab Dialogue supervened.[6] For Mark Steyn, the Battle of Poitiers had been re-enacted with – so far – the opposite results: the Muslim armies had already reached the suburbs of Paris where they burned cars and the streets of Malmö where they blocked ambulances.[7] In the years of the War on Terror, so crucial for the Islamophobia that subsequently seized swathes of Europe as an organised political force, this myth was enunciated by outlets in the Western mainstream.[8] It was then sucked up by the fighters on the streets.

There arose the English *Defence* League, with offshoots across Europe, wielding symbols from the Crusades. The EDL vowed to 'defend our land from 1,400 years of jihad' and, addressing its Muslim enemies, to 'contest your kind, as our forefathers did, relentlessly pursuing you in our quest to see all shari'a banished from our great democratic country' (this evocation of the Crusades being, of course, one of many points where the notion of 'defence' collapsed into its opposite).[9] In France, Génération Identitaire emerged as the main direct-action group practising the ideas of the Great Replacement – 'the Greenpeace of the right', in the words of one of Alain de Benoist's journals – and made headlines in 2018 when it sent a hundred activists to patrol the border in the mountains between France and Italy, marching in blue windbreakers with 'DEFEND EUROPE' written on the back, surveilling the woods with a helicopter and addressing potential immigrants with a giant banner: 'Closed border. You will not make Europe your home. No way.' But that wasnt

---

6 Ye'or, *Eurabia*, for example 9–10, 190–4.
7 Steyn, *America*, 123–4.
8 On the War on Terror and its production of Islamophobia, see for example Sherene H. Razack, *Casting Out: The Eviction of Muslims from Western Law and Politics* (Toronto: University of Toronto Press, 2008); Kundnani, *The Muslims*; Lean, *The Islamophobia*. For examples of mainstream palindefensive myth-making in this period, see for example Niall Ferguson, 'Eurabia?', *New York Times*, 4 April 2004; Frits Bolkestien in Alain Gresh, 'Malevolent Fantasy of Islam', *Le Monde Diplomatique*, mondediplo.com, August 2005; Bruce Anderson, 'We Are Destroying the Very Values Which Could Save Us in Our Battle Against Islam', *Independent*, 2 June 2008.
9 Quoted in Kundnani, *The Muslims*, 237, 239.

the group's first newsworthy action. What marked the birth of the Génération was the occupation of a mosque in the city of Poitiers.[10]

The website 'Gates of Vienna', central to the counterjihad movement, adopted the following motto: 'At the siege of Vienna in 1683 Islam seemed poised to overrun Christian Europe. We are in a new phase of a very old war.'[11] The parties of the far right could not get enough of such mottos. In the city of Vienna itself, the FPÖ arranged an opulent celebration of the 333rd anniversary of the end of the Ottoman siege under the slogan 'Defending the Occident, then as now'.[12] It also produced a booklet retelling the battle in cartoons, complete with Muslim soldiers riding around with Christian babies impaled on their lances. Thankfully, early modern superhero Heinz-Christian Strache appears on the walls of Vienna, dressed in a blue knight's gown, and teaches an enthusiastic blond boy how to 'burn a Mustafa' (that Mustafa, again).[13]

The Lega had a special fetish for the Battle of Lepanto. In October 1571, a fleet led by the Venetian Republic sailed across the Ionian Sea and trounced the Ottoman fleet in a titanic naval engagement that arrested further advances from the east into the Mediterranean. In October 2000, 429 years after that battle, Lega Nord leapt into the public eye when it descended on the small town of Lodi, where some workers of Moroccan, Senegalese and Albanian descent had filed an application for a place of worship. The Lega activists marched into the town with flags and banners copied from the Battle of Lepanto.

---

10   Karina Piser, 'The Swiftly Closing Borders of Europe', *Atlantic*, 19 December 2018; Cécile Chambraud, 'Génération identitaire poursuivie pour l'occupation de la mosquée de Poitiers', *Le Monde*, 21 October 2017. On the Great Replacement, see Davey and Julia Ebner, 'The Great', 5, 8. On 'Greenpeace of the right': Clément Martin, 'Génération Identitaire, le "Greenpeace de droite"', *Éléments*, October 2019. The group subscribed to standard French green nationalism. See for example Valeurs Actuelles, 'La remigration est la seule solution à la hauteur des défis de notre temps', YouTube, 19 November 2019.

11   gatesofvienna.net.

12   *Der Standard*, 'Wiener FPÖ feiert Ende der Türkenbelagerung', 7 September 2016.

13   *OE24*, 'FPÖ-Comic sorgt vor Wien-Wahl für Aufregung', oe24.at, 25 September 2010. Cf. Hans-Georg Betz and Susi Meret, 'Revisiting Lepanto: The Political Mobilization against Islam in Contemporary Western Europe', *Patterns of Prejudice* 43 (2009): 320.

Chanting 'No to the Islamic invasion', they reached the site of the proposed Islamic centre – not a mosque proper – and poured urine from pigs on the soil. The party continued its long march with the promise to once again throw Islam back to where it came from: the other side of *mare nostrum*.[14] On his way to power over the fate of fleeing people approaching the shores of Italy, Matteo Salvini repeatedly invoked 'the spirit of Lepanto'.[15]

Vox styled its whole political project as a new *reconquista*. To kick off the election campaign that took it to parliament, the party congregated with flags at the cave and statue in Covadonga, the little corner of Asturias where the lengthy 'reconquest', culminating in the final expulsion of Moors and Jews from Spain, is supposed to have begun in 718 or 719. 'Europe is what it is thanks to Spain – thanks to our contribution, ever since the Middle Ages, of stopping the spread and the expansion of Islam', a Vox leader said on his way to the cave.[16] The mythology of the *reconquista* also crossed the Atlantic to Brazil, where the far right rediscovered the European Middle Ages as its

---

14 Chantal Saint-Blancat and Ottavia Schmidt di Friedberg, 'Why Are Mosques a Problem? Local Politics and Fear of Islam in Northern Italy', *Journal of Ethnic and Migration Studies* 31 (2005): 1089–90; Stefano Allievi, 'Sociology of a Newcomer: Muslim Migration to Italy – Religious Visibility, Cultural and Political Reactions', *Immigrants and Minorities* 22 (2000): 151; Betz and Meret, 'Revisiting', 327, 334.

15 See for example Matteo Salvini's Twitter account, 7 October 2012 and 7 October 2015 (both anniversaries of the original 'victory against the Islamisation of Europe'); Paolo Farinella, 'Matteo Salvini: giurando su Vangelo e Costituzione, si è fregato da solo', *Il Fatto Quotidiano*, ilfattoquotidiano.it, 27 February 2018. For a typical publication by the party, see Andrea Stella, 'Lepanto la battaglia della cristianità', *Lega Nord Belluno*, 6 October 2014.

16 Miguel González, 'Abascal lanza desde Covadonga su cruzada contra separatistas y "progress"', *El País*, 13 April 2019; Iván Espinosa de los Monteros of Vox quoted in Sohail Jannessari and Darren Loucaides, 'Spain's Vox Party Hates Muslims – Except the Ones Who Fund It', *Foreign Policy*, 27 April 2019. See for example Patricia R. Blanco, 'Ni España existía ni la Reconquista es tal y como la cuenta Vox', *El Pais*, 12 April 2019; *El Mundo*, 'Un día con Santiago Abascal (Vox) en la Granada "reconquistada"', YouTube, 21 April 2019. Ironically, the start-up capital for Vox in 2013 came from the generous sponsor Mojahedin-e Khalq, the anti-Khomeinist Iranian sect (which thereby set some new record for the degeneration of Iranian diaspora politics): possibly the first major Islamophobic party in Europe to be set up with money from Muslims. See Jannessari and Loucaides, 'Spain's Vox'. But, then again, the Christian-Muslim allies during the centuries of the *reconquista* were famously flexible.

cradle. It nurtured a narrative of the white Brazilian nation as the extension of the liberation of the Iberian Peninsula from Islam. In one popular documentary called *The Last Crusade* from 2017, intellectuals of the Bolsonaro sphere – foremost among them Olavo de Carvalho – took a 'dive into the remote origins' of Brazil, namely the moment when Charles Martel halted the Muslim advance into Europe. From this moment, the Christians holding out in the kingdom of Asturias began to roll back the Moors, until, after seven gruelling centuries, they had 'only the sea left to conquer': and so Brazil was born. The function of this newfound medievalism was, of course, to specify the nation as white in origins – not indigenous, not African, by no means polymorphous.[17]

But it was in Eastern Europe that the myth of palindefence held the greatest sway. When the Fidesz government began erecting two rows of four-metre-high steel fence topped with barbed wire along Hungary's borders with Serbia and Croatia in 2015, so as to close the route migrants had used to travel to Germany and Scandinavia in search of asylum, it conjured up a millennium of bittersweet memories. No one had withstood the onslaught of Muslim armies with greater stamina than Hungary. It had paid the highest price, but always kept its fortress solid and proud. Throughout the centuries when the Ottoman Empire pushed into Europe, the Hungarian nation had been the forward defence base, without getting the recognition it deserved from other Europeans, safe in their homes: and now it was the same story all over again.[18] It was a narrative embedded in the deepest and broadest layers of Hungarian nationalist ideology. In polls conducted in 1993 and 2000, nearly 75 per cent of respondents from all walks of life agreed with the statement that 'for a thousand years, Hungary was the bulwark of the West, but we, even today, have never been thanked for this.'[19] Fidesz and Viktor Orbán

---

17   Brasil Paralelo, 'A Última Cruzada', YouTube, 20 September 2017. Cf. Paulo Pachá, 'Why the Brazilian Far Right Loves the European Middle Ages', *Pacific Standard*, 18 February 2019.

18   Viktor Glied and Norbert Pap, 'The "Christian Fortress of Hungary": The Anatomy of the Migration Crisis in Hungary', *Yearbook of Polish European Studies* 19 (2016): 140–2, 148.

19   James W. Scott, 'Hungarian Border Politics as an Anti-politics of the European Union', *Geopolitics* (2018), online first, 11.

had a ready-made legend for the so-called refugee crisis, with the potential to reinforce their position as commanders of the fort: once again, Hungary had took it upon itself to defend Europe from a Muslim invasion, and if some liberals in Brussels whimpered about the harsh methods, it was yet another sign of ingratitude.[20]

The myth could be extended to other enemies, past and present. 'We sent home the sultan with his army, the Habsburg Kaiser with his raiders and the Soviets with their comrades. Now we will send home Uncle George', Orbán thundered to a hundred thousand supporters in Budapest before the election of 2018 – 'Uncle George' being, of course, Soros (a native of the country).[21] At this point, the myth had a palpable if diffuse impact on people's lives. It provided the ideological coating for a major section of the European wall that condemned migrants to hold out as best they could on the high waves of the Mediterranean, under the bombs falling over Aleppo or on the parched fields of drought-stricken Iran. But no matter how effective the Hungarian fences were in blocking the influx, with few if any migrants in sight, the Fidesz government in 2017 resolved to upgrade the construction with thermal sensors and electric shocks to anyone touching it.[22] Measures like this turned Hungary into a utopia for much of the far right in Western Europe, a land attracting longing glances from *inter alia* Vox, which repeatedly cited it as the model for the Union. The Spanish party raised its knightly banners against 'migratory invasion', against Soros and his globalism and, of course, against Islam, all of which enemies were folded into the myth of centuries of defensive struggles uniting white Christians from Cordóba to Buda in a timeless bond.[23]

---

20 See further Céline Cantat and Prem Kumar Rajaram, 'The Politics of the Refugee Crisis in Hugary: Bordering and Ordering the Nation and Its Others', in Cecilia Menjívar, Marie Ruiz and Immanuel Ness (eds), *The Oxford Handbook of Migration Crises* (Oxford: Oxford University Press, 2018), 181–96; Scott, 'Hungarian'.
21 Witte, 'Once-Fringe'.
22 Marton Dunai, 'Hungary Builds New High-Tech Border Fence – with Few Migrants in Sight', Reuters, 2 March 2017; Cantat and Rajaram, 'The Politics'.
23 Nacho Alarcón, 'Qué es el Grupo Visegrado: El Modelo de Vox para la UE', *El Confidencial*, 21 February 2019; Beatriz Rios, 'Spanish Far-Right Joins Orban in his Defence of the Europe of Nations', *Euractiv*, 7 May 2019

Meanwhile, in Poland, the PiS drew on its own antique ideology of the nation as a *przedmurze* – bulwark – of Europe or, in the language of Catholicism, *antemurale Christianitatis*. It was constructed in the late Middle Ages and early modern era, when Polish kings and their diplomats worked tirelessly to elicit support from the Vatican and other allies in their wars against Turks and Tatars: fund us, arm us, because we are blocking the menace for the benefit of you all. Created by the top tiers of the feudal Polish state, the idea of the *antemurale* trickled down to popular layers in the seventeenth century and ripened as a myth of national nobility during the nineteenth, when nationalists in partitioned Poland longed for their lost independence.[24] The heroism of Jan Sobieski, who arrived with his Polish cavalry at the gates of Vienna and lifted the Ottoman siege at the eleventh hour, was the finest moment in this history. With the approach of the so-called refugee crisis, the PiS knew exactly what national pride to summon. And yet both Hungary and Poland must so far be considered second runners in the game of palindefensive myth-making. It was another Eastern European country that claimed first place.

While hundreds of diplomats and delegates made their way to Rio de Janeiro to hammer out the UNFCCC in the early summer of 1992, the tanks of the Serbian army and the jeeps of the Chetniks rolled ever deeper into the green valleys of Bosnia. Among the towns that fell in these weeks was Foča. A typical Bosnian population centre, with some twenty thousand inhabitants identifiable as Muslims and roughly as many as Serbs, its fate was like that of any other place subjected to occupation. The Muslims were rounded up and detained. They had their valuables taken away from them. Their houses were methodically burnt to the ground, while fire brigades protected neighbouring Serbs from the flames; every mosque was blown up and bulldozed until only gravel remained. Commanders gave orders

---

24 For extensive documentation of how the myth was constructed over these centuries, see Paul W. Knoll, 'Poland as "Antemurale Christianitatis" in the Late Middle Ages', *Catholic Historical Review* 60 (1974): 381–401; Janusz Tazbir, 'From Antemurale to Przedmurze, the History of the Term', *Odrodzenie i Reformacja w Polsce* 61 (2017): 67–87; Katarzyna Niemczyk, 'Antemurale Christianitatis? Anti-Turkish Propaganda and the True Goal of Johannes Olbracht's Crusade', *Tyragetia, Serie Nouă* 12 (2018): 31–43.

to 'shoot Muslims'. Some three thousand of them – virtually all unarmed civilians – were killed in Foča alone. Special detention centres were set up for women. For months on end, Serbian nationalist soldiers entered those centres and raped the women in groups and told them they would never again give birth to Muslim babies. In January 1994, the town was rechristened 'Srbinje', literally 'the place of the Serbs'; no traces of a Muslim presence then remained. In its investigation of the events in Foča, the Hague tribunal later observed that 'the sole reason for this treatment of the civilians was their Muslim ethnicity'.[25] Why did the Serbian nationalists behave in this way? Or, rather, what did they *think* they were doing?

'We defended Europe from Islam six hundred years ago', said Radovan Karadžić, president of the Serbian nationalist entity known as Republika Srpska: 'We are defending Europe again.'[26] During the prelude to the war, no theme loomed larger than the endurance of the Serbs on the cross where the Muslims had nailed them. More precisely, the Serbian nation entered eternity through the Battle of Kosovo Polje, or the Field of Blackbirds in 1389. There the troops from the remainders of the feudal Serbian empire met Ottoman forces and fought to the moment of bitter defeat. Legend has it that their leader, Prince Lazar, a Christ-like figure and nationalist martyr, was killed because a Muslim-loving Serbian Judas betrayed him.[27] In 1989,

---

25 United Nations International Criminal Tribunal for the Former Yugoslavia (ICTY), 'Prosecutor vs. Dragoljub Kunarac, Radomir Kovac and Zoran Vukovic: Judgement', icty.org, 21 February 2001, 205. For the events in Foča, see this document, for example 16–21, 28–31, 203–9. The best accounts of the ethnic cleansing of Bosnia are Roy Gutman, *A Witness to Genocide: The 1993 Pulitzer Prize-Winning Dispatches on the "Ethnic Cleansing" of Bosnia* (New York: Macmillan, 1993); Ed Vulliamy, *Seasons in Hell: Understanding Bosnia's War* (London: Simon & Schuster, 1994); Norman Cigar, *Genocide in Bosnia: The Politics of 'Ethnic Cleansing'* (College Station: Texas A&M University Press, 1995); Michael Sells, *The Bridge Betrayed: Religion and Genocide in Bosnia* (Berkeley: University of California Press, 1998).

26 Quoted in Cigar, *Genocide*, 65.

27 Sells, *The Bridge*, 31, 39–40, 63; Michael Sells, 'The Construction of Islam in Serbian Religious Mythology and Its Consequences', in Maya Shatzmiller (ed.), *Islam and Bosnia: Conflict Resolution and Foreign Policy in Multi-ethnic States* (Montreal: McGill-Queen's University Press, 2002), 63; Vamik D. Volkan, 'Bosnia-Herzegovina: Chosen Trauma and Its Transgenerational Transmission', in ibid., 86–97; David Bruce Macdonald, *Balkan Holocausts? Serbian and Croatian Victim-Centered Propaganda and the War in Yugoslavia* (Manchester: Manchester University Press, 2003), 232; Paul A.

actually existing socialism, including the republic of Yugoslavia, fell apart: and it also happened to be the six hundredth anniversary of the Battle of Kosovo Polje. Squeezing every drop of significance out of the occasion, Slobodan Milošević, heir apparent to the disintegrating bureaucratic apparatus in Belgrade, ordered the body of Prince Lazar exhumed from its shrine and carried by foot to monasteries, all the way down to the original battleground at the Field of Blackbirds. Venerated in macabre revival meetings, the holy relics inspired nationalist pledges such as the following: 'We will do our utmost to crush their race and descendants so completely that history will not even remember them.'[28] In a long series of mass meetings leading up to the anniversary date, feelings of unjust victimisation, of Serbs as heroes daring to fight against hopeless odds but never receiving their due – until now – acquired the delirium of a racial frenzy. The bureaucracy in Belgrade swung behind it, as did Serbian television, press outlets and intellectuals, the ideological state apparatuses hanging on every word from Milošević and, like him, baying for blood.[29]

The campaign culminated on the day of the battle, a hot 28 June 1989, when Milošević descended by helicopter at the Field of Blackbirds. Collapsing more than half a millennium into a single hyperintense moment of defence, Milošević told the crowd of between 1 and 2 million Serbs:

> Six centuries later, now, we are being again engaged in battles and are facing battles. They are not armed battles, although such things cannot be excluded yet. However, regardless of what kind of battles they are, they cannot be won without resolve, bravery, and sacrifice, without the noble qualities that were present here in the field of Kosovo in the days past.[30]

Cohen, *History and Popular Memory: The Power of Story in Moments of Crisis* (New York: Columbia University Press, 2014), 1–32; Jelka Redep, 'The Legend of Kosovo', *Oral Tradition* 6 (1991): 253–65.

28 Sells, *The Bridge*, 63.

29 See for example Christina M. Morus, 'Slobo the Redeemer: The Rhetoric of Slobodan Milosevic and the Construction of the Serbian "People"', *Southern Communication Journal* 72 (2007): 1–19.

30 Vulliamy, *Seasons*, 51; Sells, 'The Construction', 88–9; quotation from Cohen, *History*, 25.

When battles did break out, the Serbian regiments streaming into Bosnian villages crowed over their palindefensive exploits; upon cleansing a settlement of Muslims, they received medals named after the heroes of 1389. In the first concentration camps Europe had known since the Second World War, guards bawled songs in praise of those heroes and forced their prisoners to join in.[31] The late medieval bourdon laid the base for a shrill hysteria over Islamisation. Serbian nationalists incessantly bellowed reports of sharia laws, claims of Muslim backwardness, revelations of secret plans to turn the continent into an Islamic state and projections of a Muslim demographic takeover, first in the Balkans and then the rest of Europe. The Muslims were spawning into power, which was why, to quote Mark Steyn, they had to be culled. Throughout the war, Karadžić and Milošević and the other commanders never tired of telling Western Europe that it too would be overrun, unless the Serbian troops on the ground were supported and emulated.[32] The future of the continent was that of Bosnia.

And the climax came at Srebrenica, where a platoon under the command of Ratko Mladić in July 1995 netted tens of thousands of famished and desperate civilians who had fled to the enclave during earlier phases of cleansing. UN troops from the Netherlands had been assigned the protection of the 'safe area' but handed it over to the Serbian nationalists and assisted them in their work. The boys and men were placed on buses. They were thrown off at various sites around the town, blindfolded, pushed to the ground, executed and

---

31 Sells, *The Bridge*, 50, 79; Michael Sells, 'Christ Killer, Kremlin, Contagion', in Qureshi and Sells, *The New*, 358; Rezak Hukanović, 'The Evil at Omarska', *New Republic* 214, no. 7 (1996): 29.

32 See for example Cigar, *Genocide*, for example 28–9, 77, 85, 96, 100–1; Norman Cigar, 'The Nationalist Serbian Intellectuals and Islam: Defining and Eliminating a Muslim Community', in Qureshi and Sells, *The New*, for example 326–30; David Rohde, *Endgame: The Betrayal and Fall of Srebrenica, Europe's Worst Massacre since World War II* (Boulder: Westview, 1998), 214; Anthony Oberschall, 'The Manipulation of Ethnicity: From Ethnic Cooperation to Violence and War in Yugoslavia', *Ethnic and Racial Studies* 23 (2000): 982–1001; Renaud de la Brosse, *Political Propaganda and the Plan to Create a 'State for All Serbs': Consequences of Using the Media for Ultra-Nationalist Ends*, Expert Report for the Office of the Prosecutor of the International Criminal Tribunal for the Former Yugoslavia (2003), available at balkanwhitness.glypx.com; Sonja Biserko, 'The Serbian Elites and Genocide in Bosnia', *Helsinki Charter* no. 109–10 (2007): 1–19.

rolled over into mass graves. In some storehouses they were hacked to death with axes and crowbars, in others locked up in rooms where machine guns worked on the bodies for hours on end. More than eight thousand civilians were murdered in the worst massacre in Europe since the Second World War.[33] Twenty years later and counting, Bosnian Muslims still assembled in Srebrenica every July to bury the latest victims, identified through painstaking reconstruction of the disjointed corpses.

What did the Serbian nationalists think they were doing in Srebrenica? Just before marching into the enclave, on 28 June – the 606th anniversary of the Battle of Kosovo – Ratko Mladić swore to restore the glory of the 'Serbian chevaliers, who, although they had fallen in battle, stopped the great Arabic [!] flood that threatened to sink not only our nation but Europe as well'.[34] The war was of a purely defensive nature, imposed on the Serbs: 'We were forced to create a bulwark to protect ourselves from Islamic expansion towards the West. Besides the enemies we are fighting in Bosnia, there is the entire Muslim world', the butcher of Srebrenica spelled out his cosmology before getting to work.[35] His and his peers' battles became a cause célèbre for much of the European far right, including parties such as the DF and the FPÖ.[36] One aspiring intellectual marvelled at the war of liberation and imbibed its gospel to the full. She called herself Bat Ye'or.[37]

The myth of palindefence can do a great deal of political work. With its purported historical depth, it effectively separates two kinds of people: one bound by ancestry to the victims and heroes of the homeland, the other to the interlopers and usurpers from distant lands. The temporal collapse – 1389 is 1989, Poitiers is now – hypostatises a transhistorical antagonism between collectives of human

---

33 The definitive account is Rohde, *Endgame*.
34 Speech included in the hagiographic Milo Yelesievich, *Ratko Mladić: Tragic Hero* (New York: Unwritten History, 2006), 462.
35 Ibid., 439.
36 See Betz and Meret, 'Revisiting', 329–30.
37 For the influence of Serbian nationalism on Bat Ye'or, see for example Bat Ye'or, *The Decline of Eastern Christianity under Islam: From Jihad to Dhimmitude* (Madison, NJ: Fairleigh Dickinson University Press, 1996), 28; Sells, 'Christ Killer', 360–5; Carr, 'You Are', 8; Bangstad, *Anders*, 150.

beings defined by their descent. It supplies the central illusion that 'the generations which succeed one another over centuries on a reasonably stable territory' have 'handed down to each other an invariant substance'.[38] It constructs races. To the sons of Charles Martel or Prince Lazar, the myth has a rousing message: you are selected – nay, obliged – to assume our mantles and weed out the traitors and complete our work, with the beatification passing on to you at the end of the day. Nationalists can obviously take this hyperbole entirely seriously. Here they drink at their wells: 'Perhaps we are no less brave than our forefathers', Lega Nord pondered in a text on Lepanto, but only if 'we learn to think of ourselves as members of a community. A true community, with roots and a history that have shaped us and make us love one another, because we recognise each other as similar to ourselves' and the others in our towns and on our shores as similar to our foes from centuries ago.[39]

The ideological trick here, of course – so typical for nationalist interpellation – is that the brave, authentic, self-identical people is presumed to exist independently of the myth, when in fact the myth is used to call it into being. If identification with the nation came naturally, by origin and birth, nationalists would never need to say anything at all (hence every exertion on their part is a performative contradiction). People could just as well identify with their neighbourhood or gender or dietary habits or pets or class, but nationalists interpellate them as – above every other loyalty – subjects of the nation, and only insofar as they respond to the call do they *become* subjects of a nation that they know and love and will defend against adversaries with all the requisite harshness.[40] By pumping up the

---

38 Balibar in Etienne Balibar and Immanuel Wallerstein, *Race, Nation, Class: Ambiguous Identities* (London: Verso, 2010 [1991]), 86.

39 *La Padania*, 'Lepanto: La battaglia che salvo' l'Europa', cumasch.org, 7 October 2004.

40 Here we follow the theory of nationalist interpellation laid out in Maruice Charland, 'Constitutive Rhetoric: The Case of the *Peuple* Québécois', *Quarterly Journal of Speech* 73 (1987): 133–50; Alan Finlayson, 'Nationalism as Ideological Interpellation: The Case of Ulster Loyalism', *Ethnic and Racial Studies* 19 (1996): 88–112; and especially Ieva Zake, 'The Construction of National(ist) Subject: Applying the Ideas of Louis Althusser and Michel Foucault to Nationalism', *Social Thought and Research* 25 (2002): 217–46.

volume of a battle story rarely heard in Yugoslav times, Milošević and the other leaders of Serbian nationalism constituted their separate people. They hailed the Serbs as righteous victims who refused to be victims anymore, vowed to give them their rightful share of the earth: and demanded the forfeiture of the others' share.[41]

The contemporary European far right has discovered the interpellative power of the myth of palindefence. That myth clearly has an ability to play on some people's heartstrings and rally them to the cause; it can give their lives a heightened moral meaning.[42] It is also always a myth in that second sense of inventing history, even when it – unlike Mattias Karlsson of the SD – deals in memories of actual occurrences. Historical evidence suggests that there were indeed battles at Poitiers, Kosovo Polje, Lepanto, Vienna. This is not the place for a thorough investigation of just how grossly the far right has distorted the record of what happened on those and other sites. Suffice it to say that, for a start, it was a tragedy for the people living north of Poitiers that the Muslims did not win that battle. As David Levering Lewis, the African American historian and biographer of W. E. B. Du Bois, shows in his *God's Crucible: Islam and the Making of Europe, 570–1215*, life in Dar al-Islam was vastly preferable to that under the Frankish and Saxon and Slav kings in every conceivable respect: equality, religious tolerance, education, peace and civility, material standards of living. In the year 800, in Lewis's assessment, Muslim Iberia was 'at least four centuries more advanced than Western Christendom'.[43] Continued expansion to the north of Poitiers would have thrown Europe half a millennium forward in social progress: for the unromantically minded, a counterfactual to mourn.

As for Vienna, the Polish cavalry that Sobieski brought to the gates was largely made up of Tatars. They were practising Sunni Muslims.[44] And so the distortions go, but they are of course utterly trivial to the far

---

41 See the brilliant Althusserian (drawing on Charland, 'Constitutive') analysis of the Kosovo myth in Morus, 'Slobo'.
42 Cf. Zake, 'The Construction', 223–5.
43 David Levering Lewis, *God's Crucible: Islam and the Making of Europe, 570–1215* (New York: W. W. Norton, 2008), 286.
44 For a splendid demolition of the far-right myth of Vienna, see Dag Herbjørnsrud, 'The Real Battle of Vienna', *Aeon*, 24 July 2018.

right itself, for which the whole point of the exercise is to energise *defence of the nation against its present enemies*. While seeking to create an ambience of animosities rooted in the very ancient past, the primacy of the present is here absolute, the hoary stories told to serve the agendas of the day. And here we can see how palindefence is a myth in step with our times. It is the perfect far-right turn-on in a conjuncture marked by '*gatedness* as the emerging social paradigm'. It speaks to the anxieties over borders. Better than the myth of palingenesis, it matches the mania of contemporary European politics: the phenomenon of immigration. It does not live on the optimism of the high modernist zeitgeist, but on the feeling that possessions built up over many years must now be safeguarded against various threats of diminishment.[45] It binds a history of victimhood to a sense of entitlement.

If palingenesis was a myth for the rising side of the curve of capitalist history, when this mode of production still had its best years ahead of it, its golden age and great acceleration, palindefence is one for a long downturn. That includes crises generated by fossil fuels. In a moment of acute mitigation or adaptation crisis, fantasies of national renaissance seem less likely to catch on than tales of a defensive mission; connected to particular memorialised battles or not, far-right politics would be more prone to defensive than to generative postures. If there is any climate-induced migration towards Europe, it will involve people identifiable as Muslims, because potential sending countries happen to have Muslim majorities; one could imagine how believers in the myth would read it. Palindefensive politics would seem a better fit for the plots of both *American War* and *The Wall*. But there could, again, be overlaps. The contemporary far right can segue between the myths with relative ease; the device 'take our country back' – used from Trump's White House to 55 Tufton Street – combines the two. Victorious defence will presumably give the nation new life. Griffin, for one, thinks that 'moments in history when an old order seems doomed to total annihilation still create the ideal climate for palingenetic myth'.[46]

---

45 Cf. Eley, 'Fascism', 105, 109.
46 Griffin, *The Nature*, 34. Vox combined the *reconquista* motif with the meme 'Make Spain Great Again'. Adam Ramsay and Claire Provost, 'Exclusiva: un esquema de financiación coordinado internacionalmente trabaja para aupar a Vox y a la extrema derecha europea', *openDemocracy*, 26 April 2019.

Now is there any reason to consider the palindefensive myth any less *fascist* in character? It seems *prima facie* hard to find one. Several factors point to the opposite conclusion. First, the two are homologous: palindefence is but another modality of the *ur*-myth of a unity between present generations and distant ancestors of an ethnically constituted nation. As with the romances of *romanità* or *Volksgemeinschaft*, perpetual invocations of bygone battles fix a white ultra-nation with a manifest essence and destiny. Palindefence has just the same capacity to cast the fascist death spell: you deserve to be killed because you are an enemy of this nation.

Second, elements of palindefensive myth-making were present already in classical fascism. Palingenesis functioned as its 'mythical core' in a very general sense. More particularly, the myth of Judeo-Bolshevism was the pulsing rage of Nazi fascism: the idea that the Jews were plotting and spreading Communism to realise their world rule. In his remarkable study *A Specter Haunting Europe: The Myth of Judeo-Bolshevism*, Paul Hanebrink dredges up a forgotten far-right discourse from the interwar period, which likened the advances of Judeo-Bolshevism from the east to the Muslim penetration centuries earlier. In the 1920s, central European ultranationalists saw in the Jews and the Communists – one and the same gestalt – the ghosts from Poitiers and Vienna returning to take their revenge. The Polish army restaged the *antemurale Christianitatis* and plastered the country with posters of Leon Trotsky as a Jewish devil sending forth the Asian hordes of the Red Army; Hungarian would-be fascists equated revolutionaries with Ottomans; far-right Catholics in Austria used commemorations of the Battle of Vienna to turn their spears against the Jewish Marxists; French anti-Semitic writers perceived Jewish refugees in Budapest, who had fled from the Tsarist pogroms, as 'the last onslaught of Asiatic peoples', even more dangerous than the Muslims of yore.[47] Continuities can indeed exist in history.

---

47 Paul Hanebrink, *A Specter Haunting Europe: The Myth of Judeo-Bolshevism* (Cambridge, MA: Harvard University Press, 2018), for example 9, 41–4, 107 (French writers Jérôme and Jean Tharaud quoted on 275); Paul Hanebrink, 'Islam, Anti-Communism, and Christian Civilization: The Ottoman Menace in Interwar Hungary', *Austrian History Yearbook* 40 (2009): 115–16, 121–4.

With such mythical gobbets, Hitler cooked the soup for his crusade. In *Hitler's Crusade: Bolshevism, the Jews and the Myth of Conspiracy*, Lorna Waddington shows how the Nazis banged on about Germany as a shield defending not only the *Volk* but European civilisation as a whole. In the autumn of 1941, the Danish foreign minister recorded what Hitler had told him and other diplomats:

> We have been fighting this tremendous battle against the constantly onrushing East for one and a half thousand years. Earlier there were the Huns, the Tartars and the Mongolians; today it was Bolshevism that had mobilized the whole of Asia against Europe. Germany, having borne the greatest blood burden in this struggle and again bearing it this time, was fighting for the whole of Europe. If Germany had not recovered and prepared herself for this great struggle, these Bolshevik-Asiatic hordes would have rolled across Europe like a wave.[48]

Palindefensive elements in classical fascism have so far received little attention, but Hanebrink keenly observes their return in the early twenty-first century. The myth of a racial adversary once again coming to erase borders and debase the nation has returned: Judeo-Bolshevism now goes under the name of Islamisation. Where the Jew was demonised as an agent of subversion and erosion of national sovereignty, so is the Muslim today – without, as we shall see, supplanting the Jew.[49]

Third, and perhaps most important, palindefence has been the principal myth for actual fascists resolving their problems with machine guns and pistols in the recent European past. Griffin believes, as we have seen, that Anders Breivik demonstrated the hopelessly isolated position of palingenetic politics today. That belief is based on a slipshod reading of the *ideas* – supposedly the forte of Griffin and other liberal scholars of fascism – inside the mind of this conscious and wily living phenomenon.[50] Why did Breivik call his

---

48   Quoted in Lorna Waddington, *Hitler's Crusade: Bolshevism, the Jews and the Myth of Conspiracy* (London: I. B. Tauris, 2012), 205. See further for example 2, 45, 123–5, 188–9.

49   Hanebrink, *A Specter*, 274–82.

50   The standing accusation of these scholars against Marxists is, of course, that the latter do not take fascist ideas seriously.

monster manifesto *2083*? Because 2083 will be the four hundredth anniversary of the Battle of Vienna. His entire worldview was steeped in the palindefensive mindset (and it presumably remains as such, as he sits in his cell). Likewise with the slaughtering of one hundred thousand Muslims in Bosnia: this was a genocide committed under the sign of palindefence. Paxton is entirely correct in suggesting that Serbian nationalism in the 1990s came close to the 'functional equivalent' of fascism-as-force – closer than any other in Europe since 1945, because it emerged in a deep crisis (the collapse of the Eastern bloc), during which leading sections of the dominant class threw themselves into ultranationalism so as to maintain power (Milošević on the trail to the Field of Blackbirds) and thereby, with considerable mass support, ushered in a regime of systematic violence against those identified as enemies of the nation.[51] The exceptionality had, as Paxton notes, not the classical form of an end to electoral multi-party politics inside the Serbian entities. Instead it concentrated all its forces on the racialised others. In this respect too, the Bosnian genocide foreshadowed the trajectory of the European far right. The genocide for which (mainly) palingenetic anti-Semitism was responsible is common knowledge, but few remember that palindefensive Islamophobia has its own skeletons to account for; among the many differences, the former was one or two orders of magnitude larger and widely relegated palingenetic anti-Semitism to the rooms of a few boys and elderly men with dreams everyone else found sickening (at least for some time). Since Srebrenica, the politics of palindefensive Islamophobia has gained steadily greater traction in Europe.

The memory of the Bosnian genocide has been preserved by some people. On his long way to the social-democratic camp at Utøya, Anders Breivik stopped to sit at the feet of Serbian nationalist war criminals.[52] When Brenton Tarrant drove his car to the Al Noor mosque in Christchurch, he listened to an anthem from the war called 'Remove Kebab'. Lauding the hero Radovan Karadžić, the song was composed to heighten the morale of the Serbian nationalist forces; the original music video has paramilitaries singing in the hills

---

51 Paxton, *The Anatomy*, 190–1.
52 Bangstad, *Anders*, 75.

and pictures of Muslim interns in concentration camps. In his manifesto, Tarrant called himself a 'kebab removalist', a term he also had inscribed on one of his guns. On another, he had written 'Charles Martel'. Indeed, the propaganda material he posted was covered in palindefensive references, to the Crusades, Vienna 1683 and the rest of the interminable story.[53] Scholars of genocide have not forgotten either. One of them, Alex Alvarez, who engages in his own modelling of worst-case scenarios in a hotter world, believes that Bosnia provides a prototype for the organised racist violence to come.[54]

An amendment to the definition of fascism seems in order. As a set of ideas, fascism is palingenetic *or palindefensive* ultranationalism (or a combination thereof). As a real force, it is such politics coming to the fore in a conjuncture of deep crisis, and so on. It follows that if we can speak of fascist tendencies in the present, these very much include movements in the realm of ideas. It also follows that if we regard *fossil* fascism as a set of ideas, it must be an ultranationalism that somehow connects fossil fuels to palingenetic or palindefensive purposes. But why would it ever do that? We might ask why fossil fuels would be mobilised as a nationalist energy in the first place.

The Mythical Body of the Stock

Fossil fuels make up a stock of energy.[55] A cache of power buried under the ground, it is inherited from past photosynthesis, past climes and plants and slopes that have sunk into the subterranean bowels of a territory. Renewables, on the other hand – sun, wind, water, wave – belong to the flow of energy: a power that comes and goes, visits one spot and continues to another, shines and fades, blows and slackens, rushes and rests without petrifying in any particular

---

53   Hannah Ellis-Petersen, Graham Russell, Kevin Rawlinson et al., 'Christchurch Massacre: PM Confirms Children among Shooting Victims – As It Happened', *Guardian*, 15–16 March 2019; Robert Coalson, 'Christchurch Attacks: Suspect Took Inspiration from Former Yugoslavia's Ethnically Fuelled Wars', *Radio Free Europe/Radio Liberty*, rferl.org, 15 March 2019; Associated Press, 'New Zealand Shooting Gunman's Rifles Covered in White Supremacist Symbols Popular Online', CBS News, 15 March 2019; Tarrant, 'The Great', 5.

54   Alvarez, *Unstable*, for example 63–70.

55   The following paragraphs use terminology laid out in Malm, *Fossil*.

precincts. So far in history, the stock has proven infinitely more congenial to the imaginaries of nationalism. It can be apprehended as *our* coal, *our* oil, *our* gas with no existing equivalent on the renewable spectrum. 'Our sun' is, of course, an oxymoron: the sun is the common property of the solar system. It cannot be appropriated by company or country. The light it emits seems to slip out of the hands of nations as soon as it strikes the surface, whether in the form of direct radiation or as moving wind and water; *qua* fuel, it cannot be part of a national corpus. Only its relics can. Nations blessed with fossil fuels have felt the stock within them, as an ultra-deep material inheritance to which the mystique of nationalism easily sticks. The stock will always be found under the land within our borders, the flow may be for a passing moment but not the next one. The flow has a weak bond to the homeland; it is at home both nowhere and everywhere; it does not anywhere reside under any particular soil. Solar and wind are the Jew and the Muslim of energy.[56]

Fugitive flow, autochthonous stock: this explains why none of the parties we have investigated – those that take nationalism towards its logical endpoints – promote the former as a source of pride. It explains why they at most show perfunctory acceptance of the flow and more often hate it, above all wind, a major segment of any decarbonised economy. In technical, physical principle, a nationalist programme for renewables would be possible in many countries. The US could rely on domestic renewable energy for 100 per cent of its needs and export associated manufactures – panels, turbines, troughs, storage systems – to the rest of the world, but no agenda for 'energy dominance' has been outlined on such a basis. Poland could have continued to rise to the top of wind power capacity. Germany has made strides in renewable energy generation and might well do more: and the AfD wants to shut down all of it. No far-right government, coalition or president has consecrated a wind farm or solar power plant as the pedestal of the national character the way Norway has named its oil and gas fields after native heroes. None has walked out into a field of turbines or arrays and proclaimed that this source of energy will last for one thousand years – it could last forever, of course, but it would have none of

---

56 Or perhaps the Roma people of energy, ever stateless.

the rich corporeality nationalists smell in the stock. The only part of the flow that has been subjected to similar framing is waterpower, more particularly the mega-dams built across the world in the post-war decades, for rivers – fixtures of the land – have had a place in national mythology. The Nile, the Indus, the Yangtze were harnessed in projects of nationalist engineering and patriotic monumentality (in the very different context of postcolonial independence), but rivers have little left to contribute to the flow, particularly in Europe; waterpower is a small piece of the fossil-free puzzle.[57]

Outside of continent-size countries such as the US or Brazil, 100 per cent renewable energy will work best in internationally integrated grids. On a day when thick clouds cover Germany, the wind blows in Spain and the waves lap Norwegian shores. With cross-border grids pooling together flow resources from across borders, baseload and stable supplies can be guaranteed; for fairly small countries, a flow-based economy presupposes a modicum of cosmopolitanism. Nationalist fervour throws a spanner in any work to connect the material base of one country with that of another – particularly if that other country is Muslim, with all the sun one could dream of but no self-evident white reliability.[58] Europe could by now have replaced a massive chunk of its fossil fuels with solar power imported from North Africa. Technologically feasible, such schemes have so far stumbled on – among other obstacles – the fear that the Muslim partners would use their deserts to blackmail Europe or let the infrastructure be blown up by terrorists.[59] The likes

---

57 For a good survey of dam-building in the post-war era, see John McNeill, *Something New under the Sun: An Environmental History of the Twentieth-Century World* (New York: W. W. Norton, 2000), 149–91.

58 On the general logic, see Batel and Devine-Wright, 'Populism'.

59 See Karen Smith Stegen, Patrick Gilmartin and Janetta Carlucci, 'Terrorists versus the Sun: Desertec in North Africa as a Case Study for Assessing Risks to Energy Infrastructure', *Risk Management* 14 (2012): 3–26; and particularly the brilliant Luiz Enrique Vieira de Souza, Estevão Mota Gomes Ribas Lima Bosco, Alina Gilmanova Cavalcante and Leilada da Costa Ferreira, 'Postcolonial Theories Meet Energy Studies: "Institutional Orientalism" as a Barrier for Renewable Electricity Trade in the Mediterranean Region', *Energy Research and Social Science* 40 (2018): 91–100. This fear is all the more irrational for the stability of oil and gas exports to the EU from for example Algeria, Libya and Egypt, despite some very major social upheavals, as pointed out by Vieira de Souza et al., 'Postcolonial', 96.

Mythical Energies of the Far Right

of Santiago Abascal and Alexander Gauland will not hasten implementation.

On the other hand, the far right in Western countries has cultivated a special veneration of one category of workers, namely those white men who extract black fuel – coal in particular, as in the US, Poland and Germany. In the eyes of Donald Trump, the PiS and the AfD, their work is imbued with an authenticity few other categories of workers can approximate: they haul up the inner body of the nation.[60] In both the US and Germany, the renewable energy sector employs several times more people than coal, but none of the grim glamour attaches to the former, too ethereal to count. Regenerative forces come from the ground. In the US and Poland, the Trump and PiS administrations have sutured palingenetic and palindefensive figures of thought to the stock, and we can expect more of this, particularly of the defensive kind: some people will want to preserve their fossil status.[61] Some ultranationalists will try to defend or revive a way of life that really was – no fantasy here – built on the foundation of the stock.

It remains to be seen if nationalist imaginaries can become invested in the manufacturing of technology for the capture of sun and wind and wave or in the landscapes most suitable for it (straits, deserts). Until now, it would seem that among the many hurdles the transition faces is precisely *the mythical investment in the stock as the body of the nation*, upheld and amplified by the ascendant far right. Had nationalists found reason to love renewables – or had internationalists been more capable of setting the agenda – our world might not have been warming quite so rapidly.

Can the far right change its mind? The alloy appears to have entered language itself, at least English. The *Oxford Dictionary* gives two definitions of the term 'extraction': 'the action of extracting something, especially using effort or force' and 'the ethnic origin of

---

[60] See, for the American case, John Hultgren, 'Those Who Bring from the Earth: Anti-environmentalism and the Trope of the White Male Worker', *Ethics, Policy and Environment* 21 (2018): 21–5.

[61] In Poland, with a political culture steeped in victimhood and the myth of palindefence, any critique of coal extraction seems to elicit the response, *Now they are out to get us once again.*

someone's family', as in being of German extraction. Stock can mean several things, among them 'a supply or quantity of something accumulated or available for future use' and 'a person's ancestry or line of descent'. Migration, on the other hand, is 'movement of people to a new area or country in order to find work or better living conditions' or just 'movement from one part of something to another'. Either way, it flows. And no element is as quintessentially migratory as the wind. One is here reminded of Klaus Theweleit's museum-sized psychoanalytic readings of the writings and images of the *Freikorps* and their proto-fascist pals in the early Weimar Republic. Fresh from the trenches, they dreaded the Bolshevik revolution as an incoming flood or an ocean moving past German borders; from the west, meanwhile, came the horrifying spectacle of non-white soldiers – the Moroccan, Senegalese, Indochinese troops stationed by the French victors – on national soil. Both represented a nauseating flow. Men of the far right feared losing themselves in these undulating surroundings – drowning in the flood – and grasped for 'a rock amid the raging sea'. 'Nothing is to be permitted to flow', Theweleit diagnoses their craving for rigidity. 'Death to all that flows.'[62]

This explains, lastly, the structural proximity between primitive fossil capital and the far right. It does not, however, explain why nationalists in countries with no extraction can cherish fossil fuels as much as their producing brethren. Nor does it solve one remaining riddle: such fuels have oiled the wheels of globalisation. They and they alone have enabled the abstract space and time of borderless capital. With its integration in weather cycles and landscape forms, the flow could scarcely have fuelled the transcendence of natural boundaries in late capitalism and the release of industrial production from national states. One kind of cosmopolitanism is loaded with fossil fuels – namely *elite* cosmopolitanism, the plague someone like Gauland purported to combat: the rootless vagabonds move between their flats and restaurants on aeroplanes. When the popular Spanish singer Rosalía tweeted 'Fuck Vox' in November 2019, the party responded from its official Twitter account with a picture of her in an

---

[62] Klaus Theweleit, *Male Fantasies, Vol. 1, Women, Floods, Bodies, History* (Minneapolis: University of Minnesota Press, 1987 [1977]), 232, 230.

aeroplane and clarified: 'Only millionaires with private planes like you can afford not to have a homeland' – a paraphrasing of fascist philosopher Ramiro Ledesma Ramos, whom Santiago Abascal liked to quote.[63] Now if the far right hates this 'globalism', why does it not confront its energy, or at least associate it with evil rather than good? Or will it soon?

Myths of Plots and Hoaxes

The anti-climate politics of the far right draws additional force from a battery with a special capacity to light up people and keep them going for a long time: conspiracy theories. Such theories have been present since climate science first became contested. We are not here referring to primitive fossil capital pulling some strings to mangle the science and sabotage mitigation; that it has done so is not a theory, but a well-documented fact.[64] As we have seen, the initial attempts were undisguised, Exxon and the rest operating not behind the scenes but smack in the middle of them. The turn towards secrecy – as in anonymous funding – was a tactical reaction to the public outcry against the transparently self-interested denialism, which in turn triggered new rounds of embarrassing revelations. The history of the denialist ISA is one of perennial disclosures and readjustments. To the extent that it strove for seclusion, it was largely a failure. As such, it was no more of a conspiracy than the messaging of a Baptist church or a BBC show. That oil corporations lied for decades about climate change is no more mysterious than the Bush regime lying about weapons of mass destruction in Iraq; deception does exist in human history, as do, of course, plots to stab the Roman Caesar and fly aeroplanes into skyscrapers, not to speak of the more humdrum

---

63 Vox's Twitter account, 12 November 2019. Cf. *Huffington Post*, 'La (MUY) llamativa verdad escondida tras la frase de Abascal "solo los ricos pueden permitirse no tener patria"', 5 November 2019.

64 Cf. Karen M. Douglas and Robbie M. Sutton, 'Climate Change: Why the Conspiracy Theories Are Dangerous', *Bulletin of the Atomic Scientists* 71 (2015): 102; Joseph E. Uscinski, Karen Douglas and Stephan Lewandowsky, 'Climate Change Conspiracy Theories', in *Oxford Research Encyclopedia of Climate Science*, September 2017, 16–17.

machinations of police informants, marketing strategists, political front groups, companies withholding knowledge of their latest products and any number of other agents that must pursue their goals at least partly in the dark.[65] Primitive fossil capital has also acted in the shadows, as an all-too-human entity, fallible and opposable, pushing its pedestrian class interests. But if someone would claim that ExxonMobil has operated a fleet of UFOs abducting critical climate scientists and returning them to earth as muted conservatives, or that the Heartland Institute has distributed a pizza ingredient that induces paralysis, either idea would indeed count as a conspiracy theory – but these are stories not yet written and sold, unlike those working in the opposite direction.

What sets a conspiracy theory apart, then, is the belief in 'the existence of a vast, insidious, preternaturally effective international conspiratorial network designed to perpetrate acts of the most fiendish character'.[66] The network wills the destruction of all that is held dear. The power it wields is demonic, boundless in malevolence and limitless in efficacy; the conspiracy has infiltrated every nook and cranny of social life and works single-mindedly, without pausing, erring or splitting, to bring on perdition. It appears behind events that might seem random to the untrained eye, but which form a tight pattern woven by the hidden hand.[67] The truth is out there, by definition concealed – the epistemic passion of every conspiracy theory,

---

65 See particularly Jeffrey M. Bale, 'Political Paranoia v. Political Realism: On Distinguishing between Bogus Conspiracy Theories and Genuine Conspiratorial Politics', *Patterns of Prejudice* 41 (2017): 45–60; also Thomas Milan Konda, *Conspiracies of Conspiracies: How Delusions Have Overrun America* (Chicago: University of Chicago Press, 2019), 10, 13–14; Karen M. Douglas, Joseph E. Uscinski, Robbie M. Sutton et al., 'Understanding Conspiracy Theories', *Political Psychology* 40 (2019): 4; Cass R. Sunstein and Adrian Vermule, 'Conspiracy Theories: Causes and Cures', *Journal of Political Philosophy* 17 (2009): 206–8, Jesse Walker, 'What We Mean When We Say "Conspiracy Theories"', in Joseph E. Uscinski, *Conspiracy Theories and the People Who Believe Them* (New York: Oxford University Press, 2019), 53–62.

66 Richard Hofstadter, 'The Paranoid Style in American Politics', in *The Paranoid Style in American Politics and Other Essays* (Cambridge, MA: Harvard University Press), 14.

67 Drawing on ibid., 29–32; Bale, 'Political', 50–3; Konda, *Conspiracies*, 17–22; Brian L. Keeley, 'Of Conspiracy Theories', *Journal of Philosophy* 96 (1999): 116–17; S. Jonathan O'Donnell, 'Islamophobic Conspiracism and Neoliberal Subjectivity: The Inassimilable Society', *Patterns of Prejudice* 52 (2018): 9, 18.

which makes for an odd relation to evidence. To take one recent example, when a fire erupted in the Notre Dame cathedral on 15 April 2019, conspiracy theorists rushed like moths to the flames, among them Glenn Beck, star of the American conservative airwaves, who instantly knew that this was a 'World Trade Center moment'. But 'if', he added, 'this was started by Islamists, I don't think you will find out about it.' 'They', meaning the French government, 'will keep it quiet.'[68] So the absence of observable evidence for an Islamist conspiracy behind the Notre Dame fire would, for Beck, serve as so much confirmation of it. The conspiracy will erase all traces and leave invisibility as proof of its success. If anyone provides countervailing evidence, it is short-circuited into its opposite: if the French police were to present a reconstruction of the accidental fire, that would merely be a predicted *part* of the plot. All that gainsays the conspiracy can be attributed to it, if only the circles of complicity are widened to include media and academia and other channels of information. The theories seal themselves off and become immune to falsification or even buoyed by it, in a logic without equivalent in the realm of thought: these are 'the only theories for which evidence *against* them is actually construed as evidence *in favour* of them.'[69] Believers will hold on, no matter how large amounts of data are thrown against them year after year. We can now begin to see why climate denial and the capitalist class fractions propagating it had to end up in this camp.

In a first phase of innocence, climate denial does not depend on a conspiracy theory. Consider Mo Brooks, a Republican congressman from Alabama, who in May 2018 asserted that sea level rise is caused not by any warming trend, but rather by the rocks that tumble into the oceans from the California coastline and the White Cliffs of Dover. Together with soil from the mouth of rivers, these rocks force 'the sea level to rise, because now you have less space in the oceans,

---

[68] David Brennan, 'Glenn Beck Compares Notre Dame Fire to 9/11: "This Is Their World Trade Center Moment"', *Newsweek*, newsweek.com, 16 April 2019.

[69] Keeley, 'Of Conspiracy', 120. Emphasis in original. See further 120–2; Maarten Boudry and Johan Braeckman, 'How Convenient! The Epistemic Rationale of Self-Validating Belief Systems', *Philosophical Psychology* 25 (2012): 351, 354; Konda, *Conspiracies*, 93; Sunstein and Vermule, 'Conspiracy', 206–7, 210; Bale, 'Political', 48.

because the bottom is moving up'.[70] This American lawmaker could simply have been a little lost and lacking in education, his belief formed independently, without any elaborate tale of a wicked conspiracy. But a problem arises when such a belief is confronted with counterproof. If there is an overwhelming scientific consensus about global warming, and if it cannot be accepted as the result of tens of thousands of researchers pouring their findings into a common pool of knowledge, how on earth can it be accounted for? The only way out for the orthodox denier not prepared to give up his view is to postulate a conspiracy. These scientists must be colluding in the fabrication of their data, and for this they must have an ulterior motive and very great resourcefulness. Denial may not have needed conspiracy theory in, say, the mid-1980s, but the more time has passed, depositing ever-rising mountains of evidence, the harder it has become for it to maintain credibility *without attaching itself to some conspiracy theory.*[71]

Hence deniers have put forth whole libraries worth of stories about climate scientists suppressing discussion, monopolising grants, perverting the peer-review process and brainwashing students all over the world. The IPCC, Fred Singer explained in a 2008 report from the Heartland Institute, gathers at 'exotic locations' to advance its 'activist enterprise' at a safe remove from peering eyes.[72] Around this time, the tropes of the 'hoax' or the 'scam' or the 'cover-up' had become central in the denialist rhetoric; Inhofe's *The Greatest Hoax* was a run-of-the-mill title. The idea resonated with a considerable segment of the American public, the share subscribing to statements

---

70   Quoted in Scott Waldman, 'Republican Lawmaker: Rocks Tumbling into Ocean Causing Sea Level Rise', *Science*, 17 May 2018. On denial as potentially free from conspiracy theories, cf. Sunstein and Vermule, 'Conspiracy', 206; Douglas and Sutton, 'Climate', 101.

71   Stephan Lewandowsky, Gilles E. Gignac and Klaus Oberauer, 'The Role of Conspiracist Ideation and Worldviews in Predicting Rejection of Science', *PloS One* 8 (2013): 3; Stephan Lewadowsky, Klaus Oberauer and Gilles E. Gignac, 'NASA Faked the Moon Landing – Therefore (Climate) Science Is a Hoax: An Anatomy of the Motivated Rejection of Science', *Psychological Science* 24 (2013): 623; Uscinski et al., 'Climate', 21.

72   S. Fred Singer, 'Preface', in S. Fred Singer (ed.), *Nature, Not Human Activity, Rules the Climate* (Chicago: Heartland Institute, 2008), iv. See further Uscinski et al., 'Cimate', 16; Douglas and Sutton, 'Climate', 98–100; Lewandowsky et al., 'The Role', 3; Konda, *Conspiracies*, 205–6, 309.

such as 'global warming is a hoax' or 'a myth concocted by scientists' standing at around 40 per cent.[73] It obviously required a leap of faith to think that the women and men behind the satellites, observatories, laboratories, university administrations, editorial boards, research institutes, UN agencies and other units involved in the production of climate science could have been the silent, loyal members of a cabal – it would indeed have been 'the greatest hoax ever perpetrated'. Out of all these uncounted thousands, should not someone have spilled the beans?[74]

And then, in November 2009, just weeks before COP15 in Copenhagen, a dossier of one thousand emails hacked from climate scientists at the University of East Anglia was dumped on the internet and immediately hailed by denialists as the definitive proof that these scientists were indeed jiggering their data. *Investor's Business Daily* claimed that the emails showed 'attempts to conspire' and judged the scientists 'guilty of fraud', while in the *Telegraph*, James Delingpole declared that 'the conspiracy behind the Anthropogenic Global Warming myth' had been 'quite deliciously exposed'.[75] He coined the term 'Climategate' for the event. So what dark secrets exactly did those emails divulge? No fewer than nine independent investigations in the US and the UK subsequently presented their inquiries: all exonerated the scientists of any wrongdoing.[76] But, for the believers, these investigations were, of course, a part of the plot. 'Don't believe the "independent" reviews', counselled a *Wall Street Journal* column.[77] Long after the scientists had been cleared of any

---

73 Uscinski et al., 'Climate', 3; cf. Nicholas Smith and Anthony Leiserowitz, 'The Rise of Global Warming Skepticism: Exploring Affective Image Associations in the United States over Time', *Risk Analysis* 32 (2012): 1021–31.

74 As pointed out by for example Powell, *Inquisition*, 155; Uscinski et al., 'Climate', 6–7.

75 *Investor's Business Daily* quoted in Brett Jacob Bricker, 'Climategate: A Case Study in the Intersection of Facticity and Conspiracy Theory', *Communication Studies* 64 (2013): 230; James Delingpole, 'Climategate: The Final Nail in the Coffin of "Anthropogenic Global Warming"?', *Telegraph*, 20 November 2009. For a thorough reconstruction of the events, see Bricker, 'Climategate'; cf. Stephan Lewandowsky, 'Conspiratory Fascination versus Public Interest: The Case of "Climategate"', *Environmental Research Letters* 9 (2014): 1–4.

76 Uscinski et al., 'Climate' 5.

77 Patrick J. Michaels, 'The Climategate Whitewash Continues', *Wall Street Journal*, 12 July 2010.

suspicion, denialists continued to dwell on 'Climategate' as their very best case, kept alive the rumour of crooked scientists and extracted maximum fortunes from the pseudo-event: 'I think the scandal has made the opponents of energy-rationing legislation stronger and more confident', said Myron Ebell of the Competitive Enterprise Institute.[78] A fable spun around an unsubstantiated core, a conspiracy theory can build up with anything that comes its way, as its adherents have taken leave of reality – perhaps the only way explicit denial can survive in a steadily warming world. Along this way, the denialist ISA also moved further and further to the right.

If a general need for conspiratorial thinking pushed denial towards the far right, there was also a more specific content that sped up that movement: the theory that the conspiracy was perpetrated *by the left*. Climate politics carries several birthmarks from the historical moment around 1989, when actually existing socialism collapsed and – an epic coincidence – the science of anthropogenic global warming matured. Conspiracy theories, of course, accept no coincidences. This one asserts that once the Soviet Union and its allies were gone, the communists, also referred to as the socialists, the Marxists or simply the left, abandoned the dead horse and invented climate change as their next bet to dominate the world. 'Just as Marxism is giving way to markets, the political "greens" seem determined to put the world economy back into the red, using the greenhouse effect to stop unfettered market-based economic expansion', explained *Forbes* in 1989.[79] The Rio summit fit the pattern perfectly. 'The International Socialist Party, which is intent upon continuing to press countries into socialism, is now headed up by people within

---

78   David R. Baker, ' "Climategate" Fallout May Impact Legislation', *San Francisco Chronicle*, sfgate.com, 19 July 2010. Cf. for examples James Delingpole, 'Climategate 2.0', *Wall Street Journal*, 28 November 2011; and for an analysis of the impact on American public opinion, Anthony A. Leiserowitz, Edward W. Maibach, Connie Roser-Renouf et al., 'Climategate, Public Opinion, and the Loss of Trust', *American Behavioral Scientist* 57 (2012): 818–37.

79   Quoted in Dunlap and McCright, 'Sources', 245. For an identical argument two decades down the road, see Charles Krauthammer, 'The New Socialism', *Washington Post*, 11 December 2009. See further Naomi Klein, *This Changes Everything: Capitalism vs. the Climate* (London: Allen Lane, 2014), 31–5.

the United Nations', reported one early denialist from the negotiations: 'The radicals are in charge.'[80]

This is the *left climate conspiracy theory*, or the 'left climate theory' for short. It is originally a child of 1989, a product of the conjuncture when the forces of the left crumbled worldwide. Magically converting historical weakness into near-demonic power, it suggests that the left pulled off the hoax of climate change in particular and environmentalism in general to make up for its losses. In their classic account of the origins of denial, *Merchants of Doubt: How a Handful of Scientists Obscured the Truth on Issues from Tobacco Smoke to Global Warming*, Naomi Oreskes and Erik M. Conway show how Fred Singer and other pioneers graduated from the most aggressively anti-communist circles of the late Cold War. With the left climate theory, they could extend their assignment into the post-communist era, keeping up the fight against an enemy with eternal life and – appearances to the contrary – undiminished ability to jeopardise capitalism.[81] That fight became a defining mission of the denialist ISA.

Delusionary as this theory is, it has two features common to conspiracy theories: it turns the weak into the strong, and it contains a grain of truth. After the wall had fallen, some Marxists did look to the environment as their next best thing. One of them was Perry Anderson, who, in his typically Olympian survey of the post-1989 political landscape, 'The Ends of History', published in the year of Rio, acknowledged defeat. The Third International in ruins, the Second sterile, the national liberation movements exhausted – 'none of the political currents that set out to challenge capitalism in this century has morale or compass today.'[82] After this rigor mortis, could there be a second life for socialism? Anderson had glimpsed one chance: the ecological crisis, to which 'market forces contain no

---

80 Dixie Lee Ray quoted in Jacques et al., 'Organisation', 362–3.

81 Oreskes and Conway, *Merchants*, for example 213–5, 249–54. An earlier analysis of how the American right based its hostility toward environmentalism on lingering anti-communism and various conspiracy theories can be found in Andrew Rowell, *Green Backlash: Global Subversion of the Environmental Movement* (London: Routledge, 1996), particularly ch. 6.

82 Perry Anderson, 'The Ends of History', in *A Zone of Engagement* (London: Verso, 1992), 358.

solution'. The imperative of profit can only accelerate planetary spoliation, making the deduction from green to red hard to contest:

> At this higher level, where the fate of the Earth itself will be decided, do not the classical arguments of socialism for intentional democratic control of the material conditions of life stage their comeback? If there is to be, as the most prescient analysts insist, an Environmental Revolution comparable in significance only to the Industrial and Agricultural Revolutions before it, how could it be other than consciously realized – that is, planned? What else are the targets already feebly set by various national governments and international agencies?[83]

Not only planning but equality too would be a prerequisite for bringing the crisis under control. On what principle should 'the emission of carbon' be distributed among the peoples of the world, if not on equal entitlement? 'The malignant appropriation of the world's riches by a privileged minority, with which the destruction of its resources is now fatally interwoven, threatens any common solutions to the massive dangers now gathering momentum.' As revolting as this logic was to the aficionados of the market – distracted from celebrating what should have been their final victory – it had greater weight than all the rubble from the Berlin Wall, and it has only gained weight since Anderson laid it down: every postponement of the 'day of reckoning' has intensified the contradiction between capital accumulation and the life-support systems of the earth.[84] Every additional gigatonne of carbon sent into the atmosphere has made half-measures less viable. Every moment of stalling mitigation has ensured that if it ever commences, it will have to exercise the highest degree of control over the material conditions of life – first of all, over the privileged minority wasting the resources on which all others depend, notably the carbon sink of the atmosphere.

If the left climate theory seemed plausible to certain segments around 1989, then, it might well become more convincing in more intense stages of the crisis. The development of climate politics seems to obey a law of polarisation: *the higher the temperatures, the more*

---

83   Ibid., 363.
84   Ibid., 364, 375.

acute the antagonism between a left that alone stands ready to pick up the instruments for alleviating the crisis and a right that, for that very reason, refuses to contemplate it. A recursive cycle has been rolling for some time. Every year of inaction necessitates more revolutionary action the next; every threat of such action – if only of a hypothetical, tautological character – strengthens the conviction that this is a plot by the left; every strengthening of that conviction drives the right farther to the right and indurates its heart, and so on.[85] But this cycle was set in motion just as the left experienced the most epochal sapping of its forces. The ecological crisis might, Anderson noted, elevate the interest in an alternative mode of production to the biosphere itself – but it is 'accompanied by a reduction in the social capacities to fight for one'.[86] This peculiarity of the post-1989 predicament predisposes the right to fence ever more wildly: for it fights not a strong left, but the *ghost* of one.

Conspiracy theories can then go into a spin. One that made the rounds in the 2010s focused on the circumstance that Earth Day, celebrated worldwide every year on 22 April – in 2016 with the ratification of the Paris Agreement in 175 countries – happens to fall on Lenin's birthday. Moreover, the very first Earth Day, 22 April 1970, was also the centenary of Lenin's birth. 'Is Earth Day on Lenin's Birthday a Coincidence?' asked a missive from the Competitive Enterprise Institute on the occasion in 2015: 'no' was the natural answer.[87] It had a very deep meaning indeed. Another architect of

---

85 One politician who followed this cycle was Margaret Thatcher. In the 1980s, she made a name for herself as an environmentally conscious and compassionate conservative, but around the turn of the millennium she turned towards denial; in her 2003 book *Statecraft* she accused Al Gore of 'apocalyptic hyperbole', asserted that 'a new dogma about climate change has swept through the left-of-centre governing classes' and called mitigation a 'marvellous excuse for supra-national socialism'. Roger Harrabin, 'Margaret Thatcher: How PM Legitimised Green Concerns', BBC News, 8 April 2013; James West, 'How Thatcher Made the Conservative Case for Climate Action', *Grist*, 8 April 2013.

86 Anderson, 'The Ends', 366.

87 Robert J. Smith, 'Is Earth Day on Lenin's Birthday a Coincidence?', Competitive Enterprise Institute, 22 April 2015. Cf. for example David Ziemer, 'No Coincidence Earth Day Is Lenin's Birthday', *Wisconsin Law Journal*, 26 April 2010; Jack Cashill, 'It's No Accident That Earth Day Falls on Lenin's Birthday', *Sentinel*, 22 April 2017. In reality, the man who initiated Earth Day, democratic senator Gaylord Nelson, was a committed Malthusian with no discernible Leninist loyalties. Robertson, *The Malthusian*, 2, 152, 168–71.

the 'Climategate' narrative, Brian Sussman, a meteorologist-turned-conservative radio host in California, explained in 2019 why the left had smuggled its project of world domination into this particular date in the calendar:

> Environmentalists have always admired Lenin. He was the first disciple of Karl Marx to capture control of a country, and the opening act of his seven-year reign commenced with the abolition of all private property – a Marxist priority. Despite overseeing a bloody civil war, a devastated economy and a citizenry without hope, Lenin made it a priority to implement his signature decree, 'On Land'. In it he declared all forests, waters, and minerals to be the exclusive property of the state, and he demanded these resources be protected.[88]

Worries about climate change had the same terminus; hence the choice of 22 April.

Sussman presented an expanded version of the theory in his book *Eco-Tyranny: How the Left's Green Agenda Will Dismantle America*, published in 2012, at a time when Obama inspired a flurry of fresh theorising. The date of 22 April was selected back in 1970 because environmentalists acted on the mandate of Lenin and, further behind the curtain, Marx. Today they cannot stop talking about global warming. 'It's all a lie. There is no such planetary crisis. It's a concocted calamity churned out initially by Marx himself, and furthered by his modern devotees', among whom Sussman identified Rachel Carson, Paul Ehrlich, Indira Gandhi, the Rockefeller family, Al Gore (his father was friends with a Jewish-Russian businessman who ran eight drugstores from which he transferred the profits straight to the Bolshevik party), James Hansen and many others in an axis of evil that culminated, of course, in 'Barack Hussein Obama'.[89]

Sussman's most original contribution – his denial of trend and attribution and carbon vitalism seriously shopworn by now – lay in

---

88 Brian Sussman, 'Earth Day: The Real Story', *Climategate*, 16 April 2019.
89 Brian Sussman, *Eco-Tyranny: How the Left's Green Agenda Will Dismantle America* (Washington, DC: WND Books, 2012), 1, 211–12.

deriving the conspiracy from Marx himself. The founding father of communism pored over the latest reports of environmental crises and then passed on a red-green plot against mankind to generations of co-conspirators, trained to 'bewilder and lie' for the cause.[90] The more common version of the theory said that climate was a *surrogate* for original Marxism – 'Karl Marx is dead, they needed something else. CO_2 has taken the place of Marx', as the climate spokesman for the FrP said in 2012 – but Sussman rather construed it as the quiddity of the whole Marxian oeuvre, transmitted by direct line of descent.[91] This achievement rendered him a column in *Forbes*, the billionaire magazine, much besotted with his work. There he vituperated Gore and Obama for forcing their communist agenda onto Americans and 'purposefully keeping the bulk of [our] resources – oil, natural gas, minerals, timber, water – out of our reach'. The way to fight eco-tyranny was to 'demand what is rightfully ours' and 'drill it, dam it, log it'.[92] Out of such sentiments arose the 45th and first white president. Very similar sentiments invigorated another white president's quest to tear down the legacy of Lula and Dilma.

We must here recognise the exceptional status of climate conspiracy theories in the early twenty-first century. That period has so far been rather well endowed with such theories – about 9/11, vaccines, Pizzagate, Flat Earth, disappearing aeroplanes, extraterrestrial

---

90 Ibid., 113. Sussman cites Marx's writings on Justus von Liebig as well as modern eco-Marxist John Bellamy Foster. He also invents a book by Marx called *Laws of Matter* and hits some sort of new record in imaginative interpretation of Marxism: 'Marxists contend that some people are randomly spit out of their mother's womb with a better brain than most. Those with the best brains have a Darwinian authority to rule over those with the lesser brains, lest those with the deficient brains destroy the planet and kill one another.' Ibid., 5. Sussman was also a birtherist and thought that 'Islam has always sought to take over the world'. For a collection of his views, see Justin Berrier, Eric Schroeck, Chelsea Rudman and Melody Johnson, 'Who Is Fox's Latest Global Warming Expert Brian Sussman?', *Media Matters*, 28 January 2011.

91 *Verdens Gang*, 'Fremskrittspartiet: CO_2 er sosialistenes nye Karl Marx', 6 December 2012. The spokesman was Per-Willy Amundsen, a minister in the Norwegian government from 2013 to 2018.

92 Brian Sussman, 'Drill It, Dam It, Log It: How to Fight Eco-Tyranny in America', *Forbes*, May 2012. On *Forbes* being besotted with Sussman: see for example Mark Hendrickson, 'Climate Change: "Hoax" or Crime of the Century?', *Forbes*, 16 September 2012; Mark Hendrickson, 'The Palpable Politicization of Science by Global Warming Alarmists', *Forbes*, 20 September 2013.

reptilians hijacking the earth and controlling it from underground catacombs, to mention a few – but none has come anywhere near the standing of those pertaining to climate.[93] *Forbes* and the *Wall Street Journal* have never cosied up to the idea that Mossad organised the attacks on World Trade Center. No president, American or otherwise, has made the murder of Princess Diana or the suspicion of genetically modified foods a plank of his programme. Climate conspiracy theories are unique in being ubiquitous and popular *and* refuted by an ever-growing body of science *and* anchored in dominant class fractions.[94] They present established scholarly models for understanding the phenomenon with an anomaly.

Conspiracy theorists are usually portrayed as outsiders and underdogs. 'Higher levels of conspiracy thinking correlate with lower levels of education and lower levels of income', making these theories especially 'prevalent among members of low-status groups attempting to explain their status' – in short, '"conspiracy theories are for losers."'[95] Vulnerable groups wish to accuse the mighty of evil and descry their handiwork. There is an influential Marxist version of this model, based on a quip from Fredric Jameson: 'Conspiracy, one is tempted to say, is *the poor man's cognitive mapping*.'[96] The man who preaches or believes in it is poor, in the sense of being economically

---

93 For an introduction to the landscape in the late 2010s, see for example Joseph E. Uscinski, 'Down the Rabbit Hole We Go!', in Uscinski, *Conspiracy*, 1–32. But conspiracy theories gained noticeable popularity in the US already at the end of the twentieth century: see for example Peter Knight, *Conspiracy Culture: From the Kennedy Assassinations to the X-Files* (London: Routledge, 2000).

94 Cf. for example Konda, *Conspiracies*, 308–9, 313; Uscinski, 'Climate', 2, 11, 18; Josh Pasek, 'Don't Trust the Scientists! Rejecting the Scientific Consensus "Conspiracy"', in Uscinski, *Conspiracy*, 202.

95 Douglas et al., 'Understanding', 9–10, 13.

96 Jameson, 'Cognitive Mapping', in Cary Nelson and Lawrence Grossberg (eds), *Marxism and the Interpretation of Culture* (Urbana: University of Illinois Press, 1988), 357. Emphasis added. For an interpretation of the quip, see Fran Mason, 'A Poor Person's Cognitive Mapping', in Peter Knight (ed.), *Conspiracy Nation: The Politics of Paranoia in Postwar America* (New York: New York University Press, 2002), 40–56. A recent elaboration, which argues that conspiracy theories are 'revolutions of the weak' meant to relieve 'suffering' in late capitalism – focusing on cases like conspiratorial thinking in the Occupy movement, while ignoring certain more powerful instantiations – is Todor Hristov, *Impossible Knowledge: Conspiracy Theories, Power, and Truth* (Abingdon: Routledge, 2019), quotations from 54, 6.

disadvantaged and perhaps also pitiable, a little tragic and nutty, with no access to a more realistic map of the world. Jameson developed his theory of cognitive mapping in the very same conjuncture of 1989, when the prospects for a socialist world had receded and capitalism ruled triumphant. In the resulting cultural condition, the poor man is bombarded with images from afar and signals from distant markets, his life overwhelmed by the power of transnational corporations inscrutable to him. He is like the first-time visitor to a mega-city, who cannot find his way around. But just as navigating a city requires some cognitive map – to feel properly at home, one needs a mental picture of how the neighbourhoods and streets are interlinked – so this life demands some representation of the ungraspable, overbearing totality. Socialism no longer supplies one. After its demise, the poor man is thus drawn to conspiracy theories, which centre not on capital – the city no longer represented – but on secret government laboratories, the British queen, a family of bankers, aliens or some other stand-ins. Conspiracy is 'a degraded figure of the total logic of late capital'. It is the poor substitute for what used to be called class consciousness.[97]

This model has its undeniable strengths, to which we shall return later, but it does not work in the case of the left climate theory. When *Forbes* informed its readers as a matter of routine in 2013 that climate science was the bogus product of 'billions from government grants and neo-Marxist environmentalist largesse' reserved for those 'in favour of the politically correct theory', this was not a poor man speaking loudly to himself on the sidewalk.[98] This was *the rich man's cognitive mapping*. After 1989, the sharpest indictment ever put forth against the capitalist mode of production – it destroys the conditions for life as such – coincided with the sudden disappearance of the enemy that had always promoted an alternative mode, and some very rich men could only make sense of this terrain by updating their dog-eared map from a century or two of

---

97 Jameson, 'Cognitive', 356; on cognitive mapping as class consciousness, see Fredric Jameson, *Postmodernism: Or, the Cultural Logic of Late Capitalism* (London: Verso, 1991), 417–18.

98 Peter Ferrara, 'The Disgraceful Episode of Lysenkoism Brings Us Global Warming Theory', *Forbes*, 28 April 2013. Ferrara was from the Heartland Institute: see further above.

class struggle. It was, more precisely, the rich *white* man's cognitive mapping. This conspiracy theory had scant appeal in the global South – unlike some theories starring CIA and Mossad, HIV and vaccines – until Bolsonaro came along. More precisely still, it was the cognitive mapping of the rich white man who did not dare trust in capitalist climate governance, for fear that if it were to make good on its promise to address the crisis, it would turn into something else. Naomi Klein has suggested that the men of the denialist ISA correctly understood what any mitigation would imply.[99] Their great folly, of course, was to throw themselves into a fallacy similar to *post hoc ergo propter hoc*: mitigation might well take an anticapitalist form, hence the need for it must be dreamed up by the left (much in the same way some believed that because the state of Israel reaped geopolitical advantages from 9/11, it must also have masterminded the attacks).

One might say that once the science had matured, climate change became a revolutionary problem without a revolutionary subject. In the years around 1989, the environmental movement did indeed, as we have seen, turn towards justice and the left (another grain of truth in the conspiracy theory). But it was not able to challenge capitalism with anything like the power once evinced by the Third International or the national liberation movements, or even the social-democratic parties of the Second International; a lame successor, it won no Vietnam War and built no equivalent of the welfare state. The brakes it managed to put on the destructive forces of capital were decidedly weaker and more marginal, which is why the tasks it set itself have only grown ever more urgent. It remains to be seen if the climate movement surging up in the late 2010s can develop into the revolutionary subject the situation cries out for. In its absence, the anti-revolutionary thrust of fearful rich white men must target *the problem as such*, leading (some of) them to bid farewell to science and reality. The left climate theory is a heavily degraded figure of the total logic of this conjuncture, as seen from high above.

It is otherwise with chemtrails. The theory that the white stripes

---

99   Klein, *This*, 40, 43.

from aeroplanes contain chemicals sprayed by mighty elites to poison or desensitise those below, or wreak havoc with their weather or simply govern them, can be read as an extremely degraded form of subaltern ecological class consciousness: someone up there makes us suffer by polluting the sky.[100] 'Look up' is the motto of the chemtrailers. 'I never saw clouds like that as a kid. My gut and heart still tell me something's going on', one believer explained to the *Guardian* in 2017.[101] Relatively popular as it has recently become, particularly in the US, where some 5 per cent of the public had been swayed by the late 2010s, the chemtrails theory has yet to receive support from *Forbes* or the equivalent and make it into the corridors of power. Continued deadlock over the climate – not to speak of actual stratospheric aerosol injection – might widen its support base. For now, we must conclude that the most politically significant conspiracy theories of our time have been *seeded from the top*, which calls for a rethinking of the social function of such theories; and indeed, recent research, outside of Marxism, has begun to question the model of the accusatory loser. Far from expressing some ill-defined discomfort with the status quo, conspiracy theories 'may actually bolster' it, as they demonise – almost literally – those who want to do things differently. Mighty evil is at work, but it comes from outside and stands opposed to the system. These theories 'defend the social system when its legitimacy is under threat. In this respect, they join the ranks of other system-justifying processes', such as the belief that the poor get what they deserve or that women are by nature inferior.[102]

---

100 For an excellent analysis of the chemtrails phenomenon, drawing on Jameson, see James Bridle, *New Dark Age: Technology and the End of the Future* (London: Verso, 2018), 193–9, 205.

101 Cary Dunne, 'My Month with Chemtrails Conspiracy Theorists', *Guardian*, 22 May 2017.

102 Daniel Jolley, Karen M. Douglas and Robbie M. Sutton, 'Blaming a Few Bad Apples to Save a Threatened Barrel: The System-Justifying Function of Conspiracy Theories', *Political Psychology* 39 (2018), quotations from 475; cf. Christopher M. Federico, Allison L. Williams and Joseph A. Vitriol, 'The Role of System Identity Threat in Conspiracy Theory Endorsement', *European Journal of Social Psychology* 48 (2018): 927–38.

Now there is one Marxist who fathomed this logic long ago, namely Theodor Adorno, writing about the

> attitudes and opinions of all those who, for reasons of vested interests or psychological conditions, identify themselves with the existing setup. In order not to undermine their own pattern of identification, they unconsciously do not *want* to know too much and are ready to accept superficial or distorted information as long as it confirms the world in which they want to go on living.[103]

Winners may be afraid of learning about the sources of their wins, and they have never had greater reason to fear – 'knowing too much has assumed a subversive touch' on the brink of climate catastrophe.[104] This is cognition for the man on top: I cannot believe that the prevailing order, which has given me all I own, is capable of producing this terror; it must be slander from those who wish me and others like me harm. Whether or not it comes with a yarn about Lenin's birthday, climate denial has this base affective logic. A plenitude of data suggests that humankind in a warming world divides itself into left and right. Poll after poll, questionnaire after questionnaire, one psychological experiment after another has showed that people ideologically affiliated with the right, happy with free markets and private property, tend to dispute the existence of the problem – or if not, downplay its severity – whereas those on the left fret about it.[105] This

---

103 T. W. Adorno, Else Frenkel-Brunswik, Daniel J. Levinson and R. Nevitt Sanford, *The Authoritarian Personality* (London: Verso, 2019 [1950]), 662. Emphasis in original.
104 Ibid.
105 For example Irina Feygina, John J. Tost and Rachel E. Goldsmith, 'System Justification, the Denial of Global Warming, and the Possibility of "System-Sanctioned Change"', *Personality and Social Psychology Bulletin* 36 (2010): 326–8; Kirsti M. Jylhä and Nazar Akrami, 'Social Dominance Orientation and Climate Change Denial: The Role of Dominance and System Justification', *Personality and Individual Differences* 86 (2015): 108–11; Bruce Tranter and Kate Booth, 'Scepticism in a Changing Climate: A Cross-National Study', *Global Environmental Change* 33 (2015): 154–64; Aaron M. McCright, Riley E. Dunlap and Sandra T. Marquart-Pyatt, 'Political Ideology and Views about Climate Change in the European Union', *Environmental Politics* 25 (2016): 338–58; Aaron McCright, Sandra T. Marquart-Pyatt, Rachael L. Shwom et al., 'Ideology, Capitalism, and Climate: Explaining Public Views about Climate Change in the United States', *Energy Research and Social Science* 21 (2016): 180–9; Kirsti M. Jylhä, Clara Cantal,

polarity is not reduced at higher levels of education. To the contrary, it is accentuated, right-wing people showing *less* concern about global warming the better equipped they are to absorb the science. A highly educated man of the right excels in screening out information that would 'license restrictions on commerce and industry', which seems to suggest that teach-ins for the rich would be actively counterproductive.[106] They don't want to know and they know how to avoid it.

If commitment to the prevailing order trumps everything else, however, it has more foundations than class. In a seminal article from 2011, two leading experts on the denialist ISA noticed that most of its figureheads – Singer, Inhofe, Beck – were not only conservative but *white men*. Did the same pattern apply to the US public at large? Indeed, 59 per cent of conservative white men denied attribution, compared to 31 of all other adults; 65 per cent of the former believed that the media exaggerates global warming as against 30 of the latter, and so on. Rich more than poor, men more than women, whites more than non-whites rallied to the denialist pole, which could only be explained by their loyalty to a status quo that had been serving them well.[107] These findings have likewise been replicated, including on a global scale: in the early twenty-first century (again, before

---

Nazar Akrami and Taciano L. Milfont, 'Denial of Anthropogenic Climate Change: Social Dominance Orientation Helps Explain the Conservative White Male Effect in Brazil and Sweden', *Personality and Individual Differences* 98 (2016): 184–7; Matthew J. Hornsey, Emily A. Harris, Paul G. Bain and Kelly S. Fielding, 'Meta-analyses of the Determinants and Outcomes of Belief in Climate Change', *Nature Climate Change* 6 (2016): 622–6; Mark Romeo Hoffarth and Gordon Hodson, 'Green on the Outside, Red on the Inside: Perceived Environmentalist Threat as a Factor Explaining Political Polarization of Climate Change', *Journal of Environmental Sociology* 45 (2016): 40–9; Olve Krange, Bjørn P. Kaltenborn and Martin Hultman, 'Cool Dudes in Norway: Climate Change Denial among Conservative Norwegian Men', *Environmental Sociology* 5 (2019): 1–11; Stephan Lewandowsky, 'In Whose Hands the Future?', in Uscinski, *Conspiracy*, 152; Lewandowsky et al., 'Role'; Lewandowsky et al., 'NASA'; Uscinski et al., 'Climate', 2, 17, 21.

106 Dan M. Kahan, Ellen Peters, Maggie Wittlin et al., 'The Polarizing Impact of Science Literacy and Numeracy on Perceived Climate Change Risks', *Nature Climate Change* 2 (2012): 732–5; quotation from 732.

107 Aaron McCright and Riley E. Dunlap, 'Cool Dudes: The Denial of Climate Change among Conservative White Males in the United States', *Global Environmental Change* 21 (2011): 1163–72.

Bolsonaro) climate denial was a rarity in Latin America and sub-Saharan Africa, which registered high and rising levels of concern.[108] Such levels correlated negatively with GDP. If it were up to the average woman in Mozambique rather than reader of *Forbes*, truth-recognition would come easily.

All that psychic investment in the capitalist mode of production did not have to express itself in conspiracy theories – it could stick with the tumbling rocks of Dover or just sheer dismissive indifference – but they represented its sharpest edge or most sensitive antenna. For those who stood to lose most from a break with business-as-usual and were aware of it, if only subliminally, conspiracy theories about climate were attractive. The law of polarisation redoubled it. So did the mountains of evidence, pushing those who had chosen this path further and further along until they entered the far right, the natural habitat for conspiracy theories, the destination marked out by the rich man's cognitive mapping – after some crossing point, the only island where the facts of a rapidly warming world could be rigidly ignored. The fourth phase of denial was a kind of homecoming. It coincided with the rise of what has been called 'post-truth'. In the most creditable scholarly analysis to date, Lee McIntyre defines this as a condition where truth has been eclipsed and rendered politically irrelevant. Appeals to affect override respect for facts.[109] Anyone, including aspirants to the highest offices, can flourish patent untruths without paying a price for it. Whence this wantonness?

Recapitulating the history of what we have called the denialist ISA, McIntyre argues that it was organised climate denial that sowed the seeds of post-truth and made it *de rigeur* to subordinate the outer

---

108 Hanno Sandvik, 'Public Concern over Global Warming Correlates Negatively with National Wealth', *Climatic Change* 90 (2008): 333–41; So Young Kim and Yael Wolinsky-Nahmias, 'Cross-National Public Opinion on Climate Change: The Effects of Affluence and Vulnerability', *Global Environmental Politics* 14 (2014): 79–106; Susan Clayton, Patrick Devine-Wright, Paul C. Stern et al., 'Psychological Research and Global Climate Change', *Nature Climate Change* 5 (2015): 640–6; Alex Y. Lo and Alex T. Chow, 'The Relationship between Climate Change Concern and National Wealth', *Climatic Change* 131 (2015): 335–48; Stuart Capstick, Lorraine Whitmarsh, Wouter Poortinga et al., 'International Trends in Public Perceptions of Climate Change over the Past Quarter Century', *WIREs Climate Change* 6 (2015): 35–61.

109 Lee McIntyre, *Post-Truth* (Cambridge, MA: MIT Press, 2018), xiv, 5, 9.

world to one's own private gut. Cognitive bias, confirmation bias, motivated reasoning and the other psychological mechanisms that rule this condition were pioneered by 'oil interests', from their first encounter with the naked truth about fossil fuels. Decades of efforts to wave that truth away corroded political culture so thoroughly that it ended up at 'risk of being estranged from reality itself'.[110] Denial of this one problem could have such a far-reaching effect because it concerned the material base of capital accumulation, unlike, for instance, Darwinian evolution, which some on the American right also denied. Only resistance against this particular truth could radiate into the daily tweets of Trump and his copycats: if you can lie about climate, you can lie about anything – the signature procedure of the far right.[111] Hence the AfD combined its climate freakshows in the Bundestag with tirades against the 'systematic discrimination of men' and the 'islamicised federal state of Germany'.[112] If there is merit to this analysis, it implies that *by reacting to climate science with denial, fossil capital not only smothered mitigation but also midwifed the contemporary far right* or at least assisted in its intellectual legitimation. The fourth phase of denial would be a logical extension of the first.

In this phase, however, the genre of conspiracy theory underwent a metastasis. Classical practitioners were hung up on their own punctiliously assembled facts. Light in their eyes, they would produce documents proving that the World Trade Center must have come down through controlled demolition, or point to visual anomalies in the images from the moon demonstrating that NASA staged the landings in studios, or write treatises striking down on every glitch in the official account of some murder; detectives who would never stop sniffing, they based their cases on data, if only errant data. But in the 2010s, fewer of them seemed to care to. In *A Lot of People Are Saying: The New Conspiracism and the Assault on Democracy*, a piercing inquiry into the epistemology of the contemporary far right, Russell Muirhead and Nancy L. Rosenblum argue that conspiracy is

---

110   Ibid., 22, 172; see further 14, 20–34, 79, 109.
111   Ibid., 116.
112   Rensmann, 'Radical', 60.

now rather alleged by 'bare assertion'. The birtherists, for instance, composed nothing like the filmography and literature on 9/11 made by the truthers. They were content with asserting that Obama was born in Kenya, or even just alluding to the possibility. When Trump claimed that Obama had tapped his phones, or that the media concealed Islamist terrorist attacks, he didn't bother to adduce anything that looked like evidence. The 'new conspiracism' proceeds by blurting out the plot and the hoax and nothing more. It satisfies itself 'with a free-floating allegation disconnected from anything observable in the world', to which it 'pays no fealty'; instead of converting counterproof into proof, it dispenses with the question of proof altogether.[113] It is even more untethered from reality than classical conspiracy theory, of which it is a degraded figure.

How, then, does the new conspiracism establish credibility? Not by pseudoscientific validation, but by *repetition*, and more precisely virtual repetition: the veracity of a statement is measured in the number of likes, retweets, threads and views it generates. Trump was wont to back up his accusations with the high number of sympathisers who agreed with them – 'a lot of people are saying' X or Y, hence it is true. This epistemology is obviously the making of social media. With its inbuilt character limit, Twitter is a machine for bare assertion or, if you will, for the most stripped-down, contentless interpellation: 'Hey, you there!' and very little else. But not every Twitter user interpellates her followers with talk of fake news and treasons. Muirhead and Rosenblum observe that new conspiracism – unlike, or so they claim, classical conspiracy theory – is a monopoly of the right, circulated as a cheap identity card for the insiders of a threatened nation. It is performed by winners, presidents included.[114] It is also the most recent mode of climate fabulation. Inhofe managed to write a whole book on *The Greatest Hoax* and Sussman added to the library, but in the fourth phase, 'hoax' could be sufficient attestation in itself.[115] Far-right denial mostly took the form of the barest

---

113   Russell Muirhead and Nancy L. Rosenblum, *A Lot of People Are Saying: The New Conspiracism and the Assault on Democracy* (Princeton: Princeton University Press, 2019), quotations from 25–6, 123.
114   Ibid., for example 5, 32–3, 40, 49–51, 127–8, 134.
115   As argued in ibid., 107–10.

assertion – theory as meme – but it wasn't any less potent for that. It enhanced the replicability. Global heating exerts selective pressure on denialists; finding refuge in conspiracy theory was a first adaptation to the mountains of evidence, but as the space for denial became further circumscribed, taking flight in new conspiracism offered another, possibly safer way to survival.

Some of this progressing degradation was put on display by the conspiracist-in-chief when he returned to the topic of wind power. He took aim at those proposing that

> we have an economy based on wind. I never understood wind. I know wind mills very much. I've studied it better than anybody and it's very expensive . . . They manufacture tremendous – if you're into this – tremendous fumes, gases, are spewing into the atmosphere. We have a world, right? So the world is tiny compared to the universe. So tremendous amounts of fumes and everything – you talk about the carbon footprint – fumes are spewing into the air . . . And if you own a house within visions of some of these monsters, your house is worth 50 per cent of the price. They're noisy, they kill the birds. You want to see a bird graveyard, you go take a look, go under a wind mill someday, you'll see more birds than you've ever seen ever in your life [laughter and applause from the audience of young conservatives in Florida] . . . You know what they don't tell you about wind mills? After ten years, they look like hell. They start to get tired, old,

and so on.[116] Trump must have here broken some sort of sound barrier. He so completely inverted the matter at hand that his discourse became devoid of substantive content and approximated guttural noise, although listeners could make out that it maintained a thematic relation to climate and paraded absolute contempt for the problem. In a sense, denial had here completed the circle, through conspiracy theory and new conspiracism into pure burbling. Rising temperatures could not extinguish it, apparently, only drive it into new rounds of speciation. But none of the old genera died out. As

---

116   Richard Luscombe, '"I Never Understood Wind": Trump Goes on Bizarre Tirade against Wind Turbines', *Guardian*, 24 December 2019.

Muirhead and Rosenblaum suggest, the denialist ISA has rather been confirmed by the new climate conspiracism, as when the reader of a story receives a nod of recognition from a by-passer; together, they have kept the narrative alive.[117] A similar dialectic could be discerned at another point where the far right liked to convene: the theory of Cultural Marxism.

The Myth of Cultural Marxism

A spectre is haunting the far right: something it calls 'Cultural Marxism'. What is this ghoulish figure? What dark, malevolent forces does it channel into the world, and through what mediums? According to the theory, once the Russian Revolution had failed to spread to Western Europe, Lukács and Gramsci understood that some ramparts had to be removed: things like Western culture and the Christian religion.[118] The proletariat was too deeply identified with these to rise up. In the 1920s, a coterie of Marxist academics, who happened to be – a not incidental detail – of Jewish origin, headed by Adorno, Horkheimer and Herbert Marcuse, formed the Frankfurt School to begin the work of breaking the ramparts down. They mixed in a bit of Freudian theory to also dissolve the traditional family and normal sexuality. They sought to uplift women, blacks, gays and assorted other minorities; instead of open class conflict, they pursued identity politics and 'political correctness'. When the Frankfurt School relocated to the US after Hitler's seizure of power, it found itself in just the right place – next to Hollywood and the campuses – to implode the sleeping giant from within. America

---

117 Muirhead and Rosenblum, 110.
118 Scholarly studies of the theory include John E. Richardson, '"Cultural Marxism" and the British National Party: A Transnational Discourse', in Nigel Copsey and John E. Richardson (eds), *Cultures of Post-War British Fascism* (Abingdon: Routledge, 2015), 202–26; Jérôme Jamin, 'Cultural Marxism and the Radical Right', in Paul Jackson and Anton Shekhovtsov, *The Post-War Anglo-American Far Right: A Special Relationship of Hate* (London: Palgrave, 2014), 84–103; Jérôme Jamin, 'Cultural Marxism: A Survey', *Religion Compass* 12 (2018): 1–12; Tanner Mirrlees, 'The Alt-Right's Discourse of "Cultural Marxism": A Political Instrument of Intersectional Hate', *Atlantis: Critical Studies in Gender, Culture and Social Justice* 39 (2018): 49–69.

would be made communist through the spread of homosexuality, feminism, sexual liberation, atheism, the mixing of races and the hedonistic counter-culture of the 1960s. John Lennon was privy to the plot. Indeed, according to one strand of the theory, Adorno himself wrote all of the Beatles' lyrics.[119] But the central contention is that Marxism initiated a 'long march through the institutions', worming its way through schools, universities, government bureaucracies, media, film studios, the fine arts, slowly but surely, to the point where the West would come undone. Just as in the left climate theory, an epochal defeat of the left – not in 1989, but in the years after 1917 – is here transformed into a devious project of world domination.

Replete with pseudo-factual paraphernalia and laboriously connected dots, this is a most classical conspiracy theory. It issues from the oldest one – the world suffers under secretive Jewish power – and more directly from Judeo-Bolshevism and its subheading *Kulturbolschewismus*, *idée fixe* of fascists in interwar Europe.[120] The chief populariser of the later iteration was William S. Lind, a paleoconservative white man and military strategist, who brought the details up to date with the early millennium. On the right of the established American right, with his own weekly TV show, articles in the *Marine Corps Gazette* and encounters with Donald Trump, he could send Cultural Marxism into wider channels of circulation. At a conference for Holocaust deniers in 2002, he felt sufficiently at home to raise the curtain on the true identity of the Frankfurt School conspirators: 'These guys were all Jewish.'[121] Then the theory took on a life of its own.

---

119 See for example David Gerard, 'Theodor Adorno Wrote All the Beatles' Songs as a Cultural Marxist Assault on America. Possibly', *Rocknerd*, rocknerd.co.uk, 13 October 2016; *Guardian*, 'A Little Help from My Neo-Marxist Philosopher: Was Adorno the Fifth Beatle?', 10 September 2019.

120 See for example Hanebrink, *A Specter*; Waddington, *Hitler's*; and the revelatory Michael Kellogg, *The Russian Roots of Nazism: White Émigrés and the Making of National Socialism* (Cambridge: Cambridge University Press, 2005).

121 Bill Berkowitz, '"Cultural Marxism" Catching On', *Intelligence Report*, Southern Poverty Law Center, 15 August 2003. For a sample of his output, see William S. Lind, John F. Schmitt and Gary I. Wilson, 'Fourth Generation Warfare: Another Look', *Marine Corps Gazette* December (1994): 34–7; William S. Lind, 'What Is "Political Correctness"?', in William S. Lind (ed.), *"Political Correctness": A Short History of an Ideology* (Washington, DC: Free Congress Foundation, 2004), 4–9; 'The Roots of Political Correctness', *American Conservative*, theamericanconservative.com, 19 November

In the 2010s, Cultural Marxism approached the status of a metatheory of the far right, popping up high and low, as all-inclusive scoop or the barest tribalist assertion. Over at Breitbart, it was the revealed truth. Before his death, the founder of the site, Andrew Breitbart, described his discovery of it as 'my one great epiphany, my one a-ha moment where I said, "I got it – I see what exactly happened in this country."'[122] A top aide of Trump wrote a long memo on how the Frankfurt School had created a deep state 'beholden to no one', a cruel 'god bestriding the earth', now busy handing over the West not to communism but to Islam.[123] The largest party in the Netherlands, as of the 2019 elections, had the theory as an intellectual fundament. Thierry Baudet tweeted about it – no doubts or qualms: the whole cock-and-bull-story about the Frankfurt School drawing up its masterplans – with the regularity of a muscle twitch. The European Union and mass immigration ranked highest among the Marxist achievements.[124] Baudet here learned from Paul Cliteur – supervisor of his PhD thesis, law professor at Leiden, signatory to CLINTEL and top party intellectual – who in 2018 published an anthology with a white man forced to drink a chalice of poison on the cover and a title translating as *Cultural Marxism: A Spectre Is Haunting Europe*.[125] A spectre indeed. It had perverted European culture with an overdose

---

2009; Thomas Hobbes [pseudonym for William S. Lind], *Victoria: A Novel of the 4th Generation War* (Kuovola: Castalia House, 2014). On Lind's influence, cf. Chauncey DeVega, 'How "4th Generation Warfare" Helps Explain the Rise of Donald Trump', *Salon*, salon.com, 5 July 2016; Paul Rosenberg, 'Donald Trump's Weaponized Platform: A Project Three Decades in the Making', *Salon*, 16 July 2016.

122 Scott Oliver, 'Unwrapping the "Cultural Marxism" Nonsense the Alt-Right Loves', *Vice*, 23 February 2017.

123 Rich Higgs, 'POTUS & Political Warfare', May 2017, 2; document publicised in Jana Winter and Elias Groll, 'Here's the Memo That Blew Up the NSC', *Foreign Policy*, foreignpolicy.com, 19 August 2017.

124 Thierry Baudet's Twitter account, for example 31 July 2017, 12 August 2017, 19 August 2017, 20 August 2017, 18 September 2017, 6 October 2017, 1 May 2018, 30 May 2018, 1 June 2018. On 1 April 2018, for example, Baudet tweeted a recommendation of a Lind imitation: Linda Kimball, 'Cultural Marxism', *American Thinker*, 15 February 2007.

125 Paul Cliteur, Jesper Jansen and Perry Pierik, *Cultuurmarxisme: Er waart een spook door het Western* (Soesterberg: Aspekt, 2018). On Cliteur as supervisor, cf. Max van Weezel, 'Waarom FvD-huisfilosoof Paul Cliteur een complotdenker is', *Vrij Nederland*, 27 July 2018.

of compassion for oppressed groups. Intersectionality was its latest invention, with worries about ecology bundled into it.[126]

The most revered internet guru of the right, psychology professor and self-help bestseller Jordan Peterson, bought into the idea too. In November 2016, he posted on his Facebook site an article headlined 'Cultural Marxism Is Destroying America', the first sentence of which ran: 'Yet again an American city is being torn apart by black rioters.' The events in question were taking place in Charlotte, North Carolina, after the killing of a black man by a police officer, which could only be explained by all the usual suspects from Gramsci onwards having established a 'fifth column' to foment sedition.[127] A go-to theory for explaining any malaise, it worked well without evidentiary elaboration. But the far-right corners of the web overflowed with material for those who wished to learn more. Here one could find some two hundred thousand YouTube videos on the topic of Cultural Marxism.[128] At the grandiloquent pole, again, was Breivik, who began his manifesto with an excerpt from Lind, plagiarised him repeatedly and used the term 'Cultural Marxism' and its derivatives more than six hundred times ('Frankfurt School' eighty-eight times, 'Adorno' twenty-six, 'Marcuse' twenty-five).[129] If Breivik targeted a Marxist

---

126 One of the contributors accused the left of demanding an excessive amount of protection and care for the following groups: 'workers, women, children, and later also homosexuals, other sexual minorities, handicapped, indigenous people, black and brown races, people with a migration background and animals.' Those not on the list – a rather diminutive segment of humanity – were subject to reverse discrimination. Jesper Jansen, 'Femin-islam: Strange Bedfellows?', in Cliteur et al., *Cultuurmarxisme*, 257. On the ecological component of the plot, see, in the same volume, for example Derk Jan Eppink, 'De Europese Unie en het cultuurmarxisme', 142; Jan Herman Brinks, 'Over het drievoudig falen van westerse intellectuelen tussen 1917 en 2017', 102–3.

127 Jordan B. Peterson's Facebook page, 7 November 2016; Moses Apostaticus, 'Cultural Marxism Is Destroying America', *The Daily Caller*, 29 September 2016. On Peterson and his relation to the theory, see further Mirrlees, 'Alt-Right's', 61–2; Dorian Lynskey, 'How Dangerous Is Jordan B Peterson, the Rightwing Professor who "Hit a Hornets' Nest"?', *Guardian*, 7 February 2018; Charles Mudde, 'Jordan Peterson's Idea of Cultural Marxism Is Totally Intellectually Empty', *The Stranger*, 25 March 2019; as well as YouTube rants such as Jordan B. Peterson, 'Postmodernism and Cultural Marxism', *The Epoch Times*, YouTube, 6 July 2017.

128 Mirrlees, 'Alt-Right's', 55–6; this paper includes an extensive survey of Cultural Marxism on the web.

129 Breivik, *2083*. On the plagiarisation of Lind, see Chip Berlet, 'Breivik's Core Thesis Is White Christian Nationalism v. Multiculturalism', *Talk to Action*, 24 July 2011.

gathering, others could choose a Jewish one. The perpetrator of the Pittsburgh synagogue shootings in October 2018, in which eleven people were killed during Shabbat prayers – the worst attack on the Jewish community in US history – spent his ample screen time on Gab, a 'free speech' network where Cultural Marxism was a favourite talking point. There he posted pictures of Jews trying to press the African continent into the EU and organising the caravans of Central American migrants to the US. When he stormed into the synagogue, he screamed: 'All Jews must die.'[130]

More genteel in their manners, the leaders of the Sweden Democrats had to deal with yet another storm of criticism after they had, for the second time, declared that Swedish Jews would always prioritize their religion over their country. Mattias Karlsson thought the storm was whipped up by 'cosmopolitans' and 'Cultural Marxists', and in the spring of 2019, he announced that he would step down from his post as the parliamentary leader of the SD to build a new think tank with the mission to 'poke a stick into the Cultural-Marxist hornet's nest'.[131] Immonen and Halla-aho of the PS and the SS were fervent believers.[132] But these parties could still only dream of the influence exerted by their brothers in Brazil. The theory of Cultural Marxism became a pillar of the Bolsonaro government. Jair, his sons and foreign minister Araújo had their own internet guru in Olavo de Carvalho, another disgruntled white man, who, from his desk in Virginia, spoke to the nation through a thousand screens. A pipe dangling from his mouth, with an air of enviable erudition, he looked deep into the eyes of his followers and revealed to them the secrets of the Cultural Marxist plot to control Brazil. The theory was here adapted to the vitriolic demonisation of Lula's and Dilma's Workers' Party. During its years in power, de Carvalho beamed his profound

---

130   Alex Amend, 'Analyzing a Terrorist's Social Media Manifesto: The Pittsburgh Synagogue Shooter's Posts on Gab', *Southern Poverty Law Center*, 28 October 2018; Samuel Moyn, 'The Alt-Right's Favorite Meme Is 100 Years Old', *New York Times*, 13 November 2018.

131   Daniel Swedin, 'SD: extremism blir allt tydligare', *Aftonbladet*, 20 June 2018; Dick Erixon, 'Mattias Karlsson (SD) om de nya utmaningarna', *Samtiden*, 14 March 2019.

132   For example, the many texts on the topic collected on Immonen's blog: 'Kaikki blogit puheenaiheesta Kulttuurimarxismi', immonen.vapaavuoro.uusisuomi.fi, n.d; Antifasistinen Verkosto, 'What Is'.

explanations of communist depravity into the hearts of true white and Christian men and women. He had a background as an astrologist. He would not accept evidence of heliocentrism. At the centre of Cultural Marxism was the artifice of climate change, and oh, Adorno wrote all the Beatles' lyrics.[133]

In the fourth most populous nation on earth, Olavo de Carvalho attained an oracular, Rasputin-like status few intellectuals in the world came close to: Bolsonaro gave his victory speech on the election night with a book by the man in front of him. Araújo and other ministers appear to have been picked directly by de Carvalho.[134] The former credited him with having single-handedly broken the 'psycho-political control system' of Cultural Marxism, as maintained by Lula and Dilma – 'nothing short of a miracle'.[135] Here the anti-communism of the military dictatorship was revived but, unlike in the 1960s and '70s, without anything like communism on the horizon, giving it a strangely psychedelic quality. It was anti-communism on mushrooms, a degraded figure of the class struggles of the twentieth century after they had fizzled out into the crisis of the biosphere.

Protean in the extreme, Cultural Marxism can swim back and forth between classical theory and new conspiracism. It does not have to come with the full package of fantasies. Not everyone who

---

133 See his websites olavodecarvalho.org and blogdoolavo.com. Representative texts on the theory of Cultural Marxism are 'Do marxismo cultural', published first in *O Globo* and then on olavodecarvalho.org, 8 June 2002; Olavo de Carvalho, 'Marxismo cultural', olavodecarvalhofb.wordpress.com, 17 January 2019. Heliocentrism: Felipe Mora Brasil, 'Olavo tem razão. Parabéns, professor', *Veja*, 15 February 2017; Olavo de Carvalho's Facebook page, 9 August 2017. Beatles: *Folha de S. Paulo*, 'Olaveo de Carvalho diz que quem escreveu as músicas dos Beatles foi sociólogo alemão', 7 September 2019. De Carvalho as Jair Bolsonaro's guru: Iracema Amaral, ' "Guru" de Bolsonaro, Olavo de Carvalho chama parlamentares do PSL de semianalfabetos e caipiras', *Estado de Minas*, 17 January 2019.

134 Brian Winter, 'Jair Bolsonaro's Guru', *Americas Quarterly*, 17 December 2018.

135 Ernesto Araújo, 'Now We Do', *New Criterion*, January 2019. For Aráujo as the first foreign minister appointed to extirpate Cultural Marxism, including the idea of climate change, see Jonathan Watts, 'Brazil's New Foreign Minister Believes Climate Change Is a Marxist Plot', *Guardian*, 15 November 2018; Estadão Conteúdo, 'Futuro chanceler diz que vai libertar o Itamaraty do "marxismo cultural"', *Exame*, 28 November 2018; Andre Pagliarini, 'The Worst Diplomat in the World', *Jacobin*, 26 February 2019.

uses the term will also retell the Frankfurt story or know about William S. Lind, just as belief in 'Islamisation' does not require knowledge of Bat Ye'or and her elucubrations. Evasiveness can be a protective layer.[136] Baudet and Peterson maintained an air of smartness potentially vitiated by the claim that Adorno wrote all the lyrics to the Beatles' songs. Men like them could vacate the theory of narrative content and turn it into a series of formal equivalences: Cultural Marxism *is* Political Correctness *is* Multiculturalism *is* Feminism *is* Communism, all promoted by the same left which – the irreducibly conspiratorial core – controls cultural life. With its capacious accommodation of everything the far right dislikes, the theory is permeable, open to cross-fertilisation with George Soros and Eurabia and the Great Replacement. Eclecticism can ensure reproduction. But Cultural Marxism also works as a tightly packed prism, through which negative developments are rendered as epiphenomena of a single force pulling the levers of history in the wrong direction.[137] It has been called an 'instrument of intersectional hate', turned against Black Lives Matter, abortion laws, Pope Francis, Hollywood films with a trace of an anti-capitalist message (*Elysium*) or multiracial casts (*Star Wars: The Force Awakens*) – all of this and much more, at one point or another, pressed through the prism.[138]

And pressed was also, of course, the climate. The left climate theory was a product of 1989, but conjointly with Cultural Marxism, it was relaunched in a more pungent form in the 2010s. The natural step for someone like James Delingpole, the man who gave the world 'Climategate', was to transition from the *Telegraph* and *Forbes* to Breitbart, where he became an anti-climate correspondent, telling his readers that the 'scare' had been invented to 'destroy Western liberal civilisation' as part of a war started by 'the Cultural Marxists of the 1930s Frankfurt School'.[139] The take of

---

136   Cf. Konda, *Conspiracies*, 193–6.
137   Cf. Richardson, '"Cultural"', 202, 222.
138   Mirrlees, 'Alt-Right's', 55–9.
139   James Delingpole, 'Why Conservatives Will Always Lose the War on Climate Change', *Breitbart*, 9 April 2016. The Heartland Institute named a special 'freedom center' in honour of Andrew Breitbart. 'Heartland Institute Announces the Naming of the Andrew Breitbart Freedom Center', Heartland Institute, 16 March 2016.

the green nationalists would be that Cultural Marxism has arrogated ecology to itself and must be kicked out of it.[140] Much more common was a monobloc fury against both, so that someone who believed in the theory would typically also deny climate change – Jordan Peterson being another case – and denialist parties feel confirmed in their desire to revive forgotten languages.[141] The AfD denounced the Greens as 'crypto-communist decomposers of the fatherland'.[142] Into the halls of COP25 in Madrid in late 2019, Vox sent a small delegation to 'keep a close eye on the extreme left', which supposedly ran the summit as a religious service issuing new edicts for 'how we should live, what we can eat and which industries can prosper'.[143] But the truly novel element in the late 2010s was the emergence of a climate movement as a real political force, if not yet quite a revolutionary subject. It prodded conspiracy theorists to exercise their cells again. From early 2019, they focused on the person of Greta Thunberg.

If Thunberg's homeland came late to the school strikes, only joining in big numbers in the autumn of 2019, it was quicker to produce food for the thought of her enemies. After she had returned from the Extinction Rebellion actions that shut down parts of London in the spring, Jimmie Åkesson told Swedish media that 'she doesn't do this on her own. This', referring to the strikes, 'is not a campaign she has initiated spontaneously. It is obvious if you follow social media that this is not something that has spread out of sheer coincidence – it is staged.'[144] If you follow social media: supporters of the SD had a good idea about whom precisely to suspect. A few weeks later, a report

---

140   Cf. Tarrant, *Great*, 20, 58.
141   On Peterson, see above and Henry Mance, 'Jordan Peterson: "One Thing I'm Not Is Naïve"', *Financial Times*, 1 June 2018; Samuel Earle, 'Outselling the Bible', *London Review of Books Blog*, 15 March 2018; Gyrus, 'The Hero and the Dark Forest: Jordan Peterson and Environmentalism', *Dreamflesh*, October 2018.
142   Rensmann, 'Radical', 61.
143   *El Español*, 'Vox acusa a la izquierda de usar el cambio climático para cambiar "nuestro modo de vida"', elespanol.com, 2 December 2019; *Directivos Empresas*, 'El medio rural responde a los retos del cambio climático', directivosempresas.com, 12 December 2019.
144   Lotten Engbom, 'Jimmie Åkessons känga till Greta Thunberg: "Ett barn"', *Expressen*, 25 April 2019.

spread through the middle ranks of the party that Greta Thunberg was on the payroll of George Soros and chaperoned by an agent with a suitably Jewish-sounding surname. From south to north, party chapters disseminated the theory – meme, rather – on Twitter and Facebook.[145] Their leaders refrained from making the Jewish connection in public statements, but stepped up the attacks on Thunberg and felt emboldened in their denial. 'You will see that this is all a great climate hoax,' Björn Söder commented on the world's most famous Swede, by now speaking for the party consistently polling as the country's largest.[146] It readied for government rule, in alliance with the two main conservative parties. (Soon after the election in 2018, Antonia Ax:son Johnson of *Axess* broke the *cordon sanitaire* by arguing that normal bourgeois parties should accept the SD as a partner in power that would do no harm to Swedish business. The business daily agreed.[147] One year later, this was indeed the new political bloc poised to take over, Sweden moving with the plates of the continent.)

Wherever the name of Greta Thunberg travelled, the right had to make sense of her and the movement she represented, and it worked with two versions of one theory: that she was a puppet on the string of Soros or of the Marxists. The *Spectator* considered her a 'proxy' for 'those on the Left who seek to use climate alarmism to further their war on global capitalism'.[148] A Fox News pundit called her a 'mentally ill child' used as a stooge by 'the international left'. When she tweeted that 'the climate crisis is not just about the environment', since 'colonial, racist and patriarchal systems of oppression have created and fuelled it', a sprightly James Delingpole had all his suspicions

---

145   Jonas Persson, 'George Soros behind Greta Thunberg', *Nya Tider*, 24 April 2019; Kim Fredriksson and Magnus Karlsson, 'SD-politikerna sprider antisemitiska konspirationsteorier om Greta Thunberg', *Aktuellt Fokus*, 7 June 2019; Norea Dahlskog, 'SD-förening sprider konspirationsteorier om islam', *Nyheter 24*, 31 October 2019; Jimmie Larsson, 'SD-politiker spred konspirationsteori om Greta Thunberg', *Östra Småland Nyheterna*, 8 June 2019; Magnus Karlsson, 'Här sprider SD:s kommunförening antisemitiska konspirationsteorier om Greta Thunberg', *Aktuellt Fokus*, 31 October 2019.

146   Joakim Magnå, 'SD-toppens attack på Greta och klimatet: "Stor klimatbluff"', *Aftonbladet*, 19 October 2019.

147   P. M. Nilsson, 'Antonia har rätt', *Dagens Industri*, 24 September 2018.

148   Ross Clark, 'The Trouble with Greta Thunberg', *Spectator*, 23 April 2019.

confirmed: 'The teenage Climate Puppet has gone full Marxist.'[149] One could peruse articles with headlines such as 'Marxists Hope to Take Over the World by Terrifying Children about Climate Change'.[150]

On the other note, one AfD branch in Bavaria (of all provinces) created a montage with the face of Thunberg next to that of Soros and two devilish yellow eyes in the dark, under the headline 'The Power behind It'.[151] More widespread became a picture purporting to show a smiling Soros putting his arms around her. It was a doctored photo from her meeting with Al Gore. As she took the world stage, photoshopped pictures of Thunberg became a free-floating conspiracist subgenre of its own: Thunberg as the tip of an iceberg made up of Soros; Thunberg crying in the arms of an ISIS soldier; Thunberg eating lunch on a train passing by starving children, this one shared by Eduardo Bolsonaro.[152] (There was also a picture of Thunberg dressed in a t-shirt with the text 'Antifascist All Stars', proof that she was an antifa terrorist. That photo was authentic. Participating in rallies against Nazi marches in Sweden, Thunberg was a committed anti-fascist, although not a member of Antifascist Action.) The country where this theorising flied highest, however, was, true to form, Hungary.

The government of Viktor Orbán had not said much about the climate before spring 2019. But as the school strikes took off, it developed a keen interest in the issue: behold, Soros and the left have found a new stake on which to impale the Hungarian nation and leave it for refugees to consume.[153] From Fidesz media, there came a

---

149   Clark, 'The Trouble'; Justin Baragona and Maxwell Tani, 'Fox News Guest Calls Greta Thunberg "Mentally Ill Swedish Child" as Right Wing Unleashes on Climate Activist', *Daily Beast*, 24 September 2019; James Delingpole, 'Greta the Teenage Climate Puppet Goes Full Marxist', *Breitbart*, 25 November 2019.

150   John Eidson, *Life Site News*, 30 September 2019.

151   AfD Berchtesgadener Land's Facebook page, 3 March 2019.

152   See for example Ruir Pedro Antunes, 'Fact Check. Greta Thunberg é controlada por George Soros?', *Observador*, 30 December 2019; Joana Splieth, 'Nein, Greta Thunberg ist nicht die Enkelin von George Soros', *Correctiv*, 20 December 2019; Daniel Funke, 'Greta Thunberg Did Not Pose with ISIS or George Soros', *Politifact*, 27 September 2019; Paul Harper, 'Fury over Fake Photo of Thunberg Eating Lunch in Front of Poor Children', *Metro*, 27 September 2019.

153   See the excellent analysis in *Hungarian Spectrum*, 'A New Enemy: The "Neo-Bolshevik" Greens and Climate Change', hungarianspectrum.org, 15 August 2019; 'The "Perfectly Balanced" Hungarian Media and Climate Change', 24 September 2019; 'The Green Movement: A Conspiracy against Orbán's Government', 27 December 2019.

downpour of analysis of this *klímakommunizmus*. The strikers had been paid by Soros; when they marched in Budapest, their ranks were made up of 'foreigners'; duped by the 'fecal ideology' of Bolshevism, these kids would have been better off had they gone straight for the Kabbalah or the prayer to Mecca. The 'climate hysteria' itself had no scientific foundations.[154] But the novelty of 2019 was the shift in focus from science to movement, and the latter could also, in turn, be implicated in the *impacts* of climate breakdown. If this is a hoax perpetrated by enemies of the people, might they not be organising all these extreme weather events?

In 2019, one SD representative in Umeå, the largest city in Sweden's north, submitted a series of theories about the wildfires that struck the country – not in the summer, but in the exceptionally hot and dry month of April. Linking them to Notre Dame, he tried out the theory that Islamists had lit the fires, or perhaps antifa terrorists. Eventually he settled on 'a group of climate activists doing this to fan the flames of climate hysteria', the climate movement being 'a religious fanatical doomsday cult'.[155] Earlier, the grapevines of Swedish-Democratic social media had also spread the theory that the water shortages on Öland were in fact caused by asylum-seekers and, more specifically, Muslims who overconsumed water because they washed themselves five times per day before praying.[156] Trump claimed that the figure of three thousand Puerto Rican casualties from Hurricane Maria was confected by Democrats to make him look bad; the real number was in the range of '6 to 18'.[157] After the inferno in the Amazon, Jair Bolsonaro informed the world that it was Leonardo DiCaprio who had ignited it. The theory was that the American actor had paid NGOs to set the rainforest on fire and then

---

154 Megadja Gábor, 'A klímakommunizmus kísértete', *Magyar Nemzet*, magyarnemzet.hu, 1 October 2019; Orbán János Dénes, 'A klímaváltozás és a kínai verebek', *Magyar Nemzet*, 31 July 2019. See further for example Bayer Zsolt, 'Ezek őrültek vagy gazemberek? 1. rész', *Magyar Nemzet*, 26 July 2019; '2. rész', 27 July 2019.

155 Inte Rasist Men, 'SD-politiker – hemlig terrororganisation attackerar Sverige med hjälp av skogsbränder', interasistmen.se, 29 April 2019. The representative was Lars-Arne Ivert.

156 For example Thoralf Alfsson, 'Vattenbrist på Öland', thoralf.bloggplatsen.se, 5 February 2016; postings on the Facebook group 'Stå upp för Sverige', spring 2017.

157 Muirhead and Rosenblum, *A Lot*, 102.

take pictures to blacken the image of Brazil; asked for evidence, the president said that there could be 'no written plan', as 'that's not how it's done'.[158] In a rather fascinating imaginative leap, Bolsonaro and Salles were also in the habit of referring to Brazilian environmentalism as 'Shiite ecologist activism' or 'eco-Shiites', as though it was Shia Islam that had descended on the nation to deny it the riches of the Amazon.[159] Insofar as the climate movement mobilises greater strength in the years ahead, we should expect more of this to come; with or without it, the impacts will continue to feed the far-right imagination.

One of the ironies here – more eerie than amusing – is that few thinkers have analysed the profile of this sludge with greater precision than the Jewish Marxist the contemporary far right loves to hate the most, namely Adorno. The more communism 'is emptied of any specific content, the more it is being transformed into a receptacle for all kinds of hostile projections, many of them on an infantile level somehow reminiscent of evil forces in comic strips', he wrote in *The Authoritarian Personality*.[160] But the receptacle is arguably emptier today than in the 1940s. 'The less it is able to exercise a Marxist dominance over the situation, the more the dominance of Marxism is made responsible for every misfortune', noticed antifascist Richard Löwenthal – but that was in 1935.[161] As the radioactive decay from the Russian Revolution proceeds, perhaps already past half-time, the fantasy of Marxist deviltry rather appears to radiate brighter again. That paradox can be explained by the depth of revolutionary problems, for which the status quo cannot take responsibility and which do not yet have a matching subject: Marxism must be recruited as a revenant to fill the gap.[162]

---

158   Reuters, 'Brazil's Leader Falsely Blames Leonardo DiCaprio for Amazon Fires', *New York Times*, 30 November 2019. These fires were, however, as we have seen, impacts not of global warming as such, but of a politics for accelerating it.
159   Garrand, 'Parts'; Heriberto Araújo, 'Brazil: Bolsonaro's Threat to the Amazon', *Global Americans*, 29 March 2019.
160   Adorno et al., *The Authoritarian*, 723.
161   Richard Löwenthal, 'The Origins of Fascism', in Beetham, *Marxists*, 308.
162   Or, to paraphrase Voltaire, if a communist threat does not exist, it is necessary to invent it. Or perhaps, in Bloch's words, 'Marxism still encircles and fetters those who, in their ignorance, think they have overcome it subjectively.' Bloch, *Heritage*, 91.

A break with reality is forced by this situation. 'Fascist propaganda', Adorno wrote in 1946, drawing from empirical studies of far-right demagogues on the US West Coast, 'attacks bogies rather than real opponents, that is to say, it builds up an *imagery* of the Jew, or of the Communist, and tears it to pieces, without caring much how this imagery is related to reality.' He observed an 'amazing stereotypy' among the agitators he listened to. In their radio talks, sidewalk speeches and pamphlets, they repeated a small number of clichés and worked up a monotonous innuendo – 'for example, the agitator says "those dark forces, you know whom I mean", and the audience at once understands that his remarks are directed against the Jews.' The listeners are 'getting the inside dope, taken into confidence, treated as of the elite who deserve to know the lurid mysteries hidden from outsiders'.[163] It can be a gratifying experience. Addictive, it locks the insiders into a ' "closed system" of delusions' that tend to ' "run wild", that is to say, make themselves completely independent from interaction with reality'.[164] The ticket into this loop is unconditional fidelity to an unsustainable status quo. Because the chaos cannot be comprehended as the product of the system itself, it must be projected onto an external enemy, whose elimination would restore order: the anti-Semite is fuelled by a 'spiteful adherence to the existent'.[165] He subsumes initiatives to change under the general heading of foreignness. In reality, the hated outgroups are 'objectively *weaker* than the groups whom they supposedly threaten', but so it must be; had they been stronger, the haters would have the status quo in sight. From this point, there is little to rein them in. 'As soon as prejudice in any amount is allowed to enter a person's manifest ways of thinking, the scales weigh heavily in favour of an *ever-increasing expansion* of his prejudice.'[166] Of this expansion, if it continues unopposed, there will eventually be victims.

Two factors make the edges of the present conform closer to this profile than that of Adorno's own post-war years: the enormity of the

---

163 Theodor Adorno, 'Anti-semitism and Fascist Propaganda', in *The Stars Down Earth* (Abingdon: Routledge, 1994), 220–8. Emphasis in original.
164 Adorno et al., *The Authoritarian*, lx, 613.
165 Ibid., lix.
166 Ibid., 148, 629–30. First emphasis in original, second added.

crisis and the completeness of the capitalist victory over any socialist challenges (at a closer look, one and the same thing). Under their pressure, there develops that 'complex of "psychotic" thinking which appears to be a crucial characteristic of the fascist character'.[167] The orators Adorno studied in California were in the wrong place at the wrong time; the moment when they could approach real power had yet to come. He would have recognised a Baudet, an Åkesson, a de Carvalho. On the other hand, this might mean that we have less to do with post-truth or new conspiracism or any other neologism meant to keep up with the degradation than with that old *potential for fascism* that so worried Adorno.[168] And this is not, as we shall see, the last facet of our problems anticipated and illuminated with inimitable clarity by Adorno.

These, then, are some of the ideas with which the far right enters the rapidly warming world of the 2020s and beyond: palindefence and sometimes palingenesis; veneration of the stock and deprecation of the flow; the theories of the left climate conspiracy and Cultural Marxism. We have dwelt on these ideas at some length because they provide mythical energies to the far right, with which it, so to speak, crashes into this warming world. The danger of fossil fascism is the danger that such ideas become much more dominant. In this salmagundi, intellectual coherence is not the highest ideal, but nor are the component parts mutually exclusive. It is evidently possible to believe in all those things simultaneously, and they may well reinforce one another: we have to defend ourselves again; we must take what is ours out of the ground; the enemy is Marxist and Muslim and Jewish and here comes his next attack, and so on.[169] But it would be far too simplistic to say that such estrangement from reality serves only the interests of primitive fossil capital. The scandalous truth of climate breakdown throws much wider sets of privileges into question, notably those pertaining to whiteness.

---

167   Ibid., 665.
168   'All modern fascist movements', he noticed, 'have aimed at the ignorant; they have consciously manipulated the facts in a way that could lead to success only with those who were not acquainted with the facts.' Ibid., 658. For an excellent critique of the notion of a new post-truth condition, see Rune Møller Stahl, 'The Fallacy of Post-Truth', *Jacobin*, 14 December 2016.
169   As the antipode to the Frankfurt School, climate denialist Anders Breivik posited 'the Vienna school' of palindefence. Bangstad, *Anders*, 79.

9
Skin and Fuel

On 6 September 2016, nine activists from Black Lives Matter approached London City Airport in wetsuits. They boarded inflatable rafts, crossed the marina separating the airport from the suburb of Newham, stormed the runway, erected a tripod, locked themselves to it and to one another and unfolded a banner that read: 'The climate crisis is a racist crisis.' Nine thousand passengers had their travel plans disrupted. The airport complained of losses in revenue and damages to its reputation. When the activists were taken to court, the judge recognised their peacefulness and sincerity and spared them jail sentences, but she scolded them for being so confused and baffling. 'I find it rather hard to see', she demurred, 'the link between the movement which started in America and goes by name of Black Lives Matter, which as I understand protests against the treatment of the black population by the police in America. I don't see how they link to London City Airport or climate change.'[1] What had violence against black people to do with flights taking off from London? The activists pointed to a typical local environmental injustice – an airport for the wealthy located in Newham, a racially diverse suburb – and to its

---

1 Judge Elizabeth Roscoe in Patrick Sawer and Danny Boyle, 'City Airport Protest: Judge "Finds It Hard to See Link" between Runway Stunt and Black Lives Matter Campaign', *Telegraph*, 14 September 2016.

magnification on a global scale: 'Seven out of ten of the countries most affected by climate change are in sub-Saharan Africa.'[2] Whites fly, blacks die; something a white judge in London might have a hard time seeing, but for which there is some pretty solid evidence.

We have already come across the fundamental divergence between gains and losses from the fossil economy in the case of Mozambique. It is documented in a voluminous literature. To pick just one paper, half a year before the action in London, James Hansen and his long-time collaborator Makiko Sato mapped the landscape of 1°C average global warming and found that, so far, the impacts – temperatures that test human tolerance limits, heat unbearable for workers outdoors or in poorly ventilated workplaces, intensified droughts, vector-borne diseases – clustered around lower latitudes, meaning sub-Saharan Africa, the Middle East, India, Southeast Asia. But the map of causation was exactly the reverse. In Hansen's and Sato's calculation, the UK and the US were *each* responsible for more than one-quarter of cumulative emissions since 1751.[3] The concentration would be even more extreme if the gains were counted for corporations rather than nations, or if the effects of globalisation were taken into account; in the early twenty-first century, much of the emissions attributed to China stemmed from the manufacturing of goods savoured in places like the US and the UK.[4] Hansen and Sato then restated a conclusion heard innumerable times before: those 'experiencing the largest change of

---

2   This statement and more explanations of the action appear in for example Alexandra Wanjiku Kelbert, 'Climate Change Is a Racist Crisis: That's Why Black Lives Matter Closed an Airport', *Guardian*, 6 September 2016; Tara John, 'Black Lives Matter Protest at London Airport Leaves Some Confused', *Time*, 6 September 2016. Cf. Kara Thompson, 'Traffic Stops, Stopping Traffic: Race and Climate Change in the Age of Automobility', *Interdisciplinary Studies in Literature and the Environment* 24 (2017): 92–112.

3   James Hansen and Makiko Sato, 'Regional Climate Change and National Responsibilities', *Environmental Research Letters* 11 (2016): 1–9. The figure for the UK is here larger than in most calculations and probably unrealistically large (although a more qualitative measurement of the British historical responsibility might well yield a quarter or more, given that this was the country where it all began and from where it spread, very often through imperial violence).

4   As pointed out in the accompanying Steven J. Davis and Noah Diffenbaugh, 'Dislocated Interests and Climate Change', *Environmental Research Letters* 11 (2016): 1–2. It should be pointed out, however, as is well known, that *absolute* temperature increases are highest on northern latitudes.

prior normal climate bear negligible responsibility' for it.[5] What the two scientists did not note, or say out loud, however, was a facet of the pattern of which few could be entirely unaware: it mapped onto a global colour line.[6] Did that have any significance?

Frantz Fanon, in *A Dying Colonialism*, writes that the colonised person

> perceives life not as a flowering or a development of an essential productiveness, but as a permanent struggle *against an omnipresent death*. This ever-menacing death is experienced as endemic famine, unemployment, a high death rate, an inferiority complex and the absence of any hope for the future.[7]

Anti-racist students of political ecology have found in this statement a kind of prophecy. The pattern Fanon described in the 1950s has re-emerged with a vengeance precisely in those parts of the globe that experienced it during the colonial era. In the words of Romy Opperman, the phrase 'omnipresent death' now suggests a 'cumulative weight and exhaustion' in which the impoverished 'conditions of life, such as water, air, food, and labor, reduce life to a struggle for survival'.[8] Ever-menacing death now arrives via the atmosphere, and the circumstance that it descends first on those places that suffered from it during colonialism is neither accidental nor conspiratorial: it is structural. Enhanced vulnerability is the legacy of centuries of bleeding.[9]

---

5   Hansen and Sato, 'Regional', 6.

6   Cf. Leon Sealey-Huggins, '"The Climate Crisis Is a Racist Crisis": Structural Racism, Inequality, and Climate Change', in Azeezat Johnson, Remi Joseph-Salisbury and Beth Kamunge (eds), *The Fire Now: Anti-racist Scholarship in Times of Explicit Racial Violence* (London: Zed Books, 2018): 99–113; Laura Pulido, 'Racism and the Anthropocene', in Gregg Mitman, Marco Armiero and Robert Emmett (eds), *Future Remains: A Cabinet of Curiosities for the Anthropocene* (Chicago: University of Chicago Press, 2018): 116–28.

7   Frantz Fanon, *A Dying Colonialism* (New York: Grove, 1965), 128. Emphasis added.

8   Romy Opperman, 'A Permanent Struggle against an Omnipresent Death: Revisiting Environmental Racism with Frantz Fanon', *Critical Philosophy of Race* 7 (2019): 70. Cf. for example Stephanie Claire, 'Geopower: The Politics of Life and Land in Frantz Fanon's Writings', *Diacritics* 41 (2013): 60–80.

9   Cf. for example Sealey-Huggins, '"The Climate"'; Pulido, 'Racism'; Leon Sealey-Huggins, '"1.5°C to Stay Alive": Climate Change, Imperialism and Justice for the Caribbean', *Third World Quarterly* 38 (2017): 2444–63; Branwen Gruffydd Jones, 'Race in the Ontology of International Order', *Political Studies* 56 (2008): 907–27.

But the significance of the colour line is greater still, because it might well inspire *indifference* to the problem in and of itself. Racism is all about putting a lower value on some lives, dividing those worthy of living the good life from those unworthy, and if we stay with Africa, there is, counting from the beginning of the slave trade, a five-centuries-long tradition of considering black lives as the latter kind. This tradition does not necessarily live on in the desire to whip black bodies to death. It can be entirely unconscious, with no malicious intent required.[10] It can be expressed in ostensibly neutral economic rationality, such as when it is simply cheaper to locate a toxic landfill in a black neighbourhood or ship off electronic waste to Ghana, where some of the most hazardous chemicals on earth are streaming in from European phones, laptops, tablets, fridges and other devices and entering the food chains on which black people – whom the consumers will never meet – depend.[11] With or without open animus, there is, anti-racist scholars have argued, a very long tradition of treating Africa as 'the disposable trash container of the world' and 'black people as *waste*', clogging up the streets and the planet like so much human rubbish, deserving of having waste products dumped upon them.[12] In *Is Racism an Environmental Threat?*, Ghassan Hage contends that a similar perception of Muslims as waste has sunk in over the past decades.[13]

Now if the early impacts of climate change are primarily an affliction for black and non-white people, one could imagine some

---

10   As pointed out, in the context of climate change, by Milton Takei, 'Racism and Global Warming: The Need for the Richer Countries to Make Concessions to China and India', *Race, Gender and Class* 19 (2012): 134–5, 143.

11   As argued by Laura Pulido, 'Rethinking Environmental Racism: White Privilege and Urban Development in Southern California', *Annals of the Association of American Geographers* 90 (2000): 12–40; and see Peter Beaumont, 'Rotten Eggs: E-Waste from Europe Poisons Ghana's Food Chain', *Guardian*, 24 April 2019.

12   Axelle Karera, 'Blackness and the Pitfalls of Anthropocene Ethics', *Critical Philosophy of Race* 7 (2019): 44; Janine Jones, 'To Be Black, Excess, and Nonrecyclable', in Naomi Zack (ed.), *The Oxford Handbook of Philosophy and Race* (Oxford: Oxford University Press, 2017), 321. Emphasis in original. This, of course, is what environmental racism as classically conceived is all about: see for example Laura Pulido, 'Geographies of Race and Ethnicity I: White Supremacy vs White Privilege in Environmental Racism Research', *Progress in Human Geography* 39 (2015): 809–17.

13   Ghassan Hage, *Is Racism an Environmental Threat?* (Cambridge: Polity, 2017), for example 48–9, 68.

people beholden to this long tradition thinking, 'Oh, that's just another piece of bad news for the wretched of the earth. Why bother?' Jesse Chanin examined the results from an extensive survey of public attitudes in the US and found that, indeed, individuals who questioned the full humanity of people of colour were also unbothered by the state of the environment.[14] Using another method to measure the indifference, CARE International studied the ten humanitarian crises that received the least media attention in 2018 and found that climate change was implicated in most of them: frequent droughts ravaging farmland and pastures in Sudan; yet another typhoon displacing 1 million people in the Philippines; chronic malnutrition exacerbated in Chad as the eponymous lake shrunk to a pond; the corn fields of Madagascar pummelled by drought and cyclones; people in Haiti having to skip meals to survive – all off the radar north of the colour line.[15] And if European states are untroubled by the thousands drowning in the Mediterranean every year, why should they care about the millions of lives disappearing beneath the waves and desiccating in the arid heat much farther afield?[16] In 2014, Andrew Lilico, a bourgeois economist, contributor to the *Telegraph* (yes, the *Telegraph* of Climategate) and later Brexiteer, advocated adaptation as the best approach to the problem, not mitigation. When asked how the tropics would possibly adapt to 4 degrees of heating, he responded: 'I imagine tropics adapt to 4C by being wastelands with few folk living in them. Why is that not an option?'[17]

Somewhere in there is one link the judge could not see. Global heating extends the disregard for black lives 'to unprecedented levels', to quote Leon Sealey-Huggins; furthermore, such disregard might be a *sine qua non* of business-as-usual itself.[18] Laura Pulido has suggested

---

14 Jesse Chanin, 'The Effect of Symbolic Racism on Environmental Concern and Environmental Action', *Environmental Sociology* 4 (2018): 457–69.

15 CARE International, *Suffering in Silence III: The Ten Most Under-Reported Humanitarian Crises of 2018*, care.org, 21 February 2019.

16 Naomi Klein, 'Let Them Drown: The Violence of Othering in a Warming World', *London Review of Books* 38, no. 11 (2016): 13. This essay is reprinted in Klein, *On Fire*, 149–68.

17 Quoted in James Murray, 'Climate Adaptation Lobby Is Reckless, Dangerous, and (Partly) Right', *Business Green*, 1 April 2014.

18 Sealey-Huggins, ' "The Climate" ', 106.

that if the gains were coterminous with the losses, they would not justify keeping the fires alive. Only because the multitudes first sentenced to die are non-white and out of the way can combustion be allowed to continue.[19] Representatives of the global South have repeatedly sought to awaken the advanced capitalist countries to this callousness, for instance at COP15, where Lumumba Di-Aping, the eloquent and irate negotiator from Sudan, denounced the proposal from the US and its allies to let go of binding commitments and settle for voluntary pledges. It was, he charged, like 'asking Africa to sign a suicide pact, an incineration pact in order to maintain the economic dependence of a few countries [on fossil fuels]. It's a solution based on values that funnelled six million people in Europe into furnaces.' Western diplomats thought the latter analogy was beyond the pale. Ed Miliband of the UK called it a 'disgusting comparison', the Swedes 'absolutely despicable'.[20] And, of course, six years later the pact was signed.

The logic appears robust. In a world where black and brown lives matter little, and where global warming first destroys such lives, then it will not be a matter of great concern. But if there is indeed a real effect of this kind – it's a problem for non-white trash, so let's keep burning – we should expect it to be most powerful *in the early stages of warming*, up to, say, 2°C, whereas at very late stages, at 6°C and 8°C and beyond, it would presumably wane with the differentials in vulnerability. At 10°C, the blondest Swedes will be reduced to cinders too. In other words, the effect would be most politically efficacious precisely in the window in time when mitigation could make the largest difference. Everyone will be inside the same furnace and see their shared destiny only when it's far too late to do anything about it. A curve of rising disparity across the colour line and indifference among those with the power to curb the warming and diminishing opportunities for doing so must, sooner or later, pass an inflection point, beyond which there is the rest of the breakdown. This also implies that something more than indifference is at work in the early

---

19 Pulido, 'Racism', 121.
20 John Vidal and Jonathan Watts, 'Copenhagen Closes with Weak Deal That Poor Threaten to Reject', *Guardian*, 19 December 2009.

stages, namely a failure to see universal humanity in black and brown people, as if what befalls them could not eventually also sweep up the whitest of the white; as if the catastrophes could be forever contained in the lower decks of the ship where they have always played out.[21] *De te fabula narratur*, 'Of you the tale is told', Marx prefaced *Das Kapital* for the German workers; he described the conditions for the English, but told his countrymen to expect the same. Every flood in Mozambique, every drought on the Horn of Africa in recent decades has had a similar predictive tale to tell whoever would care to listen.

If some of this is in the nature of speculation, we can be positive about the far right acting out this effect to the best of its ability. When the Sweden Democrats explained, just before the wildfire explosion of 2018, that 'it will be developing countries that suffer, if any', and therefore the climate crisis was not an issue for the party, they lifted the disregard out of the unconscious. From the Finns' cartoon to Trump's tearing up of the 'nigger climate accords', most of the forces we have inspected have associated the struggle against climate change with black and other non-white people – it's for them, not us. 'Let them drown' is here not a faint, undefined propensity: it is the policy. The choice of apocalypse – the real threat to the world is *their presence among us* – aggravates it further. But the far right would scarcely be able to advance this message if it did not have a wider indifference to work on and, as it were, mobilise. We can reformulate this as a general hypothesis: *the anti-climate politics of the far right is now a phenomenon of such rank that it must stand on the shoulders of a much wider and broader set of relations* of the kind that we normally refer to as 'racism'.

Before teasing out this hypothesis, it is worth noting that racism is one of the least investigated dimensions of the climate crisis, the studies we have cited so far constituting a sub-subfield of little more than a dozen papers.[22] The attitude of the London judge is customary.

---

21 Cf. Kendi, *How To*, 132. There is nothing novel about the Anthropocene: for the victims of colonialism, worlds have come to an end for half a millennium by now. This argument is pursued in Kathryn Yusoff, *A Billion Black Anthropocenes or None* (Minneapolis: University of Minnesota Press, 2018). Cf. Donna Orange, *Climate Crisis, Psychoanalysis, and Radical Ethics* (London: Routledge, 2017), for example 54.

22 This deafening silence is noted in for example Pulido, 'Racism', 116; Nancy Tuana, 'Climate Apartheid: The Forgetting of Race in the Anthropocene', *Critical Philosophy of Race* 7 (2019): 1–31.

This is an oversight in the study not only of climate, but of racism as well. In *White Identity Politics*, Jardina argues that white Americans use their whiteness as a compass when setting their policy preferences on everything from welfare and trade to crime and presidents. But she sees one exception: 'It seems hard to imagine that white identity predicts opinion on an issue like climate change, which has not been markedly framed as harming or benefiting whites as a group.'[23] That a prominent scholar of race could so miss the plot is one indication of the research gap, and of an implicit belief in mystery. If race is a master category of the modern world in general and the American social formation in particular, as scholars of race have demonstrated, and if global heating is the most gigantic crisis the former has ever faced, as any well-informed observer should know by now, it would indeed be mysterious if the one had no bearing on the other.

Towards an Excavation of Fossilised Whiteness

Forestalling emissions cuts is in the objective interest of fossil capital. It is not in the objective interest of everyone with pale skin or all members of the white nation or white race. The far right might act as if the latter were the case, but nothing could be further from the truth: firstly, because humans with pale skin already suffer from consequences of climate breakdown and more will do so in the years ahead; secondly, because whiteness itself, in an important sense, is a fiction. The anti-climate politics of the far right does not represent the true consciousness of white people the way it arguably does so for primitive fossil capital in particular. But it has a power to speak *in the name of* white people that must be explained, and we might begin by returning to the original scene of interpellation. A police officer hails a man in the street: 'Hey, you there!' Recognising that the call is directed to him, the man turns around and faces the officer. This is Althusser's most famous example; the next, Moses at the burning bush, from which God shouts out: '"Moses! Moses!" And Moses said,

---

23 Jardina, *White*, 42. This blindness is not, of course, shared by all scholars of race. For a high level of race-climate consciousness, see Kendi, *How To*.

"Here I am."[24] In both cases, the interpellation is successful because the individuals *recognise themselves* as the ones being hailed. But such recognition cannot be taken for granted. If the man on the street has been parachuted in from his hunter-gatherer camp, where he has spent his whole life, isolated from the arm of the law, he might not understand what the police officer expects of him and fail to show him deference. If Moses had remained loyal to the palace where he grew up and kept his faith in the deities of ancient Egypt, he would not have presented himself to the burning bush. Such abortive interpellations happen all the time. When a missionary screams 'Jesus loves you!' in a town square in northern Europe, the vast majority will pass him by, even look down, because they do not recognise themselves in the call and do not feel touched in *their* souls. So what determines if an interpellation falls on deaf ears or not?

In a scintillating essay, Rebecca Kukla points out that interpellation 'only works if subjects feel that the hail recognizes who they *really, already are*'.[25] The man on the street already considers himself a subject of the law and reacts to the summons by understanding, reflexively and instantly, that the officer has seen him as such. Moses already identifies himself as Moses and therefore responds when God calls him by that name. But if a communist enters the floor of the New York Stock Exchange and starts crying out, 'Workers of the world, unite – you have nothing to lose but your chains!', the traders will shake their heads, perhaps ask the guards to throw her out, because they do not recognise themselves as workers and will (rightly) think that the communist has misidentified them. The interpellation hits the mark only when it sees the addressee as she sees herself. Now – and this is a crucial proviso – there does not have to be an *absolute* identity between the images, no exact mirror effect: but an interpellation must stir at least *some* self-recognition. Moses did not become Moses the prophet and deliverer of Israel before his first encounter with God in the burning bush, but he was *on his way there,* sufficiently self-identified as a subject of Yahweh to accept the

---

24 Althusser, *Lenin*, 118, 121.
25 Rebecca Kukla, 'Slurs, Interpellation, and Ideology', *Southern Journal of Philosophy* 56 (2018): 18. Emphasis in original.

assignment. 'The one hailed', writes Kukla, 'must in fact *come to be* (at least incrementally more) *the self she is already recognized as being*'; she must feel a 'visceral pull' to step into the shoes tailored specifically for her.[26] Interpellation, in other words, is a cumulative, sequential process that has to tie in with previous rounds of interpellations.

How far back does that process go? No human life exists prior to it: 'Ideology has always-already interpellated individuals as subjects', asserts Althusser.[27] This might sound strange, but it is rather uncontroversial. Consider a couple who has just learnt from an ultrasound scan that they will have a baby girl. From that moment on, consciously or not, they prepare to welcome the child as a girl and not a boy, and for most parents, that reception comes with a whole culture's worth of expectations and models and norms, so that the child is interpellated as a particular kind of subject already before being born. The process of the little girl recognising herself as such, standing before the mirror in a princess's dress, then comes naturally (or so it seems). The category of race often works in a similar way. A child born in Stockholm to parents born in Mogadishu will be hailed as an *invandrare*, literally 'someone who has walked in', even though she has done no such thing and is no immigrant in the proper sense of the term: but surrounding society will recruit her into that position, until the moment when she recognises that a group of young white men calling her *invandrarhora* or 'immigrant slut' might in fact be aiming their vitriol at her.[28]

In all walks of life, a universe of ideology stands ready to greet new humans. 'The Law of Culture', as Althusser sometimes calls it, 'has been lying in wait for each infant born since before his birth, and seizes him before his first cry, assigning to him his place and role, and hence his fixed destination.'[29] Now, this immediately raises a

---

26 Ibid., 13, 14. Emphases in original.
27 Althusser, *Lenin*, 119; cf. Althusser, *On the Reproduction*, 193–4.
28 Kukla develops her powerful account of slurs as a particular kind of interpellations – generic, derogatory, subordinating – in Kukla, 'Slurs', 19–31.
29 Althusser, *Lenin*, 144. Althusser, of course, here follows Lacan. Cf. Stuart Hall, 'Signification, Representation, Ideology: Althusser and the Post-Structuralist Debates', *Critical Studies in Mass Communication* 2 (1985): 109; Pascale Gillot, 'The Theory of Ideology and the Theory of the Unconscious', in Diefenbach et al., *Encountering*, 289–306.

problem. If people are ensnared in pre-existing ideologies from the moment they are born, how do they ever break out of them? People evidently retain the capacity to shut their ears to some interpellations, dispute others, renounce beliefs, change gender, convert, join an underground cell, escape from an ideology they have been trapped in.[30] But how is that possible? If successful interpellations are by definition cumulative, how can there be anything else than inertia in the sphere of ideology? Moreover, there is obviously a difference between volunteering as a far-right activist in response to recruitment efforts and succumbing to the unwanted designation *invandrare*; between rising to a call with some degree of freedom and falling under the unfreedom of classification. We shall return to both of these problems, but let us, for the moment, stay with the view of interpellation as a summons expanding on a previous summons. Interpellation is essentially *reproductive*.[31]

Kukla gives the example of school students. It is one thing for a child to be officially enrolled in a school, but quite another to have a deeply felt student identity. How does the child move from one to the other? The first day starts with the teacher reading names off a list, and when her name comes, the student raises her hand and calls back 'here!' By doing so, she recognises herself as a student *and* the teacher as someone with normative claims on her ('when I say your name, you shall respond'); as the school day unfolds, a repeated process of call and response constitutes the students 'as having this identity. It builds it into their bodies.' Day after day, they are 'inducted not just into how to respond to teachers, but into how to talk, how to dress, how to socialize, and how to spend their time. And in time they come to reproduce this student role automatically and without need for conscious reflection.'[32] The reproductive interpellations here extend over a few weeks or months, during which a child learns to be a student. At the level of the social formation, however, such processes require far longer time. The school system as such has been erected and reproduced over centuries. General subject positions into which

---

30 Cf. Rehmann, *Theories*, 152–3.
31 Rastko Močnik, 'Ideological Interpellation: Identification and Subjectivation', in Diefenbach et al., *Encountering*, 316.
32 Kukla, 'Slurs', 14.

people can be inducted – girl, *invandrare* – do not pop up over the course of a person's early life; it is in their nature to be transmitted from earlier generations. The equivalent of 'my life in school' for the social formation as a whole is rather *all the history* that has flowed into it, giving the dominant ideology its present content and depositing a base for sorting people into their positions. An interpellative movement must reckon with this past and reproduce some aspect of it. No ideological trend materialises *ex nihilo*. Someone who flies in from the outside and starts calling on people to do this or that, without paying heed to how they have seen themselves until now, will not be listened to; conversely, if people do flock around such a person, one can be certain that they recognise something in themselves – something about who they *really, already are* – in her sermons.

Now consider the interpellation of the contemporary far right, which we can, a tad crudely, sum up like this: 'Hey you, people of our great white nation! Keep burning fossil fuels! If you have them in the ground, they are yours to dig up! That talk of climate change is a hoax people like you have seen through – the real problem is all the immigrants swarming around here.' This is not the sole message the far right can muster, as we have seen, nor will it be repeated forever, but it has exerted so much visceral pull in recent years as to merit an inquiry into the history of this ideological nexus. Enough people have responded by saying 'Yes, that's us! You're right!' to compel us to ask *what has led them here*, to this conjuncture where a fire is burning and a voice from within it booming, asking them to line up as white people who will be staying white and – or even *by* – adding more fuel to the flames. What in history has taught them to step forward like this? There must be some linkage that can explain why so many people in Europe and the Americas happily appear as white burners, and it cannot be their racial character. There are, of course, no races. No discreet genetic pools separate people into such categories. Every human being shares some 99 per cent of her genetic material with every other member of that species, and out of the tiny remainder of variation, the bulk appears *within* populations from the same continents rather than between them. Human descent groups do possess distinctive sets of genes, but these map poorly onto skin colour and facial features and other popular attributes of 'races'; under similarly dark or light skin, the greatest diversity can bloom.[33] As biological units, races have no more existence than phlogiston or abominable snowmen, and as even the most cursory glance at history will show, a generic 'white' category of people is just as mythical.

People inhabiting ancient Greece and its neighbours did not think of themselves as white.[34] Early European travellers found Arabs and Indians and Chinese fully as 'white' as themselves, whereas in China, those travellers were seen as peculiarly ashen or 'pale blue' rather than 'white', a colour associated with privileged life indoors.[35] One could imagine that Swedes would go through history as the nonpareil of whiteness, but Benjamin Franklin placed them outside of that circle, as bearers of 'what we call a swarthy Complexion'. In a letter from 1901, a Minnesota lumberjack described how there were '15 white men here to 60 Swedes'; trying his best to avoid them, this poor man had to mingle with the 'beasts' when night fell. It made him depressed, particularly as the Swedes had an insufferable odour, which stemmed 'from generations of unwashed ancestors' and could never be acquired 'without the aid of heredity'.[36] Swedish experts in their turn regarded Finns as non-white members of a Mongol race.[37] The term 'Caucasian' is an established synonym for 'white' in the US

---

33   Michael J. Bamshad and Steve E. Olson, 'Does Race Exist?', *Scientific American* 289 (2003): 78–85.

34   Nell Irvin Painter, *The History of White People* (New York: W. W. Norton, 2011), ch. 1.

35   Alastair Bonnett, 'Who Was White? The Disappearance of Non-European White Identities and the Formation of European Racial Whiteness', *Ethnic and Racial Studies* 21 (1998): 1033–7.

36   Quoted in Catrin Lundström and Benjamin R. Teitelbaum, 'Nordic Whiteness: An Introduction', *Scandinavian Studies* 89 (2017): 152. There is a history of associating Scandinavians with dirt. In the 920s, the Arab traveller Ahmad Ibn Fadlan encountered what we would today call Vikings on the banks of the Volga and was shocked by their uncleanliness: 'They are Allah's dirtiest creatures: they don't clean themselves after shitting and urinating, nor after intercourse, and they do not wash their hands after a meal. They are like lost donkeys.' Quoted in Dick Harrison, 'Vikingarnas smutsiga sätt chockade Ibn Fadlan', *Svenska Dagbladet*, 31 July 2017. Cf. the theory about Muslims and groundwater on the island of Öland.

37   Anna Rastas, 'The Emergence of Race as a Social Category in Northern Europe', in Philomena Essed, Karen Farquharson, Kathryn Pillay and Elisa Joy White (eds), *Relating Worlds of Racism: Dehumanisation, Belonging, and the Normativity of European Whiteness* (Basingstoke: Palgrave Macmillan, 2019), 360.

and elsewhere; invented by a pseudo-scientist who thought Georgian women were the most beautiful on earth, the category has often included North Africans, whereas the inhabitants of the actual Caucasus region have been deemed 'black' by their Russian masters. 'Black' in the Southern slave states was any person with one-quarter or one-eighth African ancestry, until the 'one-drop rule' decreed that *one* African ancestor somewhere in the family tree was enough to make a person black. 'White' in the US long meant 'Anglo-Saxon', barely covering the French. But people of Mexican descent were considered white in the 1930s. The hovering of the Irish and Italians and Jews on the fringes of American whiteness and their later inclusion is a familiar story.[38]

In this historical snafu, no natural kinds can be spied. The classification of races has been whimsical, changing from one state or century to the next, the taxa swelling and shrinking, the criteria sometimes fanciful.[39] One and the same individual might be sorted into one box in one setting and a different one in another. Stuart Hall has described how during 'thirty years in England, I have been "hailed" or interpellated as "coloured", "West-Indian", "Negro", "black", "immigrant"', while in Jamaica, 'where I spent my youth and adolescence, I was constantly hailed as "coloured"' – which, in the Jamaican spectrum of gradations, meant *anything but* black.[40] And on the interpellations have swirled.

For all of this instability and fluidity, however, race also has a way of freezing. After centuries of colonial domination, 'white' eventually came to mean people of European stock. The Chinese and Arab elites learned to forget that they had once perceived themselves as white by virtue of their freedom from manual labour, and the English that they had once branded native workers in the London slums as foreign and all but black. After all that had happened, from the dawn of the slave trade to the consummation of the British Empire, 'white' had turned into an exclusively European designation, a talisman even the most

---

38 Painter, *The History*, for example 72–86, 360, 368.
39 Cf. Troy Duster, 'The "Morphing" Properties of Whiteness', in Birgit Brander Rasmussen, Eric Klinenberg, Irene J. Nexica and Matt Wray (eds), *The Making and Unmaking of Whiteness* (Durham: Duke University Press, 2011), 113, 133.
40 Hall, 'Signification', 108.

malodorous Swede could claim to bear, unlike, say, the most light-skinned North African.[41] The chief medium for accomplishing this separation was violence. As James Baldwin observes in his short essay 'On Being White . . . and Other Lies', Norwegians did not sit outside their cabins in Norway congratulating themselves on their wonderful whiteness, but once they came as settlers to the US, they were inducted into that category through certain rites of passage: they 'became white by slaughtering the cattle, poisoning the wells, torching the houses, massacring Native Americans, raping black women'.[42] Whiteness was constituted and acquired through some very practical acts. Then it could be passed on to subsequent generations, as a property etched into a white or rather pinkish hue of the skin.[43]

Inside the old world, a similar process took place through vicarious initiation. Agents of the European empires were baptised into whiteness by engaging in violence on the colonial frontiers, which *reflected back* upon civilians in the metropolis, who could then imagine themselves as united in whiteness and superior to those ruled overseas. When population movements changed direction in the twentieth century, as people from the peripheries found ways into the European core, the encounter with non-whites gained more immediacy for metropolitan populations. For the first time, the bodies constructed as the negation of whiteness could be seen on the streets, in the factories, in the shops and schools: the *Ausländer*, the *invandrare*, the immigrants had arrived, and these were heavily *racial* appellations. The hundreds of thousands of white Americans living in Britain were not called 'immigrants', nor were Germans who had moved to Norway or Canadians to France.[44] On the other hand, a person perceived as black might live her whole life in Denmark as an *indvandrer* in the eyes of others. 'Immigrant' and its equivalents in European languages functioned as the name for someone frozen in the position of

---

41 Bonnett, 'Who'.
42 Baldwin, 'On Being', 136.
43 See further Cheryl I. Harris, 'Whiteness as Property', *Harvard Law Review* 106 (1993): 1707–91.
44 As pointed out by for example Sally Davidson and George Shire, 'Race, Migration and Neoliberalism: How Neoliberalism Benefits from Discourses of Exclusion', *Soundings* 59 (2015): 83; cf. Balibar and Wallerstein, *Race*, 20, 221–2.

non-whiteness. Much as in the United States, the racial identities coursing and flowing through history attained a terrible solidity.[45]

When race petrifies in this manner, it does not have to sit in the skin. There is, for example, no one-to-one relationship between pigmentation and immigrant status: among the Syrians who made it to Sweden during the so-called refugee crisis, a fair number had very pale skin, but as soon as they introduced themselves to an employer or landlord as 'Khadidja' or 'Ahmed' they would be placed in the category of *invandrare* – meaning *not* Swedish – whereas Swedes of unbroken Swedish descent with chestnut skin would never have their Swedishness questioned. More particularly, Khadidja and Ahmed would frequently find themselves identified as *Muslims* merely by saying their names, and that could be damning. Just like Jews, Muslims had slid from a religious into a racial genus, presumed to have their most defining characteristics conferred to them from their source of origin. Such traits were held to be inherently *different* from the white norm and dangerous, inferior, defective, harmful or otherwise *negatively charged*.[46] Hence especially bad treatment could be reserved for them. The people in question had been racialised, although not so much on the basis of somatic features, the absence of which might elicit nervousness about the hidden, invisible Muslim or Jew; in Nazi Germany, Jewish bodies had to be marked out with yellow stars of David. But racialisation can also seize upon the quantity of eumelanin in the basal layer of the epidermis. Then essence is supposed to inhere in darkness of skin.[47]

---

45 It is ultimately this *lack* of fluidity that constitutes race as a problem in people's lives. Scholars who emphasise the perpetual construction and formation of race sometimes seem to forget this: for example, Michael Omi and Howard Winant, *Racial Formation in the United States*, 3rd ed. (Abingdon: Routledge, 2015), vii–x.

46 Cf. Teun A. van Dijk, *Elite Discourse and Racism* (Newbury Park: Sage, 1993), 20–4; George M. Fredrickson, *Racism: A Short History* (Princeton: Princeton University Press), 7–8. 140, 145, 169–70; Balibar and Wallerstein, *Race*, 22–6, 57. On racialisation of Muslims, see further for example Nasar Meer, 'The Politics of Voluntary and Involuntary Identities: Are Muslims in Britain an Ethnic, Racial or Religious Minority?', *Patterns of Prejudice* 42 (2008): 61–81; Rachel A. D. Bloul, 'Anti-discrimination Laws, Islamophobia, and Ethnicization of Muslim Identities in Europe and Australia', *Journal of Muslim Minority Affairs* 28 (2008): 7–25. On racialisation in general, see for example Rohit Barot and John Bird, 'Racialization: The Genealogy and Critique of a Concept', *Ethnic and Racial Studies* 24 (2001): 601–18. For an excellent Marxist account, see David Camfield, 'Elements of a Historical-Materialist Theory of Racism', *Historical Materialism* 24 (2016): 31–70.

47 On the instability of skin colour, see Painter, *The History*, 394–5.

Whatever characteristics a process of racialisation pounds on, the end-result is the same: race matters in people's lives. It can have lethal consequences. For a black infant at a slave plantation, a Jewish girl in a Polish ghetto or a Muslim boy in a Bosnian enclave, it branded the body with death: race is 'arguably the most violent fiction in human history', as noted by Ruth Frankenberg, but even in less blood-soaked settings, such as the Swedish labour market or the Italian ports, it might well have all-pervasive effects.[48] Inscribed in cities, seaways, restaurants, landfills and a million other sites where humans dwell, it has taken on a material reality of its own.[49] But it is always at its root a fiction, something humans have made up – and so is capital, another very tangible presence in people's lives. Neither race nor capital exists in nature. They are not 'natural phenomena, like comets or quarks' but *social artefacts*, a distinction presupposed by every critical theory of race.[50] Racists refuse to distinguish between the natural and the social. They believe that Jews or Aryans or Africans are this way or that by dint of their nature, while anti-racists see all racial classifications as mere concoctions, which can, however, become real in their own way. To say that race has no ontological reality because it is no part of nature would be to arbitrarily restrict the realm of the real. 'Social reality, though distinct from natural reality, *is nevertheless real*. Race derives its ontological status from social reality', and the circumstance that artefacts such as capital and race attach to physical entities does not alter that status.[51]

---

48 Ruth Frankenberg, 'The Mirage of Unmarked Whiteness', in Rasmussen et al., *The Making*, 72.
49 Cf. Geoff Eley, 'The Trouble with "Race": Migrancy, Cultural Difference, and the Remaking of Europe', in Rita Chin, Heide Fehrenbach, Geoff Eley and Atina Grossmann, *After the Nazi Racial State: Difference and Democracy in Germany and Europe* (Ann Arbor: University of Michigan Press, 2010), 175–7.
50 Paul C. Taylor, 'Race, Rehabilitated – Redux', *Critical Sociology* 36 (2009): 179. On distinguishing between the natural and the social, see further Malm, *The Progress*.
51 John H. McClendon III, 'On the Nature of Whiteness and the Ontology of Race', in George Yancy (ed.), *What White Looks Like: African-American Philosophers on the Whiteness Question* (New York: Routledge, 2004), 213–14. This argument is very powerfully made in ibid.; and in Stephen C. Ferguson II, 'Exploring the Matter of Race: A Materialist Philosophical Inquiry', in Zack, *The Oxford*, 263–5. See for example George Lipsitz, *The Possessive Investment in Whiteness: How White People Profit from Identity Politics*, rev. ed. (Philadelphia: Temple University Press, 2018), 2, 266; Birgit Brander Rasmussen, Eric Klinenberg, Irene J. Nexica and Matt Wray, 'Introduction', in Rasmussen et al., *The Making*, 8–9; Omi and Winant, *Racial*, 105–10; Camfield, 'Elements', 50.

The value in a piece of gold or green paper is not an intrinsic property of the object, but a result of conventions – resting, in the last instance, on a shared fiction. A child quickly learns that there is something special about that paper bill.

We can then say that race is fully entrenched *when individuals are always-already interpellated as racial subjects*. Like capital, this is a relation that has *developed over time* and must be seen as a fundamentally *historical construction*. 'We are all immersed in the waters of history, and those waters are pretty murky', notes Frankenberg: murky because hard to see through and filled with blood and mire.[52] It follows, furthermore, as Stephen C. Ferguson II reminds us, that 'race in abstraction from racism does not have any causal powers.'[53] It is racism that gives rise to race, never the other way around. Racism begins when some people enslave others and make them, as Fanon puts it, 'inferior through and through'. To crack the whip over these others and drive them into abjection, they must be rendered inferior, and racism is 'the emotional, affective, sometimes intellectual explanation of the inferiorization'.[54] Or, in the words of one classic account, when a 'historical collectivity dominates, excludes, or seeks to eliminate another on the basis of differences that it believes are hereditary and unalterable', there is racism happening, and it is this that creates racial subjects.[55]

What then is this thing called 'whiteness'? Frankenberg, building on Hall, offers the most persuasive one-sentence definition: 'Whiteness is a location of structural advantage in societies structured in racial dominance.' It is not a shortage of eumelanin in the basal layer of the epidermis, but a 'standpoint' and 'site of privilege'.[56] Not everyone with scant pigmentation enjoys or draws advantage from it, but it *exists* as such, just as any other location in this kind of society; indeed, every subordinate racial subject – the black on the

---

52 Frankenberg, 'The Mirage', 74.
53 Ferguson II, 'Exploring', 264.
54 Frantz Fanon, *Toward the African Revolution* (New York: Grove, 1988 [1952–61]), 40.
55 Fredrickson, *Racism*, 170.
56 Frankenberg, 'The Mirage', 76. Cf. Stuart Hall, 'Race, Articulation and Societies Structured in Dominance', in UNESCO (ed.), *Sociological Theories: Race and Colonialism* (Paris: UNESCO Press, 1980), 305–45.

plantation, the Muslim on the labour market – exists only in relation to it as to the glowing centre. Whiteness is a sun that always scorches.

We can then reformulate our hypothesis as a question. What is it in the historical production of whiteness that has primed people at that location to respond to anti-climate interpellations from the far right? Or, when and how did being white come to mean burning fossil fuels? No identity between energy and race can be expected, only an articulation of one to the other. We are looking for moments in history when fossil fuels have been inserted into whiteness and vice versa.[57] The premise here is that ISAs need raw material to work on. They process, refine, systematise ideology already present in crude form at a more primary level.[58] Ideology, on this view, is profoundly material: it exists not so much inside people's heads as in what they do. Religions provide plenty of examples, as when a Catholic kneels and receives mass or when a Muslim performs the *salat* and by that very ritual becomes one of the faithful. Or consider money. That phenomenon is, as Kukla remarks, ideological on account of how we use it. Someone who walks into a store with money in her pocket enters the sphere of commodity circulation, where everything can be exchanged for everything else via the universal equivalent known as money. Whether she has this in mind or not, through the transactions at the counter the customer participates in an ideology that posits money as the measure of all things – a most contingent and questionable proposition, but one that is built into the exchanges that take place every second in a commodified society. The consumer enacts the ideology and reproduces it through a practice so quotidian as to escape notice.[59] Lodged in the most primary level of capitalist property relations, this raw material can then be picked up by all sorts of what we might call second-order interpellations: 'Hey you, buy this new skin lotion, or Protect your

---

57 This draws on Hall, 'Race', 325–8, 337–40; cf. Stuart Hall, 'Gramsci's Relevance for the Study of Race and Ethnicity', *Journal of Communication Inquiry* 10 (1986): 23; Hall, 'Signification', 92, 111.

58 Cf. Rehmann, *Theories*, 154; Hall, 'Race', 341.

59 Adapted from Kukla, 'Slurs', 9–12, 18 (her primary example, however, is democratic voting). On the materiality of ideology, cf. for example Hall, 'Signification', 99–105; Wolf, 'The Problem', 249–51; Gillot, 'The Theory', 292–3.

freedom on the market, or Vote for tax cuts that will leave more money in your pocket.' We are looking for something similar in the first-order deployment of energy. It might not, however, be repeated on a daily basis like the use of money, but could rather have happened in the relatively distant past and then been sedimented in the unconscious.

Ideology is not only material, but also unconscious: when the man on the street turns to the officer and the costumer approaches the counter, the ideas folded into the act are rarely deliberated upon. Racist ideology has a special knack for inhabiting the unconscious, and most particularly when it is bequeathed from the past.[60] In one of his reflections on racism in Britain – incomprehensible without close attention to the history of empire – Hall cites Marx's adage about the tradition of dead generations weighing 'like a nightmare on the brains of the living'. Ideas from the past, he goes on, 'leave traces of their connections, long after the social relations to which they referred have disappeared. These traces *can be re-activated at a later stage*, even when the discourses have fragmented as coherent or organic ideologies.'[61] Such a later stage might well be a crisis. It has often been observed that whiteness does not, in ordinary times, speak its name; rather it appears – to whites, that is – as normality itself. No one is supposed to hear or see white privilege. It works best when the beneficiaries do not need to give it a thought, but can cash in on it while going about their daily business.[62] Something similar applies to fossil fuels: that they are good for people could long be taken as read, a doctrine so neutral and naturalised that it rarely needed to be stated in words.

---

60 As argued forcefully in Zeus Leonardo, 'Through the Multicultural Glass: Althusser, Ideology and Race Relations in Post–Civil Rights America', *Policy Futures in Education* 3 (2005): 400–12. On ideology as unconscious, see for example Althusser, *For*, 199.

61 Hall, 'Signification', 111. Emphasis added. 'It could be said, for example, that Britain's long imperial hegemony, and the intimacy of the relationship between capitalist development at home and colonial conquest overseas, laid the trace of an active racism in British popular consciousness.' Hall, 'Race', 338.

62 See for example Lipsitz, *Possessive*, 1, 119; Shannon Sullivan, 'White Privilege', in Zack, *The Oxford*, 332–4; Dalton Conley, 'Universal Freckle, or How I Learned to Be White', in Rasmussen et al., *The Making*, 38; Mab Segrest, ' "The Souls of White Folks" ', in ibid., 43; Frankenberg, 'The Mirage', 75, 81; Jardina, *White*, 7.

But in a crisis, silence will no longer do. As a conjuncture in which an ensemble of contradictions is condensed, a crisis can cause dislocations in the dominant ideology.[63] If it activates the contradictions of an entire mode of production and its relation to extra-human nature, the disturbances might be extraordinarily jolting. Nothing throws open the prisons of interpellation like a real crisis: when elements of the dominant ideology fall apart, subjects can slip through and run towards counter-apparatuses. The Russian soldiers deserting en masse in 1917 and joining the Bolsheviks form one of many unforgettable cases (analogously, missionaries know that individuals and communities in crisis are most susceptible to proselytising). Crisis is the wellspring of ideological novelty, but it is therefore also the moment when the dominant ideology *goes into overdrive* to stay in place. It develops, in Adorno's words, a '*cult of the existent*' on the furthest right.[64] There is nothing sedated about it anymore; it takes on urgency, equating the existent with existence as such and juxtaposing victory to apocalypse. Questioned and challenged, things like white privilege and fossil fuels are defended with a vehemence never seen before. The ISAs are on a war footing.[65] Some ideas kept in the collective unconscious can then be put into position and triggered.

Railways across the Bush

By posing the question this way, we have already indicated that indifference to non-white suffering is not the whole story. It is part of it, but it appears at the end of the pipe, after the distribution of the consequences of fossil fuel combustion. The paradigm of environmental justice has attuned us to inequities in the sink: but we should also be looking for articulation in the source.[66] If race is a social relation with a material reality of its own, it must be because

---

63 Following Poulantzas, *Political*, 41–2, 93–5.
64 Adorno, 'Anti-semitism', 227. Emphasis in original.
65 Cf. Althusser, *On the Reproduction*, 206, 220.
66 On the need to go beyond that paradigm, see Tuana, 'Climate', 3–4, 10–11, 18–19, 22–3; Opperman, 'A Permanent', 61.

it channels resources to some groups and away from others – in the last instance, because it is a relation between humans *as they relate to the rest of nature*. Race is not *in* or *of* nature, but neither is it some mental picture or numinous thing. It is material, which means bound up with nature, much like capital; it comes about by humans working their strange ways through a world of matter. If every critical theory of race presupposes a distinction between the natural and the social, a theory of the ecology of race also posits their interpenetration. When modern racism began in the enslavement of African people, certain bodies were forced to produce commodities for others' enjoyment and enrichment – an arrangement not of concepts, but of soil and fibre and carbohydrates and blood. White masters and black slaves were two positions in a relation for working up matter, and race must ultimately be founded on such *Stoffwechsel*, in the past or present. Like the master's house on the plantation, the location of whiteness is built on the processing of biophysical resources at the expense of others.[67] If race is a triangle, whiteness is the apex and non-whiteness and nature the base on which it stands.

One of the first to see this was Fanon. 'Race prejudice', he writes in one of the essays in *Toward the African Revolution*, 'obeys a flawless logic. A country lives, draws its substance from the exploitation of other peoples, makes those peoples inferior. Race prejudice applied to those peoples is normal.'[68] In *Black Skin, White Masks*, he notes that black bodies are associated with animals, as though they belonged to the biological world, to be subdued and procured. The white man sets himself up as the owner of these things. 'The white man wants the world; he wants it for himself alone. He finds himself predestined master of this world. He enslaves it. An acquisitive relation is established between the world and him.'[69] What he abhors most are nature and natives that run wild:

---

67 Cf., for the general argument, Gruffydd Jones, 'Race', 916–19, 924; McClendon III, 'On the Nature', 214; Taylor, 'Race', 185; on the interpenetration of the natural and the social, see further Malm, *The Progress*.

68 Fanon, *Toward*, 40–1.

69 Frantz Fanon, *Black Skin, White Masks* (London: Pluto, 2017 [1952]), 106.

Hostile nature, obstinate and fundamentally rebellious, is in fact represented in the colonies by the bush, by mosquitos, natives and fever, and colonization is a success when all this indocile nature has finally been tamed. Railways across the bush, the draining of swamps and a native population which is non-existent politically and economically *are in fact one and the same thing*,

we read in *The Wretched of the Earth*.[70] Race, on this account, is a way of ordering non-white people and nature at the bottom, dominating them, extracting what has value, discarding the rest and according the white summit some well-deserved opulence. Still under-researched, this link between the subsumption of nature and the oppression of non-whites was forged in the earliest days of the European colonial enterprise and might well persist into this day.[71]

It might not even be a link, as Fanon suggests, but rather one and the same thing. That is the gist of the argument of Ghassan Hage, who thinks that Muslims have now fallen victim to the masters' impulse to tame and control and manage everything in nature. A ferocious wildness is imputed to Muslim populations. They are likened to wolves sneaking across borders and cockroaches multiplying uncontrollably; they mark a failure of the West to domesticate the planet and are therefore so very annoying. Fences and walls must be built around them, as around animals on a farm.[72] Hage follows this homology to the letter: racism and environmental degradation has *the exact same origins* in the desire to domesticate non-human life forms. Rejecting talk of intersection or articulation, he considers these mounting problems of the twenty-first century an age-old

---

70 Frantz Fanon, *The Wretched of the Earth* (London: Penguin, 2001 [1961]), 201. Emphasis added.

71 For some recent arguments in this vein, see Stefano Harney and Fred Moten, 'Improvement and Preservation: Or, Usufruct and Use', in Gaye Theresa Johnson and Alex Lubin (eds), *Futures of Black Radicalism* (London: Verso, 2017), 83–91; David Roediger, *Class, Race, and Marxism* (London: Verso, 2017), 103–7, 123–37; Gargi Bhattacharyya, *Rethinking Racial Capitalism: Questions of Reproduction and Survival* (London: Rowman and Littlefield, 2018), for example 61–3, 68, 105.

72 Hage, *Is Racism*, for example 33–7, 45. He could have backed up his case with empirical material from the homologous agitation against wolves and Muslims in for example the AfD and the DF: see further below.

identity, one coin minted when humans first began to tame other species. Since then, there has developed a 'generalized domestication', defined as a 'mode of inhabiting the world through dominating it for the purpose of making it yield value', whether in the form of convenience or beauty or fortune. Having a picnic is one way of engaging in generalised domestication. So is sitting down in a lounge chair, adjusting it for maximum comfort, propping up the back with cushions and inadvertently 'killing an ant or mosquito that was in the way'.[73] Here the theory appears to go astray. It lets colonialism off the hook, derives present ills straight from the Neolithic Revolution and indicts the most transhistorical, innocuous acts of daily life as the mechanisms that maintain persecution of Muslims and destruction of the biosphere alike. It leaves little room for change. One cannot reasonably ask people to stop having picnics and make themselves comfortable in chairs, let alone grow plants. (Muslims do this too.) A much-needed attempt to theorise the relation between racist and ecological crises, Hage's is ultimately unsatisfactory, because it is *too general*.

One could argue that fossil fuels represent the apogee of domination of nature and *ipso facto*, given the history of the past centuries, of non-white people. But that would still be too abstract. More concretely, Naomi Klein and others have argued that the extraction of fossil fuels requires 'sacrifice zones' that tend to be inhabited by Native Americans, the Ogoni people in the Niger Delta, Arabs in Iraq and other non-white populations that must be pushed aside and deprived of their sustenance: and this cannot work unless their human worth is diminished. If they counted as full humans, seizing and poisoning their land would be inadmissible.[74] The very act of drilling sustains racism: call this *frontier racism*. It can be broadly defined as follows. When there is a commodity to be appropriated on the other side of some line in space, the lives of the population present there might have to be seriously disrupted, which demands and calls forth their 'inferiorisation', to use Fanon's term. As such, frontier

---

73 Ibid., 87, 104.
74 Klein, *This*, for example 169–7. Cf. for example Jodi Melamed, 'Racial Capitalism', *Critical Ethnic Studies* 1 (2015): 83; Pulido, 'Racism', 120.

racism obviously concerns more than fossil fuels. There could hardly be a more emblematic case than the Brazilian Amazon.

Shortly after Portuguese colonisation in the sixteenth century, the Amazon basin was plundered of spices and animal oils shipped back to Europe; in the seventeenth, the coveted goods included turtles and manatee meat; in the nineteenth, rubber for tyres. Each commodity boom required incursions into the rainforest. Here the indigenous populations had utilised a gamut of plants and animals to satisfy their needs, at rates – because those needs were limited – permitting them to regenerate. When colonial agents entered, however, they sought to extract as much of the commodities as possible to feed foreign markets and their own pockets: predation was their modus operandi. Indigenous peoples were decimated, flora and fauna killed off in cycle after cycle that emptied the basin of forms of life in the service of European capital, until the process was modified in the 1950s. Now the Brazilian state took it upon itself to mediate between the interior and the world economy. It continued in the grooves of three-centuries-old frontier racism, inflected by a new nationalist ethos: under the military dictatorship, the state conceived of the commodity frontiers in the Amazon as the advanced positions of the nation. It would brook no internal voids.[75]

A 'green desert', the Amazon was considered underdeveloped, no real part of Brazil, inhabited by primitive nature and some elusive humanoid creatures. Until 1988, every text of the constitution spoke of indigenous peoples as 'savages'.[76] The resources had to be wrested from them and the forest for the nation to fill up its territory; leaving the Amazon to its own would be to forgo fortunes, risk them being

---

75   This follows the classic Stephen Bunker, *Underdeveloping the Amazon: Extraction, Unequal Exchange, and the Failure of the Modern State* (Chicago: University of Chicago Press, 1988), for example 55–6, 60–6, 77–81, 250.

76   See for example S. H. Davis, *Victims of the Miracle: Development and the Indians of Brazil* (Cambridge: Cambridge University Press, 1977); Marcos Napolitano, *1964: Hisória Do Regime Militar Brasileiro* (São Paolo: Contexto, 2014). 'Green desert': Rodrigo Augusto Lima de Medeiros, *Decodificando a internacionalização da Amazônia em narrativas e práticas institucionais: governos da natureza no Brasil e nos EUA*, dissertation (Brasília: Universidade, 2012), 143. 'Savages': Kleger Gesteira e Matos, 'Ordem E Progresso Na Amazônia: O Discurso Militar Indigenista', unpublished master's thesis (Brasília: Universidade de Brasília, 2019), 66.

gobbled up by other nations, invite security threats – notably communist insurgencies – and let nature rule over what should be a modern state. The Yanomami terrain, for instance, was seen as a no man's land that risked departing from Brazil unless physically subsumed. The indigenous presence counted only as a vestige and nuisance. With this brand of nationalist frontier racism, the military dictatorship let loose miners, gold-diggers, rubber barons, cattle ranchers and other agents of nation and capital to cut down and dig up the Amazon. The result was an avalanche of death – by disease, displacement, torture, bombings, human-hunting with machine guns – tearing through indigenous peoples, realising Fanon's logic of railways-across-the-bush.[77] The grandest project was the Trans-Amazonian Highway, which, by drawing a path of asphalt through the rainforest, would open 'lands with no men' for the 'men with no land.'[78] These were the decades when deforestation leapt to the scale of a global concern. The Highway was never finished, but other roads intersected the basin, more thoroughly opened for capital – still very much including *foreign* capital, for which the state paved the way.

It was to this legacy Bolsonaro harked back with no remorse. He demanded that non-white minorities inside the rainforest be fully 'integrated'; he considered the setting aside of a reserve for the Yanomami in 1992 an act of national treason. As for indigenous populations, they 'do not speak our language, they do not have money, they do not have culture. They are native peoples. How did they manage to get 13 percent of the national territory?'[79] Reservations and *quilombolas* were hollows

---

77 See for example Antonio Carlos Wolkmer and Gustavo Silveira Siqueira, *XXIV Encontro Nacional Do Conpedi–UFS: História Do Direito* (Florianópolis: Conselho Nacional de Pesquisa e Pós-Graduaçâo em Direito [CONPEDI], 2015), conpedi.org.br; Bunker, *Underdeveloping*, for example 81, 84, 102–7, 82, 98, 118–22, 238–41.

78 Thaigo Oliveira Neto, 'Rodovia Transamazônia: Falência De Um Grande Projeto Geopolítico', *Revista Geonorte* no. 12 (2013): 282–98. See for example Bruce Albert, 'Terras Indigenas, Politica Ambiental E Geopolitica Militar No Desenvolvimento Da Amazonia: A Proposito Do Caso Yanomami', in Philippe Léna and Adélia Engrácia De Oliveira (eds), *Amazonia: A Fronteira Agricola Vinte Anos Depois* (Belem: Museu Paraense Emilio Goeldi, 1991), 37–58; Charles H. Wood and Marianne Schmink, 'The Military and the Environment in the Brazilian Amazon', *Journal of Political and Military Sociology* 21 (1993): 81–105.

79 Survival International, 'What Brazil's President, Jair Bolsonaro, Has Said about Brazil's Indigenous Peoples', survivalinternational.org, n.d.

in the national frame that must be penetrated, by an organ as strong as the military had once been. For the president and the broader far right, international complaints about the treatment of indigenous peoples and worries about deforestation were so many attempts to deny Brazil its sovereignty. Even the most minimal climate mitigation would undercut it more.

Just as during the military dictatorship, however, the riches of the Amazon were also uncorked for foreign capital and markets to consume, although the main commodity was now inserted into rather than plucked from the basin: cattle. In the 1990s, cattle ranching became the chief driver of deforestation in the Amazon, the dense forests razed to make way for pasture. By the early millennium, this activity alone accounted for three-quarters of all forest clearing, and once again, the boom was driven by export, an intricate web of supply chains stretching from the ranches to all corners of capitalism. Hide from cows would turn up in baseball gloves on American baseball fields or dog collars in Swedish parks. Their tallow would find its way into shaving cream, their intestines into tennis racket strings, hooves and horns into piano keys and lipsticks; the beef itself – of which Brazil became the world's largest exporter – might show up on Italian or Russian tables.[80] It was this irruption of 'cattle capitalism' the state of Lula made some attempts to hold back.[81]

Bolsonaro sought to knock down every door still closed. The trees fell on his watch to clear the way for companies like JBS and its suppliers.[82] In the summer of 2019, fires were three times more common in

---

80 Nathalie F. Walker, Sabrina A. Patel and Kemel A. B. Kalif, 'From Amazon Pasture to the High Street: Deforestation and the Brazilian Cattle Product Supply Chain', *Tropical Conservation Science* 6 (2013): 446–67. See for example Jonas Karstensen, Glen P. Peters and Robbie M. Andrew, 'Attribution of CO_2 Emissions from Brazilian Deforestation to Consumers between 1990 and 2010', *Environmental Research Letters* 8 (2013): 1–7; Florence Pendrill, U. Martin Persson, Javier Godar et al., 'Agricultural and Forestry Trade Drives Large Share of Tropical Deforestation Emissions', *Global Environmental Change* 56 (2019): 1–10.

81 Marcus Kröger, 'Deforestation, Cattle Capitalism and Neodevelopmentalism in the Chico Mendes Extractive Reserve, Brazil', *Journal of Peasant Studies* (2019), online first.

82 Dom Phillips, Daniel Camargos, Andre Campos et al., 'Revealed: Rampant Deforestation of Amazon Driven by Global Greed for Meat', *Guardian*, 2 July 2019.

zones from which meat giants sourced their beef.[83] As with Trump, then, there was nothing flukey about Bolsonaro: he rose on the crest of a very *longue durée* indeed.[84] His nationalist frontier racism was a vulgar replica of the military dictatorship, but just like that model of his, it worked as an intermediary between inner peripheries and outer cores. For all its patriotic rhetoric, the state of Bolsonaro grabbed hold of the Amazon and threw it into global circuits, connecting the frontiers to far-away stores and shelves and high streets, an infinite multiplication of the pattern from the days of colonisation. Only now it happened on the threshold of runaway climate change.

Frontier racism, then, was not a Brazilian affair, not even in its nationalist edition; it was materially integrated in the most globalised form of capitalism. At the same time, it cannot be understood outside of the particularities of *this* social formation, rather different from, say, Spain or Hungary. The articulation of race to energy and nature does not look everywhere the same, and frontier racism is still only part of the story, tied as it is to specific places. Coal first came from the heartlands of whiteness. When large-scale combustion of fossil fuels became a fact of modern life in the nineteenth century, they were extracted from places like Yorkshire and Lancashire, where no populations needed to be inferiorised as savages. If that was – as we have reason to suspect – the century when racism and fossil fuels were originally conjoined, it must have happened through some other process than frontier racism *sensu stricto*, as the share of such fuels appropriated from bushes and swamps with native populations was for a long time inconsiderable. Moreover, extraction can never have the largest ideological radius. Oil and coal and gas are hauled up from specific points in the landscape, but then they can be *consumed* virtually anywhere and fill up the space of propertied agents, rather like money. They become currencies in circulation when they are driven around – above all, as fuels for vehicles.

---

83 Alexandra Heal, Andrew Wasley, Sam Cutler and André Campos, 'Revealed: Fires Three Times More Common in Amazon Beef Farming Zones', *Guardian*, 10 December 2019.

84 On this continuity, cf. Andre Pagliarini, '"De onde? Para onde?" The Continuity Question and the Debate over Brazil's "Civil"-Military Dictatorship', *Latin American Research Review* 52 (2017): 771.

It is here, in acts of mobile *combustion*, that we shall look for a history of that primary level on which interpellations in the core can build. It is a history tied up with frontier racism, but it has spread broader and sunk deeper than acts of drilling. Focus will have to be shifted away from primitive fossil capital. An exhaustive investigation is obviously outside the scope of this work: what we shall offer are rather two vignettes. One concerns a primary vehicle in the nineteenth century, the steamboat, as deployed by the leading fossil economy of that period, the British Empire; the other, the automobile in the mid-twentieth-century US. Their complicated histories will be radically compressed to bring out some central themes. Our hope is that they can serve as signposts for further investigations.

What, Then, Has Coal to Do with the White Race?

Before 1840, steamboats had never been tested on the battlefield of a major war. But in the summer of that year, the British Empire dispatched a squadron to the coasts of Lebanon and Palestine. Tensions between the Empire and Muhammad Ali were coming to a head. Nominally a pasha under the Ottomans, Muhammad Ali had used his power base in Egypt to carve out his own Arab empire, stretching all the way from Cairo to the doorsteps of Istanbul. Perhaps an even greater offence, in the eyes of Britain, was his campaign for building a modern manufacturing industry centred on cotton.[85] The diplomatic machinery could not stand him. 'For my own part, I hate Mehemet Ali, whom I consider as nothing but an arrogant barbarian', wrote Lord Palmerston, the mighty foreign secretary, in 1839: 'I look upon his boasted civilization of Egypt as the arrantest

---

85 The historiography is vast; see for example Afaf Lutfi al-Sayyid Marsot, *Egypt in the Reign of Muhammad Ali* (Cambridge: Cambridge University Press, 1984); Jean Batou, 'Muhammad-Ali's Egypt, 1805–1848. A Command Economy in the 19th Century?', in Jean Batou (ed.), *Between Development and Underdevelopment: The Precocious Attempts at Industrialization of the Periphery, 1800–1870* (Geneva: Centre of International Economic History, 1991): 181–218; Laura Panza and Jeffrey G. Williamson, 'Did Muhammed Ali Foster Industrialization in Early Nineteenth-Century Egypt?', *Economic History Review* 68 (2015): 79–100.

humbug.'[86] London grew more belligerent by the month. 'Know', the consul-general in Alexandria warned the pasha, 'it is in the power of England to *pulverize* you.'[87] 'We must strike at once rapidly and well', the ambassador in Istanbul sent home his advice, and then 'the whole tottering fabric of what is ridiculously called the Arab Nationality will tumble to pieces.'[88] With such words ringing through the corridors of Whitehall, the Royal Navy sent four shining new steamboats towards Beirut. Their commander, Charles Napier, stood onboard the *Gorgon*, a vessel propelled by a 350-horsepower engine, with room for 380 tonnes of coal, 1,600 soldiers and six guns – 'the first true fighting steamship'.[89] It was the coal that gave it propulsive force. Short on the fuel while reconnoitring the Levantine coast, Napier sent a call to several officers: 'You must send me coal vessels here *at all costs*, because steamers without coal are useless.'[90] The coal arrived, and on 9 September 1840, the bombardment of Beirut commenced. Much of the city was destroyed. Then Napier steered south and chased the Egyptian forces from one coastal fortress to another: 'Steam', he reported back to London, 'gives us a great superiority, and we shall keep them moving.'[91] Latakia, Trablus, Saida, Sur and Haifa fell like dominoes, the enemy retreating under the relentless, unpredictable attacks. Then Napier proceeded to the decisive battle at the Palestinian port town of Akka.[92]

---

86 Quoted in C. K. Webster, *The Foreign Policy of Palmerston, 1830–41: Britain, the Liberal Movement and the Eastern Question* (London: Bell, 1951), 629.

87 Colonal Hodges quoted in William Holt Yates, *The Modern History and Condition of Egypt*, vol. 1 (London: Smith, Elder and Co., 1843), 428.

88 'Constantinople 22 March 1846: Secret Memorandum on the Syrian War of 1840–1841' (by General Jochmus), MM/SY/1-3, Broadlands Archive, University of Southampton.

89 David K. Brown, *Before the Ironclad: Development of Ship Design, Propulsion and Armament in the Royal Navy, 1815–60* (London: Conway Maritime Press, 1990), 61.

90 Letter in Major-General Elers Napier, *The Life and Correspondence of Admiral Sir Charles Napier, K.C.B., From Personal Recollections, Letters and Official Documents*, vol. 2 (London: Hurst & Blackett, 1862), 21–2. Emphasis in original.

91 Letter in Charles Napier, *The War in Syria*, vol. 1 (London: John W. Parker, 1842), 83.

92 For a blow-by-blow account of the war, see Letitia W. Ufford, *The Pasha: How Mehemet Ali Defied the West, 1839–1841* (Jefferson: McFarland, 2007). A classic account is Vernon J. Puryear, *International Economics and Diplomacy in the Near East* (Stanford: Stanford University Press, 1935).

The old crusader capital of Akka had been garrisoned by the Egyptians with thousands of soldiers. The sturdiest fortress and main depot on the coast, it was also filled with ammunition. In early November, the four steamboats reached Akka, followed some days later by sailing ships that had been detained by weak winds. The fleet was arrayed in front of the town, with the steamers in the centre so as to make maximum use of their mobility. Massive shelling commenced; the Arabs returned the fire, but the steamboats constantly shifted their positions and easily evaded the shells. Suddenly a deafening detonation ripped through the battlefield. Like a volcano eruption, a column of smoke and debris rose from within Akka. 'The smoke rested for a few moments like an immense black dome, obscuring everything.'[93] One of the steamboats had hit the great powder magazine. Correspondence between the British commanders suggests that they were aware of its position and had planned to target it.[94] The explosion concluded the battle. 'Two entire regiments were annihilated, and every living creature within the area of 60,000 square yards ceased to exist; the loss of life being variously computed from 1,200 to 2000 persons', said one dispatch to Lord Palmerston.[95] When the British soldiers entered Akka, they were greeted by utter devastation:

> Corpses of men, women, and children, blackened by the explosion of the magazine, and mutilated, in the most horrid manner, by the cannon shot, lay everywhere about, half buried among the ruins of the homes and fortifications: women were searching for the bodies of their husbands, children for their fathers.[96]

An officer from one of the steamers described seeing hands, arms and toes sticking out of the rubble. In a letter to his wife, Napier himself expressed unease: 'I went on shore at Acre [Akka] to see the

---

93 Robert Burford, *The Bombardment of St. Jean d'Acre* (London: Geo. Nichols, 1841), 8.
94 Lord Minto to Robert Stopford, 7 October 1840, ELL/216, Elliot Papers, National Maritime Museum, London.
95 Report by Colonel Charles F. Smith to Palmerston in 'Correspondence Relative to the Affairs of the Levant', in *Parliamentary Papers* (1841) VIII, 56.
96 *Tait's Edinburgh Magazine for 1841*, 'Political Register', vol. VIII (1841): 65.

havoc we had occasioned, and witnessed a sight that never can be effaced from my memory, and makes me at this time even almost shudder to think of it.'[97] Not a single man on the four steamboats was killed or wounded. Another resource, however, was exhausted in the action: coal. When silence fell over the body-strewn beaches, none of the steamers had more than one day's supply onboard, nearly all of their coal having being burnt in the pulverisation of Akka.[98]

The battle ended the war in one fell stroke. Muhammad Ali surrendered completely; his empire and cotton industry collapsed. In 1841, the success was repeated in the more famous First Opium War, in which another squadron of steamboats prised open China for trade. After these *tours de force*, the *Observer* printed an homage to steam:

> In the Chinese waters, as well as on the coast of Syria, its employment has produced results which astonished mankind; in the extreme west of Asia as well as in the extreme east – in China and in Palestine – it has finished wars which, under former circumstances, might have been protracted to infinity. But even these achievements, great and important as they are, do not seem to bear any proportion to those which are still possible to that wondrous power. Steam, even now, almost realizes the idea of military omnipotence and military omnipresence; it is everywhere, and there is no withstanding it.[99]

The *Mechanics' Magazine* was even more excited. 'Let war come to a conflict of steam-engines, and all the barbarian rabble of the world, Turks and Tartars, Arabs, Indians, Africans and Chinese, must obviously be out of question at once.'[100]

Up until the second quarter of the nineteenth century, the British Empire was built with wind. Colonies, bases, naval battles

---

97 Letter from Charles Napier to his wife, 13 November 1840, in Napier, *The Life*, 113.

98 W. Patison Hunter, *Narrative of the Late Expedition to Syria*, vol. 1 (London: Henry Colburn, 1842), 299; Robert Stopford to Lord Minto, 5 November 1840, ELL/214, Elliott Papers.

99 *Observer*, 'The Recent Victories', 28 November 1842.

100 *Mechanics' Magazine*, 'Probable Effects of the Railway System', 16 December 1837. Note that this was written before the two wars, which had a long prelude of propaganda for the military powers of steam.

and blockades were won by means of sailing ships. Yet for all its supremacy on the seas, the metropolis encountered limits on precisely this count: the reliance on wind restricted the latitude of power. Sailing upstream on rivers was a slow and arduous task, if possible at all. Imperial stations were generally located on shorelines or islands, in the nature of toeholds; on the subcontinent, in China, the Levant, West Africa and Latin America, the wind-dependent Empire mostly had to stay on the margins, unable to penetrate interiors. Steamboats promised to take it straight into the hearts of those dark continents.[101] Wind belonged to the flow of energy, coursing through the landscape and shifting in time, from one moment of gale to another of calm, often too feeble to carry a ship against the current. The stock would set the waterborne Empire free from such restrictions. Taken out of the repositories of energy from the very distant past, belonging neither to the landscape nor the weather in which the boats operated, coal allowed captains to go against the wind and the tide: this fuel was entirely in the hands of its masters.

But the rotative steam-engine that could propel vehicles such as boats required more than coal from the mines of Britain. A steam-engine was built of various metal parts rubbing against each other. This produced friction, and so the owner had to apply oil; unless regularly lubricated, the engine would break down.[102] From where, then, did the British owners of steam-engines get their oil? They got it from the Niger Delta. By far the most important lubricant in early nineteenth-century British industry was palm oil, almost exclusively imported from West Africa, predominantly the banks of the river Niger, where nuts from palm trees were turned into oil through a burdensome process performed by women. Animal fats were neither sufficient nor suitable, whereas palm oil was perfect for lubricating not only steam-engines, but also railway carriages, spinning-mules

---

101 The standard accounts here are Daniel R. Headrick, *Tools of Empire: Technology and European Imperialism in the Nineteenth Century* (Oxford: Oxford University Press, 1981); *Power over Peoples: Technology, Environments, and Western Imperialism, 1400 to the Present* (Princeton: Princeton University Press, 2012).

102 See for example John Farey, *A Treatise on the Steam Engine: Historical, Practical, and Descriptive* (London: Longman, Rees, Orme, Brown & Green, 1827), 462.

and other machines. Were it not for all that oil, those machines would have literally come to a standstill.[103]

As the adoption of steam power and the mechanisation of British industry advanced in leaps and bounds in the second quarter of the nineteenth century, the importation of palm oil exploded. From 1807 – the year when the Empire officially prohibited the slave trade – to the 1840s, the annual quantity ferried to Britain increased by a factor of more than 200.[104] Like all seagoing commerce up to that point, this trade was conducted by wind. British merchants had to dock at trading posts on the mouth of the Niger and wait for native partners to come down with the goods. The local kingdoms resisted foreign penetration, and as long as they had to trust to the wind, the British had no physical or mechanical weapon to overcome them; instead they were forced to deal with middlemen and brokers, backed up by autonomous chiefs and kings, who delivered the palm oil as and when they saw fit. When demand increased, this system turned into a source of deep frustration. British traders complained ever louder about the natives obstructing the trade, slowing it down, sometimes suspending it entirely.[105] But the solution to this embarrassing predicament came, fortuitously, in the shape of the steam-engine itself. With its power the merchants could pierce through the Delta, bypass the coastal states, reach the producers directly and pour the palm oil freely into their vessels. The brothers Richard and John Lander, the first British explorers to map the course of the Niger in 1830, laid out the vision:

> The first effect of a trade being opened will be to do away with the monopoly near the mouth of the river, which has hitherto been held by the chiefs of the lower countries. Steam-boats will penetrate up the river . . . and will defy the efforts of these monopolists to arrest

---

103   The most comprehensive studies of this trade are Kenneth Onwuka Dike, *Trade and Politics in the Niger Delta, 1830–1885: An Introduction to the Economic and Political History of Nigeria* (Oxford: Oxford University Press, 1956); Martin Lynn, *Commerce and Economic Change in West Africa: The Palm Oil Trade in the Nineteenth Century* (Cambridge: Cambridge University Press, 1997).

104   Lynn, *Commerce*, 3, 13–15, 41.

105   See for example James M'Queen, *A Geographical and Commercial View of Northern Central Africa* (Edinburgh: William Blackwood, 1821).

their progress. The steam-engine, the grandest invention of the human mind, will be a fit means of conveying civilization among these uninformed Africans, who, incapable of comprehending such a thing, will view its arrival among them with astonishment and terror.[106]

The first steam-powered expedition set out in 1832 under the leadership of the shipper Macgregor Laird. Two boats drove up the river. Naturally, the denizens of the Delta were not happy about it. In a casual tone, Laird narrated how they took shots at his boat; in one rebellious area, 'we agreed to land and burn the town as an example to the rest'.[107] Such scenes were repeated throughout the expedition. During one ambush, massive casualties were inflicted on the sailors (one of those killed was a Swede). The expedition ended in disaster, with most of the crewmen perishing in engagements or disease, but Laird had demonstrated that the Niger and its riches were indeed accessible by steamboats. He ended his two-volume account of the events with a manifesto for what might be called steam or stock imperialism: we have a power, he wrote,

bequeathed to us by the immortal Watt. By his invention every river is laid open to us, time and distance are shortened ... This power, which has only been in existence for a quarter of a century, has rendered rivers truly 'the highway of nations', and made easy what it would have been difficult, if not impossible, to accomplish without it. *We are the chief repository of it*: *our mineral wealth* [i.e. the coal reserves of Britain] and the mechanical habits of our people, give us *a superiority over all others* in the application of it.[108]

It is the privilege of bourgeois visionaries to be able to see their visions come true only by swimming with the tide of history. In the early 1850s, Laird's company established the first regular steamship

---

106 Richard Lander and John Lander, *Journal of an Expedition to Explore the Course and Termination of the Niger*, vol. 2 (New York: Harper & Brothers, 1839), 310.

107 MacGregor Laird and R. A. K. Oldfield, *Narrative of an Expedition into the Interior of Africa*, vol. 1 (London: Richard Bentley, 1837), 84.

108 Ibid., 398–9. Emphasis added.

service between Britain and the Niger Delta; soon his boats were plying up and down the Niger and back and forth between the West African coast and the metropolis. The result was a second, even more marvellous explosion in the palm oil trade.[109] Similar outflows were set off from Egypt, whose lands were converted to the production of raw cotton for British mills – the once prospering Egyptian mills derelict after Akka – accounting for more than half of supplies imported by the metropolis from 1848 onwards.[110] The trade route between Egypt and Britain was among the first to shift to steam. Driving up and down the Nile, steamboats picked up bales of raw cotton and carried them to Alexandria, whence they were shipped home; meanwhile, Egypt underwent the most extreme deindustrialisation anywhere in the nineteenth century.[111] But the fates of the peoples of the Nile and the Niger were all too representative. Across the seven seas, the peripheries of the capitalist economy were bound in the position of suppliers of raw materials to the core. That division of labour had been under construction for some time, as we have seen, but the nineteenth century was 'a period of exceptionally rapid divergence between core and periphery, and that divergence was most dramatic over the half century 1820 to 1870', observes Jeffrey G. Williamson in *Trade and Poverty: When the Third World Fell Behind*. In passing, he also notices that peoples in these southern parts of the planet were not thrilled about submitting to imperial trade, but 'the naval muscle of the industrial leaders made them comply'.[112] Those muscles were nourished by fossil fuel. It had prodigious effects on the location known as whiteness.

Before the nineteenth century, white people who lay claim to superiority tended to invoke religion. Early enslavement of Africans was justified with the 'curse of Ham', a creative interpretation of a

---

109 See Lynn, *Commerce*.
110 Calculated from figures in E. R. J. Owen, *Cotton and the Egyptian Economy, 1820–1914: A Study in Trade and Development* (Oxford: Clarendon Press, 1969), 161.
111 Jeffrey G. Williamson, *Trade and Poverty: When the Third World Fell Behind* (Cambridge, MA: MIT Press, 2013), 68.
112 Ibid., 13. For another study by a bourgeois economist that more stringently isolates the shift to steamboats in global trade as the cause of the divergence, see Luigi Pascali, 'The Wind of Change: Maritime Technology, Trade, and Economic Development', *American Economic Review* 107 (2019): 2821–54.

passage in Genesis where Ham happens upon his father Noah without clothes and is punished with a curse; black people, the assumed descendants of Ham, would have to pay for his crime – seeing the father naked – by forever being slaves.[113] But over the course of the nineteenth century, whiteness was placed on an apparently firmer footing. In *Machines as the Measure of Men: Science, Technology, and Ideologies of Western Dominance*, Michael Adas has documented the shift to another gauge of human worth: people who mastered nature stood above those who did not. It would never occur to Africans to push aside large stones that blocked a path. Indians venerated cows. On the dark continents, people lived in thatched huts, surrounded by wild animals, at the mercy of unmanaged, untapped nature; Europeans alone had cracked her secret codes and yoked her power to machines. Their material achievements set them off from all others, who were now perceived as residues from earlier stages of evolution. Essentially inhabiting another time, primitive peoples were the stunted children of mankind, so immature that they bowed down to the forces of nature and let them go to waste: they deserved to be ruled. Europeans had a higher mandate to govern them, not given by God so much as by technology, which supplanted other criteria and solidified into the insignia of superiority – and only then, after centuries of fluctuations, Adas contends, did the category of race come to a kind of rest.

Now it was the industrious whites above all others. The inferiors included the Middle East and China, previously held in some esteem for their arts and monuments of civilisation; after the wars of the early 1840s, they seemed to prostrate themselves before the machine as deeply as the Africans. At the turn of the century, a near-total intellectual consensus – trickling down to lower classes through travel accounts, adventure novels, the penny press – interpreted the near-total European dominance of the world as the manifestation of a racial character unrivalled in excellence.[114] This broad chronology

---

113 Robin Blackburn, *The Making of New World Slavery: From the Baroque to the Modern 1492–1800* (London: Verso, 2010 [1997]), for example 65–76.

114 Michael Adas, *Machines as the Measure of Men: Science, Technology, and Ideologies of Western Dominance* (Ithaca: Cornell University Press, 1989), for example 143–6, 164–5, 194–6, 213–19, 307–18, 341–2.

is, of course, found not only in Adas's work. 'Racial antagonism attained full maturity during the latter half of the nineteenth century, when the sun no longer set on British soil', Oliver Cox stated in 1948.[115] Whiteness and its surrounding negations were fixed in place during this particular century of violence from the core. Modern racism, in other words, is unthinkable without *techno-racism*.

And no technological complex was as pivotal as this one: the steamboats, the railroad cars and all the other steam-powered machines of white Europe. Travellers boarded them on triumphal processions through the colonies and sent home word of the shock and awe. Reports streamed in from all corners of the British Empire: from Vancouver Island on the Pacific coast of Canada, energetic entrepreneur Gilbert Malcolm Sproat described, in *Scenes and Studies of Savage Life*, how he and his crew first encountered Indians in 1860. A pathetic dialogue ensued. 'We do not wish to sell our land nor our water; let your friends stay in their own country', the nameless spokesman of the savages implored Sproat. He responded that 'since you do not work your land', it is 'of no use to you'; the king in London had ordered that it must be sold, and anyway the white men 'are your superiors'. The Indian would not be persuaded. 'We do not want the white man. He steals what we have. We wish to live as we are', and these were his last words as recorded in *Scenes*. Immediately afterwards, 'a civilized settlement' was built in the midst of the savages, who could not stop staring at the steam-engine.[116] One of the first thing the occupiers of Vancouver Island did was to uncover seams of coal.[117]

Sproat claimed to have spent happy years in the settlement. He criss-crossed the island and observed the habits and characters of the savages, who, he could not fail to notice, were swiftly going extinct. Why? No violence or molestation had, he claimed, been meted upon

---

115 Oliver C. Cox, *Caste, Class, and Race: A Study in Social Dynamics* (New York: Doubleday, 1948), 330.

116 Gilbert Malcolm Sproat, *Scenes and Studies of Savage Life* (London: Smith, Elder and Co., 1868), 4.

117 See Robert A. Stafford, *Scientist of Empire: Sir Roderick Murchison, Scientific Exploration and Victorian Imperialism* (Cambridge: Cambridge University Press, 1989), 73–4.

them, nor had diseases or alcoholic beverages anything to do with the matter. The savages had received wages from the settlement and on the whole had their station improved. So then why did they die in such numbers? Sproat ventured the following explanation: the mind of the savage 'was confused and his faculties surprised and stunned by the presence of machinery, steam vessels, and the active labour of civilized men; he distrusted himself, his old habits and traditions, and shrank away despondent and discouraged.' He abandoned his bows and canoes. He sank into a torpor, regressed to the stage of an animal and gave up on life: a natural outcome of 'the presence of a superior race'.[118] In this account, steam induced genocide by working out its mechanical wonders, before which the savages could only lay down and die. And that was, on the whole, a good thing, as when the sun rises in the morning.

From India, Whig intellectual Harriet Martineau shared her impressions of natives quivering before the British engines. She toured the country on a train: 'Under the western Ghauts [a mountain range on the western peninsula] the villagers come out at the sound of the steam whistle, and the babies gasp and cry as the train rushes by; and nobody denies that the railway is a wonderful thing.' The Hindus saw their world shatter, their inborn ideals – 'immutability, patience, indolence, stagnation' – crushed by the onrush of steam. Once these slow Hindus had 'hated the Mussulmans for invading with their superior energy; and now what is Mussulman energy in comparison with ours, judged by our methods of steaming by sea and land'.[119] The whistle said it all.

In the metropolis, scholars generalised such observations from the frontiers into systematic theories of races. Two papers read before the Ethnological Society of London in 1866 offer a sample. In 'Aptitudes of Races', Frederic Farrar, a reverend, polymath and friend of Charles Darwin, distinguished between three types: the 'irreclaimably savage', the 'semi-civilised' and the 'civilised races'. The first included the majority of humankind, which had never made a single

---

118 Sproat, *Scenes*, 278–9.
119 Harriet Martineau, *British Rule in India: A Historical Sketch* (London: Smith, Elder and Co., 1857), 340–1.

discovery or hit upon even one minor invention. Chattering like the monkeys in their woods, scarcely cleverer than the dogs around their huts, these races were doomed to pass out of history and leave behind no other trace than their 'actual organic remains'. Then there were the Chinese, the semi-civilised race which had once invented a few things, only to stagnate, decay and live out its 'tendency to physical obesity and mental apathy'. On top stood 'the Aryans'. 'To them belong the steam-engine, the printing-press, the ship, the lighthouse, the electric telegraph', every invention of any value, the steam-engine first and foremost; to them also 'belong the destinies of the future'.[120] All other races – black and red, brown and yellow – were giving way.

In his paper, John Crawfurd, a learned man with an eminent career as an imperial agent coursing through Asia onboard steamships, compared the European and Asiatic races. The 'natural attitude' of the European is to stand, of the Asian to sit; the former enjoys walking, the latter detests it; the former is white, the latter 'swarthy'; beauty and 'symmetry of person' decrease the farther east one travels. Women are treated as nearly equal to men in Europe and 'looked upon as but handmaids' in Asia. The European has an immeasurably rich literature. The Asian has the Qur'an. But the pith of superiority in Crawfurd's scheme was, of course, the machines powered by steam: while the European race was busy inventing them, the Asiatic races sat like immovable stones. Nothing could explain this divergence other than 'a broad innate difference, physical, intellectual, and moral', present since 'the first creation of man'.[121]

In most of these writings, the actual fuel of the steam-engine was left unmarked and unsaid. It was common knowledge that steam derived from coal; that Britain had reserves in abundance; that its mines had been open for centuries while other people sat ignorantly on top of theirs and that wherever they went, British colonisers

---

120 Frederic W. Farrar, 'Aptitudes of Races', *Transactions of the Ethnological Society of London* 5 (1867): 116, 120, 124–5.

121 John Crawfurd, 'On the Physical and Mental Characteristics of the European and Asiatic Races of Man', *Transactions of the Ethnological Society of London* 5 (1867): 59, 60, 75, 81. For more examples of techno-racism in the sign of steam, see Adas, *Machines*, 165, 186–8, 190–1, 213–14, 221–3, 233.

searched for extra coal to dig up. Sometimes the substratum of dominance was brought to the surface, as when Macgregor Laird stated that 'we are the chief repository of it: our mineral wealth', source of the superiority. Sometimes it was obliquely alluded to, at other points hidden under the belief that racial character in and of itself fuelled victory. And on occasion, an agent of the Empire would draw a straight line from subterranean base to racial summit.

John Turnbull Thomson was an average engineer, explorer and superintendent employed by various branches of imperial administration in Asia, acceding to the post of surveyor-general of New Zealand in 1856. In a lecture two decades later, he recalled how the South Island had been 'in a state of wilderness' when he arrived. Since then, the railway system had made 'rapid strides', along with foundries producing steam-engines and machinery for coal mines. Progress had a sign in the sky: 'The tall chimneys everywhere rising indicate that, in this part of the world, all the skilled trades have found a suitable and ever-increasing location.' Painting on a broader canvas, Thomson then retraced the advances of steam across America and Asia, from the Mississippi to the Yangtze, before returning home to retell an anecdote about a Maori girl killed, cooked and eaten 'by her own race. Was this not the shadow of coming events, an allegory of the certain fate of so inhuman a race?' Likewise, 'the red-coloured man has been swept off the face of the northern continent of America'; Bibles translated into his languages already gathered dust, as every potential reader was dead. After surveying this global theatre of violence, Thomson reached his punchline:

> Then, what has made the white man – or more conspicuously the Anglo-Saxon – of the Teutonic race so ubiquitously progressive and aggressive; this more especially of so recent a date? It is his humanity and science, combined with steam. And what makes steam for him? It is coal. What then has coal to do with our race? As far as we know yet, everything.[122]

---

122 Address by 'J. T. Thomson, F. R. G. S., President, in the Chair' to the New Zealand Institute, 7 April 1874, in James Hector (ed.), *Transactions and Proceedings of the New Zealand Institute, 1873* (Wellington: James Hughes), 441, 444–5.

Other Europeans wanted a share in the spoils. In *Les Lois Psychologiques de l'Évolution des Peuples*, or *The Psychology of Peoples: Its Influence on Their Evolution*, appearing in 1894, Gustave Le Bon decided that there were four types of races: the 'primitive', the 'inferior', the 'average' and the 'superior', the latter synonymous with the Europeans, so pre-eminent because they 'have discovered steam and electricity'. Among the inferior were the Ethiopians, who had only managed to degrade arts and instruments borrowed from others, 'the race being endowed with insufficient brain capacity'. Mechanical genius was the index of racial quality. Alone among the races, Europeans had an elite of 'superior men', who shouldered the task of 'synthesising all the efforts of the race', and that was how the steam-engine had come about: not as the flash of individual prodigy, but as the collective outpouring of racial intelligence. Le Bon, however, ended his work on a sombre note. The achievements of the Europeans were in jeopardy, for the rise of socialism threatened to 'facilitate the destructive invasions' from Asia.[123]

More important for subsequent developments, this steamy techno-racism was taken up by some of the most prestigious men of the US intelligentsia. In *The History of White People*, Nell Irvin Painter chronicles the ever-unfinished work of constructing whiteness, from early modern Europe to the age of Obama; focusing on the US, her account of the nineteenth century has Ralph Waldo Emerson as the central character. She dubs him 'the philosopher king of American white race theory.'[124] No one did more to codify the perception of white people as the truest face of that country. Emerson adulated the Saxon, also known as the English, from whom all that was good in America descended. He offered a full portrait of this figure in *English Traits* from 1856, an account of travels in the fatherland and extended argument for the determinative power of race – 'it is in the deep traits of race that the fortunes of nations are written', for better or for worse. 'Race in the negro is of appalling importance.' It is an equally 'controlling influence in the Jew, who, for two millenniums, under every

---

123 Gustave Le Bon, *The Psychology of Peoples: Its Influence on Their Evolution* (New York: Macmillan, 1898), 26–8, 109, 201, 227.
124 Painter, *History*, 151.

climate, has preserved the same character and employments'. Peerless in all respects are the English. Here is 'the best stock in the world, broad-fronted, broad-bottomed, best for depth, range, and equability', gorgeous and vibrant, with a manly energy inherited from Nordic ancestors. But the core of *English Traits* is a paean to the most modern prime mover. 'Steam is almost an Englishman', Emerson opens the argument, the qualification soon discarded.[125] As he could see on his journeys, England was awash with capital like no other nation, and it had steam to thank for it. Since the English acquired 'this goblin of steam, with his myriad arms, never tired, working night and day everlastingly, the amassing of property has run out of all figures.' Steam-powered machinery is 'wise, versatile, all-giving', consequently so many traits of the English; like all of their advantages, steam 'breaks out *in their race*'. The virility of the Vikings has returned. 'The old energy of the Norse race arms itself with these magnificent powers', the descent from Odin and the true Saxon blood on display in furnaces and steamships.[126]

Emerson also had a keen appreciation of the substratum. 'The English are so rich, and seem to have established *a tap-root in the bowels of the planet*, because they are constitutionally fertile and creative.'[127] Racist theories have rarely distinguished themselves for intellectual clarity, and here, as in Thomson, a paradox could be discerned: were white people so great because they used coal? Or was coal so great because it was used by white people?[128] Eliding the fine points, Emerson posited an organic unity between the fuel in the bowels of the planet, the immensity of the capital accumulated and the brilliance of the English race, the last rooted in the first and thereby indebted to it. But, conversely, the English had learnt to use coal because they were the best stock in the world.

He did notice that it stained the visage of the fatherland. 'Add the coal smoke. In the manufacturing towns, the fine soot or *blacks* darken the day, give white sheep the color of black sheep, discolour

---

125 Ralph Waldo Emerson, *English Traits* (Boston: James R. Osgood & Co., 1876 [1856]), 137, 53–4, 99.
126 Ibid., 163, 97, 165. Emphasis in original.
127 Ibid., 169. Emphasis added.
128 This recalls Plato's Euthyphro dilemma.

the human saliva, contaminate the air, poison many buildings, and corrode the monuments and buildings.' Fly in the ointment, air pollution was accompanied by another momentous change, which, however, Emerson judged to be positive. 'The enormous consumption of coal in the island is also felt in modifying the general climate.' Furthermore, 'the climate too, which was already believed to have become milder and drier by the enormous consumption of coal, is so far reached by this new action, that fogs and storms are said to disappear'.[129] Similar hopes were expressed in another Emersonian paean to coal:

> We may well call it black diamonds. Every basket is power and civilization. For coal is a portable climate. It carries the heat of the tropics to Labrador and the polar circle: and is the means of transporting itself whithersoever it is wanted. Watt and Stephenson [inventor of railways] whispered in the ear of mankind their secret, that a half-ounce of coal will draw two tons a mile, and coal carries coal, by rail and by boat, to make Canada as warm as Calcutta.[130]

All praise to the English race; in Painter's assessment, Emerson could hardly have been more successful in his efforts to whiten the US. 'He spoke for an increasingly rich and powerful American ruling class.' No matter how obtuse his ideas, 'they circulated as American orthodoxy', and this very much included his techno-racism.[131] Laymen could sip on it from periodicals such as the *Popular Science Monthly*, which, in 1876, carried a piece on 'The Apotheosis of Steam', that godlike mechanical power that had suddenly handed over the whole earth to 'the Euraryan race'.[132]

If Emerson was the king of white race theory, one of the princes was Edward A. Ross. In a paper called 'The Causes of Race Superiority', presented to the American Academy of Political and Social Science

---

129   Emerson, *English*, 45, 99.

130   R. W. Emerson, *The Conduct of Life* (Boston: James R. Osgood & Co., 1876 [1860]), 74–5.

131   Painter, *History*, 183, 187.

132   John S. Hittell, 'The Apotheosis of Steam', *Popular Science Monthly*: Vol. IX (New York: D. Appleton & Co., 1876): 430–8.

in 1901, he offered the umpteenth rating and explanation of the innermost nature of the races. This being a time in history when white racists delighted in migration, Ross argued that one race surpasses the other 'in energy' when it sends its children to foreign lands. Whites had 'superior migrancy', plus the ability to subjugate wild nature. All the barbarian rabble of the world was out of the question:

> The exploitation of nature and man by steam and machinery directed by technical knowledge, has the strongest of human forces behind it, and nothing can check its triumphant expansion over the planet. The Arab spreads the religion of Mahomet with the Koran in one hand and the sword in the other. The white man of to-day spreads his economic gospel, one hand on a Gatling [machine-gun], the other on a locomotive.[133]

Ross, however, was as pessimistic about the future as Le Bon. The steam-powered white race enthroned in the US was subjected to a 'silent replacement'. No bloodshed, no war, but 'the heavy influx of a prolific race from the Orient' threatened to finish off the whites, chiefly by giving birth to too many children. For this process, Ross coined the term 'racial suicide'.[134]

Techno-racism survived the turn of the century, but it had its critics too. When W. E. B. Du Bois in 1940 wrote an imagined dialogue between a white man and a black man, as a sort of digest of popular conceptions of race, he had the former adducing steam, factory, mine and all the wealth flowing therefrom as proof of white superiority. Du Bois's black man accepted that the white man had indeed created these things. But they stirred in him 'no envy; only regrets'. They made up a 'vast Frankenstein monster' that gave humans neither rest nor leisure, neither 'community of feeling' nor tolerance and understanding. Would the white man's creations ever further such ideals,

---

133 Edward A. Ross, 'The Causes of Race Superiority', *Annals of the American Academy of Political and Social Science* 18 (1901): 71, 74–5.
134 Ibid., 88.

then, all hail, White Imperial Industry! But it does not. It is a Beast! Its creators even do not understand it, cannot curb or guide it. They themselves are but hideous, groping hired Hands, doing their bit to oil the raging devastating machinery which kills men to make cloth, prostitutes women to rear buildings and eats little children.[135]

If the triumphal processions of steam made the riders giddy with excitement, things appeared very different from the underside. An inhabitant of Akka or what came to be known as Vancouver Island had little reason to be sanguine about the long-term effects. There is a subaltern tradition of forebodings and denunciations as rich as it is unexplored: from the very first encounter with European steamboats to the earliest days of climate science, a non-white shadow of critique followed white fetishisation. It deserves its own encyclopaedia, but we shall offer only one more example here. In his long poem 'The "Red Indian's" Penultimate Speech to the White Man', appearing in Arabic in 1992, Mahmoud Darwish, the bard of Palestine, refracts the Native American and Palestinian experiences through each other. The effect is a universal *j'accuse* for the geological epoch:

> What the stranger says is truly strange.
> He digs a well deep in the earth to bury the sky [...]
>
> Don't dig any deeper![136]

Speaking to the white man, the Native American–Palestinian pledges to 'defend the trees we wear' and 'the leaping dear', but

> soon you will erect your world on our remains,
> you will pave over the sacred places to open a road
> for the satellite moon.
> This is the age of industry,
> the age of coal and fossils to fuel your thirst for fine wine.[137]

---

135 W. E. B. Du Bois, 'Dialogue with a White Friend', in David R. Roediger (ed.), *Black on White: Black Writers on What It Means to Be White* (New York: Schocken, 1998), 36.

136 Mahmoud Darwish, *The Adam of Two Edens: Selected Poems* (Syracuse: Syracuse University Press, 2000), 136 (translated by Sargon Boulos).

137 Darwish's poem as translated by Le Trio Joubran, *The Long March*, Randana, 2018.

Or, in another rendering of Darwish's Arabic:

> This is the age of minerals, and out of coal the champagne of the strong will dawn [. . .]
>
> Where, master of white ones, do you take my people . . . and your
>   people?
> To what abyss does this robot loaded with planes and plane carriers
> take the earth, to what spacious abyss do you ascend?
> You have what you desire: the new Rome, the Sparta of technology
>   and the ideology
>   of madness,
> but as for us, we will escape from an age we haven't yet prepared our anxieties for.[138]

Now imagine that someone in, say, the year 1900 would have told Europe and its offshoots: you have to stop burning fossil fuels. Not another lump or drop can be taken out of the ground. What would the response have been? Quite possibly, a laugh in the face of the messenger, to the effect that 'those fuels have freed us from a world of back-breaking labour' (the Competitive Enterprise Institute). Any insistent campaign might well have been disparaged as an attempt at 're-primitivization' (Mark Steyn) or a 'conspiracy against the whites' (Jean-Marie Le Pen). Or perhaps someone would have opposed the ideal of 'prehistoric men with no access to technology, science, information and the wonders of modernity' (Jair Bolsonaro).[139] But these are, of course, speculations: no one demanded cessation of fossil fuel extraction and combustion in 1900 – at least not with reference to climate – and when the message eventually arrived in the early 1990s, the landscape of energy consumption had transformed beyond recognition. What reason is there to suppose that Victorian technoracism had anything to do with the reaction by then? The case could be made that the musings of a Sproat or an Emerson are as distant

---

138   Mahmoud Darwish, 'The "Red Indian's" Penultimate Speech to the White Man', *Harvard Review* 36 (2009): 158–9 (translated by Fady Joudah).

139   Tom Phillips, '"He Wants to Destroy Us": Bolsonaro Poses Gravest Threat in Decades, Amazon Tribes Say', *Guardian*, 26 July 2019.

from the perceptual world of the average far-right voter in the early twenty-first century as *The Song of Roland*.

But it could also be hypothesised that the articulation of energy and race that developed over the nineteenth century, in the most primary levels of modern capitalism, sedimented an association between whiteness and fossil fuels that wells up like magma in a time of climate breakdown. Certain defenders of such fuels might feel, on some level, questioned not only as burners but more specifically *as white people*, who for so long have had their tap-root of riches in the bowels of the planet. One can hear some of this association echoing in one of the boilerplates of white supremacy, in the US in particular: whites should be proud for having invented the modern world. In late 2018, a young, well-dressed Columbia University student entered the hall of ephemeral internet fame when he harassed some fellow black students by shouting: 'Europeans! We invented science and industry, and you want to tell us to stop because *we're so baaad*' and, getting more fired up: 'White people are the best thing that ever happened to the world. We are so amazing! I love myself! And I love white people! Fuck yeah, white men! We did everything!'[140] The far right tends to cultivate a 'producerist' notion of white people as generators of wealth and non-whites as parasites: it works hand in glove with unswerving loyalty to the productive forces (literally) in question.[141] But the continuities were perhaps most unbroken in Brexit. As Danny Dorling and Sally Tomlinson have showed in *Rule Britannia: Brexit and the End of Empire*, the right-wing assemblage that saw this project through dripped with nostalgia for imperial power. Some of the key figures embodied the persistence in their lives. (One of them, Arron Banks – of whom more shortly – spent parts of his childhood in southern Africa, where his father ran sugar estates.) With this came an order of

---

140 See for example Conor Friedersdorf, 'Probing the "White People" Rant That Roiled Columbia', *Atlantic*, 8 March 2019.

141 On producerism (a truly transnational facet of the far right) in the US context, see Daniel Martinez Hosang and Joseph E. Lowndes, *Producers, Parasites, Patriots: Race and the New Right-Wing Politics of Precarity* (Minneapolis: University of Minnesota Press, 2019), for example 25–7, 120.

priorities.[142] If one is loyal to the British Empire, one cannot be disloyal to fossil fuels.

Much further investigation would be needed to demonstrate the genealogy. It should be fairly obvious, however, that the relation between the contemporary far right, energy, climate and nature cannot be understood in abstraction from the history of modern racism and how it has related to those things. And steam was only just the beginning.

It Takes Gas to Go

To save this thing called climate, they want us to stop driving cars: thus the signal has been widely received, and not without reason. This crisis calls into question the automobile, as a privately consumed commodity with disastrous aggregate effects. No other sector has increased its emissions as fast over the past half-century as transport. With road vehicles making up the bulk, it accounted for more than half of all oil burnt in 2010, even though participation remained extremely skewed: no more than 10 per cent of humanity was responsible for 80 per cent of the kilometres travelled with motors, 'with much of the world's population hardly travelling at all'.[143] And no other nation had attached so many motors and pipes to its population as the US. In 2016, the country had 832 automotive vehicles – cars and trucks – per thousand inhabitants, to be compared with 36 in India and 39 in Africa, or 606 in Western Europe; it retained by far the highest level of dependency on this one mode of transport.[144] It has to be broken. Although some cars can be powered by electricity from sun and wind or hydrogen or potentially some other non-emitting fuel, any

---

142 Dorling and Tomlinson, *Rule*. Banks: 78.
143 R. Sims, R. Schaeffer, F. Creutzig et al., 'Transport', in O. Edenhofer, R. Pichs-Madruga, Y. Sokona et al. (eds) *Climate Change 2014: Mitigation of Climate Change. Contribution of Working Group III to the Fifth Assessment Report of the Intergovernmental Panel on Climate Change* (Cambridge: Cambridge University Press, 2014), 605–8; quotation from 606.
144 Stacy C. Davis and Robert G. Boundy, *Transportation Energy Data Book* (Oak Ridge: Oak Ridge National Laboratory, 2019), tables 3.2 and 3.6.

transition to an economy that does not trash the climate and other life-support systems would also have to involve a shift in modes: from private cars to walking, bicycling and, not the least, public transport.[145] People would have to start travelling together. From the earliest days of climate awareness, this implication has been disseminated among the republics of drivers in advanced capitalist countries, and it has run up against more than the weight of highways and parking lots and car lobbies, for drivers tend to feel with, and through, their own private vehicles. Cars give them a 'sense of *who they are* in the world'.[146] Cars, in other words, are modules of ideology, some of which has to do with race.

The car became a fixture of American life in the decades around the First World War. In 1909, there were 3.5 vehicles per thousand inhabitants, rising to 219 in 1929 (a ratio roughly six times higher than that in India and Africa as of 2016).[147] Up to the interwar period, city dwellers had moved around by foot, bicycle and public transport – streetcars or trolleys – but the lines for the latter were now infamously ripped up and scrapped at the behest of car producers. The shift to the car was, of course, predicated on abundant oil. It was not the prerogative of all Americans; some of the early manufacturers stipulated that their motorcars must not be sold to blacks even if they had the money. As one auto journal noted in the 1920s, 'illiterate, immigrant, Negro and other families' remained outside the market.[148] The racial status of the car was evident in the most iconic

---

145 See for example Sims et al., 'Transport', 619, 648; Matthew Paterson, *Automobile Politics: Ecology and Cultural Political Economy* (Cambridge: Cambridge University Press, 2007), ch. 7; Mimi Sheller, *Mobility Justice: The Politics of Movement in an Age of Extremes* (London: Verso, 2018), 84–8.

146 Paterson, *Automobile*, 223. Emphasis in original. See further Mimi Sheller, 'Automotive Emotions: Feeling the Car', *Theory, Culture and Society* 21 (2004): 221–42; Sheller, *Mobility*; Alan Walks, 'Driving Cities: Automobility, Neoliberalism, and Urban Transformation', in Alan Walks (ed.), *The Urban Political Economy and Ecology of Automobility: Driving Cities, Driving Inequality, Driving Politics* (London: Routledge, 2015), 5–13; and the classic Peter Freund and George Martin, *The Ecology of the Automobile* (Montreal: Black Rose Books, 1993), ch. 5.

147 Davis and Boundy, *Transportation*, table 3.7.

148 Paul Gilroy, 'Driving While Black', in Daniel Miller, *Car Cultures* (Oxford: Berg, 2001), 93–4; the article 'The Motor Car Has Created the Spirit of Modern America', from the journal *Auto*, 1923, quoted in Kevin Douglas Kuswa, 'Suburbification, Segregation, and the Consolidation of the Highway Machine', *Journal of Law in Society*

photograph from the Great Depression, in which Margaret Bourke-White captured an all-black line of people, queuing with empty buckets and baskets in their hands. Some of them stare impassively ahead; others look resigned and exhausted. Above them towers a billboard with a man behind the wheel of his car, his wife in the seat beside him, a girl and a boy and a dog behind, all smiling, all white. The text says 'World's Highest Standard of Living' and 'There's No Way Like the American Way'.

That way was, however, called into question during the Second World War, when the US was hit by severe shortages of rubber. Critical for everything from tanks to gas masks, the material had to be saved, and so the federal government imposed rationing. As tyres were the main source of domestic rubber consumption, this effectively curtailed the private driving of cars. The measure was followed by direct gasoline rationing, and the government forced car companies to convert their factories to war production – tanks, jeeps, fighter planes and the like – the resources of the nation redirected towards the one goal of defeating the enemy. From 1941 to 1945, the cars-to-people ratio fell from 262 to 222 per one thousand. That was the largest drop ever recorded in US history; the first and, as of 2019, only time when the preponderance of the American automobile has been thrown into doubt.[149] Masses of drivers were pushed back onto trolleys and buses. How did they respond?

In a revelatory dissertation, Sarah Frohardt-Lane has studied the repercussions in Detroit, 'the Motor City' rechristened 'the Arsenal of Democracy', where the material pressures of the war were acutely felt. Half a million black migrants streamed into the converted factories. Like other residents, these African American workers had to rely on public transport to get to their jobs. The trolleys and buses filled up with black passengers, alongside the white ones who could no longer take their cars (Detroit, as a northern city, didn't have Jim

---

31 (2002): 44. On the long gestation of US suburbia and the role of transport technologies in it, see Kenneth T. Jackson, *Crabgrass Frontier: The Suburbanization of the United States* (New York: Oxford University Press, 1985).

149   Sarah K. Frohardt-Lane, *Race, Public Transit, and Automobility in World War II Detroit*, dissertation (University of Illinois, 2011), 12–13, 123–8, 137–8; Davis and Boundy, *Transportation*, table 3.7.

Crow laws). Not only were the shared vehicles overcrowded – 'jammed to the doors with people', as one letter-writer complained – but black and white bodies jostled against each other like never before. They competed for seats and waited at stops as packed buses passed them by. Black people – women not the least – were employed as operators of trolleys, conductors and bus drivers, charged with moving Detroiters of all colours around. The face of transport changed: and many whites resented it violently.[150]

When rationing began to bite in the summer of 1942, 'racial disturbances' on Detroit's public transport lines exploded. The city's police department received a dozen reports per week, but the real figure was in all probability higher. The vast majority of incidents occurred between blacks and whites: a white rider would imagine that a black brushed against him on purpose and throw out a racial slur. A tussle over a vacant seat would escalate into a stabbing. There were rumours of black Detroiters organising 'bump clubs' to deliberately push into whites at fixed dates and places, and fears of black men sexually assaulting white women. Repeated inquiries yielded no foundations for such claims, which appear to have expressed intense white revulsion and anxiety over physical contact with black bodies. The sight of African American personnel – women to boot – seems to have been particularly provocative, suggesting an undermining of white authority: 'I don't ride with niggers', a black bus driver would often hear.[151] The disturbances reached a pitch of intensity in the so-called race riots of June 1943, a two-day-long orgy of killing during which white mobs stormed trolleys or cut their wires, dragged out black passengers, beat or stabbed them, or simply drove their cars to trolley stops and shot those waiting.[152]

After the mayhem, public inquiries descended on Detroit to understand what had happened. They heard whites bemoan a city invaded. 'Negro migrations into communities which cannot absorb them, either on account of their physical limitations or cultural background', should

---

150 Frohardt-Lane, *Race*, for example 7–9, 19–27, 30–5; quotation from 33.
151 Ibid., 35–65; quotation from 45.
152 Ibid., 68–93. Of the thirty-four killed, twenty-four were black, seventeen of whom were killed by police eagerly assisting the white gangs.

be terminated, the US attorney general recommended: 'It would seem pretty clear that no more Negroes should move to Detroit.'[153] More particularly, a torrent of complaints about blacks overtaking public transport and whites hamstrung by the rationing gushed forth. 'One gets in a [street]car – a colored man or colored woman in every seat', one fifty-nine-year-old white woman protested in a letter to the mayor. 'We rather stand then sit with them, not only we folk but plenty more. I have lived in Detroit thirty-eight years [and] in my home thirty-four and the last few years we have seen this sort of thing grow worse and worse. Folk that have their own car don't have to contend with this sort of thing.'[154] 'The colored people is placing themselves better than the white', wrote another white woman. She had an exit strategy: all those coloured people could be evaded in recreational areas outside the city. But, she pointed out despondently, 'it takes gas to go'.[155]

Peace released the pent-up frustration. In the days after Japan surrendered, white car-owners tore up their ration coupons, rushed to the nearest stations to buy as much gasoline as they could and drove around and out of the city for the mere pleasure of it. 'Fill 'er up!' was the motto of peacetime Detroit. The city authorities fomented and satisfied the desire with a master plan for new highways, expressways and parking lots, while public transport – never furnished with resources to accommodate the wartime passengers – was left to decay.[156] The Second World War, Frohardt-Lane concludes, 'was a major environmental opportunity lost'. It could have turned the US away from the private car towards public transport, but white aversion to travelling with black people foreclosed such a shift, which in turn 'dramatically constrained possibilities for addressing the environmental consequences of automobile use'.[157] Detroit was, in this respect, a microcosm of the nation.

Fresh into the peace, in 1947 Congress authorised 37,000 miles of new highways. That was a mere rehearsal for the 1956 Federal-Aid Highway Act, which initiated the largest single construction project

---

153 Francis Biddle quoted in ibid., 93. Note the term 'cultural background'.
154 Mrs B. Hyde quoted in ibid., 1.
155 Mrs. M. Halic quoted in ibid., 96. See further 90–5.
156 Ibid., 102–9, 149–56, 169, 187.
157 Ibid., 13, 11.

in human history: ribbons of highways or 'freeways' – roads exclusively devoted to cars – laid out between states, within cities and to new neighbourhoods around them.[158] A smooth way was paved to suburbia, where whites could isolate themselves from black masses. In the decades following the war, fresh waves of migration brought African Americans to the north, away from lynch mobs and Jim Crow laws and towards the promise of decently paid work, but for every two non-whites who moved into cities, three whites moved out. They settled in the suburbs, kept vanilla by means of discriminatory lending practices, unspoken agreements and explicit covenants against 'Negro invasion'. The flipside was, of course, the ghetto. To clear the way for the roads, bulldozers and wrecking-balls smashed into African American districts, bisected them, reduced them to junkyards – later to be filled with incinerators and dump sites – and accelerated the vicious circle that constituted the black ghettos in the post-war era.[159] As affluent whites withdrew into monochrome serenity, they left behind the urban rot from which the image of the menacing black man could grow, the one space inconceivable without the other. Suburbia became the prime site of twentieth-century American whiteness, literally a *location* where light skin denoted the 'world's highest standard of living' and a safe distance from squalor and crime. Whites went to school and married and consumed alongside other whites, with blacks relegated to a place somewhere else, best escaped and forgotten. The material mirror of segregation made race *look real*.[160]

But segregation could not have been built without 'the highway machine' – which is to say not only the concrete and asphalt, but also

---

158 Amy L. Scott, 'Cities and Suburbs in the Eisenhower Era', in Chester J. Pach (ed.), *A Companion to Dwight D. Eisenhower* (Chichester: John Wiley & Sons, 2017), 117; Simon Pirani, *Burning Up: A Global History of Fossil Fuel Consumption* (London: Pluto, 2018), 84; Jackson, *Crabgrass*, 248–50.

159 Scott, 'Cities', 114–15, 118–19, 123–7; Painter, *History*, 371–2; Lipsitz, *Possessive*, 6–8, 151; Marshall Berman, *All That Is Solid Melts into Air: The Experience of Modernity* (London: Verso, 2010 [1982]), 291–9, 307–8.

160 Martha R. Mahoney, 'Segregation, Whiteness, and Transformation', *University of Pennsylvania Law Review* 143 (1995): 1659; David Theo Goldberg, 'The New Segregation', *Race and Society* 1 (1998): 16, 25; Douglas Kuswa, 'Suburbification', 46–51; Duster, 'The "Morphing"', 117, 129; Scott, 'Cities', 124; Lipsitz, *Possessive*, 7; Omi and Winant, *Racial*, 121.

the automobile as such.[161] That vehicle alone enabled whites to drive past the ghettos, without ever having to sit next to someone black. 'Our cities', said a report from the Eisenhower administration in 1955, 'have spread into suburbs, dependent on the automobile for their existence. The automobile has restored a way of life in which the individual may live in a friendly neighborhood', words not difficult to decipher (by then, the cars-to-people ratio had rebounded to 380).[162] Seen from the opposite side, with Paul Gilroy, 'white flight from urban centres was not just accomplished by means of the automobile, it was premised on it.'[163] Or, with Sikivu Hutchinson, 'the internal combustion engine's post-war resurgence played a key role in transforming black neighborhoods into poster children for American "inner city pathology".'[164] It afforded both the means and the motive for escaping non-white destitution. Compact living and travelling gave way to infinite sprawl, such as in Atlanta, where the greatest distance between two points in the city area had reached 137 kilometres by the first decade of the twenty-first century, compared to 37 in Barcelona, a city with similar population and wealth. Atlanta had eleven times higher CO_2 emissions from transport per capita; in the Catalan capital, every fifth trip was made on foot, whereas in the Georgian the share was too small to be registered.[165]

Switching a city like Atlanta to other modes of transport now seems like an exceedingly difficult undertaking. There is a Metropolitan Atlanta Rapid Transit Authority, abbreviated MARTA, which, however, has been colloquially known as 'Moving Africans Rapidly Through Atlanta'. It was established in the 1960s, at the same time that many whites departed the city centre in the wake of

---

161   Douglas Kuswa, 'Suburbification'.
162   Eisenhower's Advisory Committee on a National Highway Program, also known as the Clay Commission Report, quoted in Scott, 'Cities', 117 (emphasis added); Davis and Boundy, *Transportation*, table 3.7.
163   Gilroy, 'Driving', 94.
164   Sikivu Hutchinson, 'Waiting for the Bus', *Social Text* 63 (2000): 113. See further Sikivu Hutchinson, *Imagining Transit: Race, Gender, and Transportation Politics in Los Angeles* (New York: Peter Lang, 2003), for example 3–10, 40, 79–80, 93, 180; Rebecca J. Kinney, 'The Auto-mobility of *Gran Torino*'s American Immigrant Dream: Cars, Class and Whiteness in Detroit's Post-industrial Cityscape', *Race and Class* 57 (2015): 58–9; Thompson, 'Traffic', 102; Jackson, *Crabgrass*, 241–2, 275–6, 285–9.
165   Pirani, *Burning*, 45.

civil rights struggles. White suburbs then resisted extension of MARTA lines, citing fears that black passengers would venture out of the city on those buses and trains to steal TVs and depress property values.[166] In the northeastern outpost of Gwinnett County, residents voted overwhelmingly against connecting with MARTA in 1971 and then again in 1990: 'That place [central Atlanta] has a reputation for murder and rape. The wrong people. We don't need 'em, we don't want 'em', one man explained at a packed public meeting before the latter referendum. Instead more highways were built for the outliers in the race to emit.[167] Cause and consequence, the underlying attitude broke into local media in 2000, when a reporter accompanied a young white baseball star as he drove his SUV down a multi-lane freeway and expressed his repugnance at the mere thought of public transport. He would never set foot in the New York subway:

> Imagine having to take the Number 7 train to the ballpark, looking like you're riding through Beirut next to some kid with purple hair next to some queer with AIDS right next to some dude who just got out of jail for the fourth time right next to some 20-year-old mom with four kids. It's depressing.[168]

The comments caused an uproar, but the baseball player received support from fans, and the Georgia Highway Contractors Association took the opportunity to run TV ads against what it called 'radical environmentalists' promoting public transport. It showed dilapidated apartment blocks and black people disembarking a bus. A voice-over warned that the rights of Atlantans to drive and live where they wanted were in peril.[169] Then in early 2019, the suburbs were

---

166 Jason Henderson, 'Secessionist Automobility: Racism, Anti-urbanism, and the Politics of Automobility in Atlanta, Georgia', *International Journal of Urban and Regional Research* 30 (2006): 298–300.

167 Doug Monroe, 'Where It All Went Wrong', *Atlanta Magazine*, 1 August 2012; Jenny Jarvie, 'After Decades of Resistance, Atlanta's Booming Suburbs Face a Historic Vote on Public Transit', *Los Angeles Times*, latimes.com, 18 March 2019; Laura Bliss, 'Atlanta's Big Transit Vote Is a Referendum on Race', *CityLab*, 15 March 2019.

168 Quoted in Henderson, 'Secessionist', 297.

169 Ibid., 300.

again importuned to rejoin the city and accept MARTA; as the gateway to integration, Gwinnett County had a third referendum. The naysayers recirculated the classical arguments – trains and buses would be used by 'illegals' aka immigrants; 'we don't want more riff-raff'; rails are an antique technology; 'we just want value for our money' – and carried the day.[170] For the third time, the County voted to keep its distance and stay behind the wheel. By now, it was also a de facto referendum on the climate crisis.

Drawing on the Atlanta case, Jason Henderson has proposed the term 'secessionist automobility' for the process and practice of *seceding from the city of others by means of the car*.[171] Owners can physically separate themselves from perceived foreignness, gender deviance and non-white menace, coil up in their private cocoons and maintain necessary links to the city by commuting. Known as a means for bringing people from place A to B, the car also keeps people X apart from Y, working as wall as much as bridge. But this is to say, of course, that secession depends not only on the car, whose shell would be immobile on its own. It takes gas to go: the white automobile and suburb from the very start presupposed superabundant petroleum products. Ideology did not always mask this fact. In his germinal *Lifeblood: Oil, Freedom, and the Forces of Capital*, Matthew Huber studies hundreds of petroleum advertisements from the 1950s and finds little visual variation: 'Man works, women stay at home, and *everyone is white* and lives in a single-family home with a garage, children, and sometimes pets.'[172] As in

---

170 Bliss, 'Atlanta's'; Jarvie, 'After'; Tyler Estep, 'Fears about MARTA in Suburbs Familiar but Changing', *Atlanta Journal-Constitution*, 21 February 2019; Tyler Estep, 'Gwinnett MARTA Voting Analysis: 5 Takeaways You Need to Know', *Atlanta Journal-Constitution*, 22 March 2019.

171 Henderson, 'Secessionist', and 'The Politics of Mobility in the South: A Commentary on Sprawl, Automobility, and the Gulf Oil Spill', *Southeastern Geographer* 51 (2011): 641–9. Roads through the Amazon would rather seem to represent *integrationist* or *subsumptionist* automobility. An excellent Marxist-Sartrean analysis of the car is Dag Østerberg, *Arkitektur og sosiologi i Oslo: en sosio-materiell fortolkning* (Oslo: Pax, 1998).

172 Matthew T. Huber, *Lifeblood: Oil, Freedom, and the Forces of Capital* (Minneapolis: University of Minnesota Press, 2013), 85. Emphasis in original. Cf. for example Henderson, 'The Politics', 646; George A. Gonzalez, *Urban Sprawl, Global Warming, and the Empire of Capital* (Albany: SUNY Press, 2009).

Margaret Bourke-White's smiling family, but with the blacks out of sight.

'Race', writes Mimi Sheller, 'is a performance of differential mobilities.'[173] People sorted into the privileged location move fast and freely, in and out as they wish, across and beyond borders in pursuit of their interests, while the others are stuck, blocked, hemmed in. The history of race in the US is one long history of differential mobilities: whites crossing the sea to shackle black bodies and hold them on plantations; Jim Crow laws fixing 'Negroes' in certain compartments; mass incarceration keeping an African American population behind bars.[174] 'The highway machine', writes Kevin Douglas Kuswa, reproduced a pattern where 'the have-nots become the move-nots, resigned to remain within a crowded cage contrasted with the adjacent freedom of superhighways and airports.'[175] And first-class mobility has its special energy carriers. Just as the steam-engine enabled white colonisers of the British Empire to enter the dark continents – chase the enemy down a coast, penetrate a river basin – so the internal combustion engine allowed white denizens of the US to exit congregations of black people. Go out and take: withdraw and consume; in both centuries, the stock of energy underwrote the concentration of resources in white locations. But while the steam-engine required a collective operation – no one drove his own railway across the bush – the automobile was a singularly *private* vehicle. It was filled by one man and his family. An extension of the home, the owner selected his company as he would his guests. The gasoline-fired engine sent the individual on his own burst through space, inside a bubble of steel, unbuckled from masses of human bodies: the bourgeois subject in a shell.[176]

Secessionist automobility realises one particular spatial modality of the car. It has reached an apotheosis of sorts in the SUV. The

---

173 Sheller, *Mobility*, 57. Cf. Kendi, *How To*, 166–9.

174 As pointed out in Cotten Seiler, '"So That We as a Race Might Have Something Authentic to Travel By": African American Automobility and Cold War Liberalism', *American Quarterly* 58 (2006): 1093; Cotten Seiler, 'The Significance of Race to Transport History', *Journal of Transport History* 28 (2007): 308.

175 Douglas Kuswa, 'Suburbification', 55.

176 John Urry, 'Inhabiting the Car', *Sociological Review* 54 (2006): 17–31; Huber, *Lifeblood*, for example 43, 74, 112–13; Paterson, *Automobile*, for example 18, 51; Henderson, 'Secessionist', 301.

phenomenal popularity of this vehicle owes much to 'the sense of inviolability that a couple tonnes of steel and fiberglass can instill'. Speaking on behalf of the SUV owners, Don Mitchell describes them as so many almighty atoms:

> Cocooned in a sealed chamber, behind tinted glass, with the temperature fully controlled, and the GPS system tracking, and sometimes dictating, our every turn, our every stop and start, we are radically isolated from each other, able to communicate only through the false connectedness of the cell phone. We ride high and sovereign; we are masters of space; we are safe against all who might intrude, all who might stand in our way (and against the weather, too).[177]

By the twenty-first century, the SUV had become the vehicle of choice for secessionist automobility, the chariot that brought the owner safely to the suburb and the conveyance of yet more rings of sprawl.[178] It had a round of disastrous aggregate effects all of its own. After electricity generation, the largest driver of increasing global CO_2 emissions between 2010 and 2019 was the ballooning SUV fleet – not transport in general, nor car traffic, but specifically the rising number of SUVs rolling down the roads and carrying their bulky, hulking frames by combusting extravagant amounts of gas.[179] Rippling out from American suburbia, SUVs first filled up North Atlantic car dealerships and then spread to other markets, manufacturers scurrying to push and cash in on the trend. It might have been the most gratuitously destructive paroxysm of automobility so far in history. Could it have arisen from any other context than white American secessionism?

---

177 Don Mitchell, 'The SUV Model of Citizenship: Floating Bubbles, Buffer Zones, and the Rise of the "Purely Atomic" Individual', *Political Geography* 24 (2005): 96. This tendency inhered in the automobile already before the rise of the SUV: see for example Freund and Martin, *The Ecology*, 87, and further below.

178 This is briefly noted in the case of the rich's secession into their own enclaves in Britain in Rowland Atkinson, 'Padding the Bunker: Strategies of Middle-Class Disaffiliation and Colonisation in the City', *Urban Studies* 43 (2006): 829.

179 Laura Cozzi and Apostolos Petropoulus, 'Growing Preference for SUVs Challenge Emissions Reductions in Passenger Car Markets', International Energy Agency, iea.org, 15 October 2019.

But by now neither the SUV nor the motorcar was, of course, the exclusive possession of white people. No car dealer in his right mind would turn away a non-white customer with enough purchasing power. Just before the financial crisis of 2008 struck the US, 76 per cent of African American households owned at least one car. The figure for white households, however, was 93 per cent. Blacks were nearly 3.5 times more likely to lack access to a car than whites and almost 6 times more likely to use public transport; in urban areas, 54 per cent of passengers on buses, subways and commuter trains were African American or Latino.[180] To some, it would still seem to be 'a colored man or colored woman in every seat'.

What of Europe? Here the equivalent to black migration to the north, across the Mason-Dixon line, is non-white emigration to the north of the Mediterranean. As in the US, one original pull factor was the demand for labour-power. Once the domestic reserve armies had been drained by the post-war boom, European capital turned towards southern shores and called up people who would later be recognised as Muslims to the points of production; Belgium, for instance, recruited young male workers from Morocco, Tunisia and Turkey to its coal mines, steel plants, construction sites and auto factories.[181] When the boom came to a halt in the early 1970s, those workers were the first to be dismissed. But they had built lives in their adopted homelands, stayed put and were joined by families and relatives. Amid the capitalist downswing, a series of protracted wars stretched through the southern rim – from Algeria to Somalia, Lebanon to Kurdistan, Iraq and Iran and Afghanistan – and activated a whole set of push factors, refugees scrambling to the north for safety. These were some of the processes that rooted non-white populations in European cores once overwhelmingly, if not quite spotlessly, white.

---

180   Robert D. Bullard, Glenn S. Johnson and Angel O. Torres, 'Transportation Matters: Stranded on the Side of the Road before and after Disasters Strike', in Robert D. Bullard and Beverly Wright (eds), *Race, Place, and Environmental Justice after Katrina* (Boulder: Westview Press, 2009), 66–9.

181   Hassan Bousetta and Laure-Annes Bernes, 'Muslims in the EU: Cities Report: Belgium', EU Monitoring and Advocacy Program and Open Society Institute, 2007, 11–12.

And gradually, patterns of segregation resembling those in the US became facts of life in European cities, due to the same kind of white flight: 'immigrant areas' in inner cities and housing estates were vacated by white residents, who concentrated in upscale neighbourhoods and outer suburbs. Darkening in their complexion, the former places earned a reputation as crowded, crime-infested sinkholes, with poor schools, no jobs, shabby cars.[182] In Ghent, a Belgian city maintaining a good deal of industrial production, people of Moroccan and Turkish descent were crammed into the nineteenth-century districts in the centre. In 2015, a team of researchers reported the following sentiments from whites who had relocated to the suburbs: 'I moved away because of the immigration of allochthons, just like my parents.' 'Too many blacks. Beware, I am not a racist, but at my age [seventy-three] I don't feel safe among them.' 'It used to be a nice neighbourhood, but now it has become Little Turkey. Friday evening, it's all mosque hustle there.' One man offered a testimony from behind the wheel: 'That's already a foreign country there. In the past, it was not like that, but now it's terrible. Actually, we drove through this week and my wife told me "I really don't want to live here".'[183] Folk that had their own car did not have to contend with that sort of thing.

If secessionist automobility is still an under-researched phenomenon in the US, however, it is almost completely unstudied in Europe. What role the car has played in segregating this continent is simply not known. Public transport is woven into the urban fabric to a far greater extent than in the US; the typical European city still looks more like Barcelona than Atlanta, car dependence is less extreme, the average cars-to-people ratio about a fourth lower. But dim traces of

---

182 See for example Anne Power, 'Social Inequality, Disadvantaged Neighbourhoods and Transport Deprivation: An Assessment of the Historical Influence of Housing Policies', *Journal of Transport Geography* 21 (2012): 39–48.

183 Nick Schuermans, Bruone Meeus and Pascal De Decker, 'Geographies of Whiteness and Wealth: White, Middle Class Discourses on Segregation and Social Mix in Flanders, Belgium', *Journal of Urban Affairs* 37 (2015): 486–9. On the general pattern in Belgium, cf. Lex Veldboer, Reinout Kleinhans and Jan Willem Duyvendak, 'The Diversified Neighbourhood in Western Europe: How Do Countries Deal with the Spatial Distribution of Economic and Cultural Differences?', *Journal of International Migration and Integration* 3 (2002): 48–50.

the same patterns have surfaced in some research. Interviews with suburbanites in Berlin suggest that they prefer to travel by car because it relieves them of encounters with strangers. From Milan, one study describes public transport as a space of face-to-face interaction with the whole panoply of human characters, enjoyed by some and disliked by others. 'You see the weirdest people, the most incredible races, I mean . . . people I'd expect to see in Bangkok or Manila, but I find them in Milan', said one man in his sixties, explaining his preference for public transport before going into particulars: 'It fascinates me, notably black people: their clothes, their hair, their way of carrying themselves, their behaviour, it makes me really curious, that's all.'[184] The study did not catch the opposite preference on tape. From general surveys, we know that European cities have become less compact over the past half century, dispersed suburbs now regularly housing more residents than the centres; that the car has been instrumental to the trend; that the effects on emissions and land use and social cohesion have been broadly similar to those in the US; that car-dependent suburbs have become citadels of the right.[185] As in so many other respects, the old world has drawn closer to the new.

But the logic of secessionist automobility is most often discernible only via proxy indicators, as in Sweden, a country where the car occupied an unusually central position in early twenty-first-century climate politics. Because it had no fossil fuels of its own and imported a limited amount to generate electricity, the share of emissions emanating from transport here stood at one-third, above the global average of one-fourth; any transition in Sweden would mean the final retirement of much of the car fleet. Use of the vehicle was unevenly distributed. On an average day, one-fourth of the

---

184 Giulio Mattioli, 'Moving through the City with Strangers? Public Transport as a Significant Type of Urban Public Space', in Timothy Shortell and Evrick Brown, *Walking in the European City: Quotidian Mobility and Urban Ethnography* (Abingdon: Routledge, 2016), 57–74 (quotation from 66; unpublished Berlin study referenced on 62).

185 For example Power, 'Social'; Anne Power, 'Regional Politics of an Urban Age: Can Europe's Former Industrial Cities Create a New Industrial Economy to Combat Climate Change and Social Unravelling?', *Palgrave Communications* 4 (2018): 6–7, 11; Walks, 'Driving', 19; Alan Walks, 'Driving the Vote? Automobility, Ideology, and Political Partisanship', in Walks, *The Urban*, 199–220.

population – predominantly outside of cities – accounted for nine-tenths of the travelled kilometres.[186] So-called *invandrare* tended to congregate within cities. They had lower levels of car ownership.[187] The cities had developed patterns of hyper-segregation: white people with high incomes 'choose to live closely together with others in the same category', whereas 'people with origins outside of Europe and in Muslim countries are over-represented in low-income neighbourhoods, without noticeable alterations over periods of residence.'[188] Descent itself appeared to lock them in place. Colour lines cut through Swedish cities as if some invisible Jim Crow hand had painted them, with tunnels and highways amply serving the brighter sides.

One could study Malmö, a compact post-industrial city, probably the most stigmatised in northern Europe, subject of a never-ending stream of rumours about Muslim takeover and apocalyptic criminality. In 2001 – coincidentally close to the baseballer uproar in Atlanta – the microprocesses of hyper-segregation in Malmö were laid bare by two pieces of investigative journalism. One representative of the municipal housing company related the experience of allocating apartments to immigrants in, 'well, let's call them "white" areas – it doesn't work at all. And that's because of the "whites" who live there. If there is someone dark-skinned from Iran or Iraq moving in with lots of children, they will be bullied', and hence the company had learned to concentrate the non-whites in separate areas.[189] In the second incident, city authorities had dared to offer accommodation to two families from Rosengård, the most maligned of all districts, in

---

186 Lena Winslott Hiselius and Lena Smidfelt Rosqvist, 'Segmentation of the Current Levels of Passenger Mileage by Car in the Light of Sustainability Targets: The Swedish Case', *Journal of Cleaner Production* 182 (2018): 331–7.

187 Jonathan Rokem and Laura Vaughan, 'Geographies of Ethnic Segregation in Stockholm: The Role of Mobility and Co-Presence in Shaping the "Diverse" City', *Urban Studies* (2018), online first.

188 Roger Andersson, 'Etnisk och socioekonomisk segregation i Sverige 1990–1998', in *Välfärdens förutsättningar: Arbetsmarknad, demografi och segregation*, SOU 2000:37, 262; *Det blågula glashuset: Struktural diskriminering i Sverige*, SOU 2005:56, 329. These public inquiries into structural racism in Sweden were flukes of a short-lived political conjuncture in the first years of the millennium; after the SD entered parliament, discussion of the issue was utterly squelched.

189 As quoted in *Det blågula*, 332.

an empty day-care centre in the wealthy suburb of Bunkeflostrand. The families in question had 'expressed a strong desire to get away from Rosengård' and 'come into contact with Swedish children'. Revolt broke out in the villas. Thirty-six of the forty-eight households closest to the vacant building signed a petition against the 'experiment', and a particularly committed couple composed its own that read: 'Bunkeflostrand is one of the last outposts near Malmö to escape to if you want to avoid Arabs around the corner.' Were the relocation to go ahead, the couple would 'move southwards to areas protected by like-minded', a reference to the ultra-rich outer suburbs, to which columns of affluent whites seceded around the turn of the millennium, voting not so much with their feet as with their steering wheels.[190]

In 2019, one of these suburbs produced a promotional video that once again put the process in the spotlight. It begins with a blond woman and her daughter running the gauntlet on the streets of Möllevången in Malmö – the only central neighbourhood in a major Swedish city, as of the late 2010s, that retained racial diversity and a working-class character, incidentally also the one urban district in the country that voted most heavily for the Left Party in the 2018 elections (45 per cent of the votes). A gang of vaguely immigrant-looking men in hoodies taunt the woman and her daughter and throw a bottle at them. These scenes are filmed in black and white. But light, warm and yellow as rapeseed, falls on the family as it flees to Staffanstorp, where, the voice-over advertises, 'you can feel like one of the team'. Here everyone is white. Everyone? When the film was shot, the authorities of Staffanstorp removed the actual owner of a kiosk – an immigrant himself – and replaced him with a white man who could act the part and sell ice cream to the white family. Safe in all-smiling, all-white Staffanstorp, the woman drives through her new life in an SUV. It is a Porsche Cayenne.[191]

---

190 Ola Rothenborg, 'Villaägare säger nej till muslimer', *Dagens Nyheter*, 5 June 2001. For all things Malmö, see Ståle Holgersen, *Staden och kapitalet: Malmö i krisernas tid* (Göteborg: Daidalos, 2017); for residential segregation, Ann Rodenstedt, *Living in the Calm and Safe Part of the City: The Socio-Spatial Reproduction of Upper-Middle Class Neighbourhoods in Malmö*, dissertation (Uppsala University, 2014).

191 'Informationsfilm om Staffanstorp', YouTube, 11 November 2019.

We can then propose another hypothesis. *Climate change has burst onto the scene of European politics just as secessionist automobility attracts a segment of the white population as never before*, a coincidence that forms one of many invisible shoals on which mitigation efforts run aground. It is another murky pool of history where the far right is blooming. If it remains to be mapped out as such, we know for a fact that *the far right is the most strident defender of the car*, as we have seen in some cases already, from the Trump administration loosening fuel-efficiency standards to Björn Höcke's branch of the AfD demanding the removal of speed limits. After tucking away their boots and bomber jackets, the SD had few extra-parliamentary branches. But one of them was 'SD Motor', an association for members – virtually exclusively men – practising motorsports and feeling the need to promote various aspects of car culture. The chair of SD Motor, Josef Fransson, was also for a long time the party's climate spokesman, responsible for much policy formation on the issue.[192] During the election campaign of 2018, chairman Åkesson again and again expressed his disdain for a programme of subsidies for electrical bicycles, in a tone that suggested that this was the sissiest mode of transportation ever invented: real Swedish men drive cars. In their main staging area in the southern countryside that defined itself in opposition to the foreign country of Malmö, the Sweden Democrats lined up their EPA tractors and SD Motor went to bat for the automobile. 'To drive and own a car is freedom', was a typical phrase; or, 'the car is our chance to live in our own *hembygd*', the Swedish equivalent of *Heimat*, roughly 'native soil'.[193] Party ideologist Richard Jomshof, the man who warned that Sweden was about to perish because of Islamisation, attributed the most recent successes of the SD to Swedes (that is, white ones) leaving Malmö and other cities to get away from the immigrants, moving to the countryside, in the process becoming dependent on the car and appreciating the SD for combatting high gasoline prices.

192 sdmotor.se; on Fransson's role, see Baas, 'SD-politik'; Hultman et al., 'The Far', 124–6. He preceded Kinnunen on the post.
193 Anna-Lena Lodenius and Mats Wingborg, *Slaget om svenskheten: Ta debatten med Sverigedemokraterna* (Stockholm: Premiss, 2009), 207–8.

Those departing the Muslim-infested cities flocked to the only party 'trying to instill some sobriety into the climate debates'.[194] It takes gas to go.

The FrP fashioned itself not only the oil party, but the car party of Norway. It always supported a road to be built somewhere. During its time in government, road construction over the Norwegian hillsides sped up, still financed through tolls. The FrP catered to the republic of drivers with these roads, but it failed to free them of the tolls, and for this perceived betrayal, voters punished it with the worst result in two decades in the local elections of 2019, while a new party to the right – so far of a single-issue character: 'People's Action: No to More Road Tolls' – ate into its support base.[195] The Finns wanted to make gasoline cheaper and considered car driving a national right. 'Recognising all the frustrations of congestion, driving a decent car remains the most fun you can have sitting down', tweeted Roger Helmer of UKIP.[196] Vox stressed the 'need to respect the freedom of using private means of transport'. As soon as the coalition of that party and the conservatives had seized Madrid from the left in the 2019 elections, it scrapped – a first in Europe – a major low-emissions zone limiting car traffic.[197] (A conservative spokesperson had branded the zone 'pure communism'.)[198]

And over in Germany, the country with the largest car fleet in Europe, the AfD sensed where the wind was blowing. As part of its 2019 anti-climate offensive, it produced a one-hour YouTube

---

194 Johan Delby, '"Många flyttar för att de inte uppskattar invandringen"', *Dagens Samhälle*, 13 June 2019.

195 For example Kjetil Gillesvik and Anders Haga, 'Bompengepartiet med ny sjokkmåling', *Bergens Tidende*, bt.no, 15 March 2019; Daniel Boffey, 'Driven to Despair: Road Toll Charges Take Centre Stage in Norway Vote', *Guardian*, 8 September 2019. It might be recalled that FrP itself started out as an anti-tax party – the wheel keeps rolling.

196 Roger Helmer's Twitter account, 10 October 2018.

197 Rocío Monasterio (Vox Madrid) in Marta R. Domingo and Tatiana G. Rivas, 'Rocío Monasterio: "Llevaría el Orgullo Gay y fiestas similares a la Casa de Campo"', *ABC Madrid*, 28 January 2019; Arthur Neslen, 'Madrid Could Become First European City to Scrap Low-Emissions Zone', *Guardian*, 31 May 2019; Sam Jones, 'Madrid's New Rightwing Council Suspends Low-Emissions Zone', *Guardian*, 18 June 2019.

198 Isabel Díaz Ayuso from Partido Popular in *El Plural*, 'Twitter se ríe del PP por tildar de "comunista" el Madrid Central de Carmena', elplural.com, 7 December 2018.

documentary called *Dieselmord im Ökowahn* – roughly, 'Ecological Delusion Murdering Diesel'. Deploring the recent bans on old diesel cars in some cities, the film defended automobility as a mode of travel that can never be replaced by public transport. 'If you want to drive a car, you should drive a car' – 'mobility is a fundamental right', the automobile the armour of 'self-determined life', restrictions on driving tantamount to expropriation. But the film ended on an positive note: Germans can keep away from these murderous delusions and enjoy their life on four wheels out in the *Heimat*.[199] In the late 1990s, activists coined the slogan *kein mensch ist illegal* – 'no human being is illegal' – which later spread across Europe, wherever movements sprung up in solidarity with refugees. In 2018 and 2019, the AfD countered with its own campaign: *kein SUV ist illegal* and *kein diesel ist illegal*, the posters this time showing white hands on the fuel pumps. By these substitutions, the party conveyed a bundle of messages: some (non-white) human beings are indeed illegal; cars in general and SUVs in particular should go free from all intereference; these machines are worth more than the humans in question, the former negating the latter.

As a rule, then, the far right objected to *any* limits on the use of cars, across the varying terrains of social and racial formations. 'Racism is like a Cadillac; they make a new model every year', Malcolm X is said to have quipped.[200] It would indeed appear that the car has provided some raw material for the far right in the early twenty-first century. But as we shall soon see, the history runs deeper still, for fascism itself was born in a combustion engine.

The Possessive Investment in the Substratum of Whiteness

Ideologies, says Althusser's simplest definition, are systems of representation. Images, myths, ideas and concepts are ordered within their frames. In ideologies, people express the way they live in imaginary form – but, Kukla updates Althusser, not by blowing free-floating

---

[199] AfD, 'Dieselmord im Ökowahn – Die Diesel-Dokumentation', YouTube, 29 September 2019.

[200] For example George Lipsitz, '"Swing Low, Sweet Cadillac": White Supremacy, Antiblack Racism, and the New Historicism', *American Literary History* 7 (1995): 701.

bubbles of ideas; rather, 'ideologies are built into practices and the material environment'. As representations, they are not posterior to people's lives, the way a developed photograph is to the scene captured. Present at beginning and end, ideologies 'play a role in *constituting* reality'.[201] Their general mechanism is, of course, for both philosophers, interpellation, but this seems to privilege the act of shouting, as though ideologies always come pumping at high volume – 'Hey, you there!', 'Moses!', 'Heave-ho!', 'Fall in line!' Kukla points out, however, that 'there need be no literal speaker who issues the interpellative hail.' She gives the example of public bathroom signs asking visitors to enter the toilet of their gender. The police can demand that a man stops by flashing blue lights at him. God also beckoned to the Israelites as a cloud of pillar and fire, and merely by showing up with a crown or a motorcade or some other emblem of authority, a ruler can demand that his subjects clear the way.[202]

If we open for the possibility of interpellation as a summons without sound, a process of recruitment that does not have to yell or speak in imperatives, we can indeed think of it as the general mechanism of ideology: when a couple glide through an immigrant neighbourhood in Ghent, the practice of car-driving and the material environment of the segregated city interpellate them as whites, even before a word is said. Or, imagine Harriet Martineau arriving in Calcutta in the 1850s, resting in a hotel and receiving refreshments from native servants and then entering a travel agency with English signs on the door. Ticket in hand, she would have boarded the train and rolled out on the Indian countryside. At every step on this trip, she must have moved through a hall of interpellations asking her to perform certain acts *as a white person* and, recognising herself in the entreaties and offers, more fully come to be just that. The steam-whistle shrieking in the villages would have been the exclamation mark, the babies' cries in response to the noise the audible proof of English superiority as a system of representation enacted on the rails.

But then there are also very vocal and verbal interpellations: a late nineteenth-century ethnologist lecturing on the Aryan race, a

---

201   Althusser, *For*; 198, 200; Kukla, 'Slurs', 9. Emphasis in original.
202   Kukla, 'Slurs', 15; Wolf, 'The Problem', 249, 252.

far-right party in the early twenty-first demonstrating against imminent Eurabian apocalypse. We seem compelled to introduce distinctions between first- and second-order interpellations, primary and derivative, materially embedded and fully politicised. One could imagine a scale from the inaudible to the ear-splitting, but other senses than sound would have to be included. The substantive point, following Althusser, is that the dominant ideology is realised in the ISAs, but 'it comes from elsewhere' – or, following Poulantzas, 'ideological relations always have roots which go beyond' those apparatuses, and the roots '*always consist in relations of power*'.[203]

This is obviously not to suggest that everyone travelling on a steamboat or driving a car by that very act becomes a white supremacist, any more than every piece of lithium is spontaneously transformed into batteries. But some raw material there must be in the ground. Indeed, the far right is *particularly* dependent on such pre-existing material. One could listen to someone who should know: 'The appeal to feelings is the mainstay of nationalism and puts it in an advantageous position vis-à-vis other ideologies focusing on the political organisation of society, making proposals and offers based on cool and rational considerations.' This is not a scholar of fascism writing, but Santiago Abascal. Honest in his own way, he continues: 'The mobilising power of national feelings and identities, *often irrational*, is greater than that of any other political and moral principle.'[204] The audience of the far right learns nothing new; if anything, it learns to unlearn rational thinking. But it must already be able to hum the tune. As observed by Adorno, nationalists can play on its 'psychological instruments' only if there is a 'susceptibility' among their listeners.[205] Because both enriched their Marxism with psychoanalysis, Althusser and Adorno were on the same page here, but the latter also studied actual fascists and noted how they, more than any other propagandists, just as Abascal acknowledges, seek to conjure political sympathy by pandering to the rawest prior emotions. The far

---

[203] Althusser, *Lenin*, 126; Poulantzas, *State*, 37. Emphasis added.
[204] González, 'Abascal'. Emphasis added. For the typical scholarly argument, see Jason Stanley, *How Fascism Works: The Politics of Us and Them* (New York: Random House, 2019), 68, 71.
[205] Adorno et al., *The Authoritarian*, 619.

right does not try to transcend the present with appeals to reason; it reaches out with cues and stimuli. It aims at 'winning people over by *playing on their unconscious mechanisms* rather than by presenting ideas and arguments'.[206] Paradoxically, this very irrationality of the far right requires it to stay close to the prevailing order. It can only be effective if there is raw material in the unconscious its interpellations can connect to. Such material is a product of history, and it can become more readily available in a crisis.

We have glimpsed into two reservoirs of fossilised whiteness: moments in history when the location of whiteness has stabilised through the combustion of fossil fuels. Like every construction of race, they were founded on violence, incidents like the pulverisation of Akka and the 'race riots' in Detroit issuing in the emergent patterns. Racist ideologies were built into and reconfirmed by such acts of aggression, through a characteristic mix of fiction and fact; steam power accorded substantial advantages to their owners, but it was not an agent of extinction on its own. No First Nations ever expired just by feeling awe and paralysis at the sight of an engine. In fact, the very perception of steam as the epitome of English or European or white ingenuity could only come about through the suppression of certain realities: most concretely, all those engines would have screeched to a halt without the palm oil squeezed from the hands of Nigerian women. That material was no exception. As Joseph Inikori has shown in *Africans and the Industrial Revolution in England*, the achievements on which the 'civilised races' so prided themselves would scarcely have come about without the raw materials, overseas markets and, of course, slave labour extracted from Africa and its diaspora.[207] European technomass was the embodiment of biophysical resources transferred from some locations and into others, but it was fetishised as a set of magic artefacts with

---

206 Adorno, 'Anti-semitism', 223, 228, 219. Emphasis in original. See further Theodor W. Adorno, 'Freudian Theory and the Pattern of Fascist Propaganda', in Andrew Arato and Hike Gebhardt (eds), *The Essential Frankfurt School Reader* (Oxford: Blackwell, 1978), 121–5, 134.

207 Joseph Inikori, *Africans and the Industrial Revolution in England: A Study in International Trade and Economic Development* (Cambridge: Cambridge University Press, 2002).

powers of their own.[208] All traces of the black sources of steam and its accessories were expunged. From that primary reification, it was but a short step into the second-order interpellations of techno-racism: the idea that steam manifested nothing so much as *the power of the white race* that owned it.

The automobile was no different from the steam-engine in this regard. It was because Japan seized the rubber plantations in Malaya and the US lost 97 per cent of its supplies that private cars had to be parked for four years. The fleets and infrastructures of motorcars would have been impossible without a global reordering of biophysical resources across colour lines. The SUV, the highest stage of antisocial transport thus far, only intensified the consumptive relations between peripheries and core: the greater the flows of rubber and bauxite and cobalt and oil, the more the owner could fancy himself armoured in solitude.[209] In the grip of extreme fetishism, he imagined the vehicle as endowed with powers of its own and an immanent freedom from others.

The ideology of automobility – secessionist by extension – has a radius far greater than anything at the point of extraction. The car is the 'ur-commodity', writes Gilroy, the arch-product of twentieth-century fossil capital and the only machine that people live privately within while on the move.[210] Like no other commodity – think, by way of comparison, of a hammer, a radio, a tennis racket, a lamp – it carries its driver through life and takes him to his right location by combusting the stock, internally; no other commodity has as much energy *and* racial significance. If the steam-engine was the motor of differential mobility in the nineteenth century and the internal combustion engine that in the twentieth, however, only the latter has come into direct conflict with the imperatives of mitigating climate change. Here the automobile is significant not only for its emissions,

---

208 Following Alf Hornborg, for example *The Power of the Machine: Global Inequalities of Economy, Technology, and Environment* (Walnut Creek: Rowman & Littlefield, 2001); *Global Ecology and Unequal Exchange: Fetishism in a Zero-Sum World* (London: Routledge, 2011); *Global Magic: Technologies of Appropriation from Ancient Rome to Wall Street* (Basingstoke: Palgrave Macmillan, 2016).
209 Frohardt-Lane, *Race*, 123–4; Mitchell, 'The SUV', 96–7.
210 Gilroy, 'Driving', 89.

but just as much for its ideology of anti-collectivity. The motorist wants to be *left alone*, preferably have the road to himself as he speeds forth. 'We are now, truly', Mitchell writes, continuing the SUV-owners' anthem, 'the liberal, autonomous subject. We own ourselves and no one can intrude upon us without our permission.'[211] The car, in other words, exudes the ideology *most detrimental to any efforts to cut emissions* – an ideology of form as much as content, abiding no rationing or accommodation of foreign others.

But if we are broadly right about the function of the automobile in segregating the US and Europe, this ideology should also be expected to veer further to the right. George Lipsitz has theorised 'the possessive investment in whiteness' – the tendency for privileged whites to hold on to their position and defend its perks – and if the automobile has been a material substratum of whiteness, just as the steam-engine once was, then we should expect some of that possessive investment *to be transferred to the substratum itself*.[212] Far-right interpellations would then be reproductive, reactive – as in countering a threat – and very loud indeed. The danger of fossil fascism is the danger of a reboot of these older articulations of race and energy, on the way down the slope of crisis.

Poitiers by Petroleum

The steamboat and the motorcar were not the sole technologies to make the white burner. Nor did the fossilisation of whiteness come to a restful end with either. On the road, the white driver bumped into an unforeseen obstacle in late 1973: from one month to the next, the price of crude oil shot up from $24 to $55 a barrel. It had hovered around $20 for the entire post-war period, allowing the US and Western Europe to treat mass consumption of petroleum as a given. Now motorists were sweating in motionless queues to gas stations; President Nixon advised Americans to drive less and switched off the lights at Washington monuments, while the Swedish government

---

211   Mitchell, 'The SUV', 97.
212   Lipsitz, *Possessive*.

distributed ration coupons. Someone had, all of a sudden, grabbed the white burners by the throat. Who?

In Said's *Covering Islam*, this is the watershed of modern Islamophobia. Before the oil crisis, Islam and Muslims-as-defined-by-their-Muslimness scarcely figured in Western politics and media, but now 'we', as in we Westerners, 'could no longer drive our cars the way we used to; oil was much more expensive; our comforts and habits seemed to be undergoing a radical and most unwelcome change.' A notion of creeping Muslim mastery took hold. 'The Muslim world seemed once more on the verge of repeating its early conquests' – Poitiers at the gas station – and 'the whole west seemed to shudder'.[213] One Swedish ethnologist has recalled 'a father-in-law *in spe*, who under the first days of the oil crisis swore over what "those Arabs were up to". The shock of the sudden threat to his love for driving disabled him from explaining what exactly "those Arabs" were up to, but whatever it was, they had no right to do it.'[214] The first crisis of world capitalism to be centred on a fossil fuel was also the first to be attributed to Muslims.

In late 1973 and the following few years, when the price of crude parked around $60, 'Muslims' were roughly synonymous with 'Arabs' in the imagination of the cursing drivers. This changed in the second shock, the Iranian Revolution, which doubled the price again, up to $125 per barrel in early 1980, an even tighter grip on the pumps – and these were no mere Arabs. The marching masses in Tehran were cast as pure Muslims defined by a congenitally violent and frenzied Muslimness. To add insult to injury, they stormed the American embassy and held fifty-two soldiers and diplomats hostage for 444 days, parading them blindfolded, treating them fairly well – no torture was ever recorded – and for the US, this was the height of

---

213 Said, *Covering*, 36–7, 5. See further for example Melani McAlister, *Epic Encounters: Culture, Media, and US Interests in the Middle East since 1945* (Berkeley: University of California Press, 2005), 134–6; Peter Gottschalk and Gabriel Greenberg, *Islamophobia: Making Muslims the Enemy* (Lanham: Rowman & Littlefield, 2008), for example 3, 37–8, 112–23; Malcolm Brown, 'Comparative Analysis of Mainstream Discourses, Media Narratives and Representations of Islam in Britain and France prior to 9/11', *Journal of Muslim Minority Affairs* 26 (2006): 299–300; Huber, *Lifeblood*, 106–8.

214 Magnus Berg, *Hudud: En essä om populärorientalismens bruksvärde och världsbild* (Stockholm: Carlssons, 2000), 9.

excruciating national humiliation. The ABC network gave the drama rolling coverage under the tag 'America Held Hostage'. On one estimate, one-fifth of all TV broadcasting in the US in 1980 concerned this crisis, which transfixed audiences: a trauma for the empire.[215] It has not recovered from it. When Trump in his brinkmanship with Iran in early 2020 tweeted the threat to bomb fifty-two cultural sites in the country, he meant to pick one for every American man held hostage forty years prior.[216] But there appears to be a wider psychological scar from the oil crises of the long 1970s: *the privilege of fossil fuel consumption is threatened by foreigners who wish us harm.*

This, at least, is the premise for the Eurabia theory. It was Bat Ye'or's intellectual achievement to turn the situation – Muslims denying us the fuel we want – into a project for the extinction of Europe. One could think that the far right should have rather reacted by developing a narrative of Muslims selling us the oil, oil heating up the planet, the planet including our national homes, and propose a transition to renewables. But insofar as energy independence from Muslims has been promoted, it has been on the basis of domestic fossil fuels, as with Hamm and Trump. If Islamophobia would have had precedence over fossilophilia – if that can be a word – the far right might have made different choices. But the sense lingering after the oil crises seems to have been one of entrenched and armoured entitlement. Not many years separated that moment from the emergence of the climate crisis.

You Remind Me of Something

So the car was the locus of the encounter with the Muslim oil crisis, but cars have not, of course, remained lily-white. In 2016, the US still reigned supreme with 20 per cent of the world's registered motor vehicles; China came second with 15 (although this figure included trucks and buses).[217] Within the US, black people had developed a car

---

215 Said, *Covering*, xxii, 25; McAlister, *Epic*, 198–208, 233.
216 On Trump's abiding obsession with the hostage crisis, see David A. Graham, 'The Iranian Humiliation Trump Is Trying to Avenge', *Atlantic*, 7 January 2010.
217 Davis and Boundy, *Transportation*, tables 3.2–3.

culture of their own. Already in wartime Detroit, the inner space of the motorcar had an alluring safety from racist harassment. African Americans had special reasons to appreciate unhindered movement from one place to another.[218] Victims of differential mobility will strive to get into the fast lane; as Gilroy has noted in a melancholy critique of black automobility, precisely the original exclusion of African Americans from the car and all it represented made them prone to consume it as an object of compensatory prestige. The liberation movements of the 1960s and 1970s and their cultures cared little for it: 'You may not have a car at all, but remember brothers and sisters, you can still stand tall', sang William DeVaughn in 'Be Thankful for What You Got' from 1974. But in 1995, R. Kelly recorded the song 'You Remind Me of Something', where the object of sexual desire is (like) a car: 'You remind me of a jeep, I want to ride it', and 'Girl you look just like my car, I want to wax it, and something like my bank account, I want to spend it, baby.'[219]

For Gilroy, the valorisation of the car in popular black culture is but one symptom of the collapse of the emancipatory projects of the 1968 conjuncture, and much the same could be said for the global South. But the urge to get into the driving seat can, of course, be found earlier. 'Coal! Coal! Coal! That is the one thing needful for me', Muhammad Ali is reported to have exclaimed after the war of 1840.[220] As these lines are written, the largest coal-fired power plant in the Middle East and possibly also on the African continent is being constructed in Abdel Fattah al-Sisi's Egypt (with assistance from Chinese engineering firms). It will have a capacity larger than Poland's Bełchatów, although the coal will be imported, chiefly from South Africa.[221] As for Nigeria, 1960 was the year of formal independence from the British Empire: and of the discovery of oil.

As Nigeria became independent, Fanon put the finishing touches to *The Wretched of the Earth*. In the last chapter, he called upon his 'comrades' in the Third World to stop emulating the ways of white folks.

218 Frohardt-Lane, *Race*, 10–11; see further Seiler, ' "So That We" '.
219 Gilroy, 'Driving', 87–8, 91, 95.
220 Andrew Archibald Paton, *A History of the Egyptian Revolution, from the Period of the Mameluks to the Death of Mohammed Ali* (London: Trübner & Co, 1863), 239.
221 See for example the tag for 'Hamrawein' at *Energy Egypt*.

Europe now lives at such a mad, reckless pace that she has shaken off all guidance and reason, and she is running headlong into the abyss; we would do well to avoid it with all possible speed . . . Let us decide not to imitate Europe; let us combine our muscles and our brains in a new direction . . . Humanity is waiting for something other from us than such an imitation, which would be almost an obscene caricature.[222]

One of the greatest ecological tragedies in modern history must be that Fanon's admonition went unheeded. But that does not appear to have broken off fossil substrata from the location of whiteness. The generalisation of 'European achievements, European techniques and the European style' rather seems to have corroborated their superiority, just as Fanon expected.[223] One could turn to another student of whiteness, Bolívar Echeverría, who observes it through the optic of a Latin American reception of the Frankfurt School and sketches a theory of its ecology: an accident of history, whiteness is based on the violent diffusion of technologies for dominating nature. Because their original owners happened to be light of skin, it has become the badge of capitalist modernity. To be modern is to be white, if only in an acquired sense. When the waters are calm and the cruise safe, certain people with darker skins can be invited onboard, provided that they demonstrate irreproachable capitalist and modern credentials; Echeverría mentions Michael Jackson, Condoleezza Rice, the figure of Uncle Tom; one could add R. Kelly and al-Sisi. More open and inclusive, this *mestizo* version of whiteness validates the exemplarity of the white way of domination. But if a crisis breaks out, it might relapse into exclusion, a strict criterion for racial whiteness reasserting itself. Echeverría's main example is German fascism.[224]

In the case of the climate crisis, the generalisation of Europe gives the far right another opportunity. It allows it to blame the non-whites when in the mood for recognising that there is indeed a problem. Thus the top candidate of the SD in the elections to the European

---

222   Fanon, *Wretched*, 252, 254.
223   Ibid., 254.
224   Bolívar Echeverría, *Modernity and "Whiteness"* (Cambridge: Polity, 2019 [2010]), 40–51.

parliament in 2019 declared that the average temperature on earth had stood still for eighteen years, that there was unwarranted 'climate hysteria', that Europe should commit to no climate policies *and* that it was the responsibility of the *Chinese* to reduce CO_2 emissions.[225] Trump, the AfD, the PS and others have made the same move: global heating is a hoax, and the ones heating up the planet are the Chinese, the Indians or some other faraway bunch. For the parties running fastest into the abyss, white is always right, the nation never wrong.

The Gender and Art of Ecological Sadism

Now if race is at the heart of the far right, so is gender. From Poland to Brazil, the parties and presidents that swept to power in the 2010s adopted remarkably similar postures of anti-feminism, the same lines bouncing word for word through their galaxy. Something called 'gender ideology' threatened to undermine the Polish as well as the Brazilian family and nation; a catch-all term for perceived challenges to the norm of husband and wife and rose-cheeked children, it became flesh in the endangered child. Our boys risk being forced to wear dresses, said the state-led campaign against 'gender ideology' in Poland.[226] Our kids are being indoctrinated into homosexuality, alleged one film clip that circulated widely during Bolsonaro's presidential campaign. More precisely, the Workers' Party had distributed baby bottles with penis-shaped nipples in day-care centres for infants to suck on. Thus they would be turned into gay people.[227] In the days

---

225 SVT, 'Toppkandidaterna: Peter Lundgren', 16 May 2019.

226 Eileen Boris, 'Gender Troubles, Redux', *Women's History Review*, online first (2019): 3; cf. Weronika Grzebalska and Andrea Pető, 'The Gendered Modus Operandi of the Illiberal Transformation in Hungary and Poland', *Women's Studies International Forum* 68 (2018): 166. For a genealogy of the theory of 'gender ideology', Elizabeth S. Corredor, 'Unpacking "Gender Ideology" and the Global Right's Antigender Countermovement', *Signs* 44 (2019): 613–38.

227 See for example Gilmar Lopes, 'É verdade que o PT de Haddad distribui mamdeira erótica nas escolas?', *E-Farsas*, 27 September 2018. The Brazilian campaign must be put in the context of the virulent misogyny in the attacks against and impeachment of Dilma: see Joseph Jay Sosa, 'Subversive, Mother, Killjoy: Sexism against Dilma Rousseff and the Social Imaginary of Brazil's Rightward Turn', *Signs* 44 (2019): 717–41.

after Bolsonaro's inauguration, one minister made it clear that from now on 'girls will be princesses and boys will be princes', while a televangelist close to the president told a cautionary tale about Sweden, where 'gender ideology' had made it mandatory for schoolchildren to wear gender-neutral orange uniforms, and if any parents dissented, they were deported to isolated islands in the North Atlantic.[228] The theory of 'gender ideology' was apocalyptic: this enemy too was 'destroying the very foundations of our civilization', in PiS lingo; the family is 'the basis of the nation's survival', according to the Fundamental Law promulgated by Fidesz. The theory was intensely conspiratorial, 'gender ideology' framed as foreign to the nation, imposed from the outside, part of the masterplans of Soros or the Frankfurt School. It was based on denial of yet more layers of reality. 'Glass ceilings exist only in women's heads', said one representative of the PiS government.[229] It called for a strong white man to root out the weeds.

In the panics about the wrong sort of babies being born, gender and race met again in the bedroom. '"New Germans?" We'll do it ourselves', said the heading of one AfD poster for the 2017 election, positioned above a woman, white, smiling and pregnant. Only fertile women and vigorous men could beat back the Muslim conquest.[230] German men, Björn Höcke explained, needed to regain their forceful and proud masculinity; nature had equipped them with 'feistiness, wisdom and leadership', as against the 'intuition, gentleness and dedication' of their women.[231] In analogy with green nationalism, quite a few far-right personalities chose to play the cards of gender equality and LGBT rights against immigrants in general and Islam in particular. That too was a vaguely insincere exception to the rule. The

---

228 Henrik Brandão Jönsson, 'Brasiliansk TV-präst: "Sverige tvingar skolelever att gå i orange uniform"', *Dagens Nyheter*, 18 January 2019.

229 Quotations from Grzebalska and Pető, 'The Gendered', 166–8.

230 See further Barbara Umrath, 'A Feminist Reading of the Frankfurt School's Studies on Authoritarianism and Its Relevance for Understanding Authoritarian Tendencies in Germany Today', *South Atlantic Quarterly* 117 (2018): 861–78; Banu Gökariksel, Christopher Neubert and Sara Smith, 'Demographic Fever Dreams: Fragile Masculinity and Population Politics in the Rise of the Far Right', *Signs* 44 (2019): 561–87.

231 Quoted in Oliver Kreuger, 'Der Militante Maskulinismus', *Die Tageszeitung*, taz.de, 24 March 2019.

myth of palindefence tended to evoke a past where men were real men and women not yet seduced by feminism and call for the warrior-man to yet again defend his woman from the *other* man, as on the AfD slave market. And just as with race, this, of course, had ramifications in ecology.

There is an even longer history of women being associated with nature and animality, but as Carolyn Merchant has shown in her foundational *Death of Nature: Women, Ecology, and the Scientific Revolution*, the workings of that association altered as Britain developed what we now recognise as the first fossil economy. In pre-fossil times, organic nature was held in some respect and even considered inviolable, rather like a mother. When extraction took off, such inhibitions had to be thrown to the side, and instead there emerged the ideal of subjection through entrance: 'There are still laid up in the womb of nature many secrets of excellent use', Francis Bacon wrote excitedly. Miners were 'searching into the bowels of nature', for truth and fortune were 'hid in certain deep mines and caves'. Digging up coal was coded as a masculine activity, a penetration of the resistant female body, nature now less of a Madonna than a whore – a 'common harlot', as suggested by Bacon, who exhorted English men to 'lay hold of her and capture her'.[232] Since then, drilling for fossil fuels has frequently been staged as a performance of muscular masculinity, and so has their burning: 'The explosive power of combustion could be crudely equated with virility', notes Cara Daggett.[233] Cars and men go a long way back.[234] Walking and bicycling have something effeminate about them, not to mention being bound to rails; whereas the stock is rigid or spurts, the flow is more like an uncatchable wave.[235] Giving up on the stock would, for some, it seems, mean emasculation.

And so climate denial also builds on fossil substrata of masculinity and centuries of interpellation of men, turned into a roar in the mouth of, for instance, Jordan Peterson, the self-proclaimed defender of the man under feminist siege:

---

232 Quoted in Carolyn Merchant, *The Death of Nature: Women, Ecology, and the Scientific Revolution* (San Francisco: HarperCollins, 1990 [1980]), 169–71.
233 Daggett, 'Petro-masculinity', 20.
234 See for example Paterson, *Automobile*, 47–8.
235 Cf. Theweleit, *Male*, for example 244, 283, 403, 429–30.

Most of the global warming posturing is a masquerade for anti-capitalists to have a go at the Western patriarchy. That's partly why the climate change thing for me is a contentious issue, because you can't trust the players. You can't trust the data because there is too much ideology involved.[236]

We have seen Pascal Bruckner shiver at a report supposedly claiming that men exhale more carbon than women. As a matter of fact, there are scientific studies indicating that men in advanced capitalist countries drive longer, eat more meat and work in the dirtiest industries – steel, oil, construction, auto – although some of that driving and working also deliver material benefits to the women of their households.[237] Patriarchy might indeed have something to do with the problem. In accordance with the logic of system justification, denial is an overwhelmingly male thing.[238] The anti-climate politics of the far right here again appears as the edge of a deeper structure. It represents a brutalised version of 'industrial masculinity' – or, following Daggett, a 'hypermasculinity' or even 'petro-masculinity', although the prefix 'petro' omits coal from the mix; perhaps we should also speak of fossilised masculinities.[239] When the manhood of cars and coal and big bloody steaks is in dispute, one reaction will be bellicose reassertion of it. The cornered man will strike back, overreact.

Daggett gives the example of 'rolling coal'. A subcultural trend that took off during Obama's second term, its basic practice is to retrofit a

---

236   In Greg Callaghan, 'Right-Winger? Not Me, Says Alt-Right Darling Jordan Peterson', *Sydney Morning Herald*, 21 April 2018.

237   For example Marjorie Griffin Cohen, 'Gendered Emissions: Counting Greenhouse Gas Emissions by Gender and Why It Matters', *Alternate Routes* 25 (2014): 55–80; Annica Kronsell, Lena Smidfelt Rosqvist and Lena Winslott Hiselius, 'Achieving Climate Objectives in Transport Policy by Including Women and Challenging Gender Norms: The Swedish Case', *International Journal of Sustainable Transportation* 10 (2016): 703–11.

238   See Rachel E. Goldsmith, Irina Feygina and John T. Jost, 'The Gender Gap in Environmental Attitudes: A System Justification Perspective', in Margaret Alston and Kerri Whittenbury (eds), *Research, Action and Policy: Addressing the Gendered Impacts of Climate Change* (Dordrecht: Springer, 2013): 159–71, and references in footnote 105, page 294.

239   On industrial masculinity, see Martin Hultman and Paul M. Pulé, *Ecological Masculinities: Theoretical Foundations and Practical Guidance* (London: Routledge, 2018), 40–4. The gender equivalent of capitalist climate governance is 'ecomodern masculinity' – picture Elon Musk in his Tesla car – as theorised in ibid., 45–51.

diesel truck so that extra fuel floods the engine. The driver then releases the half-burnt gas – surplus to the requirements of the vehicle – in the form of giant clouds of sooty fume. Thus he can spew smoke or 'roll coal' on whomever he wants to target. Not the cash-strapped proletariat, practitioners might spend thousands of dollars on adding smokestacks and further modifying fuel systems to maximise the exhaust just for the sake of it; if on social media, they will share pictures and videos of their accomplishments. At the peak of the trend, just before Trump came to power, bicyclists and drivers of hybrid or electric cars were common targets on US streets. Compilations of 'hot babes' getting bursts of smoke in their faces earned the genre the label 'pollution porn'. 'It's an adrenaline rush', one roller explained to the *New York Times*. 'The air sucks anyway' and 'smoke's pretty. I like seeing it', said another.[240] But whatever other statement they delivered, rolling coal was fundamentally anti-climate pageantry: 'That's my way of giving them a finger. You want clean air and a tiny carbon footprint? Well, screw you,' in the words of one seller of stack kits in Wisconsin.[241] As of 2020, one YouTube clip called 'Rolling Coal on Protesters Compilation: Black Lives Matter, Trump Haters, Tree Huggers' had garnered nearly 5 million views. Dudes driving their pickups up to African Americans, feminists and environmentalists and laughing as they bellow smoke on them: intersectional hatred performed through supererogatory combustion.

Something of the same attitude was on display in the video produced by the Finns – perhaps the European far-right party most skilled in visual propaganda – for the 2019 national election, in which it was a hair's breadth from becoming the largest party. It purports to be a film about the state of Finland. Over an image of a sunrise over Helsinki, a deep, manly voice tells the story of a nation once inhabited by happy people who had paid heavily for their independence and loved their homeland. Then their leaders, appearing as men in villainous masks, betrayed them, trampled on their memories, robbed them with taxes and 'opened the gates to a flood of people'

---

240   Elizabeth Kulze and Eric Eyges, '"Rollin' Coal" Is Pollution Porn for Dudes with Pickup Trucks', *Vocativ*, 16 June 2014; Hiroko Tabuchi, '"Rolling Coal" in Diesel Trucks, to Rebel and Provoke', *New York Times*, 4 September 2016.
241   David Weigel, 'Rolling Coal', *Slate*, 3 July 2014.

represented by the face of a black man. A young blond woman screams as she is taken away to be raped. Regular Finns lose their jobs and end up homeless. Then the peripeteia arrives: a middle-aged man slams his fist into the breakfast table as he reads the news and the voice deepens one notch, to the sound of dramatic strings: 'Their anger began to form into an energy' – *energia* in sonorous Finnish – 'that was thick and black.' From half a dozen apartment buildings, enormous plumes of smoke rise up, as though every Finnish household has become a power plant. The sky is a black ceiling filled with the exhaust from the people's rage. 'One night, this dark energy reached its critical mass.' The crust of the earth itself opens up and a monster emerges from the crack – 'the embodiment of being pissed off' – his body sooty, smouldering, glowing like a mass of coal. 'The defender of the people' overpowers the masked leaders and seizes their limousine. At that point, he breathes out a cloud of sparks.[242]

What is going on here? More than climate denial: perhaps climate *refusal*, as Daggett suggests; an angry, incandescent reinvestment in the fossil economy, no longer bothering to pretend otherwise. China Miéville has written about 'social sadism', a bourgeois culture of revelling in the misfortunes of the poor and laughing at them – a 'deliberate, invested, public or at least semi-public cruelty'. Perhaps what we have here is a concordant *ecological sadism*. It extracts pleasure – emotional, sexual, aesthetic, political pleasure – from the despoliation of nature. It makes ordinary fossil fuel consumption a boisterous and braying infliction of cruelty on the planet's body. The volume has been gradually turned up in the early millennium: at his last G8 summit in 2008, George Bush said 'goodbye from the world's biggest polluter' and then 'punched the air while grinning widely.'[243] Stars of the US right could, Miéville notices, chant 'Drill baby drill!' and call for their followers to 'rape the planet – it's yours. That's our job: drilling, mining and stripping.'[244] In the burning

---

242 Verkkomedia Perussuomalaiset, 'V niin kuin ketutus', YouTube, 20 March 2019.
243 Robert Winnett and Urmee Khan, 'President George Bush: "Goodbye from the World's Biggest Polluter"', *Telegraph*, 9 July 2008.
244 China Miéville, 'On Social Sadism', *Salvage* no. 2 (November 2015), 19, 23, 40. 'Rape the planet': Ann Coulter, see also for example Heather Digby Parton, 'Ann Coulter Has Fallen from Grace – and the Reason Why Is Terrifying', *Salon*, 29 June 2015.

summer of 2019, Jair Bolsonaro claimed that the Amazon is 'like a virgin that every pervert from the outside lusts for' – and in the next breath, 'the Amazon is ours, not yours'. The patent logic was that the Amazon was for Bolsonaro and men of his kin to fuck.[245] Asserted in this manner, the destruction appears to have value of its own, in excess of the exchange-value it generates.

Something of the same ecological sadism was, again, on display in the far-right bullying of Greta Thunberg. In April 2019, she gave her typical speech to the European Parliament: 'I want you to panic. I want you to act as if the house was on fire', she began. Four minutes into the speech, she relayed some basic info on the collapse of natural systems, with 'up to 200 species going extinct every day.' At this point her voice faltered. 'Erosion of fertile topsoil, deforestation of our great forests . . .' – now she cried – 'toxic air pollution, loss of insects and wildlife, the acidification of our oceans . . . these are all disastrous trends accelerated by a way of life that we in the financially fortunate parts of the world see as our right to simply carry on with.' She took a pause. She swallowed hard; a speaker handed her a glass of water. The day after, Thierry Baudet tweeted a picture of Thunberg in tears and wrote: 'Haha.'[246] He laughed at her. When she crossed the Atlantic Ocean in a sailing yacht, Arron Banks, the primary donor to UKIP and its breaking-point Brexit campaign, tweeted his wishes: 'Freak yachting accidents do happen in August . . .'[247] He wanted her to drown. Bolsonaro called her a 'brat'; Trump mocked her as – among other things – 'a very happy young girl looking forward to a bright and wonderful future. So nice to see!'[248] And in her homeland, inevitably, one man was convicted of having threatened to rape Greta Thunberg and kick her with steel-toe boots and shoot

---

245 Letícia Casado and Ernesto Londoño, 'Under Brazil's Far-Right Leader, Amazon Protections Slashed and Forests Fall', *New York Times*, 28 July 2019.

246 Thierry Baudet's Twitter account, 17 April 2019.

247 *Telegraph*, 'Arron Banks under Fire for Greta Thunberg "Joke" about "Freak Yachting Accidents"', 15 August 2019. For this man's many achievements in life, see Ed Caesar, 'The Chaotic Triumph of Arron Banks, the "Bad Boy of Brexit"', *New Yorker*, 18 March 2019.

248 Bianca Britton, 'Greta Thunberg Labeled a "Brat" by Brazilian President Jair Bolsonaro', CNN, 11 December 2019; Aaron Rupar, 'Trump's Tweet about Greta Thunberg Is One of His Ugliest Yet', *Vox*, 24 September 2019.

her with a gun and kill her mother too if she didn't publicly announce that it is good to fly.[249] No serious observer could miss the gender dimension of the desire to shoot this particular messenger down: a girl telling the men in power that their time is up; a spokeswoman of a planet that can't take them anymore. Inflicting various forms of cruelty on her became a source of pleasure. If such sadism was indeed astir, it would be the far right returning to the roots of fascism.

---

249 Rebecka Jonsson, 'Sparreholmsbo dödshotade Greta Thunberg', *Ekuriren*, ekuriren.se, 19 December 2019.

10
For the Love of the Machine

Fascism might be the only ideology in modern history with a fixed place and date of birth. On 23 March 1919, some one hundred people gathered in the meeting room of the Milan Industrial and Commercial Alliance to found the Fasci di Combattimento and 'declare war on socialism'.[1] One of the men present described Mussolini entering the hall with a truncheon in his belt: 'The commander had arrived. How could we any longer doubt our victory?' The particular founder who wrote these words was given command over some of the squadrons, and three weeks later, the time had come for their first action. The police in Milan had dissolved a rally of workers and killed one in their ranks. In response, trade unions called for a general strike on 15 April. On the morning of that day, the streets and piazzas lay quiet. Decorated officers and war veterans joined the column of the Fasci, swelling to perhaps one thousand men, marching through the town in search of the striking workers, revolvers at the ready. After several false alarms, they finally caught sight of their prey: 'Here comes the march, with women in front, holding high a picture of Lenin and carrying the red flag. – A defiant column, with confident, rapid steps moves forward. It comes to a halt behind the line of carabinieri. We

---

1 Paxton, *The Anatomy*, 5.

are on one side, they are on the other.'[2] The carabinieri gave way and the fascists charged ahead, swinging their sticks and clubs and firing their pistols into the mass, felling bodies left and right for about an hour, before proceeding to the building of *Avanti!* The main socialist publication in Italy, *Avanti!* had been leading the agitation against Mussolini and his followers; now they streamed into the offices and set them on fire. Fascism was born in this ecstatic action, as described by one of its original founders and most fertile minds: the poet and propagandist Filippo Tommaso Marinetti. His writings serve as a kind of double helix for the relation between fascism and fossil fuels.

Ten years before the founding of the Fasci and the raid on *Avanti!*, Marinetti penned his famous Futurist Manifesto. It has, in the words of Walter Benjamin, 'the merit of clarity'.[3] In the preamble, the author narrates a scene where he is driving his car at maximum speed, when suddenly a pair of cyclists come towards him and gesticulate that he is on the wrong side of the road. 'Their stupid uncertainty was in my way . . . How ridiculous! What a nuisance!' Marinetti crashes into a ditch, but he manages to revive his car and formulate the eleven points of the manifesto, the first of which promises to declaim a song 'about the use of energy'. The fourth point reads:

> We believe that this wonderful world has been further enriched by a new beauty, the beauty of speed. A racing car, its bonnet decked out with exhaust pipes like serpents with galvanic breath . . . a roaring motorcar, which seems to race on like machine-gun fire, is more beautiful than the Winged Victory of Samothrace.

The manifesto goes on to 'sing the praises of the man behind the steering wheel'. The eleventh and last point sings 'of intrepid steamships that sniff out the horizon' and 'of railway stations, voraciously

---

2 F. T. Marinetti, *Critical Writings* (New York: Farrar, Straus & Giroux, 2006), 287, 293.

3 Walter Benjamin, *Selected Writings, Vol. 3, 1935–1938* (Cambridge, MA: Harvard University Press, 2002), 121. Although Benjamin is here referring to another of Marinetti's manifestos.

devouring smoke-belching serpents' and 'of workshops hanging from the clouds by their twisted threads of smoke'.[4]

With this text, Marinetti set the scene for his futurist poetics and fascist politics, held together by the worship of the machine – not an instrument for achieving this or that end, but a dazzling end in itself. 'Combustion engines and rubber tires are divine', he would proclaim, not in jest. 'Gasoline is divine. So is religious ecstasy inspired by one hundred horsepower.'[5] The motorcar for Marinetti was a cult object, the burnt fuel an elixir that lent novel vitality to man. A typical dithyramb of his read:

> Vehement god of a race of steel,
> space-intoxicated Automobile,
> stamping with anguish, champing at the bit!
> O formidable Japanese monster with eyes like a forge,
> fed on fire and mineral oils,
> hungry for horizons and sidereal spools,
> I unleash your heart of diabolic puff-puffs[6]

Straddling the nineteenth and twentieth centuries, Marinetti also idolised the steamboat and the locomotive; indeed, every kind of engine seems to have put him into a trance. The religious and erotic aspects of fetishism were here fully at one. 'We', Marinetti wrote,

> are promoting love of the machine – that love we first saw lighting up the faces of engine drivers, scorched and filthy with coal dust though they were. Have you ever watched an engine driver lovingly washing the great powerful body of his engine? He uses the same little acts of tenderness and close familiarity as the lover when caressing his beloved.[7]

---
4 Marinetti, *Critical*, 12–14.
5 Ibid., 256.
6 F. T. Marinetti, *Selected Poems and Related Prose* (New Haven: Yale University Press, 2002), 38.
7 Marinetti, *Critical*, 85.

In Marinetti's utopia, the embrace between man and machine would be so complete as to produce one body. There would appear a post-human race, an immortal mechanical or cyborg species, in which all impulses to goodness and affection would be abolished. Harnessed to 'the all-conquering zeal of the motor', it would 'quite naturally be cruel'.[8] Anyone who stood in the way would have to suffer its fury. 'One must persecute, lash, torture all those who sin against speed.'[9] Whence the overwhelming force? 'The bowels of the earth', Marinetti sang, 'will disgorge the monster of speed.'[10]

The violent destruction of nature was here something to pine for. Marinetti poured his scorn on rivers and valleys; he wanted to see the mountains decapitated; he dreamed of filling the meadows with billboards and piercing the hills with trains and making the Danube run in a straight line at 300 kilometres per hour. Nature for him was an enemy that had to be wrestled to the ground and paved over at once. The finest evidence of progress and apex of beauty was the sight of smoke:

> We have great centres which are aflame day and night, breathing their huge fires all over the open countryside. We have soaked with our sweat a whole forest of immense mill chimneys, whose capitals of stretching smoke hold up our sky, which wishes to be seen as nothing but a vast factory ceiling,

although Marinetti, in a rare moment of humility, admitted that 'we don't as yet know what to do with the rebellion of imprisoned gases that are angrily tossing about beneath the leaden knees of the atmosphere'.[11] Perhaps he imagined there would be a way to crush that rebellion too.

---

8 Ibid., 46, 86.
9 Quoted in Anne Bowler, 'Politics as Art: Italian Futurism and Fascism', *Theory and Society* 20 (1991): 774.
10 Quoted in Fernando Esposito, *Fascism, Aviation and Mythical Modernity* (Basingstoke: Palgrave Macmillan, 2015), 298.
11 Marinetti, *Critical*, 170, 224. Cf. for example 11–12, 44–5, 163, 224–5, 245, 253–4; Marinetti, *Selected*, 16, 44. See further Andrew Hewitt, *Fascist Modernism: Aesthetics, Politics, and the Avant-Garde* (Stanford, CA: Stanford University Press, 1993), 143–8; Marja Härmänmaa, 'Futurism and Nature: The Death of the Great Pan?', in Günther Berghaus (ed.), *Futurism and the Technological Imagination* (Leiden: Brill, 2016), 337–60.

Nature was not the only thing Marinetti disliked; he avowedly scorned feminism and women. He was horrified by Italian women leaving their households to work in public. But there would be no danger in giving them the vote, for they were in 'an absolute state of inferiority from the point of view of intelligence and character'. Taking misogyny to characteristically dizzying heights, Marinetti argued that young boys should be kept separate from girls so as to not be infected by their softness and fantasised about a form of parthenogenesis in which men could bypass 'the inefficient vulva' and beget sons by themselves. He strained himself to combine such passionate homoeroticism with equally passionate homophobia. One reason to execrate communism was its sheer queerness: 'If I were a Communist, I should be worrying about the next war between homosexuals and lesbians, who will then join forces against normal men.' But normal men loved not so much women as their machines, the objects to which all lust had to be transferred. The woman is a creature tied to nature, 'whose dreamlike mane of hair extends into the forest'; like the rest of nature, she 'impedes the march of men'.[12]

Marinetti dreamed of a mechanical existence free from the botheration of nature and women.[13] The vehicle that seemed to allow for the most breath-taking of such liberation was the aeroplane. In the First World War, aeroplanes rode highest, above the anonymous mass of soldiers, like knights duelling in the sky. If trench warfare represented a faceless horror show, pilots were heroic men on winged horses, who projected their power with a hand on the joystick; blending bourgeois individuality with patriotic duty, they were the martial aristocracy of the new age.[14] Italian fighter pilots performing sorties elicited Marinetti's 'limitless joy' – the 'joy that one belongs to the same race as those magnificent fliers, who now have the know-how to keep absolute control of the skies'. He had to try it himself. One of his manifestos begins with him 'in an airplane, sitting on the fuel

---

12 Marinetti, *Critical*, 55–6, 342. See further Barbara Spackman, *Fascist Virilities: Rhetoric, Ideology, and Social Fantasy in Italy* (Minneapolis: University of Minnesota Press, 1996), 8–14, 52–5, 74; Härmänmaa, 'Futurism', 339; Bowler, 'Politics', 772–5.
13 Hewitt, *Fascist*, 150.
14 Esposito, *Fascism*, for example 184–230, 269–70.

tank', having coitus with the machine 'above the mighty chimney stacks of Milan'. His poems could holler – 'Hurrah! No more contact with the filthy earth!' – and praise life in one's habitat of choice: 'I leap from branch to branch / through the great illusory forests of smoke / that rises from the factories.' The smoky sky was superior to the ground like man to woman, the verticality confirming the dominance. Marinetti declared his 'hatred of the earth (vertical mysticism)' and stated his 'longing to break free', once and for all, of this pathetic planet.[15]

The men on the march, however, also expressed their love for a higher deity: the nation. If the machine was a fetish object for Marinetti, it was placed on the altar of the nation, or – a synonym – the race. As it happened, the Italian race had been blessed with a very special mechanical genius. Marinetti preached an 'ultraviolent nationalism', which 'means above all else fortifying national industry and commerce and intensifying the development of our intrinsic qualities as a race in the forward march of our victory over competing races'.[16] This race must hatch new members at speed; while waiting for the miracle of parthenogenesis, it had to maximise its number of 'ardent males and inseminated females. Fecundity, for a race like ours, is its indispensable defence.' Marinetti held democracy in contempt. Parliaments were places of 'negro drumtalk, and the flapping of windmills'.[17] He glorified the colonisation of Libya. A witness to the aerial bombing of that country in 1911 – the first deployment of aeroplanes in war – he considered it 'the most beautiful aesthetic spectacle of my life'. He derided people so dim as to be unmoved by the sight of technology, '*like the Arabs who looked on indifferently at the first airplanes in the skies over Tripoli*'.[18] He is probably most famous for his yearning for war. Naturally, he hated Marxism, as much for its threats to manliness as for its ideals of universal peace and the splitting of classes; workers ought to know their place and

---

15 Ibid. 308, 324; Marinetti, *Critical*, 107, 258; Marinetti, *Selected*, 39, 44; Ernest Ialongo, 'Filippo Tommaso Marinetti: The Futurist as Fascist, 1929–37', *Journal of Modern Italian Studies* 18 (2013): 403.
16 Marinetti, *Critical*, 105, 421. See further Ialongo, 'Filippo', 394–5, 400, 405–7.
17 Quotations from Spackman, *Fascist*, 12; Marinetti, *Critical*, 57.
18 Conversi, 'Art', 92; Marinetti, *Critical*, 121. Emphasis in original.

defer to the elite in charge. 'Increase inequality among mankind', was one of his watchwords. All Marinetti wanted was for the dominant order to realise its inner potential: he craved 'an Italy of big business, of huge markets, rich and great in herself', only with 'more locomotives, more factory chimneys'.[19] Business-as-usual – just more of it, and faster.

For all his bluster, Marinetti reveals a sort of DNA code of fascism. The basic strands are all there in his early writings: anti-Marxism, anti-feminism, racism, imperialism, the preoccupation with fertility, the promotion of inequality, accolades to violence and a giddy myth of palingenesis.[20] With some of his exaggerations, he would find few followers. There were no fascist programmes to bypass the inefficient vulva, and Marinetti's desire to see the leaden weight of Italian history and tradition exploded – to the point of demanding that the 'stinking little canals' of Venice be filled in and replaced with 'factories plumed with smoke' – jarred with nationalist sensibilities.[21] But in most other respects, he was a loud mouthpiece for classical fascism, not the least in his views of the machine. In *Fascist Virilities: Rhetoric, Ideology, and Social Fantasy in Italy*, Barbara Spackman has argued that Marinetti's early writings prefigured fascism by centring everything on the virtue of virility, the one point condensing all other 'interpellations – religious, political, familial, racial, and gender'.[22] But virility here rather seems derivative of the machine. It is not because they are virile that Marinetti's men use machines: only *if* and *when* they take hold of the steering wheel and put the pedal to the metal do they really become so.

---

19   Marinetti, *Critical*, 355, 240. See further Daniele Conversi, 'Art, Nationalism and War: Political Futurism in Italy (1909–1944)', *Sociology Compass* 3 (2009): 97; Ialongo, 'Filippo', 395.

20   On the palingenetic aspects and Marinetti's influence on Mussolini, see Roger Griffin, 'The Multiplication of Man: Futurism's Technolatry Viewed through the Lens of Modernity', in Berghaus, *Futurism*, 77–99; Griffin, *A Fascist*, 7; cf. Jorge Dagnino, 'The Myth of the New Man in Italian Fascist Ideology', *Fascism* 5 (2016): 130–48. On Marinetti's futurism prefiguring other themes of fascism, cf. Bowler, 'Politics', 776, 785.

21   Quoted in Ialongo, 'Filippo', 405. Marinetti, one could say, was more genetic than palin. Mussolini balanced the components better, with his combination of futurist modernism and *romanitá*. On the latter, see Romke Visser, 'Fascist Doctrine and the Cult of the *Romanitá*', *Journal of Contemporary History* 27 (1992): 5–22.

22   Spackman, *Fascist*, xi; cf. for example 2, 52.

We shall argue, in contradistinction to Spackman, that Marinetti's ideas and classical fascism rather gained a distinctive unity through a particular relation to the machine, as a nodal point in which the ensemble of interpellations came together.[23] Whatever enemies got Marinetti riled up – workers and Bolsheviks, women and lesbians, Arabs and Turks, pacifists and internationalists and democrats – his response was invariably to unleash the violence of the productive forces against them. The machine, for him, was the omnibus vehicle for upholding and increasing inequality between humans, because *it had the material power to dominate and destroy*. Clearly this did not apply to any old mechanical device. Marinetti was not turned on by a windmill or a trolley. It was the machine that exploded into action by burning its fuel, crushing underfoot its opponents and shooting man through time and space that intoxicated him: fascism was born in the love of combustion.

Now, Marinetti might have been a maniac, but the northern Italian bourgeoisie harboured feelings similar to his. It came late to modern industry. In 1900, Italy barely produced three hundred cars a year, fewer than in France by two orders of magnitude. Then the race began: by 1909, year of the Futurist Manifesto, Turin had twenty car manufacturers, Milan fifteen; Fiat, the largest among them, produced some 1,500 cars per year. When Marinetti bought his 60 horsepower Fiat and set out on the country roads, he partook in a general craze for racing cars that symbolised the belated arrival of capitalist technology *and he would tolerate no hurdles in their way*.[24] He and his futurists soon-to-be fascists positioned themselves as the shock troops of fossil capital. For them, striking workers and gesticulating bicyclists were vexations that had to be cleared off for the energy to burn freely; indeed, Marinetti defined speed, his favourite magnitude, as 'disdain for obstacles'.[25] The verbs that structured his language were 'unleash', 'release', 'let loose', all the better when

---

23 A similar argument about the place of the machine is advanced in Hewitt, *Fascist*, 134–7, 142–6.

24 Tim Benton, 'Dreams of Machines: Futurism and l'Esprit Nouveau', *Journal of Design History* 3 (1990): 19–23.

25 Esposito, *Fascism*, 311.

accompanied by force – 'Release the breaks! . . . You can't? . . . Smash them then! . . . Let the engine's pulse centuple!'[26] This was a poetics for capital in general. More particularly, it was a poetics for the capitalist class of northern Italy, so intolerant of brakes and limits because the task of adapting the country 'to the latest bourgeois technology still had to be accomplished' – that is, because it came late, and because it had a rising working class and a limited colonial base to grapple with.[27]

Marinetti marched one step ahead of this class, truncheon in hand, precisely because he made the machine more than a tool of accumulation. He was not primarily beholden to this or that bank account or rate of profit (although he did inherit a fabulous fortune from his father, a Piedmontese businessman and lawyer, by which he funded the futurist art empire). Art theorist as he was, he turned the machine into the supreme aesthetic object, and as Benjamin argued in his reading of Marinetti, this move was constitutive of fascism itself: this '*aestheticizing of political life*' and, we should add, of the materiality of the productive forces.[28] So fundamental to the advances of capitalist industry, the machine was here raised to a new level – no longer a bearer of exchange-value, but itself the highest value. It consummated *l'art pour l'art*, production for its own sake and, particularly when it came to nature, destruction for the sake of destruction, and with such a political programme, fascism truly became the avant-garde of fossil capital in the two main countries where it formed. It was exceptionally committed to the task of winning the race in record time.

Hellish Fire Can Keep Going

If the above interpretation rested on Marinetti only, it would, of course, be exceedingly weak. But he had a kindred German soul in Ernst Jünger. In 1914, at the age of nineteen, this man rushed to the

---

26 Marinetti, *Selected*, 38–9.
27 Rosenberg, 'Fascism', 170.
28 Benjamin, *Selected*, 121. Emphasis in original.

front, to return four years later with multiple wounds and decorations and the stuff for a stellar literary career. The war was the *ne plus ultra* of human existence. Nothing compared to the thrill of shells exploding and machine guns rattling and fellow men hurtling themselves against cannons. All began in 'clouds of gas and smoke'.[29] 'To live means to kill', announced Jünger; to live on the edge and slay enemy soldiers conferred special qualities on men: hardened in the 'storm of steel', they came out on the other side as a metallic nobility.[30] Marinetti, who fought on the opposite side of the First World War, could not have agreed more; war, he would endlessly repeat, is 'the sole cleanser of the world', or the sole 'hygiene' or 'purgatory', baptising the best by fire and burning away the unfit.[31]

Jünger elaborated this killer instinct into a romance of what he called 'total mobilisation'. Modern war was not the act of an individual soldier or platoon or even army. It had to mobilise the sum total of the productive forces and transmit them as one giant current of destruction onto the battlefield: 'All energies of great industrialised countries with their factory centres, transportation capabilities and armies of machines had erupted' in the *Produktionskrieg*. The blaze shot forth from the crater's vent, but its power rose from within the material base:

> Somewhere deep from the rear one by one supply carts are rolling on shiny rails, in all factories driving belts are humming, there is feverish mining for coal and ore in all mines, blast furnaces are on fire day and night in industrial zones. Supply is well maintained, hellish fire can keep going as long as it has to.[32]

Extensively branched circuits of energy were projected into the theatre of war, where they detonated in a larger-than-life spectacle of combustion. The warriors of the new age were machine operators,

---

29 Ernst Jünger, *Storm of Steel* (London: Penguin, 2004 [1920]), 83.
30 Nikolaus Wachsmann, 'Marching under the Swastika? Ernst Jünger and National Socialism, 1918–33', *Journal of Contemporary History* 33 (1998): 575; Ernst Jünger, *Interwar Articles* (n. p.: Wewelsburg Archives, 2017, archive.org), 49.
31 Marinetti, *Critical*, for example 14, 49, 52, 61.
32 Jünger, *Interwar*, 28, 36; cf. *Storm*, 69.

attacking from inside tanks and aeroplanes and artillery pieces with ever greater power. Total mobilisation was the art of tightly co-ordinating the transformation of energy – or, as Jünger articulated its general formula, 'the growing conversion of life into energy', suggesting that life itself must be dissipated – and the type of energy he had in mind is obvious to any twenty-first-century reader.[33] War was so wonderful because it concentrated the burning of fossil fuels in one central pyre. 'This is the material. Vast industrial districts with coal mine headframes and the nocturnal glow of blast furnaces before the eyes', Jünger would write and rewrite in ever more rapturous prose; 'all coal and steel, oil, explosives and electricity, manned by specialised positions from admiral to boiler heater', working towards the point of purposeless destruction, which, for him, was in itself the greatest purpose: he felt 'unbelievably lucky' at having seen it firsthand.[34] He took pleasure in the cultic event. Much like Marinetti, he conceived of war as an aesthetic and metaphysical rite, a moment of pulverisation in which the machine proved its prowess to the heavens.

But what if – god forbid – there was no war going on? What joy could there be in peacetime? Luckily, the notion of total mobilisation implied that some things in ordinary life had sheen too. In 1926, at a point in life when he was as close to Adolf Hitler as he would ever be, their bromance intense, the writer dedicating his book *Fire and Blood* to the 'national leader', Ernst Jünger published an essay entitled 'The Machine'. There he depicted the world of machines as enchanted. Some people on other continents live among colourful birds in virgin forests, but we grow up amid caravans of machines – 'a magical landscape that surpasses any fantasies of *One Thousand and One Nights*. Here we feel at home. It would not be a mistake to say that we live in a fairy world.' Anyone who suggested that the machine was to blame for spiritual ills was profoundly mistaken; only man himself was at fault, for not realising that 'higher

---

33   Jünger, *Interwar*, 126. Cf. Bogdan Costea and Kostas Amiridis, 'Ernst Jünger, Total Mobilisation and the Work of War', *Organization* 24 (2017): 479–83.

34   Esposito, *Fascism*, 254; Jeffrey Herf, *Reactionary Modernism: Technology, Culture, and Politics in Weimar and the Third Reich* (Cambridge: Cambridge University Press, 1984), 83; Jünger, *Interwar*, 29.

and more serious goals' were to be found *in* the machine. The error lay in viewing it 'as a mere means of production', when it was the deepest source of mystery and exaltation men could ever know. Critics should also keep in mind

> that the motion of machines is of a coercive character. It runs over whomever stands in its way, becoming a means of destruction. Any protests will crash against its steel shell, like the protest of English factory workers who revolted against the use of the first steam-engines.[35]

There could be no alternative but to side with the machine and take delight in it. The exhilaration of the moving front line had its counterpart in the unstoppable advance of civil machinery. Jünger himself had little difficulty in finding the magical moments, when a frisson akin to that of the battlefield would run though him:

> There are such moments when we truly are proud of that, which is called progress. Let's remember the intoxicating joy that overwhelms the modern man at the sight of his own creations burning gigantic amounts of energy in the sky above the metropolises.[36]

In the times of Weimar, Jünger never stopped haranguing his many readers on these themes. He came up with a sublime vision of 'work', defined as the dissipation of ever greater quantities of energy: work 'is like fire, consuming and transforming everything combustible'; if we throw ourselves into it unreservedly, 'we begin to get a feeling for the high temperatures' and the 'monstrous furnace'.[37] One must partake in the rite of this everyday miracle. 'Its blood is the fuel

---

35 Jünger, *Interwar*, 42, 44 (translation slightly corrected). Hitler and Jünger in 1926: Wachsmann, 'Marching', 581. On Jünger's influence on the far right and relation to Hitler in those critical years, see also Elliot Y. Neaman, *A Dubious Past: Ernst Jünger and the Politics of Literature after Nazism* (Berkeley: University of California Press, 1999), 31–5.
36 Jünger, *Interwar*, 42–3.
37 Ernst Jünger, *The Worker: Dominion and Form* (Evanston: Northwestern University Press, 2017 [1932]), 56–7; cf 109. The 'worker' in Jünger's title, then, had nothing to do with the proletariat, but everything to do with the man who knows how to burn fuels.

that drives the wheels and smolders at their axles'; blood brothers are those who intermingle through that fuel.[38] Jünger would paint vast murals of bewitching industrial landscapes, the streams of automobiles outshining any waterfall and shoal of fish in the ocean, and underscore, as in *Fire and Blood*, that 'yes, the machine is beautiful'. That beauty always communicated with the *Fronterlebnis*, or 'battlefield experience', the square to which Jünger returned again and again; war was the continuation of industry by other means *and vice versa*, and if people could only open their eyes to this presence of death in their daily life, they would see it shine.[39]

This perpetual warring rested, of course, on a primordial war against nature. That was the enemy that must be vanquished before any other. Jünger preached *Herrschaft* over nature on a telluric scale and delighted in the sight of wild landscapes retreating before the blitz of technology. The sadism tinged with masochism, he noticed that 'the machines are not only directed against nature but against us as well'.[40] Love them all the more for that: Jünger sought unconditional, devotional surrender to the violence of the machine in every sphere of life and death; at the point of production, 'merciless discipline' and 'motorised pace of work' must be savoured.[41] With his pen, Jünger intervened, like Marinetti, in the class war raging through the period. On the one side stood machine and nation; on the other, the masses and nature: 'The German "*Volk*" is erected on the foundations of subjugated nature. It is this that distinguishes it from the "mass", which whips up nature to flood and rebellion', as Klaus Theweleit encapsulates the worldview.[42] The *Volk* would be victorious if it learned to desire its fate, as marked out by the machine. All cathexis must be directed onto it. Painful or sweet, 'there is no way out, no sideways, no backward; what matters, rather,

---

38   Ibid., 27.
39   Quotation from Herf, *Reactionary*, 79. Cf. Klaus Theweleit, *Male Fantasies, Vol. 2, Male Bodies: Psychoanalyzing the White Terror*, 197–9; Brigitte Werneburg and Christopher Phillips, 'Ernst Jünger and the Transformed World', *October* 62 (1992): 42–64.
40   Herf, *Reactionary*, 88. On *Herrschaft* over nature, see Jünger, *The Worker*, for example 31, 58, 136, 140.
41   Jünger, *Interwar*, 106, 88; see further *The Worker*, for example 6, 41, 47, 57, 103, 125.
42   Theweleit, *Male Fantasies, Vol. 2*, 94.

is to *increase the force and speed of the processes by which we have been gripped.*'[43]

Women, needless to say, did not exist in this universe.[44] Men alone brought forth the new; to them belonged the *Herrschaft*, hottest and most intense in the front-line fraternity. 'What is being born' in the parthenogenesis of combustion 'is the essence of nationalism, a new relation to the elemental, to Mother Earth, whose soil has been blasted away in the rekindled fires of material battles': out of crater and pit, the phallic stock of the machine-race.[45] This would be 'a whole new race, energy incarnate', the 'masters of explosive and flame'.[46] But it was also very recognisably white and German.

With these writings, brought to the German public during the critical years of Weimar, from the pages of the NSDAP daily *Völkischer Beobachter* and other far-right publications, Ernst Jünger contributed to the rise of Nazism. His main deed was to weld together nationalism and the machine. Before the First World War, the German right had been sunk in one long daydream about fields and forests, organic nature and the grand cosmos, soulful patrician landscapes unsullied by engines and other modern barbarities; the nation would be redeemed when the machine vanished. Jünger stood all of this on its head. In *Reactionary Modernism: Technology, Culture, and Politics in Weimar and the Third Reich*, the work that has defined the field of fascism and technology studies, Jeffrey Herf pinpoints an ideological constellation named in the title of the book. It combined selective absorption of some yields of modernity – its most efficient technologies – with a repudiation of all other modern values. It wanted to charge ahead with the machine, but go back towards stricter hierarchies, and the former move would catalyse the latter. Herf's account has its problems, as we shall see, but he provides firm evidence of the

---

43 Jünger, *The Worker*, 125. Emphasis added.
44 He could refer to war as 'the male form of procreation'. Quoted in Andreas Huyssen, 'Fortifying the Heart – Totally: Ernst Jünger's Armored Texts', *New German Critique* 59 (1993): 10.
45 Quoted in Theweleit, *Male Fantasies, Vol. 2*, 88. Jünger is a protagonist of *Male Fantasies*.
46 Quoted in ibid., 159.

break effected by the 'reactionary modernists', among whom Jünger was the thinker *primus inter pares*. 'German youth owe a debt above all to Ernst Jünger for the fact that technology is no longer a problem for them', one Nazi commentator thanked him in 1934. Before Jünger opened their eyes, they lived in the 'nightmare' of anti-technological impotence; now they – 'the German youth', a euphemism, of course, for the Nazis – 'have made his beautiful confessions to technics born from fire and blood their own. They live in harmony with it.'[47] They had come to love the machine of combustion, and without that breakthrough, *the material force of Nazism would have been impossible*.

In his agitation against sentimental drivel about land and ruddy farmers, Jünger had a pragmatic argument to make. In these tough times, a nation would never be strong without learning how to burn things. Germany had lost the war because it had not understood the logic of total mobilisation, failed to build up its industrial capacity and imagined that spirit could make up for brute matter. 'Success was guaranteed to those who in the shortest time sent out trains with tons of explosives or supplemented production with coal and steel' – the lesson nationalists must take to heart before the next round. 'The foremost duty of the nationalist state', Jünger informed the readers of *Völkischer Beobachter*, 'will be the creation of a powerful army, equipped with the most modern technical means.'[48] Every fetter on such build-up would have to be unsentimentally smashed.

The great conjuring trick of Jünger, however, which enabled him to touch the German right in the depth of its romantic soul, was to *displace the mysticism of national lifeforces onto the machine*.[49] The feelings of blind love for a community bound in blood did not have to be forsaken, only transferred to the moment of mass combustion. Effeminate fantasies of a pre-industrial mist could be exchanged for the hypermasculine fog of gases, on whose beauty Jünger built his entire career, as he tried to put into words something that had to be

---

47   Wolf Dieter Mueller quoted in Herf, *Reactionary*, 80.
48   Jünger, *Interwar*, 29, 96; cf. Herf, *Reactionary*, 3, 85–7, 95, 100.
49   'We have to transfer what lies inside us onto the machine.' Quoted in Costea and Amiridis, 'Ernst', 480.

experienced; language, he constantly repeated, could not do justice to this scenery.⁵⁰ The magic of the flaming machine must be lived to be felt. Jünger thereby took technology one step further from social relations, to a plane of hyper-fetishisation where the productive-destructive forces had a 'mysterious and compelling claim' on people, who must find the highest pleasure in submitting to their laws. For those who tuned into this frequency in the Weimar cacophony of interpellations, liberation of the German machine meant more than profit or strategic strength. It represented the regeneration of the very soul of the nation, and anyone who opposed it was therefore *an enemy of that soul* – the shock troops moving into position.⁵¹ Once power had been seized, nothing would be allowed to slow down the hellish fire.

Fossil Behemoth One

In March 1923, Benito Mussolini inaugurated the construction of a road from Milan to the Alpine lakes, to be completed the next year: the first motorway in world history. Here no vehicles other than automobiles would be allowed to move. There would be no zigzagging bicyclists, disrespectful pedestrians or slow animal-drawn carts, only the car in splendid isolation, on straight and smooth lanes built to serve its needs. In *Driving Modernity: Technology, Experts, Politics, and Fascist Motorways, 1922–1943*, Massimo Moraglio tells the story of how the network of *autostrada* constructed in northern Italy in the 1920s introduced limited-access highways and revolutionised automobility. The driver would take his car to a gate next to a tollhouse. A uniformed inspector would step forth, greet the car in military style, collect the toll and open the gate for the driver, who could then gun the engine and accelerate to maximum speed – the first and, for a few years, the only place in Europe

---

50 If there was a language whose capacity Jünger fully trusted, it was that of technology itself: a language that does not feign representation, but directly expresses the user's power. On this aspect of his thought, see Walter H. Sokel, 'The "Postmodernism" of Ernst Jünger in His Proto-Fascist Stage', *New German Critique* 59 (1993): 33–40.

51 Drawing on Herf, *Reactionary*, 24, 32, 71–3, 83–4, 88, 94, 101.

where automobiles could be exercised to their full capacity, without the usual impediments.[52]

The Milanese bourgeoisie began to promote the concept immediately after the First World War. The most industrious tycoon, one Piero Puricelli, owner of a company in the road and quarry business, published a pamphlet on the 'physical martyrdom' inflicted on motorists when they had to share roads with 'undisciplined wagons'. Motorways would remove the pain. 'Industry and modern commerce employ the automobile on the road in the same way that calculating machines are employed in offices. Such are the times', Puricelli reckoned, but it was only after the fascist seizure of power that these times came into their own.[53] Mussolini and his local authorities fast-tracked the projects; while initiatives and funding were usually private, the fascists made good use of them in their propaganda, the motorways incarnating the consolidation of hierarchy, the defeat of the socialist challenge and the rebirth of the Italian nation. In the 1920s, the regime blazoned the arrival of the bourgeois family that could drive to its villa outside of town, relieved of the inconveniences of tramways and trains. It attracted enormous international interest. Technicians and engineers, businessmen and lobbyists, politicians and journalists from the US, the rest of Europe and even as far afield as Egypt and China flocked to the motorways outside Milan. Fairs and associations turned them into the living proof of the benefits of automobility and inspired similar schemes around the world. Puricelli, now a de facto ambassador for Mussolini, kicked off discussions over road development in Finland and won contracts for motorways in Brazil, but the country quickest on the uptake was, of course, as we shall see, Germany.[54]

When forming his first cabinet, Mussolini included one minister from Fiat. This company had been gravely threatened by the factory occupations during the Biennio Rosso and thanked its lucky stars for

---

52   Massimo Moraglio, *Driving Modernity: Technology, Experts, Politics, and Fascist Motorways, 1922–1943* (New York: Berghahn, 2017), for example 1, 4, 14–16, 41–5, 54, 57, 77. There were various precedents for these roads – particularly in the US – but no unification of them in a system of interurban motorways before Mussolini. Ibid., 164–9.
53   Puricelli quoted in ibid., 44–5.
54   Ibid., for example 5–7, 10, 13–14, 51–2, 70–5, 164, 169.

fascist reimposition of discipline. Workers hoisting the red flag over factories were not the sole trouble prompting Italian capitalists to look for an exceptional solution in the years after the war; a critical shortage of coal made them cast longing glances at Turkish regions rich in the fuel. We shall see how this particular appetite was satisfied. As for protecting private ownership of the means of production, fascism fulfilled all its promises; a company like Fiat enjoyed the Roaring Twenties, capturing 90 per cent of the Italian auto market and basking in profits. Mussolini continuously reassured capital of the new order by appearing behind the wheel of a racing car or, better still, descending from an aeroplane.[55]

Mussolini fancied himself the chief pilot of the nation. As Il Duce, he had several planes on standby for his exclusive use. He enjoyed piloting as well as flying in comfort; his pet plane was equipped with a bar. He liked to be photographed in a leather flying jacket, goggles on his forehead, leaning out of a cockpit with a confident mien. In August 1919, five months after the raid on *Avanti!*, he intoned '*Volare!*' – To fly! – '*Volare!* To fly over the ghastly, continual trench skirmishes that are the stuff of daily life. *Volare!* To fly for the beauty of flight, almost *l'art pour l'art*'. In a speech before a gathering of aviators in 1923, Mussolini was keen to acknowledge that 'not everyone can fly; it is not even desirable that everyone should. Flying must remain the privilege of an aristocracy; but everyone should have the desire to fly, everyone should long for it.'[56] He vowed to be an impresario of aviation and as such he acted.

Fascist Italy was at the forefront of aviation. It posted a series of spectacular records, such as when Italo Balbo – who had first made a name for himself as a *squadrista*, particularly effective in establishing

---

55 Franco Amatori, 'The Fascist Regime and Big Business: The Fiat and Montecatini Cases', in Harold James and Jakob Tanner (eds), *Enterprise in the Period of Fascism in Europe* (Aldershot: Ashgate, 2002), 65–9; Roland Sarti, *Fascism and the Industrial Leadership in Italy, 1919–1940* (Berkeley: University of California Press, 1971), 9, 17–20, 26–7, 37, 43, 113, 136.

56 Quoted in Esposito, *Fascism*, 273, 334–5. Mussolini's flying habits: Federico Caprotti, 'Overcoming Distance and Space through Technology: Record Aviation Linking Fascist Italy with South America', *Space and Culture* 14 (2011): 337; Federico Caprotti, 'Technology and Geographical Imaginations: Representing Aviation in 1930s Italy', *Journal of Cultural Geography* 25 (2008): 193–4.

the reign of terror over workers and peasants in the Po valley – led sixty-one seaplanes over the Mediterranean in 1928, the largest mass flight to that date. In 1931, he took fourteen planes in formation all the way to Rio de Janeiro and became a world celebrity. Two years later, he orchestrated another transatlantic mass flight, this time to the US, where tens or even hundreds of thousands heard him proclaim that 'Italy is no longer plebeian but is the army of civilisation *en route* in the world' (the city of Chicago named a street after him; it is still there).[57] As eyes had peered at the Italian motorways in admiration, now they turned to the skies.

Rarely if ever has flying been so exalted in ideology. In his extensive study *Fascism, Aviation and Mythical Modernity*, Fernando Esposito charts how Italian aviation built up into an aesthetic and political hysteria. Marinetti served the regime with his artistic projects of 'aeropainting' and 'theatres of the sky' – filling it with pirouetting planes and coloured smoke – and authorities regularly organised flying spectacles. Such events served as metaphors for the fascist order: on the ground, a mass of onlookers gaping at the sky; somewhere up there, invisible to the eye, the pilot in charge. The public was asked to trust in his extraordinary control and identify with him as it would with fate.[58] A fresh frenzy erupted in 1935, when the Italian air force wreaked revenge on Ethiopia – that stubbornly independent African kingdom that had routed an Italian army in 1896 – by showering its people in poison gas. Much like the British steamboats one century earlier, the aeroplanes penetrated the interior of Ethiopia with crushing superiority; afterwards, they were represented as the machines that tamed wild African nature.[59] They were totems of fascism. 'The pilot really knows what it means to govern. Hence there appears to be a necessary, inner spiritual affinity between aviation and fascism. Every aviator is born a fascist', one of many state-sanctioned publications on the topic affirmed. 'Whoever

---

57 Quoted in Esposito, *Fascism*, 345; see further 324–7; Caprotti, 'Overcoming'; Ialongo, 'Filippo', 403.
58 Here following Caprotti, 'Technology', 196.
59 Federico Caprotti, 'Visuality, Hybridity, and Colonialism: Imagining Ethiopia through Colonial Aviation, 1935–1940', *Annals of the Association of American Geographers* 101 (2011): 380–403.

does not expand, whoever stands still, will submit and become enslaved. Flying is the weapon most identified with the race', explained another. Following Marinetti, one lieutenant claimed that just as marble was the classical material and Gothicism the style of the Medieval period, so 'flight is the characteristic of our fast-flowing and metallic era', in which 'man lives, *burning*, in the turbulence of history.'[60]

Fossil Behemoth Two

During his first weeks as *Reichskanzler*, Adolf Hitler turned down a number of invitations to address ceremonies and gatherings. But he jumped at the opportunity to open the International Automobile Exhibition in Berlin on 11 February 1933. Standing before the potentates of the car industry – bald old men in dark suits, listening intently, heads tilted – he unveiled some new national priorities; they did not include railroads, which belonged to the past. With their fixed schedules and routes, they had brought 'individual liberty in transport to an end'. The automobile had given man a mode of travelling that obeyed solely 'one's own orders', respecting 'not the timetable, but man's will'. It was akin to the aeroplane, also powered by a motor. Kissing the feet of the car capitalists, Hitler lauded them and their underlings for producing 'real masterpieces of precision as well as aesthetic beauty' and pledged to do everything in his might to advance 'today's most important industry' – more specifically, he would relieve it of tax burdens, promote motorsports and launch a programme of road construction. Leaving a satisfied audience, the Führer had time for one more public appearance that day: boarding a plane to the town of Kassel, he inaugurated the Adolf Hitler House with the declaration that 'the age of international solidarity is over. The national solidarity of the German Volk will take its place!'[61]

---

60 Quotations from Esposito, *Fascism*, 3, 325; Caprotti, 'Overcoming', 336. Emphasis added.

61 Bernhard Rieger, *The People's Car: A Global History of the Volkswagen Beetle* (Cambridge, MA: Harvard University Press, 2013), 54; Max Domarus, *Hitler: Speeches and Proclamations, Vol. 1, 1934–1934* (Wauconda: Bolchazy-Carducci, 1990), 251–2.

The auto industry had languished in the doldrums of Weimar. Weak domestic demand, poor if any roads, battles with cocky unions on the shop floors conspired to depress motorisation to a trifle: 1 car per 242 inhabitants in 1927, as against 1 per 44 in the UK and France and 1 per 5 in the US. An embattled corporation such as Daimler-Benz was quick to see the promise of the NSDAP. In early 1931, it began to place conspicuous advertisements in the *Völkischer Beobachter*, even though – or rather because – it was cash-strapped and ceased advertising with the social democrats. Daimler-Benz outfitted Hitler and his entourage with shiny new Mercedes cars. Representatives of the corporation later boasted that they helped 'motorise the movement', and the Mercedes star became the vehicular mirror of the swastika: the motorcades of the Führer, gliding through the well-manicured forests of Sieg Heil salutes, would invariably be spearheaded by it. There is a scene in Leni Riefenstahl's *Triumph of the Will* where Hitler and Robert Ley, head of the Deutsche Arbeitsfront (DAF), the Nazi institution responsible for the control of labour, slowly pass by a parked Mercedes car; for a brief moment, the star on the hood and the swastikas on the arms seem to blend, before the two men inspect the ranks of obedient, sharp-chinned Aryan workers.[62]

And the Third Reich was off to a flying start: in April 1933, newly purchased vehicles powered by internal combustion engines were exempted from annual car tax charges. Costs for drivers' licences were cut and all speed limits abolished. From spring 1934, it was up to the driver to drive as he wished. The automobile lobby had long been asking for precisely these stimuli. To demonstrate its commitment, the Third Reich sponsored the racing teams maintained by Daimler-Benz and overnight turned Germany into the world leader in sports, raking in Grand Prix titles and speed records, interpreted by a rapt press as so many proofs of the superiority of German power. When all three places on the podium in Monaco were taken by its drivers in 1936, one

---

62 R. J. Overy, *War and Economy in the Third Reich* (Oxford: Clarendon, 2002 [1994]), 69; Rieger, *The People's*, 12–13, 31; Bernard P. Bellon, *Mercedes in Peace and War: German Automobile Workers, 1903–1945* (New York: Columbia University Press, 1990), 218–19, 232.

reporter exulted: 'Germany has won a hot battle in a superior manner – now we advance to further battles.' Car capitalists had already responded in kind in the months after Hitler took the helm, when the concord between the NSDAP and German capital was sealed with munificent donations from the latter; in mid-1933, the National Association of the Automobile Industry poured money into the party coffers.[63] The enrichment was reciprocal. During the first year of the Third Reich, more than twice as many cars were sold as in the preceding year; in 1934, three times as many were manufactured as in 1932 – sign of a recovery 'far more rapid and extensive than in virtually any other branch of industry'.[64] In 1935, the motor industry reached a position as the largest manufacturing employer. As for Daimler-Benz, capacity utilisation fluctuated around a fourth and a third in the early 1930s, but hit 95 per cent in 1935 and 97 in 1938. Not only was demand buoyed, but the unions were graciously obliterated. Daimler-Benz was in a state of bliss: *Herr im Hause* – masters in the house – after decades of intrusions by socialists.'[65] It counted a 'pleasant increase in profits' following the seizure of power.[66]

Then Hitler unveiled his truly ambitious projects for motorisation: first 'the people's car', or *Volkswagen*, the most highly prized composite of race and commodity, above the *Volksmotorraeder* (motorbike), *Volkstraktor* (tractor), *Volksfernseher* (television), *Volksempfaenger* (radio) and a list of other items in which the *Volk* would take material form. The idea was to have industry produce a cheap and compact car for the people. Back at the International Automobile Exhibition in Berlin in March 1934 – an annual festivity in his calendar – Hitler explained that the state 'desires that the German *Volk* take an animated interest in motorized vehicles', and it was incumbent on the

---

63   Bellon, *Mercedes*, 221; Rieger, *The People's*, 47–8, 51–2, 56 (reporter quoted on 48); Neil Gregor, *Daimler-Benz in the Third Reich* (New Haven: Yale University Press, 1998), 37. On Hitler's obsessive love for motorsports and its regenerative impact on the German Volk, see further Max Domarus, *Hitler: Speeches and Proclamations, Vol. 2, 1935–1938* (Wauconda: Bolchazy-Carducci, 1992), 1018–19.

64   Gregor, *Daimler-Benz*, 37.

65   Bellon, *Mercedes*, 232 (figures on the corporation from 220); further Overy, *War*, 71, 78–9.

66   Overy, *War*, 7; Bellon, *Mercedes*, 219.

assembled capitalists to make room for it.[67] But most of them balked at cranking out mass cars to rock-bottom prices. Hitler had to enlist the illustrious engineer Ferdinand Porsche – his confidant in automotive matters, gladly accepting the assignment to produce the Volkswagen – and transfer the project from the private sphere to Ley's DAF. The first factory would be built under Porsche's direction, with financial resources the DAF had confiscated from the trade unions after May Day 1933. It would be modelled on Ford's factories, which Porsche had admired in Detroit, but it would be larger – the largest in the world. Seizing assets from crushed labour to construct a top-of-the-line car factory under Porsche: German fascism in power. The prototype Beetle was put on display at the Berlin exhibition in 1938, and in his speech of that year, the Führer expected ordinary Germans to put away their savings for a future purchase of the car 'out of their love for and pleasure in automobiles'. As is well known, however, it was only in post-fascist conditions that Volkswagen came into its own.[68]

Hitler had better luck with the roads. Contrary to popular myth, these were not built to serve military planning primarily, nor to alleviate unemployment, on which they had negligible effect.[69] Instead, the roots of the *Reichsautobahnen* were twofold. Having set up a lobby in 1925 for the purpose, car capitalists and motor enthusiasts sought to persuade successive Weimar governments to construct a network of toll roads for automobiles only, from Lübeck in the north to Basel in the south. They made little headway: during the Weimar years, cars were frowned upon as hobby-horses of the bourgeoisie. A typical opinion was that 'roads for the enjoyment of rich people' were uncalled for in these trying circumstances.[70] If Weimar Germany

---

67 Domarus, *Hitler, Vol. 1*, 440.
68 For all things *Volkswagen*, see Rieger's excellent *The People's*. Cf. Wolfgang König, 'Adolf Hitler vs. Henry Ford: The Volkswagen, the Role of America as a Model, and the Failure of a Nazi Consumer Society', *German Studies Review* 27 (2004): 249–68. The union assets: Adam Tooze, *The Wages of Destruction: The Making and Breaking of the Nazi Economy* (London: Penguin, 2007), 147. Hitler in 1938: Domarus, *Hitler, Vol. 2*, 1018.
69 Thomas Zeller, *Driving Germany: The Landscape of the German Autobahn, 1930–1970* (New York: Bergahn, 2007), 1, 56, 59.
70 Werner Sombart quoted in ibid., 49; see further 47–9.

was poor in roads, it was all the better furnished with public transport – extensive, rapid, reliable – further reducing the perceived need for rolling out carpets for cars. Furthermore, the turnover in cabinets and ministers was so high as to leave the lobbyists without levers. The Third Reich instantly shattered those constraints. Maps and plans from the mid-1920s were elevated to a national priority, heightened further by the inspiration from Italy – not to be left unmatched – and by the fascist aura of the automobile.[71] In the autobahn, the material interests of hitherto circumscribed fossil capital were released through the ideological zeal of fascism. The road model condensed the logic: on the lanes of the *Volk*, there would be no 'bicyclists, horse-drawn carts, herds of sheep', as one observer put it, nor any other kinds of obstacles. As the laissez-faire produced a virtual slaughter of German drivers and pedestrians, the Third Reich found itself compelled to re-impose speed restrictions on some roads in May 1939. But the autobahn remained free.[72]

The Puricelli-Mussolini system was here replicated on a grander scale. After eight years in existence, the Nazi state had laid out a road network seven times longer than what the Italians had achieved in the first twelve years of their reign. Germany overtook Italy as the frontrunner; Hitler in a Mercedes speeding down his own motorway straight as a die made for flashy images, renewing the momentum for similar projects in France, Holland, Denmark, Poland.[73] But the propaganda rang loudest on the home front, where the autobahn cemented the order and extended it beyond the visible horizon. All senses were bombarded with the roads: there were autobahn paintings, autobahn novels, collectible autobahn pictures in cigarette packs, radio dispatches from the latest inaugurations and some

---

71   Richard J. Evans, *The Third Reich in History and Memory* (London: Little, Brown, 2015), 181; Zeller, *Driving*, 49–50, 56; Johan Schot and Vioncent Lagendijk, 'Technocratic Internationalism in the Interwar Years: Building Europe on Motorways and Electricity Networks', *Journal of Modern European History* 6 (2008): 200–7; Richard Vahrenkamp, 'Motorization and Autobahn Projects in Germany in the Interbellum', *TST: Transportes, Servicios y Telecomunicaciones* 18 (2010): 146–70.

72   Rieger, *The People's*, 55–6; Evans, *The Third*, 182; observer quoted in Zeller, *Driving*, 129.

73   James D. Shand, 'The *Reichsautobahn*: Symbol for the Third Reich', *Journal of Contemporary History* 19 (1984): 189–200; Moraglio, *Driving*, 138–45.

twenty state-produced films with titles such as *Open Road, Roads of the Future, On Germany's New Highways, Fast Roads* and *Roads Are Fun*. A picture book entitled *The Experience of the Reichsautobahnen* captured its subject: 'Run, my car, run! Like lightning the autobahn now flashes far through the valley from this height here'; Germany is 'wonderfully laid out' for the driver, who learns 'how rich we are and how much of it [the nation] we still need to conquer for ourselves'. In *Driving Germany: The Landscape of the German Autobahn, 1930–1970*, Thomas Zeller observes that in this genre of state propaganda, 'speed itself became the intoxicating goal'.[74] This was probably not because the Nazi hacks had read Marinetti into the wee hours; rather, both expressed a machine idolatry generic to fascism.

'Next to the automobile, the aeroplane was central to the Nazi project of modernity', notes one recent Hitler biographer.[75] The Führer would not miss a chance to descend from the skies. In the opening scenes of *Triumph of the Will*, an aeroplane is gliding for several minutes above the clouds, before its cruciform shadow darts over the town of Nuremberg and comes to a rest at the airport, where the Führer emerges before the adoring crowd. 'It is the way in which it stands above things. It is its remoteness from everything small and petty,' one Nazi wrote, explaining the appeal of the plane.[76] Hitler opined that 'in a few years nobody would even consider undertaking travel of more than 500 kilometres in anything other than an aeroplane' and ordered his armies of engineers to catch up and overtake rivals in this department too. A pet project of his was the Tempelhof airport, located smack in the middle of Berlin and refurbished in record time with a stupendous terminal and arc-shaped aircraft hangars. Before the war broke out, Tempelhof had indeed become one of the world's busiest airports.[77] But the greatest strides were

---

74   Zeller, *Driving*, 161. Quotation from same page; further 62–4. Cf. Paul B. Jaskot, 'Building the Nazi Economy: Adam Tooze and a Cultural Critique of Hitler's Plans for War', *Historical Materialism* 22 (2014): 317–18.

75   Brendan Simms, *Hitler: Only the World Was Enough* (London: Allen Lane, 2019), 219. Analytically, this book is a disaster; see further below.

76   Esposito, *Fascism*, 349.

77   Simms, *Hitler*, 220; on Tempelhof, see also for example Reinhard Mohr, 'The Mother of all Airports', *Der Spiegel*, 25 April 2008.

made not in civilian but in military aviation, a focal point of the rearmaments drive, the build-up of the Luftwaffe outpacing most if not all other industrial ventures. Out of scrap and obsolete workshops, the Nazi state assembled a formidable complex for the production of futuristic aircrafts and aero-engines – here handsomely assisted by Daimler-Benz and BMW – and posted a series of breakthroughs: the first full-metal plane, the first turbojet engine, the first fighter jet to take to the skies. The German millions might be a generation away from frequent flying, but the spectacles of the Luftwaffe mandated their imaginary identification with a *Volk* on wings.[78] If nothing else, they could drive on the autobahn, whose designers sought to make the experience as similar to flying as possible, with bends and curves and bridges producing the effect of cruising over the landscape.[79]

How did the promotion of these technologies fit into the broader political economy of Nazism? First of all, it joined the Third Reich and German capital in wedlock. The former restored order in the factories, so the latter could churn out commodities unobstructed; the latter delivered machines to the former, so it could stabilise and wage its wars.[80] The two continuously reconfirmed their union through the development of productive-destructive forces, premised on the defeat of labour: 'You must learn to love the machine like a bride', Robert Ley coached the workers at Siemens in 1933.[81] The machine would weld together workers and their superiors in one harmonious racial body unriven by class divides. At the same time, the most passionate love affairs – with the automobile and the aeroplane – had something distinctly conventional about them. Alongside motorbikes, tractors and various household appliances, these things were simply the accessories of capitalist modernity. There was a latent

---

78 Daniel Uziel, *Arming the Luftwaffe: The German Aviation Industry in World War II* (Jefferson: McFarland, 2012), for example 7–10, 60–1, 263; Tooze, *Wages*, 125–30, 165, 620.

79 Benjamin Steininger, *Raum-Maschine Reichsautobahn: Zur Dynamik eines bekannt/unbekannten Bauwerks* (Berlin: Kadmos, 2005), 103–28.

80 Here following Franz Neumann, *Behemoth: The Structure and Practice of National Socialism, 1933–1944* (Chicago: Ivan R. Dee, 2009 [1944]), for example 361; cf. Tooze, *Wages*, 114.

81 Tim Mason, *Nazism, Fascism and the Working Class* (Cambridge: Cambridge University Press, 1995), 80; on the subjugation of labour, see further for example 86–8, 93–4, 102–3.

enthusiasm for them in parts of the German population, adroitly activated by the Nazi state.[82] Longing eyes had already during Weimer fastened on the American way of life. Car capitalists had envied the levels of motorisation and methods of production, as had Hitler in *Mein Kampf*; to him, American automobility was the epitome of the life worth striving for.[83] Two historians, Adam Tooze and Brendan Simms, have recently given this component of Nazism its due emphasis, but muddied the picture by suggesting that the US and its capitalism – not the Soviet Union and Judeo-Bolshevism – were somehow Hitler's archenemies. But as their own material so volubly testifies, the US with its cars and parkways and commodities for the affluent consumer was his *ideal*.[84] It was what he wanted Germany to also become.

The speech delivered by the Führer to the 1936 edition of the Berlin auto exhibition tied together some of these themes in the Hitlerian outlook. Proud of the speedy achievements, he now offered a retrospective on the shortcomings of Weimar. The automobile industry – 'called upon to put its unique and characteristic stamp on today's age' – had been left floundering because Marxists ran the nation. They believed in a 'theory of equality, that it was necessary for the human race to become a race of primitives' and aimed at 'proletarianizing the standard of living for all so as to arrive at a level shared by as many as possible'. Hence they shunned the luxury carriage and tried to strangle it with taxes and traffic police. But in fact, Hitler argued, progress can never be collective in nature. It always 'originates with a very few individuals' and is only 'possible given respect for individual creative power' and the great feat of the Third Reich

---

82   Cf. Adam Tooze, 'The Sense of a Vacuum', *Historical Materialism* 22 (2014): 359; Zeller, *Driving*, 241.

83   See for example Zeller, *Driving*, 48; Simms, *Hitler*, 92–3, 121, 219.

84   Tooze's US-centric account of Nazi aggression is debunked in Dylan Riley, 'The Third Reich as Rouge Regime: Adam Tooze's *Wages of Destruction*', *Historical Materialism* 22 (2014): 330–50; Perry Anderson, 'Situationism à l'envers?', *New Left Review* 119 (2019): 47–93. As for Simms, rarely has a historical study in such a flippant manner presented so much empirical material contradicting its own main thesis. It is demolished in Richard Evans, 'Problematic Portraits', *Guardian*, 27 September 2019, and in a long footnote in Anderson, 'Situationism', 64. 'Ideal society' is the term used by Simms, *Hitler*, 78. See further for example 89, 220, 294, 553.

was to discard collectivist inhibitions. Now the *Volk* felt free to take the driver's seat. Under the Führer, there was 'complete liberty for the *Volk* to make use of it [the car], not only in terms of legislative liberty, but above all in terms of psychological liberty'. It could begin its ascent towards 'the brilliant example of America'. In the Soviet Union, meanwhile, the ratio of cars to people was even lower than in Weimar: the primitive morass Germany would have descended further into were it not for the Nazis uplifting the nation.[85]

Clearly, Hitler harboured resentment against the US, as against an elder brother with a lustrous career. But he dedicated to it nothing like the murderous fury and will to annihilation reserved for the Judeo-Bolshevik enemy in the east. The reaction against the two principal blockages of capitalist expansion in the Weimar period – powerful labour, drawn to the pole of revolution; truncated territory, without the colonies worthy of a superpower – was concentrated towards this cardinal point. To the east lay the source of subversion. To the east also lay the vast spaces that would give Germany the empire its dominant classes had been hungering after since the days of Bismarck. Victory in this direction passed through 'total mobilisation', the Jüngerian concept turned into a tagline by Joseph Goebbels. The machine, in other words, was the corporeal form of the great unshackling the Nazis sought to undertake: it embodied the forces of the nation, before the battlefield, on and beyond it. Only by producing as much top-class modern technology as possible could the *Volk* prevail and prosper, and it followed that any maudlin critique of it was anti-national. Hitler made as much clear; at no time did he betray any soft spot for nature. He was himself, in Herf's assessment, the greatest practitioner of 'reactionary modernism'. 'He insisted that the Germans must succeed in the battle against nature in order to win in the battle among nations and races.'[86]

And the Nazis would not have been Nazis if they did not soak this embrace of technology in racism: in their view, the white race had an unmatched rapport with the machine. 'All the results of art, science,

---

85 Domarus, *Hitler*, Vol. 2, 752–5.
86 Herf, *Reactionary*, 194.

and technology that we see before us today', Hitler observed in *Mein Kampf*, 'are almost exclusively the creative product of the Aryans.' The Moors, for instance, had used horses for thousands of years, but their time was over now. The Aryans had the motorcar.[87] The circumstance that two German inventors, Benz and Daimler, had first created a vehicle with wheels, seats and an internal combustion engine proved that Germany was the homeland of the Aryan car, but the most brilliant example of its full development was, again, America: after the natives had been pushed aside, it 'criss-crossed the area with a system of roads and railways, until finally the white race controlled the entire continent, which today represents a cornerstone of the white race'.[88] The white race had established this dominion by killing non-whites without pangs of conscience. When it took North America, Latin America, India, any other part of the globe, it exercised 'an extraordinarily brutal right to dominate [*Herrenrecht*]'. It could not afford to act otherwise. Only if cannons stayed affixed to the engines – as in the nineteenth century, when the British Empire ruled the seas – could the white race maintain a world-system of 'gigantic central world factories' on its own side and 'huge markets and sources of raw materials' on the other. Some weaklings put their trust in the market alone. In his famed speech to the crème de la crème of capitalists in Düsseldorf in January 1932, Hitler exposed their dishonesty: 'All I can answer is: that is not the spirit which opened up the world to the white race', nor would it open up the world to Germany (at which point the transcript says 'Hear, hear!').[89] One must here credit Hitler with a sense of historical realism. His outspokenness put the secret of techno-racism out in the open: technology, after all, was white because built from the bones of non-white people. He just wanted to affirm this.

Techno-racism was law in the Third Reich; an exemplary text was *Die Dynamische Wirtschaft*, written in 1936 by Fritz Nonnenbruch, economics editor of the *Völkischer Beobachter* and one of the main – if rather unoriginal – NSDAP writers on the economy. He studded

---

87 Quotation from Adolf Hitler, *Mein Kampf*, 290; Moors and motorcars, 294–5.
88 Hitler in Simms, *Hitler*, 91.
89 Domarus, *Hitler, Vol. 1*, 96–7, 102. Cf. Hitler, *Hitler's Second*, 102.

his book with criticisms of hitherto existing capitalism for not fulfilling its own promises. Speculators and rentiers – Jews, that is – had detained the forces of technology, but thankfully their power had now been broken. Nazism thus 'corresponds to the technical age. It runs capitalism as the engine [*Kapitalismus als Motor*], uses its dynamic energies, but fixes the transmission belt' or, better, 'shifts the gears'.[90] It revved up the existing order. Productive, non-Jewish capitalists – those who owned the machinery – had an enduring role to play here, Nonnenbruch made plain; any Marxist notion of the worker producing unremunerated surplus-value must be expurgated. The added value came from the machine itself. It would be all the speedier valorised now that the racial character of capital could play out freely: 'The entrepreneurs who are driven by the pleasure of technology are following the racial spirit.' Harking back to nineteenth-century teachings, Nonnenbruch defined *Rassenlehre*, or the systems of racial classifications, as 'an outflow of our Nordic will to technology', although race was not the sole essence to be expressed here. 'Technology', Nonnenbruch swaggered, 'is the field in which our manliness is shown.'[91]

The ruminations of a Marinetti or a Jünger were here trimmed into a catechism of the Nazi state. Nonnenbruch derived the mastery of technology from the front-line experience, and invested it with all the mystical qualities hitherto associated with rural nature: the flight, the car journey, the blast furnaces were 'far more romantic' than any such outmoded refuges from reality. Most romantic of all, because most expressive of the racial will to technology, were the former two. 'It is wonderful that the youth are so lively to hurry themselves to the airplane. It is wonderful that cars are bought in great numbers', because this shows that 'the *Volk* wants technology' and is ready to pay for the experience of it. Imbuing the machine with these racial

---

90   The words here are '*legt aber die Transmissionsriemen um*'. Fritz Nonnenbruch, *Der Dynamische Wirtschaft* (München: Zentralverlag der NSDAP, 1936), 43. Cf. 123, 134.

91   Ibid., 125, 129, 139, 151–3. Nazism 'is based on a race that joyfully resonates with the forces of technology on a large scale. It endorses the technology as a whole. Its economic policy, which ensures the development of technology, is the highest expression of this resonance.' Ibid., 150.

values was a way to upgrade it even further: not just a dead thing that functioned efficiently and generated profit, it materialised the good and strong *Volk*: building the autobahn 'is much more than the construction of traffic routes. It is much more, just like the car is more than just a means of transport.'[92] Neither was for Jews. The foil for this torrent of techno-racist thinking was, of course, always the Jews: the white race knew how to build machines, but Jews switched off their power. 'The Aryan Prometheus had brought the world the fire of creativity – technology – the gift of the gods. But Jehovah had cut his throat and the Jewish vulture tore out his liver', NSDAP cadres learned in 1938.[93] In the same year, Jews were banned from driving on the roads of the nation. A masterstroke of Nazi anti-Semitism was its conceptualisation of Jews as the evil geniuses of both Weimar blockages: behind internal sedition and external confinement, and hence the people that must bear the brunt of materialised Aryan power.

Not much, however, was coherent in Nazi techno-racism. Were whites creative or just good murderers? Was the innovativeness specifically German or common for all whites? What about the fuels that made the machines move? The white *Volk* could accomplish little without black fuel, and here its masters faced a predicament: German territory held no oil fields. Nor did it have rubber for tyres. The one raw material it possessed in abundance was coal, lignite in particular. In the second half of the nineteenth century, German manufactures therefore learned to squeeze all manner of strange drops out of the black rock, mixed with water or wood, soil and air and other ubiquitous inputs: the foundation for their world-leading chemical industry. The almost alchemical art consisted in transmuting coal into other substances, synthetic materials that could replace the natural ones Germany had to import. Even by the standards of the day, coal thus came to occupy a disproportionate place in this

---

92 Ibid., 153, 138, 143.
93 Emil Maier-Dorn in John C. Guse, 'Nazi Technical Thought Revisited', *History and Technology* 26 (2010): 14; further for example Herf, *Reactionary*, 204–5; Shand, 'The *Reichsautobahn*', 197; Michael Allen, 'The Puzzle of Nazi Modernism: Modern Technology and Ideological Consensus in an SS Factory at Auschwitz', *Technology and Culture* 37 (1996): 527–71.

nation – nine-tenths of fuel consumption during Weimar, one-third of the tonnage shipped on railways, the strategic base for the industrial apex: the hat out of which chemists pulled dyes, drugs, fibres, light metals.[94] In 1926, they cracked the code of turning coal into petroleum. By heating up lignite to around 500°C and adding a catalyst, they could distil a liquid brew; after further treatment in a refinery, it would be passable gasoline for internal combustion engines. The process was known as 'hydrogenation' and first implemented at commercial scale at the Leuna plant of IG Farben. The largest private corporation in Europe between 1925 and 1945, IG Farben was a colossus with feet of coal, its holdings of lignite the basis for streams of gases, explosives, mineral acids, pharmaceuticals, rayon, as well as oil from 1927 and soon also synthetic rubber. No other company would mean as much to the Third Reich.[95]

The NSDAP fell in love with hydrogenation. It held out the prospect of a home-brewed fuel for cars and aeroplanes, tanks and fighter jets and other machines with motors. Hitler was determined not to repeat the mistake from the previous war – remaining dependent on foreign fuel supplies, easily blockaded – and would accept no limits to synthetic oil production.[96] And IG Farben had reasons to fall into Nazi arms. As the world's largest chemical factory, the Leuna plant had immense capital sunk into it, but in 1932 it ran at a fifth of capacity. Burdened by underutilised mines, beset by uncertain markets, beleaguered by cheap oil – world prices falling under the Depression – and restive labour, IG Farben cried out for a state to save it. Only government guarantees for sales, prices and social order could shelter the investments. And the Nazis, of course, offered just that: during the year after the *Machtergreifung*, all the wishes of IG Farben were fulfilled. In a pact signed by the parties, the corporation committed

---

94 Peter Hayes, *Industry and Ideology: IG Farben in the Nazi Era* (Cambridge: Cambridge University Press, 2001 [1987]), 1; John Gillingham, *Industry and Politics in the Third Reich: Ruhr Coal, Hitler and Europe* (London: Methuen, 1985), 5, 10, 49; Tooze, *Wages*, 50, 342.

95 Hayes, *Industry*, xxi, 10, 16–20, 36–7; Gillingham, *Industry*, 19–20.

96 Arnold Krammer, 'Fueling the Third Reich', *Technology and Culture* 19 (1978): 394–422; Hayes, *Industry*, for example 66, 139–41; Gillingham, *Industry*, 49; Dietrich Eichholtz, *War for Oil: The Nazi Quest for an Oil Empire* (Washington, DC: Potomac Books, 2012), xvi.

to tripling gasoline production at Leuna over the next twelve months, while the state pledged to ward off foreign oil with tariffs, purchase any unsold output and guarantee a 5 per cent return on investments. Further bolstered by motorisation (IG Farben had long recommended a road network), the mines and plants were suddenly stretched to capacity, new hydrogenation facilities planned, profits spiking. In late 1933, the founder of IG Farben, Carl Bosch, best known for the Haber-Bosch process, penned an article headlined 'Where There Is a Will, There Is a Way'. He relished the 'wave of new trust and new confidence' coming from Hitler.[97] More than a betrothal, this arrangement between the Nazi state and IG Farben quickly evolved into a symbiosis that would last until defeat did them apart.

But this was not a relation involving one corporation only: the coal sector in general was well disposed to the NSDAP. It suffered the same distempers in late Weimar – deflated prices due to cutthroat competition, over-capacity, demanding unions – and welcomed the exceptional state.[98] Figures from the sector were overrepresented on the roster of pro-Nazi capitalists. One of the main emissaries to facilitate the *Machtergreifung* was Albert Voegler, chair of RWE – the corporation that still operates lignite mines in the Rhineland, among them Hambach – who then manoeuvred to advance hydrogenation.[99] The most notorious private patron of the NSDAP, Fritz Thyssen, had earned his fortunes in 'a metallurgic concern with a coal basis'.[100] In the decisive round of business donations in spring 1933, auto, electrical and machinery firms supplied one-sixth of the money, while the rest came from the coal-steel-chemistry complex.[101] Some observers at the time noticed the pattern. In his classical *Behemoth: The Structure and Practice of National Socialism, 1933–1944* – the first edition published in 1942 – Franz Neumann, the Jewish Marxist from Katowice, honed in on coal. The technological advances in German

---

97 Hayes, *Industry*, 37–47, 58, 64–8, 109–19; cf. Tooze, *Wages*, 116–18.
98 Gillingham, *Industry*, for example 28, 32.
99 Tooze, *Wages*, 111, 117, 122.
100 Neumann, *Behemoth*, 278.
101 Hayes, *Industry*, 85–6.

industry were founded on this one fuel, but the installations for its various transubstantiations were extraordinarily expensive. As a result, capital leapt to embrace a state that vowed to hedge it against losses.[102]

The behemoth, as Neumann saw it, amalgamated the Nazi state with the mightiest capital from coal. The former ensured expanding markets for the synthetic materials – civilian at first, notably cars guzzling synthetic gasoline, but before long centred on rearmament – held off natural imports and helped organise industry so as to quench ruinous competition.[103] German coal and chemical manufacturers had been pioneers of the cartel form, but the agreements tended to come apart during the crisis. The Third Reich revived cartelisation; small coal producers and retailers who undercut prices would henceforth be sanctioned out of existence. In 1934, the ten largest lignite corporations were pooled together in a combine called Braunkohle-Benzin, or Brabag, and instructed to multiply their synthetic gasoline facilities.[104] These developments merely enhanced the centrality of coal. Pumping gasoline out of the mines was the precondition for motorisation and blitzkrieg; from the same seams came tyres for the Wehrmacht and Luftwaffe; new light metals were conjured up for more advanced engines, the military-industrial dynamism of the Reich swirling around this axis. Trade and finance were demoted in Nazi capitalism, and in their place, a particular set of productive forces was enthroned: 'The vertical combine from coal (or lignite) to manufacturing is the type which best expresses the industrial leadership.'[105] Given its strategic importance, one might think that the Nazi state should have nationalised the coal sector, but instead it held up the carrot of glistening profits and trusted in private

---

102 Neumann, *Behemoth*, for example 277–82, 288–91.

103 Cf. Overy, *War*, 86. The Nazi state here created a virtuous cycle in the relation between its highly versatile primitive fossil capital and fossil capital in general: the autobahn and other pro-auto measures stimulated key branches of the latter, which increased demand for the products of the former, in itself advanced by state policies, and so on. Cf. ibid., 72, 756.

104 Neumann, *Behemoth*, 15, 261–8, 281–3, 291–2; Gillingham, *Industry*, 5, 34–5, 133–5; Tooze, *Wages*, 111–12, 119–20.

105 Neumann, *Behemoth*, 392.

enterprise to deliver the goods. 'Profits and more profits are the motive power.'[106] The spine of the behemoth, we can then conclude, was primitive fossil capital. 'Every day sees how this machinery becomes an instrument of destruction rather than welfare', Neumann despaired from his exile.[107]

This anatomy only became more pronounced in the later phases of the Third Reich. In 1936, Hitler shifted gears once more with the announcement of the Four-Year Plan, a crash programme for making Germany self-sufficient in key raw materials – oil, rubber, fats, textiles, copper . . . – and ready it for war. The office for realising the Plan was handed over to managers and technicians from IG Farben.[108] New hydrogenation units cropped up and mining accelerated, coal output increasing by 70 per cent over the seven first years of the Reich.[109] But this was still far from enough. In fact, fuel shortages crippled Germany in the mid-1930s, on the eve of the war and during every major campaign. The totally mobilised economy had an insatiable appetite for coal and its derivatives, some of which also had to be diverted to the southern ally; instead of conquering its own reserves, Italy depended on German shipments. One bottleneck was the shortage of labour-power, and so the coal sector became the first to experiment with slave labour: in 1941, there were already seventy thousand foreign workers forced into German mines.[110] The oil situation was even more critical. It called for a two-pronged offensive: further expansion of hydrogenation capacity, and conquest of actual oil fields. In 1940, Germany wrested control over Romania – the

---

106   Ibid., 354. Cf. 280, 298; Tooze, *Wages*, 256 ('with private ownership still the norm, the profit motive remained the ultimate inducement to production'); Ian Kershaw, *The Nazi Dictatorship: Problems and Perspectives of Interpretation* (London: Bloomsbury, 2015 [2000]), 66; on not nationalising the coal sector, Overy, *War*, 16.
107   Neumann, *Behemoth*, 472. More recent scholarship has corroborated the substance of Neumann's account: in particular Tooze, *Wages*; Hayes, *Industry*.
108   Tooze, *Wages*, for example 225–30, 252–3, 308, 443; Hayes, *Industry*, for example 164–183, 337–9; Kershaw, *The Nazi*, 70–3. Other coal interests, notably from the Rhineland, were here sidelined, as they had been less aggressive in developing hydrogenation capacity. Gillingham, *Industry*, 51, 164.
109   Gillingham, *Industry*, 58.
110   Tooze reconstructs the recurring fuel crises in detail in *Wages*. On the supplies to Italy, see for example Gillingham, *Industry*, 122; on slave labour, 123–5; Tooze, *Wages*, 417.

fourth largest oil producer at the time – and turned it into a 'gas station' that carried it through the early blitzes.[111] Minor spoils were picked up in Austria, Galicia, Estonia, but the real prize was the Caucasus and the Persian Gulf, the path to which passed through the Soviet Union. Operation Barbarossa would open the gates to the oil of the Orient.[112]

A rush of excitement ran through the dominant classes of Nazi Germany: at long last, the Judeo-Bolshevik enemy would be done away with and the resources it bestrode freed up. Operation Barbarossa consolidated the symbiosis, as state and capital threw themselves with combined forces into the life-and-death struggle in the east.[113] The anticipated oil booty was to be grabbed by Kontinentale Öl, or Konti, a combine formed in 1940 out of assets from IG Farben, Brabag, the four biggest German oil companies and a slew of banks. Hermann Göring presided over it. Neumann considered it 'a model for the new ruling class': a corporate entity working for private profit, through the advancing war machine. Konti had as its sole purpose 'the exploitation of oil in the conquered territories', to be incorporated into the '"big enterprises in the oil and coal industry"' – a world under the caterpillar tracks of primitive fossil capital, the future in sight when Barbarossa got underway.[114] As for Hitler, he was 'absolutely obsessed' with plundering coal and oil.[115]

But the war in the east did not, of course, pan out. The Wehrmacht came within reach of the Caucasian fields but no further. Frightful fuel shortages soon reappeared. A stressed Führer would blurt out lines such as 'whoever loses the oil loses the war' and 'if, due to the shortage of coking coal the output of the steel industry cannot be

---

111 Hermann Neubacher quoted in Eichholtz, *War*, 27; on Romania, see further 4–5, 14–15, 19–30, 34.

112 Ibid., xiv, 5, 9–13, 62–4, 76–80; Tooze, *Wages*, for example 308–9, 459–63, 469–70, 487, 665–6; Neumann, *Behemoth*, 328.

113 Eichholtz, *War*, 46; Riley, 'The Third Reich', 345; Kershaw, *The Nazi*, 75–7; Tooze, *Wages*, 561–2, 571.

114 Neumann, *Behemoth*, 396, 277, 358 (the latter quotation from *Deutsche Volkswirt*). Cf. Eichholtz, *War*, 28, 35–40, 67.

115 Ernest Mandel, *The Meaning of the Second World War* (London: Verso, 1986), 49.

raised as planned, then the war is lost.'[116] Confounding his diagnosis, it was the Soviet Union that proved the superior fossil economy, capable of hurling endless columns of well-oiled, fully tanked machines at the invader. Up against the Allies, Hitler fulfilled the scenario he dreaded most: inferiority vis-à-vis those who controlled continents of raw materials. The master of the Middle Eastern fields, the United Kingdom, imported 20 million tonnes of oil in 1944, nine times the maximum amount the Third Reich ever attained. War – here Jünger got it right – is an exercise in mass dissipation of biophysical resources, in which the winner must excel; conversely, once you are forced to produce 'less steel or less coal, you will inevitably end up producing fewer and fewer guns and airplanes'.[117] The hollow in Nazi techno-racism was exposed. Left to itself, the *Volk* really had no striking force, and the various miracle technologies in which Hitler and his holdouts invested their last hopes merely brought this home. 'The German ruling clique drove towards war because they were excluded from a position of imperial power', reflected Adorno. 'But in their exclusion lay the reason for the blind and clumsy provincialism that made' the Nazis 'uncompetitive and their war a gamble.'[118] Instead the full force of the Nazi machinery was turned inwards, against defenceless targets: above all, the Jews.

Death by Coal

When the Wehrmacht rolled into Poland in September 1939, IG Farben was not far behind. It looked for more coal and suitable sites

---

116   Quotations from Simms, *Hitler*, 470; Tooze, *Wages*, 574. In July 1941, less than two weeks into Operation Barbarossa, Hitler had this outburst in private: 'With coal we know when the reserves shrink: cavities are opened. When it comes to oil, we cannot say whether the rooms will fill up again in our invisible reservoirs. Man may be the most dangerous microbe imaginable: he takes out the whole Earth without asking whether it may be a of vital importance for another realm of life.' Adolf Hitler and Heinrich Heim, *Monologe im Führer-Hauptquartier 1941–1944*, ed. Werner Jochmann (Hamburg: Albrecht Knaus, 1980), 39. These comments might be as close as Hitler came to a critique of fossil fuels.

117   Mandel, *The Meaning*, 51. 20 million tons: Tooze, *Wages*, 412; see further for example 411, 454–5, 639–40.

118   Theodor Adorno, *Minima Moralia* (London: Verso, 2005 [1951]), 106.

for factories. In late 1940, its scouts developed an interest in the small village of Auschwitz, located in the midst of the Upper Silesian coal basin. A string of mines surrounded the village, which also sported three watersheds and a railway juncture: ideal for a new generation of chemical plants. With the assistance of the SS, IG Farben acquired its first mine, dormant for the previous two decades, which became known as Fürstengrube, or 'the Pit of the Ruler'. This and half a dozen other mines would feed an industrial zone producing synthetic rubber, gasoline, plastics, methanol, iso-octane and other compounds in high military demand: the single largest investment by the largest European corporation in the 1940s, by now wallowing in profits. Coal was the chief attraction of Auschwitz, although in 1940 it already had the additional benefit of an SS-run prison that might provide some future labour-power. The location decision of IG Farben set in motion the process by which it evolved into something rather more exceptional than a prison compound. Before Auschwitz became an extermination camp, it arose as a complex for extracting and processing coal; the two activities then hooked up and proceeded in lockstep.[119]

From the perspective of the victims, it could seem as though they were being reduced to coal and ash and smoke, the lines between human and fuel systematically blurred. One survivor, Benjamin Jacobs, describes in his memoir how he 'knew that in Auschwitz Jews were turned to ashes'. In front of the gates, 'I felt as if I were standing on hot coals . . . The billows of smoke rose from the chimneys as the sky brightened. Our brothers in the other group were also led away, soon to be silenced.' Jacobs and his companions were loaded onto trucks. They saw a band of people marching off: 'Their clothes were dirty, and they wore striped miners' lamps on their heads. They were on their way to work. I was struck by the paradox: the coal they

---

119 Franciszek Piper, 'The System of Prisoner Exploitation', in Yisrael Gutman and Michael Berenbaum (eds), *Anatomy of the Auschwitz Death Camp* (Bloomington: Indiana University Press, 1994), 37, 41–3; Raul Hilberg, 'Auschwitz and the Final Solution', in ibid., 81–3; Robert-Jan van Pelt, 'A Site in Search of a Mission', in ibid., 109, 117–18; Jean-Claude Pressac with Robert-Jan van Pelt, 'The Machinery of Mass Murder at Auschwitz', in ibid., 196; Hayes, *Industry*, xiii–xvi, 347–60; Florian Schmaltz, *The Buna/Monowitz Concentration Camp* (Frankfurt am Main: J. W. Goethe-Universität and Fritz Bauer Institut Frankfurt, 2010), 1–6; Tooze, *Wages*, 443–6.

mined might have been used to move the trains that carried us here.'[120] The railways deprecated by Nazi ideology were eminently suitable for ferrying Jews from across Europe; for the purposes of the SS, the railway juncture of Auschwitz recommended the site.[121]

Jacobs was taken to Fürstengrube. Sent into the mine as a slave, he would dig coal for the IG Farben rubber factory, a kind of work he, like most sharing his fate, had never done before. The novice miners were forced into the shafts without protective clothing and had to contend not only with the usual hazards, but with guards setting dogs on them and shooting at them; life expectancy in the Fürstengrube subcamp was four to six weeks.[122] 'When the elevator stopped and the doors opened, a thick coal fume greeted us. The air lacked oxygen and was full of coal dust. We could hardly see ahead of us.' At midday, 'we were all weary and tired, and as black as chimney sweeps. Coal dust had settled in our mouths and noses and had covered our skin.' Jacobs narrates a descent into coal, a dissolution in the fuel: it 'baked into our skins'; we 'could not wash it off. My eyelids looked as if they were coated with mascara. Papa and Josek didn't look any better. It won't be long before all of us are black Mussulmen, I thought.'[123] *Muselmann*, meaning 'Muslim', was the derogatory term for concentration camp inmates who moved as though in a coma, famished and apathetic, resigned to their impending death (this being at a time when Islam in the West was associated with submissiveness). Jacobs the Jewish coal slave felt himself becoming a black Muslim.

But Jacobs had the rare fortune of surviving; experienced in dentistry, he was set up as a camp dentist and relieved of mining. When the Nazi guards set the last prisoners in Fürstengrube on fire in their barracks, days before the Red Army arrived, he was no longer there.[124] He had been moved to a site where 'we came to a large group

---

120   Benjamin Jacobs, *The Dentist of Auschwitz: A Memoir* (Lexington: University Press of Kentucky, 1995), 119, 121. 'The camp was draped in the black of the rising smoke.' Ibid., 129–30.

121   Hilberg, 'Auschwitz', 83.

122   Shmuel Krakowski, 'The Satellite Camps', in Gutman and Berenbaum, *Anatomy*, 54–5; Piper, 'The System', 55; Hayes, *Industry*, 359–60.

123   Jacobs, *The Dentist*, 137.

124   Hayes, *Industry*, 168.

of prisoners, who were enclosed behind an eight-meter-high wire fence covered with iron bars that was normally used to store coal... Some inmates from Fürstengrube had arrived before us. The remnants of coal were everywhere.'[125] Some of that coal was used to cremate bodies. Coke was the main fuel in the crematoria.[126] One young Polish-Jewish man who survived the Dora camp gave his story:

> We were told to go down to the coal storage place, load up the coal to be taken to the crematorium. Dora camp was built on a hillside and the crematorium happened to be at the top of the hill. I remember pushing the wheelbarrow full of coal, up this slope towards the crematorium, arriving there and actually walking into the crematorium and seeing it in action – seeing the bodies lying there, waiting to be put into ovens and seeing bodies in the oven being burnt. I did that a couple of times, and by the third time I was very weakened and so tired. I was halfway up that hill pushing the enormous load of coal and I just slumped onto the ground and started crying.[127]

Burning a hated people with coal: could the fascist affair with technology have had a more fitting finale?

A Western Wall of Race and Arms

If the link between advanced fossil-fuelled technologies and classical fascism would still be considered insufficiently established, we could turn to countries outside the Axis. The frontman of British fascism, Oswald Mosley, had a brief career in the elite unit called the Royal Flying Corps during the First World War and, proud as a peacock, never tired of championing national aviation in the standard fascist

---

125 Jacobs, *The Dentist*, 169.

126 See Franciszek Piper, 'Gas Chambers and Crematoria', in Gutman and Berenbaum, *Anatomy*, 158, 168–9; Pressac with van Pelt, 'The Machinery', 187–91.

127 Jan Imich in Lyn Smith and the Imperial War Museum (eds), *Forgotten Voices of the Holocaust: True Stories of Survival – from Men, Women and Children Who Were There* (London: Random House, 2005), 227

terms.[128] A group of his followers took the machine into the waters. Founding a speedboat club, they drove through the rivers and lakes of Suffolk on boats built in Mussolini's Italy and groomed in races and regattas in the Third Reich.[129] As for the US, it seems, in the light of the above, hard to brush off as inconsequential the fact that one of the two most prominent peddlers of anti-Semitism and sympathy for the Nazis was Henry Ford. The magnate of the assembly line and pater of mass production, churning out *the* central commodity, Ford has been used to periodise capitalist history – Fordism/post-Fordism – and is also (sometimes) remembered for the long excerpts from the *Protocols* in his newspaper, the swastika pins distributed in his offices, the signs greeting workers exiting his factories – 'Jews Teach Communism', 'Jews Control the Press' – and the slave labour requested for the Ford-Werke in Cologne, the services that plant gave to the Nazi war machine and the profits it raked in.[130] It is even harder to brush it off when we consider that the second most prominent peddler was Charles Lindbergh.

When he made the first solo flight across the Atlantic in 1927, carried by nearly 2,000 litres of fuel, Charles Lindbergh became an idol. The worshipping swept up everyone from the tens of thousands who rushed to his plane on a strip outside Paris and hoisted him in the air to the *New York Times* (regularly referring to him as 'the hero of the Atlantic') and Mussolini (who congratulated the 'superman' for having 'taken space by assault and subjugated it').[131] Soon after, Lindbergh drifted into the far right. Several visits to the Third Reich imprinted in him the value of order and discipline. A week after the *Kristallnacht* pogroms, he stood on a Paris train platform and spotted a group of Jewish refugees trying to make it to the US: they 'gave me a strange feeling of pity and disgust. These people are bound to cause trouble if many of them go to America', he noted in his diary.

---

128 Colin Cook, 'A Fascist Memory: Oswald Mosley and the Myth of the Airman', *European Review of History* 4 (1997): 147–62.

129 Andrew Martin Mitchell, *Fascism in East Anglia: The British Union of Fascists in Norfolk, Suffolk and Essex, 1933–1940*, dissertation (University of Sheffield, 1999), 184.

130 See Max Wallace, *The American Axis: Henry Ford, Charles Lindbergh, and the Rise of the Third Reich* (New York: St Martin's Press, 2003), for example 7–16, 68, 137, 326–41.

131 Ibid., 73–5.

'Whenever the Jewish percentage of the population becomes too high, a reaction seems to invariably occur.'[132]

When war broke out, the pilot and the car czar led the America First Committee, the group campaigning for the US to stay outside the fray, from which Donald Trump would later take his slogan (as in 'An America First Energy Plan'). Lindbergh believed that Jewish interests were manipulating the US to intervene in the war. He found it annoying that Britain kept up its resistance and much preferred an alliance with the Third Reich; instead of a suicidal internecine battle, the 'white race' ought to unite against its external enemies.[133] These enemies came from the east:

> Oriental guns are turning westward, Asia presses towards us on the Russian border, all foreign races stir restlessly. It is time to turn from our quarrels and to build our White ramparts again . . . It is our turn to guard our heritage from Mongol and Persian and Moor, before we become engulfed in a limitless foreign sea.[134]

There ought to be constructed 'a Western Wall of race and arms' to hold back the foreigners. Such a pan-white wall would need a French army and an English fleet as well as a German air force, plus all the material the US could muster. Germany would have a special task allotted to it. This nation would be the 'essential' guardian of the eastern border of civilisation; 'she alone', Lindbergh wrote in 1940, 'can either dam the Asiatic hordes or form the spearhead of their penetration into Europe'.[135] What then had aviation to do with this global war for the white race? As far as Lindbergh knew, everything:

> Aviation seems almost a gift from heaven to those Western nations who were already the leaders of their era, strengthening their leadership, their confidence, their dominance over other peoples. It is a tool

---

132 Ibid., 382.
133 Ibid., 209, 381.
134 Charles Lindbergh, 'Aviation, Geography, and Race', *Reader's Digest*, November 1939, 66.
135 Roger Butterfield, 'Lindbergh: A Stubborn Young Man of Strange Ideas Becomes a Leader of Wartime Opposition', *Life*, 11 August 1941, 65.

specially shaped for Western hands, a scientific art which others only copy in a mediocre fashion, another barrier between the teeming millions of Asia and the Grecian inheritance of Europe – one of those priceless possessions which permit the White race to live at all in a pressing sea of Yellow, Black, and Brown.[136]

A manifesto, if you will, for secessionist aviation.

Lindbergh went on to lay out airline routes around the world. He contributed to the design and development of aircraft technology.[137] He was not, of course, the only individual in the US to be on affable terms with Nazis and then, after the tides of war had turned, reinvent himself as just another contributor to American progress. The ancestor of ExxonMobil, Standard Oil, worked closely with IG Farben on hydrogenation. Among US companies, only General Motors had a greater stake in the Third Reich.[138] Another businessman directed the building of its third largest oil refinery, which specialised in processing imported oil into high-octane gasoline for fighter jets. This man expressed his admiration for the fascist states in the 1930s and wished for the US to become like them. He was so enamoured with the German way of life that he employed a nanny and brought her back to his Kansas compound; a fanatical Nazi, dressed in starched white uniform, she established a reign of terror over his two young boys. She read stories to them about unruly children punished by having their thumbs cut off or being burnt to death. She compelled them to produce bowel movements at a specified time every morning, on pain of forced injections of enema or castor oil. The terrified boys would grow up to become known as the Koch brothers.[139]

---

136 Lindbergh, 'Aviation', 64.
137 Leonard S. Reich, 'From the Spirit of St. Louis to the SST: Charles Lindbergh, Technology, and Environment', *Technology and Culture* 36 (1995): 352. This particular paper is an untroubled hagiography of Lindbergh, focusing on his later expressions of compassion for the natural environment.
138 Hayes, *Industry*, for example 37–8, 335–6; Tooze, *Wages*, 132–3.
139 Mayer, *Dark*, 35–40. Of these first brothers out of four, Frederick and Charles, the former would be succeeded by David as one of the two Koch brothers heading the business empire.

On the Forward Rush

Towards the end of *Reactionary Modernism*, Herf backs away from the implications of his findings and cordons off Germany in a separate enclave. The phenomenon, we now learn, was peculiarly German, found nowhere else, with no causal relationship to capitalism. Offering the techno version of the *Sonderweg* thesis, Herf argues that Germany followed a special path away from capitalist modernity and suffered a deficit of modern values and mindsets and for this reason developed 'reactionary modernism', endemic of irrational German culture.[140] He never bothers to compare fascist Germany to fascist Italy. But we have two cases, not one, a duality that in itself contravenes the presumed singularity of the German path.[141] Moreover, if we agree with Herf that a signal feature of the current he examines was the fetishisation of technology, the removal of the machine from social relations and its elevation to a higher sphere where it could be imaginatively infused with some racial essence, then we must conclude that the Germans merely followed the example set by the British and the Americans. They copied their ideas *in extenso*. The techno-racism expounded by Marinetti, Jünger, Nonnenbruch, Lindbergh and other fascists was the creation and consensus of the literati in the greatest liberal empire of the nineteenth century, closely followed by that of the twentieth. There is a line running from the writings of a Macgregor Laird and a Harriet Martineau, not to mention Ralph Waldo Emerson or Edward A. Ross, to 'The Machine' and *Triumph of the Will*.

The defining idea of techno-racism – our white race has the greatest facility with advanced technology – required the seedbed of the global division of labour completed in the nineteenth century. No one would have got into his head that whites had a knack for machines in the second or the fourteenth, before the modern divergence. No artisanal guild could have inspired it. A technological system of local or even national scale could not sustain the illusion: only when the world economy actually sucked biophysical resources into the core

---

140 Herf, *Reactionary*, 217–19, 228, 233–4.
141 Cf. Moraglio, *Driving*, 9; Esposito, *Fascism*, 21–6.

could it take root; only when the deep base of technological progress was hidden from view, because of its location in remote peripheries, could the fetish be detached from them and instead attached to something like the white race. That separation – as Hitler himself intuited, in his twisted way – was accomplished by 'extraordinarily brutal' force. The British Empire led this pack. There is a line running, if only crookedly, from the pulverisation of Akka and the burnt-down villages on the river Niger to the invasion of Ethiopia and the later fascist atrocities: various points of deployment of fossil-fuelled technologies in the domination and destruction of non-white enemies.

From the time of the steamboat, the cutting edge of this violence had to be fossil. The Middle East, China, West Africa could not have been as thoroughly penetrated with sailing boats; the Italian pilots could not have dropped their poison from kites; the German cars and tanks could not have been dragged by horses (when they were, it was a sign of the chronic fuel shortages); the crematoriums burnt coal to burn people and would hardly have attained such concentrated efficiency with wood. In *The Origins of Nazi Violence*, Ernesto Traverso flips the *Sonderweg* thesis on its head to argue that Nazism was no deviation from capitalist modernity, but rather 'a *unique synthesis* of a vast range of modes of domination and extermination already tried out separately in the course of modern Western history'. He lists prisons, factories, abattoirs, the massacres of native populations, the procedures of a dehumanised bureaucracy, the mass production of commodities switched into one of corpses. To this 'concatenation of elements' that streamed into the Nazi kiln, we can now add fossil fuels.[142] If we consider that form of energy integral to capitalist modernity – a substratum not given much thought, until recently; 'all that is solid melts into air' read purely as metaphor – the perennial debates over whether fascism was modern or anti-modern can be settled. It was hyperfossil.

Fascists were not the only ones who drove cars and planes and burnt coal, obviously, just as others ran prisons and slaughtered

---

142 Enzo Traverso, *The Origins of Nazi Violence* (New York: The New Press, 2003), 151, 149. Emphasis in original.

natives. Anti-fascists engaged in the former activities too. In 1930, an Italian activist in French exile performed his own daredevil solo flight, crossed the Alps and dropped 150,000 anti-fascist leaflets over Milan, before crash-landing – just barely surviving – on the way back.[143] But fascists alone turned these machines into fetishes *in which their ideology was condensed* and made combustion a sacrament. They aestheticised political life, in Benjamin's analysis, by aestheticising the domination and destruction powered by divine gasoline and associated fuels; put differently, they aestheticised *power as power*.[144] What drove them to such extremes? According to the final paragraphs of 'The Work of Art in the Age of Mechanical Reproduction', where Benjamin sketched his well-known theory, 'all efforts to aestheticize politics culminate in one point. That one point is war', but this happens because the productive forces have been blocked from their natural outlet.[145] Capitalist property relations have prevented them from taking a healthy course, and so 'the increase in technological means, in speed, in sources of energy will press toward an unnatural use' – the inferno of human self-annihilation. 'Instead of deploying power stations across the land, society deploys manpower in the form of armies. Instead of promoting air traffic, it promotes traffic in shells.'[146] But we can now recognise these as false dichotomies. Fascists promoted *all of the above.* Shells and armies were not substitutes for traffic and power stations, but their cognates. The productive forces did not revert into agents of planetary destruction because they were barred from coming home, but *because they were accelerated so freely and furiously on the way there.*

Often the most observant critic of the machine, Benjamin here absolved it of destructivity and imagined that more power stations and air traffic would come with peace. In this text, as in some others,

---

143 Caprotti, 'Technology', 189.
144 Cf. Malm, *Fossil*, 17–19.
145 Benjamin, *Selected*, 121. Emphasis omitted. For other instances of Benjaminian productive force determinism and optimism *and* the opposite positions, see Esther Leslie, *Walter Benjamin: Overpowering Conformism* (London: Pluto, 2000), for example 35–6, 53–4, 92, 176.
146 Benjamin, *Selected*, 121–2. Cf. Ansgar Hillach, Jerold Wikoff and Ulf Zimmerman, 'The Aesthetics of Politics: Walter Benjamin's "Theories of German Fascism"', *New German Critique* 17 (1979): 103–4.

he was under the sway of the productive force determinism of the Comintern: the belief that technology was a force for good held back by monopoly capitalism and its most decayed form – fascism – in particular. As Poulantzas later pointed out, this belief was hopelessly out of tune with empirical realities. Mussolini and Hitler in no way halted the onward march of the productive forces: 'Fascism is *not* a *backward turn*, but rather a *forward rush*.'[147] The forwardness of that rush received retroactive confirmation after the war.

In 1934, fascist authorities devised a new plan for covering Italy with a network of motorways extending to a total of 6,850 kilometres. It made up the template for post-war construction, and 'curiously enough', notes Moraglio in *Driving Modernity*, the Italian network of motorways in 2017 measured 6,850 kilometres.[148] It might have taken eight decades before that fascist plan was realised, but Fiat became Italy's pre-eminent mass producer of automobiles in the 1950s.[149] It was the boost from the Third Reich that 'marked Daimler-Benz's rise to world industrial power'.[150] The Volkswagen project was revived by the British occupation authorities (after an embarrassing interlude in 1948 when Wolfsburg, the model city created for production of the car, gave revanchist NSDAP die-hards 68 per cent of the votes for the city council, a perseverance unheard of anywhere else in the Western sector). Soon the Beetles rolled off the conveyor belts and became 'virtually omnipresent' in the Federal Republic. The Nazis formulated the promise of motorisation, came up with the name and developed the prototype; under gentle guidance from the British, West Germany made it a reality. Volkswagen also became the bestselling small car in the US, while a special niche was occupied by the spin-off brand Porsche.[151]

The autobahn network has been called 'the most enduring of the Third Reich's material legacies': it still carries cars through the nation

---

147 Poulantzas, *Fascism*, 100. Emphases in original.
148 Moraglio, *Driving*, 142.
149 Amatori, 'The Fascist', 63.
150 Bellon, *Mercedes*, 219.
151 Rieger, *The People's*, 126. Wolfsburg: 94–5. For more on the Nazi and post-Nazi continuities as manifested in the Volkswagen success story, see for example 155, 327.

and retains an international reputation for the high quality of the roads and giddy absence of speed limits.[152] Ahead of its time, it took some decades for the autobahn to fill up; when it did, it was as 'corridors' of Western democracy.[153] Tempelhof offered an architectural blueprint for airports around the world: it has been crowned 'the mother of all airports'.[154] Jet engines and other aerospace innovations found their way into planes girding the globe. Hydrogenation was first picked up by the South African apartheid regime – another state in need of oil import substitution – and then spread farther afield, to the US as well as the globalising People's Republic of China. Some of the synthetic material industries stayed put. In the early twenty-first century, the rubber factory at Oświęcim, Polish for Auschwitz – the one for which Benjamin Jacobs dug coal – was the third largest of its kind in Europe. At least two of the world's leading tyre manufacturers sourced their rubber from it.[155]

The alacrity with which capitalism absorbed these particular gifts from Nazism lacked equivalents elsewhere. In no other realm of Nazi activity – jurisprudence, art, education, military and paramilitary organisations, party life – was there anything remotely similar to the continuity and uptake of fossil-fuelled technologies. There were problems of incomplete de-Nazification in other areas, to be sure, but not that gilded stamp of approval, suggesting that Nazism and non-fascist capitalism had a particular elective affinity on this point. Needless to say, it was repressed in the post-war era. The Nazi origins of the technologies in question were conveniently forgotten, or rationalised away (as in the idea that Volkswagen or the autobahn represented the isolatable *good* sides of the Third Reich).[156] 'As the pyramids record the history of the Pharaohs, so too the motorways will remind the German people forever of the most extraordinary figure in its history', one journalist wrote when the autobahn first

---

152 Shand, 'The Reichsautobahn', 189.
153 Zeller, *Driving*, 2; see further for example 171, 185, 240.
154 Jess Smee, 'Last Call for Berlin's Tempelhof Airport', *Guardian*, 31 October 2008.
155 Tooze, *Wages*, 446.
156 See Evans, *The Third*, 187; Rieger, *The People's*, for example 147, 304–5; Zeller, *Driving*, 62.

opened: pyramids indeed, but without the reminiscence.[157] One might be tempted to speculate that some of this repressed past has lately been attempting a return, perhaps in Germany in particular.

The desperate efforts of liberal historians such as Herf to immunise capitalism from the theory and practice of the Nazi machine are, we can then conclude, just that. As for Benjamin, he got it more right in other essays, such as 'Theories of German Fascism', a reading of Ernst Jünger that begins with a line from the far-right journal *Action Française*: 'L'automobile, c'est la guerre.'[158] Here war is not the alternative to the domination of nature, but its corollary. The insight was sharpened by Benjamin's associates in the Frankfurt School, notably Adorno, who thought that the most important psychological condition for Auschwitz was the pursuit of 'one's own advantage before all else' and 'the inability to identify with others'. 'Similar behavior can be observed in innumerable automobile drivers, who are ready to run someone over if they have the green light on their side.'[159] Or, 'which driver is not tempted, merely by the power of his engine, to wipe out the vermin of the street, pedestrians, children and cyclists? The movements machines demand of their users already have the violent, hard-hitting, unresting jerkiness of fascist maltreatment.'[160] This is the kind of whopping claim that, for critics of Adorno, seem to suggest an implausibly straight arrow of causation from the mundanities of capitalist modernity to mass murder. But the actual automobile adoration of the original fascists, its echoes on the far right today

---

157 Roger Griffin, 'Notes Towards the Definition of Fascist Culture: The Prospects for Synergy Between Marxist and Liberal Heuristics', *Culture, Theory and Critique* 42 (1999): 114.

158 Walter Benjamin, *Selected Writings, Vol. 2, Part I: 1927–1930* (Cambridge, MA: Harvard University Press, 2005), 312. 'Deeply imbued with its own depravity, technology gave shape to the apocalyptic face of nature and reduced nature to silence.' Ibid., 319.

159 Theodor W. Adorno, *Critical Models: Interventions and Catchwords* (New York: Columbia University Press, 2005), 201, 371. Marcuse had a similar eye for the place of the car in proto-fascist psychology: 'The average man hardly cares for any living being with the intensity and persistence he shows for his automobile. The machine that is adored is no longer dead matter but becomes something like a human being.' Herbert Marcuse, 'Some Social Implications of Modern Technology', in *Technology, War and Fascism: Collected Papers of Herbert Marcuse*, vol. 1 (London: Routledge, 1998), 47.

160 Adorno, *Minima*, 43–4.

and the aggregate effects of the private car suggest that he was on to something.

What exactly? It might be objected that fascists were not especially dedicated to fossil fuels and the things they powered, but rather craved modern technology in general, which just happened to have this energetic basis. But that would merely be to concede a point. On the other hand, it would be facile and obviously fallacious to posit any one-to-one correspondence between these productive forces and fascism, not only because anti-fascists utilised them too, but because *everyone* did (and still does). European integration was founded on the Coal and Steel Community. That did not make the Maastricht Treaty fascist. Nor can it be maintained that the forces themselves called for fascists: it was not the Mercedes car, as a mute physical object, that pulled the Führer into the driving seat. Fascists ran to their beloved machines, not the other way around. Why did they? It might have had something to do with the pursuit of 'one's own advantage before all else', and more precisely with the sense that those machines *inflated the power of the self above others* in a way their older alternatives did not: fascists were rarely aroused by bicycles or subways. They appeared to have found congenial the power of the engine to wipe out the vermin, or just race past and rise above them.[161]

We might then hypothesise a sequence of fascist love for fossil fuels: devotion to the national stock (visible in the Nazi programme for fuel autarky), appreciation of the mighty technologies predicated on it (vehicles having the place of honour) and lust for combustion as a socio-chemical reaction for dominating and destroying enemies – three moments corresponding to primitive fossil capital, fossil capital in general and fossil capital as a totality. But it might be too neat, and it is certainly not exhaustive. We have seen a great number of fascist motifs hitched to the machine: discipline, the end of class conflict, hierarchy, virility, war, the domination of nature as a virtue in itself. More can be added.

---

161 This implies that these machines lent themselves to fascist drives, with or without the presence of organised fascists. 'An observer of West German road traffic in 1957 noticed that "at the steering wheel, we yield to the apparent increase of our power, which inflates our 'ego'. Enormous forces obey at the touch of a finger; obedience is complete".' Rieger, *The People's*, 178.

If this goes some way towards synthesising the fascist love for the machine, however, we are still left with the question of why it took hold of some nations – only two in full – and not in others. Fossil fuels might have been necessary and conducive conditions, but they were clearly not sufficient. The answer is not hard to come by. The dominant classes of the US and the UK had scant need for a Lindbergh or a Mosley, because they were so much securer in their seats and unconstrained in space. They could afford to relax some. Roosevelt and Churchill had their own limousines and road adventures, for sure, but they were not quite as gung-ho about them as Mussolini and Hitler, nor did liberal thinkers at home in their countries – say, a John Dewey or a Karl Popper – ever develop any machine mysticism. Extreme aestheticisation appeared in two up-and-coming capitalist countries, where further expansion was blocked by the fait accompli of existent colonial spheres and thrown into doubt by the backwash from the Russian Revolution: here, fanaticism about fossil-fuelled technologies had a function to perform. The countries most successful and unbounded in their development had no commensurate need for explosive, violent unfettering. Classical fascism was a reaction *against limits*. It was *fossil* fascism in a different sense: assaulting those limits with the material force of fossil-fuelled technologies and breaking out of a generalised crisis which, in itself, had nothing to do with their ecological fallout. It was fossil fascism in the genetic not the defensive mode.

It follows, for the near future, that limits – of another nature: ecologically determined, global in scope, tightening over time – might trigger a reaction with some family resemblance. The danger of fossil fascism is the danger that actors of the far right will again position themselves as the shock troops of fossil capital, but on the other side of the historical curve. They will resist the imposition of purposively suffocating fetters. Indeed, this has, as we have seen, already happened: the far right has run towards the historical task of defending fossil fuels and their technologies in their terminal crisis. Perhaps we should have seen this coming. In the light of the genealogies we have sketched, there is nothing surprising about the grandchildren of classical fascism returning to protect the automobile and the aeroplane and the burning of coal against meddlesome hands. In other respects, however, the situation is one of epochal reverse. The conditions in which the far right operates today could hardly be more different from the interwar period, while at the same time being bound by some recognisable similarities. We should expect fossil fascisation to deviate massively from the antecedent. It will not be IG Farben and Fürstengrube in the 2020s. What makes for the central difference?

Fossilised Proletariat?

The starkest contrast between the interwar period and the early twenty-first century pertains to the class struggle. The working class, to put it simply, is a shadow of its former self. This is not the place for an in-depth review of the fraught question of proletarian support for the far right; the data is contradictory and contested – for both periods – and interpretations tainted by confused theories of class. But some things have come to indubitable light. The idea that Trump and Brexit, two of the main steps towards the right in recent years, were carried by an aggrieved working class – also referred to as 'the losers', 'the rustbelts', 'the left-behind' – has been thoroughly debunked: in both cases, affluent voters tipped the scales.[162] On the other hand, we know that the very long march of the contemporary far right in Europe began with it winning over layers of the working class from the left. The French, as so often, set the example, which then spread to northern Europe, the region where the allegiance of the class to its social-democratic parties had long been the most absolute and abiding. The DF, the FrP, the PS each bit off chunks of the red constituencies and presented itself, not without foundation in electoral statistics, as 'the new working-class party'.[163] Sweden was perhaps the last

---

162 See for example Gurminder K. Bhambra, 'Brexit, Trump, and "Methodological Whiteness": On the Misrecognition of Race and Class', *British Journal of Sociology* 68 (2017): 214–32; Jardina, *White*; Mutz, 'Status'; Dorling and Tomlinson, *Rule*.

163 See Jens Rydgren (ed.), *Class Politics and the Radical Right* (London: Routledge, 2013). The same trends marked the early rise of the far right in central-western Europe: see for example Anthony J. McGann and Herbert Kitschelt, 'The Radical Right in the Alps: Evolution of Support for the Swiss SVP and Austrian FPÖ', *Party Politics* 11 (2005): 152.

domino to fall. In 2018, the country that held out longest in class voting buckled too; for the first time, more workers voted right than left, and a growing share went for the farthest right.[164] Something similar happened in Germany.

A select group of old billionaires helped the AfD get off the ground and turn rightwards.[165] In the 2017 election, the typical voter profile seemed to fit the Trump/Brexit pattern – wealthier than the average, not objectively precarious – but in the following years, the party moved over to one side of the cleavage of united Germany: it stood tallest in the east, in provinces with manufacturing industries and low incomes.[166] It also made inroads into the unionised core of the class. The leading German labour sociologist, Klaus Dörre, pointed to the emergence of a far-right pole in the factory councils of the automobile industry and offered the following statement from one union activist, who had just led a strike to victory:

> Refugees should – at least that's my opinion – they should have to leave ... I wouldn't have a problem if they re-opened Buchenwald, put a barbed-wire fence around it, and then it's them in there and us out here. And no one has to deal with the other one. And leave them in there for as long – I mean, with normal humane treatment and all – but to put it bluntly, until they are deported.[167]

This was, in Dörre's judgement, an extreme expression of a common mood. Indeed, he reported 'an astonishing affinity for violence'

---

164   See Jens Rydgren and Sara van der Meiden, 'The Radical Right and the End of Swedish Exceptionalism', *European Political Science* 18 (2019): 439–55.

165   Melanie Amann, Sven Becker and Sven Röbel, 'A Billionaire Backer and the Murky Finances of the AfD', *Der Spiegel*, 20 November 2018.

166   See for example Siri, 'The Alternative', 142; Hansen and Olsen, 'Flesh', 12, 15; Christian Franz, Marcel Fratzscher and Alexander S. Kritikos, 'German Right-Wing Party AfD Finds More Support in Rural Areas with Aging Populations', Deutsches Institut für Wirtschaftsforschung, *DIW Weekly Report* 8 (2018): 69–79.

167   Klaus Dörre, 'A Right-Wing Workers' Movement? Impressions from Germany', *Global Labour Journal* 9 (2018): 342. Long before the AfD, Dörre and his colleagues identified a 'raw material' of far-right sympathies in the German working class: Klaus Dörre, Klaus Kraemer and Frederic Speidel, 'The Increasing Precariousness of the Employment Society: Driving Force for a New Right Wing Populism?', *International Journal of Action Research* 2 (2006): 98–128.

among his informants, the lines between so-called populism and plain old Nazism difficult to detect.[168]

On the other hand, again, there is research suggesting that the far right in Europe has been seated in or near state power by the most well-off layers in the most prosperous societies in history in periods of boom rather than bust.[169] The most reasonable conclusion from this mixed picture seems to be that the contemporary far right is *a transclass phenomenon*. Like a *Volkspartei* should, it attracts support from all classes and strata.[170] This raises two immediate questions. First, what's in it for the working class? Second, can classical Marxist theories of fascism survive this situation? If we accept that fascist tendencies are at work in contemporary capitalist societies, that fascisation is happening and proto-fascist politics are gaining ground, without – as of the late 2010s – any full-blown fascism being instituted anywhere, so that comparisons with the interwar period are legitimate, we have a conundrum at hand. How could capitalism give rise to this phenomenon, when there are no gravediggers in action?

'The Commies or us, that is the question', wrote Goebbels in his diary on 12 August 1932.[171] Classical Marxist theories pivoted on this choice. Whatever their affiliation, favoured analogies and preferred periodisation, they agreed on the core point: labour was *too strong* for capital to put up with, which was why it employed the fascist steamroller. 'Fascism', according to one restatement of the axiom, 'is there to solve the problem of working-class hostility to capitalism.'[172] No such organised problem exists today. One would have to bring out a massive searchlight to find any explicit, formalised hostility to capitalism in the working classes of the countries under study (Brazil and France perhaps being the exceptions). No one could argue that the dominant classes are rattled or profit rates squeezed by overly strong labour. No recent scenes of workers' militias on the streets torment bourgeois brains; no trade unions force

---

168   Dörre, 'A Right-Wing', 343.
169   Frank Mols and Jolanda Jetten, *The Wealth Paradox: Economic Prosperity and the Hardening of Attitudes* (Cambridge: Cambridge University Press, 2017).
170   Cf. Mudde, *The Far*, 80.
171   Hett, *The Death*, 154.
172   Renton, *Fascism*, 101.

employers to engage in baneful wage arbitration; no Commies wait around the corner of the next recession. The working class had already suffered defeats, granted, in both Italy and Germany at the time of fascist takeover.[173] But the defeat today is of another magnitude. Now the far right rises on the back of a decomposed working class, whose organisational infrastructure has been atrophying, class consciousness fraying, combativeness petering out (France again the partial exception) and autonomous political culture withering away since the 1980s, the fragmentation inside and outside the workplace proceeding apace.[174] If fascism has the 'defining ambition to crush the organised working class', it would seem to have nothing to do in the present.[175]

But if we slacken the axioms of classical theory, we can allow for a reverse scenario to play out. Fascisation happens because of the *extreme political weakness* of labour, as it approaches or falls below pre–First International levels. It is not the unnerving presence but the total absence of both revolutionary and steadily reformist working-class politics that now sets the slide in motion. One of the mechanisms has been noticed frequently, to the point of truism: when workers are no longer interpellated as comrades in a class with a mission, the far right fills the vacuum and interpellates them as members of a white nation. This can work only for a segment – and a shrinking one at that – of the working class in Western Europe, whose most precarious and exploited strata rather tend to become more multicoloured and thereby less available for white nationalism: but the damage will have already been done. Dishabituated of class politics, some of the class will have joined the forces of the nation, as the only collectivity at their fingertips.[176] Dörre contends that workers who have lost hope of ever wresting back some of the

---

173 Cf. Clara Zetkin above on fascism as the straw the masses grasped at after losing hope in 'the revolutionary proletarian class', which had *just* been defeated as the Biennio Rosso ended. Insofar as the present far-right wave is repeating that sequence, it has been drawn out over half a century rather than a couple of years (or one decade, in the case of interwar Germany).

174 Cf. Fekete, *Europe's*, 43.

175 Renton, *Fascism*, 101.

176 Cf. Neil Davidson and Richard Saull, 'Neoliberalism and the Far-Right: A Contradictory Embrace', *Critical Sociology* 43 (2017): 711, 716.

surplus-value from the class enemy switch to going after refugees. At last *they* can be beaten.[177] The far right, then, to repeat the phrase from Clara Zetkin, is '*a refuge for the politically homeless*'; as any homelessness, this one is the result of losses – weakness, not strength.[178]

A country that illustrates this predicament with special clarity is, again, Poland. David Ost, the authority on the Polish labour movement, has traced the paradoxes of its transition: after 1989, society became riven by class exploitation. Just when class politics was most needed, however, 'class talk was dead', buried under half a century of Stalinism and its afterlives, unavailable as a conductor of popular mobilisation. Instead the PiS called up the workers and promised them less suffering when 'they would be governed by "real Poles"'; and the party swept traditional proletarian redoubts like steel, auto and, of course, coal.[179] The apparent absence of class then came to be filled by another absence, namely that of Muslims. Monika Bobako, the Polish Marxist who has done the most to theorise the country's Islamophobia, has pointed out that if the left was weak in the rest of Europe in the 2010s, it was virtually non-existent here. There was 'no alternative vehicle for the expression of frustration and anger', nothing to drive such feelings towards a target that was more real.

> Nationalist ideology in Poland, as well as the 'Islamophobia without Muslims' that has been incorporated into it, can be interpreted as a useful instrument of affirmation of national pride and identity that is in fact a way to *channel social disaffection away from the dominating structures of economic power*,

---

177 Dörre, 'A Right-Wing', 341. The convergence between conservative and social-democratic parties in the centre is another aspect of the same process, crucial for opening the space for the far right in, for instance, Sweden: see Rydgren and van der Meiden, 'The Radical', 447–8.

178 Zetkin, *Fighting*, 80. Emphasis in original.

179 David Ost, 'Workers and the Radical Right in Poland', *International Labor and Working-Class History* 93 (2018): 118–19. For a collection of studies of this and related processes in Eastern Europe, see Don Kalb and Gábor Halmai (eds), *Headlines of Nation, Subtexts of Class: Working-Class Populism and the Return of the Repressed in Neoliberal Europe* (New York: Berghahn, 2011).

because the path leading up to them has been so utterly blocked.[180] We can then propose a provisional theorem: minimal space for the left produces a maximal break with reality. Poland has merely followed one route into this space; countries that were spared Stalinism and instead passed through social democracy into neoliberalism ended up there too. This is where the cognitive mapping of the rich can meet that of the poor, in a cross-class alliance of otherworldliness, the one partner unable to believe anything bad about his own system, the other to catch sight of its totality. Both will throw their mud against aliens.

What, then, is the place of fossil fuels in this conjuncture of working-class history? The epochal political defeat does not necessarily – so far – translate into material impoverishment in absolute terms. The class may have some things to lose. More importantly, it may have the *perception* of resources slipping out of its hands unless firmly clung onto. As Dörre and others have documented, significant layers of the European working class have come to see themselves as standing in the middle of society: unable to climb any further, looking down on others who are worse off, aware that a fall is possible.[181] This, ironically, has some resemblance to the petty-bourgeois insecurities considered most prone to evolve into fascist sympathies in classical theory. But shop owners and clerks in Weimar Germany did not come out of a century of self-conscious and aspirational class politics, on a par with the modern labour movement; now, the precarity is what comes after socialism. We might then expect spirited resentment against attempts to *also* take away certain resources – those consolation prizes, as it were, for living under the permanently deepening domination of capital. Of what nature might they be?

In another conjuncture of working-class history, in West Germany of the early 1960s, Adorno observed that

---

180 Bobako, 'Semi-Peripheral', 459, emphasis added. Cf. Davidson and Saull, 'Neoliberalism', 717.

181 Dörre, 'A Right-Wing', 340–1; Valentina Ausserladschleider, 'Beyond Economic Insecurity and Cultural Backlash: Economic Nationalism and the Rise of the Far Right', *Sociology Compass* (2019), online first; Oliver Nachtwey, *Germany's Hidden Crisis: Social Decline in the Heart of Europe* (London: Verso, 2018), 200.

the lot of workers today is actually no longer as it was in the classical analyses of Marx and Engels, that, simply stated, the proletarians today genuinely have more to lose than their chains, namely their small car or motorcycle as well, generally speaking – leaving aside the question of whether these cars and motorcycles are perhaps a sublimated form of chains –

and then hastened to add that to condemn such gadgets would be perverse. 'If the workers do indeed have more to lose than their chains, then that may be painful for the theory, but it is initially very good for the workers.'[182] Cars and motorcycles and their appurtenances may bring relief from pressing needs, even joy. Otherwise they would give rise to no attachment. In Adorno's view, they compensate for a general powerlessness, a sense of being unable to ever budge the structure, which leaves workers 'focused on what is close, what is directly in front of them' – commodities – which in turn stiffens the 'system-immanence of proletarian consciousness'.[183] If such a cycle of consumptive resignation could be traced in the years just before 1968, we should expect it to be rather more thoroughgoing three decades after 1989 and counting. This implies that, crudely put, *some fractions of the working class would be susceptible to fossil fascism in a mitigation crisis*, as determined as anyone to hold on to what they have got.

Indeed, the *weaker* the class becomes politically, the greater such susceptibility should be. This applies not only to the sphere of consumption (fossil capital in general) but also that of work (primitive fossil capital). After the defeat, coal has acquired a nostalgic quality as the base for a working-class pride and power that no longer seem to exist.[184] What is still left of it, organised labour may defend tooth and nail: hence the de facto alliance between the AfD and some German trade unions in the struggle over the coal phase-out. In

---

182 Theodor W. Adorno, *Philosophical Elements of a Theory of Society* (Cambridge: Polity, 2019 [2008]), 38, 50.

183 Ibid., 44, 54. For working-class attachment to cars and related commodities, see Huber, *Lifeblood*.

184 On this historical role of coal, see Timothy Mitchell, *Carbon Democracy: Political Power in the Age of Oil* (London: Verso, 2011).

Poland, coal has just those connotations, plus a material support function, no fewer than 70 per cent of households depending on it for winter comfort.[185] Little surprise if some would react to talk of climate with 'so you will deprive us of this too?', or that the Finns could score a point about the left trying to 'take the sausage from the mouth of workers'. For some time, this has been the strategic challenge for the climate movement, as it seeks desperately to interpellate workers as beneficiaries of a transition – a just one, a Green New Deal, a Green Industrial Revolution or the like.[186] Waning class-consciousness and ossifying system-immanence are what make that message so hard to sell. And if these factors might prime some elements of the class for fossil fascism on the mitigation side, one could only imagine their susceptibility in a serious adaptation crisis, when the wages of whiteness could become very real and material indeed.[187] What elements? There is no way around the fact that closest to the far right are those who self-identify as white and male.

A theory of fossil fascism – divinatory or creeping – would thus turn the classical Marxist theory on its head. It would give shorter thrift to what went for the classical *liberal* theory of fascism, as a deviation from capitalist modernity à la *Sonderweg* – if only Germany (the hottest case) had been fully capitalist like the UK or the US, the train wreck would not have happened. Today that thesis would be laughably implausible. The *Kaiserreich* left the nation with enough feudal impurities and premodern warts to make the idea of a *Sonderweg* command the attention of a generation of historians, who gave the empirical lie to it.[188] As the far right enters the 2020s – a

---

185 Z. Łukaszewski, 'Węgiel tak, smog nie – Swiadomosc i odpowiedzialnosc', in W. Bialy, H. Badura and A. Czerwinska-Lubszczyk (eds), *Systemy wspomagania produkcji gornictwo-perspektywy i zagrozenia: Wegiel, tania czysta energia i miejsca pracy* (P.A. Nova, 2018), 488.

186 See here in particular the recent work of Matt Huber, for example 'Ecological Politics for the Working Class', *Catalyst* 3 (2019).

187 Cf. Harris, 'Whiteness', 1758–60.

188 For a summary, see Eley, 'What'; Eley, *Nazism*. There is one Marxist theory that comes perilously close to this thesis: Ernst Bloch's idea of the non-contemporaneity of fascism. Germany is the land of 'unsurmounted remnants of older economic being and consciousness', of 'elements of ancient society which have not yet died' – a land not 'formed and balanced' by the 'capitalist Ratio'. Bloch, *Heritage*, 106–9. Although Bloch argued that modern capital made good use of this dross, his version of Marxist theory

decade of which we can only know one thing: there will be deepening crises – from a position of greater strength than at any point since 1945, no one could argue that its assembly areas are stuck in the pre-capitalist past. No one could point to any *Junkerdom* still infiltrating the army. No quasi-medieval hierarchies rule the schools in Denmark or the US, and so on: every candidate country for fossil fascism is *more consummately capitalist than it has ever been* and perhaps can ever become; the deficit that caused original fascism, on the liberal theory, is a surplus breaching every levee.

Paradoxically, the flaws of the classical Marxist and liberal theories here mirror each other. Both imply that if capitalism is the only game in town, it cannot produce fascism. Without factors to negate it – working towards a post-capitalist mode of production in the former case, tying it to its feudal predecessor in the latter – it would be serenely non-fascist. But if fossil fascism is now a danger, it is because *capitalism has been left to its own devices*, unthreatened by overthrow, unadulterated, uninhibitedly productive of ecological crisis. The only theory that holds clues for this scenario is that of the Frankfurt School: 'I would say that in comparison with the Germany of 1933 the decisive cause of fascism, namely the concentration of economic and administrative power on the one side and complete impotence on the other side, has progressed' (Adorno).[189] The fully capitalist earth radiates disaster triumphant.

It follows that in twenty-first-century fascism, *race has primacy over class*. The duty of Italian and German fascism was indeed to crush the organised working class; on its first day on the job, it jailed and killed the cadres of communism and social democracy. But if fascism were to assume power in the decades ahead, the immediate target would be racially defined – in Europe, immigrants, Muslims or

---

might – his reputation to the contrary – be the least-suited of all for the present. For him, fascism resulted from '*the prevented technological blessing, the prevented new society with which the old one is pregnant in its forces of production*.' Ibid., 113. Emphasis in original. It was a primitivism of a nearly Muslim kind. 'Rebellions of older strata against civilization have been known in this demonic form only in the Orient up to now, above all in the Mohammedan sector.' Ibid., 103.

189   Adorno, *Critical*, 305.

other earmarked non-white populations.[190] This is not, of course, to suggest that classical fascism was not racist – 'in Bavaria the fascist program is exhausted by the phrase "beat up the Jews"', Zetkin noted in 1923 – or that it could return without dominant class content.[191] But precisely for the differences outlined, the hammer would not *need* to fall first on labour, but would rather hit enemies defined by descent. This might change: a sudden upswing in class struggle would alter the calculus, as would a climate movement as frightening as the Commies once were. But on *current* trends, it will be race first. If Marinetti and his friends inaugurated fascism in the age of communism by smashing the skulls of women carrying the red flag, the model for post-communist times was laid down when the Serbian nationalist troops rolled into towns like Foča and Srebrenica. Border zones will be hotspots of fascist politics in this century, curving back into the interior of what could be whitened nations.[192]

So far, we have presumed that any such fascism would be anti-ecological in character, as it would preclude mitigation or foil the adaptation of those in greatest need. Lines from the history of fascism point in that direction. But what if it would rather appear as an actual *solution*?

---

190 As early as 1971, Ernest Mandel made this prophecy: 'The seeds of a potential new fascism are present in the disease, consciously induced in some imperialist countries, of xenophobic and racist mentality (against the blacks, against the coloreds, against the immigrant workers, against the Arabs, etc.).' Mandel, 'Introduction', 37.

191 Zetkin, *Fighting*, 82. On the other hand, weird as it is, 255 pages into his study of classical fascism, Poulantzas inserts the parenthetical: 'I cannot of course start examining racism here.' Poulantzas, *Fascism*, 255. No one could even possibly write a book about the twenty-first-century far right by focusing *exclusively* on its class dynamics. Likewise, one paragraph is devoted to the problem of racism and anti-Semitism *on the very last two pages* of the collection of classic texts in Beetham, *Marxists*, 356–7; no one could wait that long today without making a fool of herself. Such a class-centrist neglecting of racism was the worst blind spot of the original theories (amply rectified by the Frankfurt School), but it speaks to *real epochal differences* in the balance of class forces.

192 Cf. Alvarez, *Unstable*, 130; Klein, *On Fire*, 50.

Blood over Soil

There was also a green side of the brown. Italian fascism never bothered much about it, not even in rhetoric; for Mussolini, nature was barren until reclaimed by man, wilderness disobedient and natural parks unneeded.[193] But Germany was rather more complex. It is time to break with those

> who speak of dominating the earth; all of that must be brought back into perspective. There is nothing particular about man. He is but a part of this world. In the face of a good storm, he can do nothing. He cannot even predict it. He does not even know how a fly is made – as disagreeable as it may be, it is a marvel – or how a flower is organized. Man must relearn how to see the world with worshipful respect,

Heinrich Himmler, *Reichsführer* of the SS, orated in 1942, fresh from planning the Final Solution.[194] 'We recognize that separating humanity from nature, from the whole of life, leads to humankind's own destruction,' a professor of botany explained the outlook of the Third Reich. Hitler lent his own authority to 'the imperative to preserve the German landscape'.[195] Such rhetoric tapped into rich veins of thinking about the relation between nature and *Volk*. In the decades around the turn of the century, conservationists, zoologists, back-to-nature romantics, weekend-hikers on a flight from spiritless modernity, nationalists worried about class conflict and other Germans who did not feel at home among the machines created a ferment of

---

193 See Marco Armiero, 'Green Rhetoric in Blackshirts: Italian Fascism and the Environment', *Environment and History* 19 (2013): 283–311; Marco Armiero, 'Introduction: Fascism and Nature', *Modern Italy* 19 (2014): 241–5; Wilko Graf von Hardenberg, 'A Nation's Parks: Failure and Success in Fascist Nature Conservation', *Modern Italy* 19 (2014): 275–85.

194 Quoted in Johann Chapoutot, *The Law of Blood: Thinking and Acting Like a Nazi* (Cambridge, MA: Harvard University Press, 2018), 27. Cf. Hitler, *Mein*, 245, 285–9.

195 Ernst Lehmann quoted in Janet Biehl and Peter Staudenmaier, *Ecofascism Revisited: Lessons from the German Experience* (Porsgrunn: New Compass Press, 2011), 13; Hitler in Frank Uekoetter, *The Green and the Brown: A History of Conservation in Nazi Germany* (Cambridge: Cambridge University Press, 2006), 41.

ideas about the special bond between their nation and its environment. Germans were rooted in the soil and drew spiritual sustenance from rivers and mountains. Letting those treasures go to waste would lead to the asphyxiation of the *Volk*.[196] Alongside Malthus, this is the source of green nationalism.

It could then seem, at a first look, that the German fascists took seriously the task of protecting nature. At a closer look, it is apparent that when they spoke of nature, they had something else in mind than a sphere of existence to be saved from pillage. As Johann Chapoutot has recently showed in his comprehensive reconstruction of their ideology, *The Law of Blood: Thinking and Acting as a Nazi*, they rather wanted the Germans to think of themselves as subjects to the laws of nature. People were no different from animals or plants. Every living being fought others for living space. Germans should learn to behave like a race in nature, namely the Aryans, the apex predators in the struggle of all against all, not to be crossbred with rats. Extreme monists and holists, the Nazis refused to acknowledge any discontinuity between the social and the natural: Jews, with their abstract and dialectical reason, distinguished between the two, but Aryans lived them as a oneness, in which everything had to abide by the same unforgiving laws.[197] When Himmler inveighed against the domination of the earth, what he meant was that humans should not be so arrogant as to believe that they lived apart from the war of all against all – a one-sided view of biology, to be sure: the social Darwinism that suffused Nazi thinking.

Nature, then, should not be loved or protected so much as submitted to. It was the highest 'law-making authority', from which the

---

196 See for example Franz-Josef Brüggemeier, Mark Cioc and Thomas Zeller, 'Introduction', in Franz-Josef Brüggemeier, Mark Cioc and Thomas Zeller (eds), *How Green Were the Nazis? Nature, Environment, and Nation in the Third Reich* (Athens: Ohio University Press, 2005), 5–7; Charles Closmann, 'Legalizing a *Volksgemeinschaft*: Nazi Germany's Reich Nature Protection Law of 1935', in ibid., 23–4; Thomas Lekan, '"It Shall Be the Whole Landscape!" The Reich Nature Protection Law and Regional Planning in the Third Reich', in ibid., 77; Biehl and Staudenmaier, *Ecofascism*, 15–26; Uekoetter, *The Green*, 22–5.

197 Chapoutot, *The Law*, for example 20–31, 36–47, 151–66, 245, 314, 413–14.

Nazis purported to derive their policies.[198] Nature forced the races to wage pitiless war, whether they liked to or not. Nature dictated expansion to the east, prohibited racial mixing and decreed that Jews and other *Untermenschen* should be killed, just as when 'one animal devours another' (Hitler).[199] Nature abhorred equality. It demanded of German mothers that they fill their wombs and populate the earth with hundreds of millions of offspring, or the space would be taken by faster-multiplying races. (Racist demography is always apocalyptic: 'Little by little, Germanic blood is being drained from this earth', Hitler lamented in 1922; 'our people is dying', warned one demographer in 1934; 'Germans to be exterminated within two generations', screamed the *Völkischer Beobachter* in 1941).[200] Could anything like *ecology* emerge from this stew? It might occasionally sound like it. Man's attack on 'the iron logic of Nature' must 'lead to his own doom', Hitler would write, with a ring of an environmental Jeremiah to it (he here discussed the consequences of miscegenation).[201] The Nazis were wont to liken the *Volk* to a tree or a forest (to drive home a point about race). What was completely missing from them, however, was any perception of *social drivers of environmental degradation* or any identification with natural phenomena disrupted by human action – what we would today consider the barest minima of ecological politics.[202]

But there were claims that Jews mistreated nature. Germans had a Nordic closeness to nature, whereas Jews came from 'the hothouse atmosphere' of the oriental deserts, with no roots in the soil, on a quest 'to devastate the lakes and forests of verdant Europe'.[203] These

---

198 Johann Chapoutot, 'The Nazis and Nature: Protectors or Predators?', *Vingtième Siècle: Revue d'Historie*, no. 113 (2012), 5. Cf. Karl Ditt, 'The Perception and Conservation of Nature in the Third Reich', *Planning Perspectives* 15 (2000): 166; Chapoutot, *The Law*, 159–61, 314; Stanley, *How*, 79–80. 'Nature itself in its inexorable logic makes the decision', is the kind of sentence repeated *ad nauseam* in *Mein Kampf*. Hitler, *Mein*, 511.

199 Chapoutot, *The Law*, 156.

200 Quotations from ibid., 179, 178, 384.

201 Hitler, *Mein*, 287.

202 As argued in Chapoutot, 'The Nazis', 2–3; on the use of organicist metaphors, see 4–5. Cf. Biehl and Staudenmaier, *Ecofascism*, 15–17, 22–3, 41–2.

203 Chapoutot first quoting Bernhard Kummer and then referring to the views of Richard Darré (on whom more below) in *The Law*, 105, 101. See further Gesine Gerhard,

nomads and speculators could only tear things down. *Holzjuden*, or 'wood Jews', had cut 'the last of the strong oaks and the last of the beautiful walnut trees', one conservationist complained in 1937.[204] Spitting at nature and its highest creature, the Jews demonstrated their cruelty on defenceless animals: campaigns against *shechita*, or ritual slaughter, had long been dear to the heart of German and other anti-Semites, and the NSDAP protested wildly against this 'torture'. *Der ewige Jude* spent ten out of seventy minutes on a grisly scene of two cows bleeding to death at the hands of a knife-wielding Jew. 'Every animal has a right to live', Himmler told his masseur, and then immediately added: 'It's a way of seeing I particularly admire in our ancestors.' He repeated the point in October 1943, in a speech to the SS leaders, by now busy implementing the Final Solution: theirs was 'the only nation of the world with a decent attitude towards animals'.[205] The function of this compassion for animals is rather transparent. It was one more rod with which to beat the Jews.

All of these ideas came together in the doctrine of *Blut und Boden*, as formulated most fully by Richard Darré, the specialist in genetic livestock selection who took his skills to the post of agriculture minister. He elaborated on the distinction between rooted settlers and rootless nomads. The former had mixed their blood with the soil and become one with it, whereas the latter roamed the earth, striking no roots anywhere, blowing like locusts into the cities, creating no culture of their own but living off others: Aryans versus Jews.[206] In the late 2010s, Rassemblement National went on a trip down this memory lane. Darré was not only the original theorist of blood and soil and nomadism; since the end of the Third Reich, he has

---

'Breeding Pigs and People for the Third Reich: Richard Walter Darré's Agrarian Ideology', in Brüggemeier et al., *How Green*, 131; Mark Bassin, 'Blood or Soil? The *Völkisch* Movement, the Nazis, and the Legacy of *Geopolitik*', in ibid., 209; Closmann, 'Legalizing', 32; Lekan, ' "It Shall" ', 82.

204 Hans Stadler in Uekoetter, *The Green*, 38.
205 Chapoutot, *The Law*, 29; Uekoetter, *The Green*, 57.
206 The most comprehensive study of Darré's philosophy in English is Anna Bramwell, *Blood and Soil: Richard Walter Darré and Hitler's "Green Party"* (Abbotsbrook: Kensall Press, 1985), but it is tainted by sympathies with the object of study, and its central thesis has been irredeemably savaged by later historiography. See further below; for one rebuttal, Piers H. G. Stephens, 'Blood, not Soil: Anna Bramwell and the Myth of "Hitler's Green Party" ', *Organization and Environment* 14 (2001): 173–87.

sometimes been regarded as the voice of a reputed green Nazism. Blood, however, enjoyed unquestionable precedence over soil in his scheme. Land for him was the medium of the race, impregnated with its inner qualities or 'spiritual force'; on its own, it had no value or meaning.[207] He wanted his race of farmers to 'control the environment' and 'master' it. He also believed that free trade pulled up the peasants by the roots, worried much about flagging German birth rates and thought they could be revived if the *Volk* reconnected with the soil, but he did in fact not – as he pretended at his Nuremberg trial – engage in considerate landscape protection. He took up the burning core of Nazism with as much zest as anyone: 'We Germans are leaders in the field of airships and the most modern commercial aircraft. We Germans build the most modern roads in the world', and so forth.[208] Not even the Nazi theorist with the greenest reputation could avoid flipping brown naturalism into a project for actual domination of nature.

The same turnabout is on display in *Alles Leben ist Kampf*, a propaganda film from 1937 that opens with stags locking horns, bison flexing their muscles and trees stretching their branches to the sun. It then moves on to humans: gardeners removing weeds, mentally ill children looking foolish at a hospital. *Alles Leben ist Kampf* justified the Nazi sterilisation programmes by picturing humans as organisms that must eliminate weaker elements to survive. In the final scenes, the disabled are gone and the viewers treated to the utopia of healthy children playing on the beach, a steamboat ruling the waves, men constructing roads, factory chimneys spewing smoke, the obligatory fighter pilot and Nazi soldiers throwing model planes in the air: the human stags in action. 'She' – nature – 'puts living creatures on this globe and watches the free play of forces. She then confers the master's rights on her favorite child, the strongest in courage and industry',

---

207 Quoted in Bassin, 'Blood', 216; cf. Stephens, 'Blood', 178–9; Dritt, 'The Perception', 166–7. Thomas Rohkrämer, 'Martin Heidegger, National Socialism, and Environmentalism', in Brüggemeier et al., *How Green*, 196. See further Bramwell, *Blood*, for example 191–2; Gerhard, 'Breeding', 131–9; and cf. Thomas Zeller, 'Molding the Landscape of Nazi Environmentalism: Alwin Seifert and the Third Reich', in Brüggemeier et al., *How Green*, 147, 150, 155, 161. We here leave out the whole phenomenon of Heidegger.

208 Quoted in Tooze, *Wages*, 199; see further 172–4; Uekoetter, *Green*, 203.

Hitler philosophised in *Mein Kampf*. The Aryan 'is the Prometheus of mankind' who 'caused man to climb the path to mastery over the other beings of this earth.'[209] If this was as close as one could come to ecology in the Nazi doxa, what about the praxis?

Some brown steps did have a green tint. Within ten months after Hitler had become chancellor, Germany had promulgated three laws against vivisection and the brutal treatment of animals.[210] In 1935, there came a sweeping Nature Protection Law, which set out to shield rural landscapes from development and convinced conservationists to hitch their wagon to Nazism; their community had never been particularly close to the NSDAP – nor in active opposition to it – but now tried to make the most of the promising statutes. A rather high number of nature reserves were announced. Around the same time, the Third Reich adopted principles for forest management that ruled out clear-cutting and mandated a mixture of trees of different ages and species. Hitler and Himmler were, very famously, vegetarians. Some Nazis were smitten by organic farming, and the SS experimented with biodynamic agriculture. Himmler considered generating biogas from the faeces of concentration camp inmates. The guards of Dachau ran an organic herb garden.[211] Among these measures, however, those of any ecological significance lay in shambles within months or a couple of years, unceremoniously ignored, revoked and demolished when higher interests were at stake.

The Nazis quickly realised that vivisection was instrumental for military research – gassing, not the least – and handed out blank permits for all sorts of experiments. Animals fared no better under the Third Reich than under Weimar, and considerably worse once the war began; the Wehrmacht was a voracious consumer of horses and dogs. A frail construction from the start, the Nature Protection Law had only been on the books for one year when the Third Reich

---

209 Hitler, *Mein*, 134, 290. 'Soil exists for people which possesses the force to take it and the industry to cultivate it.' Ibid., 134.
210 Uekoetter, *The Green*, 55.
211 Closmann, 'Legalizing'; Gerhard, 'Breeding', 138; Zeller, 'Molding', 158–9; Uekoetter, *The Green*, for example 12, 27–30, 40, 44, 61–70, 142; Biehler and Staudenmaier, *Ecofascism*, 119–24; Michael Imort, '"Eternal Forest – Eternal *Volk*": The Rhetoric and Reality of National Socialist Forest Policy', in Brüggemeier et al., *How Green*, 43–72.

leapt into the Four-Year Plan of frenetic resource harvesting. 'No square meter of German soil shall remain uncultivated', thundered the Führer; 'there must not be any restrictions for areas used for the purposes of industry', declared the Ministry of Commerce.[212] The drive for 'inner colonisation' included draining swamps, regulating rivers, constructing dams and turning millions of hectares of 'wasteland' into arable farms, with no skimping on the pesticides or chemical fertilizers; schemes for organic farming were shelved, worries about soil exhaustion and depleted water tables suppressed. Now conservationists cried silently over the spread of 'monotonous steppeland'.[213] Forests were cut down at an ever higher rate, exceeding 150 per cent of the sustainable yield already by 1935. And then there was the autobahn. Friends of nature grumped about a 'road-building psychosis'. A group of designers with *Heimat* ideals, called 'landscape advocates', imagined that they were embedding the motorways in the landscape, drawing the curves gently, adorning the concrete and asphalt with local stone and freshly planted native flora: perhaps the original greenwashers. Their advice was neglected nearly all the time, and Hitler would not hear of valuable forest groves slowing down his roads. If any peacetime project rode roughshod over conservation, it was the autobahn.[214]

This assault on nature sped into a blitz with the advent of war. All restraints were thrown off on the eastern front, when the German war machine expelled Poles and Russians who, it was said, had no proper connection to the land. Half-wildlings, they had left their landscapes in a degenerate condition, replete with untouched marshes and wetlands. The Aryans would now convert the 'impenetrable thicket' and 'endless primeval forests' into the 'planned harmony of fields' and 'well-thought-out villages' typical for

---

212 Uekoetter, *The Green*, 31, 147.
213 Westphalian nature protection officer Wilhelm Münker in Lekan, '"It Shall"', 92; see further Gerhard, 'Breeding', 136–8; Chapoutot, 'The Nazis', 6–7.
214 For the details, see Zeller, *Driving*, for example 109–16, 143–9, 171–2, 241–2; and further Uekoetter, *The Green*, 9, 32, 71–2, 79; Guse, 'Nazi', 10–11. A strikingly naïve interpretation of the 'environmentalism' of the autobahn is William H. Rollins, 'Whose Landscape? Technology, Fascism, and Environmentalism on the National Socialist Autobahn', *Annals of the Association of American Geographers* 85 (1995): 494–520. 'Road-building psychosis': ibid., 504.

Germany, as the SS booklet *Der Untermensch* heralded in 1942; there would be motorways as well.[215] The Germans laid railways across the bush with a vehemence. Encountering nature reserves under red flags appears to have triggered a particular rage: 'Wherever the Germans came across *zapovedniki* [the sprawling archipelago of reserves created after 1917], they inflicted sadistic carnage' – butchering the bison, shooting the rare birds by firing squads.[216]

The Nazis appear to have developed a particular belligerence against mountains. Already in the early 1930s, they had produced propaganda films about mountains as the most intractable space of nature, which the Aryans had to master with explosive materials; *Berge in Flammen*, or 'Mountains in Flames', showed armed heroes digging, bombing, mining and blowing them up. From 1943 onwards, they made this fantasy a reality. To beat back the Allies' advantage in the air, the Nazis prepared to sink underground aeroplane factories into mountains, which first had to be disembowelled. Hundreds of thousands of prisoners were forced to gouge out the mountains and, when needed, decapitate them. Chapoutot relates the case of Wapelsberg, 'the jewel in the crown of the Thuringia Mountains' – incidentally the province of Björn Höcke – whose crest was levelled to allow for the take-off and landing of jet-propelled aeroplanes. Images show prisoners from camps such as Dora reduced to flesh and bone after months of hacking in the mountains. 'What can be read into those images of nature destroyed and men emaciated', Chapoutot asks, 'if not that they were treated identically and shared a common destiny?'[217] Nazism 'burns, it

---

215 Quoted in Joachim Wolschke-Bulmahn, 'Violence as the Basis of National Socialist Landscape Planning in the "Annexed Eastern Areas"', in Brüggemeier et al., *How Green*, 252. See further for example Chapoutot, *The Law*, 249–62, 328–77; Tooze, *Wages*, 469–76; Uekoetter, *Green*, 81, 158–60.

216 Douglas R. Weiner, *A Little Corner of Freedom: Russian Nature Protection from Stalin to Gorbachev* (Berkeley: University of California Press, 1999), 58.

217 Chapoutot, 'The Nazis', 11. In this article, Chapoutot executes the most concentrated demolition of the myth of green Nazism. The Nazis were the first 'radical environmentalists in charge of a state', writes Anna Bramwell in *Ecology in the 20th Century: A History* (New Haven: Yale University Press, 1989), 11. Nothing could be further from the truth, as we have seen; and here we have only summarised a vast recent historiography.

consumes products and it consumes energy'; in its essence, Nazism is fully accelerated combustion.[218]

The environmental shortcomings of Nazism did not represent a common human failure to meet lofty ideals. They were inscribed in the genetic code of fascism, both as a set of ideas and a real historical force. The machine won over nature, the spirit of Marinetti and Jünger over mountain and land *and it could not have ended otherwise.* Beyond the extremist fossilophilia, the reasons are at least fivefold. First, since fascism closed its eyes (to put it mildly) to inequalities, it could not register any socially determined processes harmful to nature; to that problematic, it was by definition blind. Second, blaming the racial enemy for environmental woes was bound to have zero positive effect, simply because no correlation or causation existed (*Holzjuden* did not in fact cut the oaks, and so on). All it could do was take the blame off any actual culprits. Third, whatever closeness to nature the Nazis managed to construct was programmed to be parochial, reserved for whites only and perversely ineffectual (herb garden in Dachau). Fourth, because it accorded absolute primacy to nation and race, it had to mobilise all resources in their service, with no compunctions about ecologically detrimental effects. Under the one purpose of arming the *Volk* with maximum force, Nazi naturalism ineluctably wrote out recipes for how to cut up nature.

And last but not least, because fascism-as-force presupposed an alliance with dominant classes, it could not afford to start questioning the productive forces but had to present itself as their best accelerator. When Adolf Hitler in 1927 composed his first text after *Mein Kampf*, a booklet to be exclusively distributed to the top industrialists in Germany, he promised to wield 'the might of the sword' on behalf of their businesses, respect 'the strength and genius of individual personality' – codeword for private property – and give the economy '*the opportunity to survive and develop freely*'.[219] He could not have written about the need to phase out substances and practices

---

218 Chapoutot quoted in Catherine Vincent, 'La protection de la nature permettait aux nazis de justifier la violence et le crime', *Le Monde*, 4 October 2019.

219 'The Road to Resurgence', included in full in Henry Ashby Turner Jr., 'Hitler's Secret Pamphlet for Industrialists, 1927', *Journal of Modern History* 40 (1968): 367, 369, 372. Emphasis in original.

profitable but damaging to ecosystems. It would have put off the only classes on whose arms fascism could swing into power. There were indeed a few odd greenish Nazis on the 'left' wing of the party, who mused about breaking up industrial cities, resettling the population in pseudo-medieval villages and shifting from coal to water power: and they were brusquely sidelined, so as not to alienate capital.[220] Is there any reason to believe that green nationalism could perform better in this century?

Border over Ecology

When assessing the prospects for an ecological far right, scholars are naturally drawn to the record of Nazism. For a long time, there lingered a scent of nature romanticism from the ruins of the Third Reich, leading some to expect the far right to protect the environment again, much as it did under Hitler. Indeed, one can still find students of the topic – who are not historians – alleging that Nazism 'integrally linked' race and conservation, and that ecology is *at the core of* nationalism, now as then.[221] To some extent, this impression is a leftover from the pre-climate era, the still relatively innocent years when ecology could be perceived as that little box of reality down there, where you find things like vivisection laws and organic herb gardens. Because Nazis placed some items in that box, they could – with some effort – be regarded as green. But, we know now, one first has to look at what they did with some rather more commonplace things like coal and cars and aeroplanes. It was in this area, conterminous with the entire field of Nazi operations, that they showed their true colours. In the biodynamic katzenjammer, far-right climate denial is a strange and surprising aberration; on the

---

220 Primarily Gottfried Feder and his small circle: see Guse, 'Nazi', 5–9. On the link between these 'greens' and the 'left' of the NSDAP, cf. Bramwell, *Blood*, for example 15.

221 Bernhard Forchtner, 'Far-Right Articulations of the Natural Environment', in Forchtner, *The Far*, 7. This is the analytical point of departure for most of the contributions of the volume, and its most serious weakness. See further for example Voss, 'The Ecological', 178; Kyriazi, 'The Environmental', 184; Forchtner and Özvatan, 'Beyond', 217–20.

latter view – more holistic, if you will – it has come very naturally indeed.²²²

Some have had an interest in bestowing green diplomas on Nazism. They fall into three categories: liberals and conservatives wishing to smear environmentalism, self-proclaimed eco-fascists and – numerically the least important group – progressive environmentalists who fought so hard to drive reactionaries out of the movement in the 1980s and 1990s that they overinterpreted the greenness of Nazism, so as to get the best shot across the bow.²²³ All of this is out of sync with the historiography, which has conclusively refuted the myth of green Nazism – even in miniature boxes – and demonstrated that no ecological Nazism could possibly have existed, any more than there could have been an egalitarian or a democratic one. What did exist was a measure of nature-related *rhetoric*, with counterpoints in other matters. Ernst Bloch witnessed how Nazis sometimes posed as anti-capitalists and coarsely plagiarised the labour movement. 'They stole the colour red', inserting the swastika; 'then they stole the street', marching under a forest of flags; they even had the audacity to call themselves a workers' party. 'The Nazi was creative, so to speak, only in the embezzlement at all prices with which he employed revolutionary slogans to the opposite effect.'²²⁴ Agents of green window-dressing achieved much the same. 'The enemies of technics', Marcuse wrote of them, 'readily join forces with a terroristic technocracy.'²²⁵ Now here is a trick to be repeated.

So far in this century, indications are that green nationalists will replicate all five moments that turned well-meaning conservationist

---

222 This sense of climate denial as aberration characterises much of the work of Forchtner and colleagues.

223 For the latter, see, above all, Biehl and Staudenmaier, *Ecofascism*. The classical liberal scarecrow is Luc Ferry, *The New Ecological Order* (Chicago: University of Chicago Press, 1995). The Nazis established 'the idea that the natural world is *worthy of respect in and of itself*, independent of all human considerations'. Ibid., 98. Emphasis in original. On contemporary eco-fascists invoking Darré and other heroes of Nazism, see Sarah Manavis, 'Eco-Fascism: The Ideology Marrying Environmentalism and White Supremacy Online', *New Statesman*, 21 September 2018; cf. Sam Adler-Bell, 'Why White Supremacists Are Hooked on Green Living', *New Republic*, newrepublic.com, 24 September 2019.

224 Bloch, *Heritage*, 64–5, 117. Cf. Neumann, *Behemoth*, 187–93, 216–17.

225 Marcuse, 'Some', 63.

Nazis – insofar as any such existed – into their opposites. The border, as we have seen, is as much of a blinder as blood. It prevents the believer from seeing the drivers of climate breakdown, which have nothing to do with immigration. Second, a racial enemy can, as we have also seen, only be the worst possible substitute for fossil capital. The El Paso killer came close to admitting this in his manifesto:

> The decimation of the environment is creating a massive burden for future generations. Corporations are heading the destruction of our environment by shamelessly overharvesting resources . . . The government is unwilling to tackle these issues beyond empty promises since they are owned by corporations . . . I just want to say that I love this country, but god damn most of y'all are just too stubborn to change your lifestyle. So the next logical step is to decrease the number of people in America using resources. If we can get rid of enough people, then our way of life can become more sustainable,

Patrick Wood Crusius reasoned before he walked into Walmart with a semi-automatic rifle and gunned down twenty-two people, the worst massacre of Latinos in modern US history. He even called his manifesto *The Inconvenient Truth*.[226] The logic of it seemed to be that corporations ought to be bridled, but since this is unachievable – due to their capture of the state apparatus – he would do the next best thing and slaughter immigrants. From the standpoint of fossil capital, this was a harmless vent, to match the switch from class to race hatred. In an environmental assessment, it was a pool of blood and body parts near the border.

Fortification of borders has so far produced a less than zero positive effect on nature. When they were barricaded in Central and Eastern Europe to shut down the so-called refugee crisis, the only tangible impacts were disastrous: spirals of razor wire blocking the path of wildlife. In Slovenia, the border fence against Croatia

---

226 'The Inconvenient Truth' by Patrick Wood Crusius, posted by Louis Proyect as 'The Manifesto of the El Paso White Supremacist Killer', louisproyect.org, 4 August 2019.

– erected to snuff out a trickle of Middle Eastern refugees – cut up a treasure-trove of lingering biodiversity where bear, lynx, wolves and deer had been roaming freely; now their habitats were dissected, the long-term viability of their populations in doubt. The structures on the US border with Mexico obstructed the movement of animals as much as people.[227] Beyond human suffering, fortified borders are poison for wildlife, particularly when animals seek to adapt to higher temperatures by migrating north. In 1990, fifteen countries in the world had fences or walls on their borders; in 2016, some seventy did and more were raring to go and there was no sign that this border-building psychosis had anything but detrimental consequences for nature – it cut no emissions, saved no species, cleaned up no plastic waste; the barriers made no contact with sources of pollution.

But there is room for environmental concern in the far right – of a distinctly parochial, herb-garden-in-Dachau variety. A vision of biodiversity has a place in it. In fact, whereas the far-right parties in the European Parliament formed one bloc against climate, they voted overwhelmingly in support of initiatives on biodiversity.[228] Nouvelle Écologie called for France to 'consider biodiversity a form of national wealth'. Amid its war on climate politics, even the AfD found a slot for 'protection of our landscapes and native animal species'.[229] This is the environmental issue most amenable to thematic harmonisation with the love of the white nation. Just as nature demands that species stay in their habitats – no walruses in the desert – and prohibits their hybridisation – ants cannot mate with parrots – so there must be one people per nation; updated for the age of ethnopluralism, this is

---

227 See for example John D. C. Linnell, Arie Trouwborst, Luigi Boitani et al., 'Border Security Fencing and Wildlife: The End of the Transboundary Paradigm in Eurasia?', *PLoS Biology* 14 (2016): 1–13; Boštjan Pokorny, Katarina Flajšman, Laura Centore et al., 'Border Fence: A New Ecological Obstacle for Wildlife in Southeast Europe', *European Journal of Wildlife Research* 63 (2017): 1–6; Reece Jones, *Violent Borders: Refugees and the Right to Move* (London: Verso, 2017), 140–3.

228 Schaller and Carius, *Convenient*, 12, 16.

229 Collectif Nouvelle Écologie, 'Nos 21 propositions pour une écologie patriote au XXIe siècle', actu-environnement.com, n.d; AfD, 'Schöne grüne welt?', 20 September 2019.

the equivalent of Nazi social Darwinism, the latest attempt to deduce racial order from a monist ontology.[230]

As an ecological project, it is unctuous, to say the least. If the far right cared so much about the survival of animals, it would demand that border walls come down. The use of biodiversity as a cipher for intra-species differentiation shines through in the hostility to invasive species, and in the sudden turnabout when wolves come into view. *Canis lupus* was eradicated in Germany in 1850, but after 2000, it returned across the Polish border, gaining a foothold in the form of some eight hundred individuals in seventy packs – far too many for the taste of the AfD, which made their hunting a demand to parallel 'minus immigration' in the east. 'A few wolves are fine', said Karsten Hilse, but then 'the population grows and the animals lose their shyness. They could take small children as prey.'[231] Germany suffered an 'infestation of blood-suckers and parasites' that would soon 'eat the flesh off the bones' of its people, claimed one AfD politician in Chemnitz with the surname Goebel.[232] The rest of the living world is accorded value if it is expedient for the defenders of the white nation, which means that it can be revoked at a moment's notice.

And yet the potential for sincerity seems greatest when the far right speaks about biodiversity. Imagine it sponsors a bill that actually saves a native frog in a forest threatened by logging. Then the benefits accrue to the nation. With climate, everything is different. Imagine a country like Germany or the US halving its emissions in five years: the benefits from such a programme would be *evenly distributed* throughout the climate system, as palpable in Addis Ababa or Baluchistan as in Aachen and California, and those who would gain most, relative to a

---

230 See Bernhard Forchtner, 'Nation, Nature, Purity: Extreme-Right Biodiversity in Germany', *Patterns of Prejudice* (2019), online first.

231 Sebastian Kositz, 'Haben Sie nur ein Thema, Herr Hilse?', *Sächsische Zeitung*, 18 January 2018.

232 Thomas Goebel, quoted in Sara Peschke, 'Wer hat Angst worm bösen Wolf?', *Süddeutsche Zeitung*, 15 August 2018. See for example Michael Haselrieder and Joachim Bartz, 'Der Wolf im Wahlkampf der AfD in Sachsen', *ZDF*, 20 August 2019; Markus Wehner, 'Die AfD will den Wolf jagen', *Frankfurter Allgemeine Zeitung*, 20 January 2020. A brilliant analysis of far-right lupophobia, which marks a break with the wolf romanticism of the Nazis, is Christoffer Kølvraa, 'Wolves in sheep's Clothing? The Danish Far Right and "Wild Nature"', in Forchtner, *The Far*, 107–20.

business-as-usual trajectory, would probably be non-white. The sole conceivable foundation for valuing mitigation as such seems to be *solidarity* – including, as solidarity once used to mean, a self-interest reflected in the common good and the fate of the worst off. Could the far right ever ascend to such a viewpoint? Is a universalist fascism conceptually possible? Adaptation, again, is different: a sea wall exclusively benefits those behind it. Addressing the environmental problem at scale and at its root appears to demand a subject of another political constitution.

Fourth, nothing as yet suggests that nationalists are ready to renounce any energy that could fill the nation with power; as with the Nazis, their naturalism is likely to flip into chimneys and jets as long as these can be had. And, last but not least, a transition would require confrontation with dominant classes. But the far right comes to power *with* dominant classes, not against them, buttressing not subtracting from their rule: 'Race always looks the way business needs it' (Bloch).[233] Could it look the way a rising capitalist class fraction based on the flow would need it? This is speculation beyond the horizon of current trends, all of which point to green nationalism as a repetition: just as there was blood over soil, it will be border over ecology. The environmentalist rhetoric of the far right will have an ersatz accent.

But this still leaves us with the question of determinants for swings within the diversity of tactics – why, for instance, did the French far right in the 2010s adopt a 'patriotic ecology' that feigned mitigation ambitions, while their German neighbour stuck its head deeper in the sand? The location of lignite riches may have been one factor, insufficient on its own (Vox and the SD took the AfD line without coal). Three others may be mentioned. Germany, but not France, had a mass climate movement with an anti-capitalist vanguard in Ende Gelände, with which the far right locked horns. France, but not Germany, had a decades-long tradition of an agrarian nationalism that pitied the farmer sacrificed to globalisation. The French Great

---

233 Bloch, *Heritage*, 85. Cf. Richard Saull, 'The Origins and Persistence of the Far-Right: Capital, Class and the Pathologies of Liberal Policies', in Saul et al., *The Longue*, 28.

Replacement theory, lastly, may have lent itself to a renewal of the theory of nomadism. From this divergence, it is difficult to extract projections. The presence of a massive and militant climate movement, for instance, could make the far right go either way: trying to steal the colour green, or washing off any taint of it.

But it might still be too convenient to conclude that ecological fascism is a *contradictio in adiecto*, the 'eco' in eco-fascism predetermined to always be a sham. One scenario remains to be considered. Imagine the German state in the year 2025 closes every last mine, converts the entire auto industry to the production of buses, trains, solar panels and rounds up all *Ausländer* in reopened concentration camps. Or, imagine that a Sweden under the rule of the SD institutes a five-year plan for halving emissions *and* repatriates anyone descended from post-1970 immigrants. We are here straining the imagination. But if such scenarios were to come to pass, the 'eco' would have substance to it: there would be actual mitigation. Besides Malthusianism and *Blut und Boden*, one tributary could be canalised into such a project, namely sheer brutal authoritarianism. In the climate debate up to this point, this position might have received the least representation, but it has been drafted in *The Climate Change Challenge and the Failure of Democracy*, a book written in 2007 by Australian scholars David Shearman and Joseph Wayne Smith and since mercifully forgotten.

Shearman and Smith reject the Marxist contention that capitalism is the source of this crisis. Instead they assign the blame to democracy. Now is the time to realise that 'freedom is not the most fundamental value and is merely one value among others. Survival strikes us as a much more basic value.' As global heating puts the survival of the human species in question, it has to rediscover its true nature: rigid hierarchy. 'The human brain is hard-wired for authoritarianism, for dominance, and submission' (just look at the apes). More precisely, Shearman and Smith advocate a fusion of feudalism and the one-party state – but without any planned economy – headed by a 'an altruistic, able, authoritarian leader, versed in science and personal skills', backed up by a class of 'philosopher kings or ecoelites' trained since childhood – 'as in Sparta' – to steer the world through the heat. We also learn that female brains are geared towards raising children; that 'black rap songs' expressing 'desires to murder white people' should be banned; that Islam is torpedoing the Western world by demographic means.[234] A similar utopia has cropped up in Hervé Juvin's proposal for a 'Council for Life' with nine members fully authorised to ensure human survival. Here it would be the stags without the smoke.

Of all the scenarios considered, this – an ecological fascism to equal fossil fascism in substantiality – would require the most profound rupture with current trends. But one small step in this direction might have been taken in Austria in the first days of 2020. After the FPÖ had disgraced itself in a corruption scandal, the ÖVP-FPÖ government dissolved, national elections were held, the Greens rode high on the Fridays for Future and eventually, after long negotiations, an ÖVP-Green government was formed. The conservatives of the ÖVP had by then photocopied the Islamophobia of the FPÖ, effectively becoming the mainstream far right. After sealing the agreement with the Greens, Chancellor Sebastian Kurz exclaimed: 'We succeeded in uniting the best of both worlds. It is possible to protect the climate and borders.' The government programme included a plan to source 100 per cent of its electricity from renewables by 2030 *and* ban the *hijab* on schoolgirls aged up to fourteen, lower the price for public transport *and* put asylum-seekers in 'preventive detention' (and lower corporate taxes.)[235] Follow that line into a mitigation crisis on a razor's edge, and ecological fascism is no longer unimaginable.

To assess its likelihood, one must ponder two questions. Will fossil capital defend itself to the bitterest end? So far, indications are that it will indeed. Will it draw on the unique resources of the far right for the purpose? That too: it remains hard to imagine a transposition whereby fossil capital would fight its rearguard battles *from the left* while the right executed its destruction. Green parties in Europe, however, can evidently not be trusted with keeping left. Non-socialist,

---

234 David Shearman and Joseph Wayne Smith, *The Climate Change Challenge and the Failure of Democracy* (Westport: Praeger, 2007), 130–4, 13, 141, 111.

235 BBC News, 'Austria Backs Green Agenda with New Coalition Deal', 2 January 2020; Benjamin Opratko, 'Austria's Green Party Will Pay a High Price for Its Dangerous Alliance with the Right', *Guardian*, 9 January 2020.

foreign to the labour movement, they could be the catalysts for something like ecological fascism, if they were to gain further support from popular climate mobilisations without red-green outlets and, with their consent, lean further to the right. But we have also seen the degenerated social democracy of Denmark take a similar step. As society careens rightwards, more collapses and cave-ins along the way cannot be excluded. The progressive climate movement would then have to wage war on at least two fronts. It has liked to argue that the transition is an opportunity to rebuild society from the bottom up, because it demands a total and extreme material makeover; conversely, if fascism were to be built into that transition, it would stand on some solid footing.[236] It would be a world saved for sustainable dystopia. At the time of this writing, the odds for such eco-fascism might be slightly lower than for eco-socialism, but markedly higher than for fossil fascism in its various shapes, including those of green nationalism, which, as practised so far, tend to be but another form of denial.

---

236 The classical argument is Klein, *This*.

11

Death Holds the Steering Wheel

Denial comes in many forms. Imagine that your neighbour beats his wife badly every Saturday. Each Sunday morning, you wake up and think: what a wonderful neighbourhood this is, peaceful and prosperous, a blessing to live in! If someone asks whether you heard strange sounds yesterday evening, you shake your head vigorously. Or you might respond that some couples behave that way, fighting it out with fists and tableware – it is just one way of conducting an argument. They seem happy enough when he's not drunk. Or you might recognise to yourself and others that there is grave violence inflicted on that woman and it ought to stop, but then you go about your daily life, month after month, and you listen to the muffled cries without acting – or perhaps you slip in the business card of a therapist through the letter slot, or talk to another neighbour who is also content just talking about the matter, and even if the assaults continue and you glimpse the woman in a state of physical collapse, you imagine that you have done your part.[1]

In *States of Denial: Knowing about Atrocities and Suffering*, Stanley Cohen introduces a basic tripartite taxonomy of denial: literal, interpretive and implicatory. If someone asserts that a bad thing does not happen and is not true, her denial is literal; if she accepts that it

---

1 Adapted from Cohen, *States*, 7, 51–2.

happens but gives it a lower degree of meaning – rewriting the event, obfuscating the effect, exculpating the perpetrator – it is interpretive. But the most insidious form is perhaps the third. Here the facts and gravity of the matter are accepted, but not *acted* upon. Knowledge is not an issue. The harm is fully acknowledged, but the obligation to intervene is suppressed through one cognitive technique or other.[2] Building on Cohen, Kari Marie Norgaard has argued, in *Living in Denial: Climate Change, Emotions, and Everyday Life*, that implicatory denial has been the general response to the climate crisis in advanced capitalist countries. Her study is based on observations in Norway during one warm winter, when the normally snow-covered mountains lie bare. Her informants do not doubt the distressing reality of climate change – it is all too obvious, and their government has vowed to address it since day one – but they find myriad ways of evading it, passing over it, pretending that they deal with it as they go about their daily lives with nugatory modifications; when it comes to doing nothing, there is no lack of ingenuity.[3] But on climate, denial cannot continue in this implicatory mode forever.

The kind of denial we have been primarily concerned with here – that propounded by the denialist ISA and most of the far right – would seem to fall into the two first categories, which we can, for simplicity's sake, merge under the rubric 'literal'. Given its rise to prominence and power in the last decade, we then face an enigma. Why would implicatory denial revert to the stage of literality? If European countries have been 'living in denial' for decades while *claiming* to be aware of the problem, why would sections of them start rallying to the most blatantly ignorant version? Global heating has a determinate temporality: absent interventions, it does not stagnate or fluctuate around an average level of ailments, but deteriorates towards catastrophic breakdown. Rudimentary knowledge contains awareness of this non-linear declension. Unlike the many atrocities Cohen writes about, here denial concerns not something in the past (a history of slavery, an old genocide) nor a body of misery that permanently deforms the present (homelessness, human rights

---

2   Ibid., for example 7–9, 22.
3   Norgaard, *Living*.

violations), but a spiral that connects past, present and future and only speeds up, through non-negotiable mechanisms of a biogeochemical and physical nature, unless mitigated at the source. This confers a particular instability on implicatory denial. Contradictions build up inside it as in a pressure cooker, and one way out is to say: well so let us go on with business-as-usual, phew, for the problem does not exist (or is not that bad, unlike immigration).

This might give some relief. One German commentator noticed as much when the AfD turned its lather from refugees to climate: 'The AfD is changing from the fear party to the relief party. By denying anthropogenic climate change, it frees people from the pressure to change. Flying, meat, the internal combustion engine – no problem at all!'[4] The far right offers people the opportunity to live as they always have without the mounting psychic conflicts of implicatory denial in an ever warmer, more degraded world. Norgaard's book ends before the surge of the far right in Europe, so it treats literal denial as an American crotchet, expecting continued lip service to rule the rest of humanity, but it might also be the case that *reversion from implicatory to literal denial makes sense with rising temperatures*. It is perhaps easier to live like Donald Trump if you think like him and not like Al Gore: and if so, it becomes comparatively easier by every warming day.

Implicatory denial would then *engender the far-right solution*. By turning up the fire under the pressure cooker, it invites the denialists to return and blow off some steam. Here, it is important to remember another particularity of the climate crisis: it is not a state-directed atrocity, in the vein of a government torturing dissidents or locking up an ethnic minority in camps. Cohen's prototypical cases are about such immediate, face-to-face violence. When global warming was first brought to the governments of advanced capitalist countries, it incriminated not this or that authoritarian state, *but the material substratum for the capitalist mode of production* as it actually existed and had operated for more than a century. Science did not say: there is evil in napalm and machetes; put them away. It said: there is evil in automobiles and aeroplanes and coal-fired power plants and a

---

4 Hillje, 'Wider'.

hundred other slow-burning fuses – and how could the governments of ordinary capitalist states respond to that?

In *The State in Capitalist Society: An Analysis of the Western System of Power*, Ralph Miliband notes that capitalist classes distinguish themselves from previous exploiting classes in history in that they do not (normally) govern directly, as did pharaonic kings or feudal lords. Governments do it for them. The holders of office will not say so, of course; they will say that growth and a healthy business environment are necessary for any other policy goals and simply assume that servicing capitalist class interests is for the general good. That commitment is so deep, so internalised and unquestioned that it not only 'colours the specific response' to a problem but the very 'mode of perception itself'. Now if a problem comes along that requires these states to act in 'fundamental opposition' to capitalist class interests, their reflexive reaction will be some form of misperception and reluctance to intervene – even if the problem were to be 'a vast criminal organisation'.[5] Writing in 1969, Miliband had no idea of just how vast and immanent it could be. But he put his finger on the structural inability of capitalist states to process information such as that about the climate and their determination to maintain business-as-usual at any cost. The many climate-related irrationalities we have inspected in this book are *spin-offs from that foundational move*.

And the move continues, without deceleration. On any minimal definition, it is irrational. Rationality, writes Adorno, 'is the organ of adjustment to reality', and simple empirical data suggest that the dominant classes do not have such an organ.[6] At the height of the extreme summer of 2018, for example, the International Energy Agency reported that more capital again flowed into fossil fuels worldwide. Not only did investments continue, but *their share relative to renewables expanded*, so that 59 per cent of all capital streaming into the production of energy now ended up in oil and gas and coal.[7] Exactly three decades after Hansen's testimony in the sweltering heat, it would have been irrational to operate any facilities for

---

5 Ralph Miliband, *The State in Capitalist Society: An Analysis of the Western System of Power* (London: Quarter, 1976 [1969]), 67, 69, 194.
6 Adorno, *Philosophical*, 5.
7 International Energy Agency (IEA), *World Energy Investment 2018*, iea.org.

burning fossil fuels. It would have been even more irrational to build new ones. It would have been the height of irrationality to increase the rate of investment – but what could be said about a scenario where that rate knocked out parallel investments in renewable energy, in the year 2018?

The prime mover of this spiral movement is capital accumulation. Enjoying halcyon days, oil and gas companies were responsible for the greatest boom in 2018. No less than 94 per cent of global energy investment was financed 'from capital incorporated into a company's balance sheet or from private individuals' own assets' – that is, from profits already made, a cycle otherwise known as self-expanding value. The companies were so aflush in money that they could throw it into ever greater capacity.[8] The compulsive logic at work was summed up by an Australian team of scientists: 'As long as it remains financially optimal, fossil fuel producers have incentives to exploit their reserves rapidly and continue exploration, in direct conflict with other efforts to mitigate climate change.'[9] In direct conflict with any other ambitions they might have voiced, capitalist states expedite this logic, through state-owned enterprises – Petrobras and Equinor come to mind – or by greasing the wheels of private ones. As of this writing, *all* behave in this fashion, whether they pay homage to the science or not. Not one of them has initiated the liquidation of primitive fossil capital. Quite a few still profess awareness of the greatest challenge humanity has ever faced and yada yada yada: Justin Trudeau of Canada, for instance, the soi-disant climate hero who told a gathering of oil and gas executives in Houston in 2017 that 'no country would find 173 billion barrels of oil in the ground and just leave them there. [Applause.] The resource will be developed. Our job is to ensure this is done responsibly, safely and sustainably.'[10] For a capitalist state, that is indeed the job description.

---

8 Ibid., for example 12–14, 119–20, 125, 132, 138; quotation from 119.
9 Saphira A. C. Rekker, Katherine R. O'Brien, Jacquelyn E. Humphrey and Andrew C. Pascale, 'Comparing Extraction Rates of Fossil Fuel Producers against Global Climate Goals', *Nature Climate Change* 8 (2018): 489.
10 Jeremy Berke, '"No Country Would Find 173 Billion Barrels of Oil and Just Leave it in the Ground": Justin Trudeau Gets a Standing Ovation at an Energy Conference in Texas', *Business Insider*, 11 March 2017.

Implicatory denial, in other words, is the psychology of capitalist climate governance – 'the rich, convoluted and ever-increasing vocabulary for bridging the moral and psychic gap between what you know and what you do', with Cohen.[11] Greenwashing is its own form of post-truth.[12] When ExxonMobil came around to climate avowal and in the fall of 2018 plastered cities with ads for its petty laboratories experimenting in algae biofuels – 'the low-emission fuel of the future', to be produced by a 'fat, fit, fantastic green machine' – it was a tactic for sheltering the core business in a thin green varnish.[13] One lie was coated in another, the original deception not rectified but compounded. The denialist far right here represents less of a deviation than a reconciliation. *The break with reality is effected in the political economy*; the far right proposes to bring official rhetoric along. In Europe, it likes to peddle the line 'we say what you are thinking'; when it makes climate denial literal, it could adapt it to 'we think and say what you are doing'.[14] Norway, with its world records in hypocrisy, is the holy land of this rapport.

And the far right performs a similar act on immigration: it is not that the borders of Europe are open and the far right wants them closed – they are already closed and have been so for some time, and it wants to be *upfront* about this, fill in any tiny remaining gaps and deport those who have squeezed in. Borders enshrine a material ideology of assessing a person's right to be present and protected from harm *on the basis of his or her origins*. The far right turns that principle into a cosmovision.[15] It can strengthen its belief with a walk or car drive through the average European city, which will quickly reveal that non-white people are de facto ascribed less value as

---

11   Cohen, *States*, 9.
12   As argued by Laura Stegemann and Marinus Ossewaarde, 'A Sustainable Myth: A Neo-Gramscian Perspective on the Populist and Post-Truth Tendencies of the European Green Growth Discourse', *Energy Research and Social Science* 43 (2018): 25–32.
13   See for example ExxonMobil, 'Creating Tomorrow's Fuel from Unexpected Sources', energyfactor.exxonmobil.eu, 19 July 2018; ExxonMobil, 'Advanced Biofuels', corporate.exxonmobil.com, n.d.; Chris Goodall, 'Is the FT Right to Carry Misleading Advertisements from Exxon?', *Carbon Commentary*, 26 October 2018.
14   Cf. Tybring-Gjedde of FrP: 'We shouldn't speak with a forked tongue! Be clear – be climate sceptics, because that's what the people are!'
15   Cf. Jones, *Violent*.

human beings. Some political scientists have theorised the permanented presence of far-right parties in European politics as a 'normal pathology', but, in an incisive paper, Cas Mudde has turned this around: they rather represent a *pathological normality*. They are 'in line with key tenets of mainstream ideologies' and advocate their 'radicalisation'. Or, as Althusser would have it: the ideology of the far right does not merely describe the real relation; it expresses it *as a will*.[16] This goes for climate and for race.

Is there anything new here? 'For more than 30 years, the tendency has been emerging among the masses of the advanced industrial countries to surrender themselves to the politics of disaster instead of pursuing their rational interests and, chief of all, that of their own survival.'[17] From this tendency, generations of irrationalities appear to be spawned. They are not hatched in dysfunctional brains; proto-fascist individuals are not ill, not clinically psychotic, but 'often even better "adjusted" than the non-prejudiced ones'.[18] Among the messages in a bottle Adorno sent out like an armada, this one is not the least disconcerting: the break with reality is *caused by reality itself* and then reacts back upon it. Under the conditions of a fossil economy, the rational thing to do is turn on the coal stove, take the car to work, fly to Thailand for holiday, buy some shares in an oil company. The *totality* is irrational. It cannot adjust to the reality it produces and so breaks off from it, one way or another, in a flight that eventually sweeps up individuals too. 'People are inevitably as irrational as the world in which they live.'[19] Or, 'the tendency toward pathological opinion must be derived from normal opinion', when what goes for normal is 'the state of the world rushing towards catastrophe'.[20] We

---

16 Cas Mudde, 'The Populist Radical Right: A Pathological Normalcy', *West European Politics* 33 (2010): 1175, 1178; Althusser, *For*, 198.

17 Theodor Adorno, 'Sociology and Psychology', *New Left Review* series 1, no. 46 (1967): 67.

18 Adorno et al., *The Authoritarian*, 748.

19 Ibid., lii. Cf. Adorno, *Philosophical*, 81, 126–8.

20 Adorno, *Critical*, 106, 13. 'It is truly startling to think about how much in society is flagrantly irrational and how much we avoid really wanting to know this.' Sally Weintrobe, 'The Difficult Problem of Anxiety in Thinking about Climate Change', in Sally Weintrobe (ed.), *Engaging with Climate Change: Psychoanalytic and Interdisciplinary Perspectives* (Hove: Routledge, 2013), 40.

can particularise this diagnosis and say that after the onset of the climate crisis, *the reproduction of fossil capital as such secretes ideologies of denial and other irrational pathologies.*

Bourgeois Optimism Kisses Reality Off

The most mainstream of liberal ideology has played its part well here. In the 2010s, one noticeable instance was the spate of neo-optimism. A brigade of bourgeois intellectuals made the case that the world was improving in a manner nothing short of supercalifragilisticexpialidocious. Book shelves in airports and kiosks filled up with titles like *Factfulness: Ten Reasons We're Wrong about the World – and Why Things Are Better Than You Think* by the late Hans Rosling, the main Swedish non-fiction export product, who really did believe that the world was becoming a better place every passing day. How did the realities of climate breakdown fit into that belief? Not at all, of course. The section on the issue in *Factfulness* starts with the line: 'When people tell me we must act now, it makes me hesitate.' Rosling grants that the problem is real – 'as real as Ebola was in 2014' – but the thrust of his argument is an assault on climate activists. They deliberately deceive the public. They are 'doomsday prophets' in the habit of making 'exaggerated or unsupported claims' and spreading 'far-fetched, unproven hypotheses'. They pretend that climate causes HIV and shark attacks. On *this* issue, you have to 'remember the uncertainty in the data'. If you look with the sober eyes of this high priest of liberal knowledge, you will realise that global heating is not that bad – sure, a problem, just like Ebola once was: a brief dip in the ever-rising curves of capitalist modernity, in which every decent Westerner should continue to swim.[21] Bill Gates handed out 4 million copies of the book as gifts to American students in 2018. Right on the heels of that year's summer, the Swedish government distributed it to *every* student in the last year of high school, alongside instructions to

---

21 Hans Rosling, *Factfulness: Ten Reasons We're Wrong about the World – and Why Things Are Better Than You Think* (London: Sceptre, 2018), 228–34.

teachers for how to convey the message: talk about an ideological state apparatus.[22]

On climate, if not on immigration, Rosling would have made a good advisor to the SD. His Danish predecessor Bjørn Lomborg was much loved by the DF.[23] In the UK, one brick in the 55 Tufton Street wall was Matt Ridley, conservative hereditary peer, coal mine owner, advisor of the Global Warming Policy Foundation and vice president of Vote Leave. He produced *The Rational Optimist: How Prosperity Evolves* – 'a blast on the vuvuzela of common sense', according to the endorsement from Boris Johnson.[24] The lullaby continued in Steven Pinker's *Enlightenment Now: The Case for Reason, Science, Humanism, and Progress*, which begins treatment of the ticklish subject with token nods to the science (for which Pinker is supposedly making his case). Indeed, 'humanity has never faced a problem like it', the Harvard liberal admits of the climate crisis. But then he presents the following contentions: populations around the world are 'better protected against storms, floods, and droughts' than they have ever been; each year of GDP growth yields even better sea walls; the world economy is undergoing a 'natural development' of de-carbonisation; emissions are flattening and falling; trying to make polluters pay is 'punitive aggression'; if people are told that the problem can be solved by geoengineering, they are more likely to believe in its existence.[25] This way of reasoning – 'reason in the service of unreason' – has a good deal of similarities with Freud's anecdote of the borrowed

---

22 Elin Joleby, 'Gates skänker Rosling-bok till studenter', *Svenska Dagbladet*, 6 June 2018; Emma Olsson, 'Gymnasister får Factfulness', *Svensk Bokhandel*, 26 September 2018. The Swedish bourgeoisie has a foible for neo-optimism: other pen-pushers of the genre, representing the uppermost strata of this class and seeing the climate crisis with eyes like Rosling's, include Stefan Fölster, Anders Bolling and Johan Norberg. The latter is a senior fellow at the Cato Institute, founded by Charles Koch, a promoter of 'Climategate' and publisher of denialist reports with titles like *Apocalypse Not* and *Climate of Fear*. See Mayer, *Dark*, for example 257, 270.

23 Pia Kjaersgaard, 'Giv os proportionerne tilbage', Dansk Folkeparti, 3 August 2009. On the role of Lomborg and bourgeois optimism in the denialist ISA, i.e. Oreskes and Conway, *Merchants*, 256–60; Powell, *Inquisition*, 83–6; Mann and Toles, *Madhouse*, 113–14.

24 For a continuous update of Ridley's curriculum vitae, see *DeSmog UK*, 'Matt Ridley', n.d.

25 Steven Pinker, *Enlightenment Now: The Case for Reason, Science, Humanism, and Progress* (New York: Viking, 2018), 137–54, 382.

kettle.[26] A man is accused of returning a borrowed kettle in a damaged condition. In the first instance, he responds that the kettle was in fact unblemished when he gave it back; in the second, that it had a hole already when he borrowed it; in the third, that he never borrowed it at all.[27]

Massively popular among well-off readers, the neo-optimists accepted the assignment of defending bourgeois civilisation against the accusation that it burns the earth, the nub still being that *no radical intervention is needed* – it would just cause unnecessary misery. Bringing emissions first to half and then to zero would, according to Pinker, 'require forgoing electricity, heating, cement, steel, paper, travel, and affordable food and clothing'.[28] In a world without emissions, people must go hungry and naked and freeze to death, the minister of enlightenment reports (while also alleging that capitalism is de-carbonising anyway), in words that do not fall far from those of Mark Steyn or Pascal Bruckner or for that matter Donald Trump.[29] The bourgeoisie shall be remembered for producing not only presidents like Donald Trump, but intellectuals who, in the midst of climate breakdown and the sixth mass extinction, proclaimed that 'everything is amazing'.[30] No matter that nature crumbled, the neo-optimists would always 'affirm the continuous perfection and endless progress of humankind' – oh, but that was Marinetti. 'Have faith in progress, which is always right, even when it is wrong.'[31]

---

26 Adorno, *Critical*, 108.
27 Sigmund Freud, *The Interpretation of Dreams* (Oxford: Oxford University Press, 2008 [1899]), 95–6.
28 Pinker, *Enlightenment*, 141.
29 Trump anticipating the results of a Green New Deal: 'No planes. No energy. When the wind stops blowing, that's the end of your electric [sic]. "Let's hurry up. Darling, darling, is the wind blowing today? I'd like to watch television, darling."' David Smith, 'Democratic Oversight Is "Bullshit": Trump Goes Off-Script at CPAC', *Guardian*, 2 March 2019. Trump responding to the presence of Greta Thunberg at the World Economic Forum: 'This is not a time for pessimism – this is a time for optimism. Fear and doubt is not a good thought process . . . This is a time for tremendous hope and joy and optimism . . . We will never let radical socialists wreck our economy.' CNBC, 'Watch President Donald Trump's Full Speech at the Davos World Economic Forum', YouTube, 21 January 2020.
30 Pinker, *Enlightenment*, 283.
31 Marinetti, *Critical*, 53, 251. On Marinetti's optimism, cf. Härmänmaa, 'Futurism', 342.

Futures of Denial

How long can this go on? We have seen predictions that 'outright climate denial' – the literal kind – will become 'less and less viable' (this from 2012) and we have seen them miss the mark. And now we hear them again. 'If straight-up denial seemed a viable strategy then' – back in the days of Anders Breivik – 'nine years later (with six of those years among the ten hottest ever recorded) it is less so today', Naomi Klein suggests in *On Fire: The (Burning) Case for a Green New Deal* from 2019. She divines that 'climate science will no longer be denied' and thinks the far right will have no choice but to turn to green nationalism.[32] Apart from the empirical circumstance that the most psychotic forms of denial exploded after the 1°C threshold was crossed in 2015, there are a number of problems with this forecast.

To posit a causal relationship between the catastrophic degree of global heating and far-right recognition of it is to bet on rationality. The underlying assumption appears to be that when denialists see their own houses burning, they will have to open their eyes – somewhat like the expectation that if racists only were to hang out with blacks and Muslims, they would come to their senses. Feel-good stories occasionally take place in real life. But it is more the rule than the exception that fantasies and prejudices of this character 'make themselves completely independent from interaction with reality', since their primary function is to serve the psychological needs of their bearers. They are founded on an abandonment of the reality principle and formed in detachment from the object; nothing guarantees that experience will disturb them.[33]

Wildfires appear to furnish strong cases in point. 'The Amazon is not burning, not burning at all', said Ernesto Araújo on 4 September 2019.[34] Just as Vox scrapped the low-emissions zone in Madrid, firefighters were struggling to contain the worst wildfires in decades around that city. The party then proceeded to block an

---

32 Klein, *On Fire*, 47.
33 Adorno et al., *The Authoritarian*, quotation from 613; see further for example xlv, 149, 608–19, 627, 973.
34 CNN, 'Brazilian FM: The Amazon Is Not Burning'.

official declaration in the Spanish parliament that would have supported the victims of such fires, due to its 'ideological content' (a formulation about climate change exacerbating the problem).[35] When California burned again in August 2018, Donald Trump blamed environmental regulations, and when Australia suffered an inferno in late 2019 topping that in the Amazon – at least fifteen times the area incinerated – an army of ministers and politicians and pundits from the right lined up to deny any climate connection, other than possibly – the now obligatory theory – that climate activists did the deed.[36] All these events struck panic and rage in masses of people. But not in the far right. To assume that it will adjust to reality as a matter of course is a little too charitable, particularly as the capacity for unmooring shows no sign of diminution; instead, digital life and new conspiracism distend it.[37]

On the other hand, as we have seen, the gathering clouds of smoke have pushed some parts of the far right towards accepting trend, attribution and impact and instead denying the actual drivers. Denial, Cohen points out, is always partial. It is in the nature of it that some

---

35 Stephen Burgen, 'Firefighters Battle Forest Blaze in Central Spain', *Guardian*, 30 June 2019; *El Periódico de la Energia*, 'A Vox no le gustan las palabras "cambio climático" por se de "progres"', elperiodicodelaenergia.com, 27 August 2019.

36 Trump: Matthew Sparke and Daniel Bessner, 'Reaction, Resilience, and the Trumpist Behemoth: Environmental Management from "Hoax" to Techniques of Domination', *Annals of the American Association of Geographers* 109 (2019): 538. Australia: see for example Paul Karp, 'Australian Natural Disasters Minister David Littleproud: "I Don't Know If Climate Change Is Manmade"', *Guardian*, 10 September 2019; Graham Redfearn, 'Tony Abbott, Former Australian PM, Tells Israeli Radio the World Is "in the Grip of a Climate Cult"', *Guardian*, 3 January 2020; Ketan Joshi, 'Something Else Is Out of Control in Australia: Climate Disaster Denialism', *Guardian*, 8 January 2010; Ketan Joshi, 'Australians Are Ready to Break Out of the Cycle of Climate Change Denial', *Foreign Policy*, 14 January 2020; Michelle Crowther, 'Can the Catastrophic Fires Bring Some Sanity to Australian Climate Politics?', *Vox*, 10 January 2020.

37 This tendency has of course been noticed by many. For just one example, see Robert Samuels, *Freud for the Twenty-First Century: The Science of Everyday Life* (Basingstoke: Palgrave, 2019), 20. Even brief online contact with climate conspiracy theories seem capable of stimulating departure from reality: Sander van der Linden, 'The Conspiracy-Effect: Exposure to Conspiracy Theories (about Global Warming) Decreases Pro-Social Behaviour and Science Acceptance', *Personality and Individual Differences* 87 (2015): 171.

information is registered.[38] The parameters of denial can then shift, as it is indeed likely to do in later stages of this crisis: yes, global warming as such exists, but it is not the cause of *this* extreme weather event; it does not induce migration; there is no emergency; people like Greta are being hysterical – secondary and tertiary forms that seem to have a future. They already have a past. Back in 2010, the Koch brothers sponsored a museum exhibition arguing that *if* global warming is a reality, humans will adapt to it by developing 'short, compact bodies' or 'curved spines' so they can live in underground cities.[39] Yes, the problem exists, and we can do nothing about it – a variant tried out by the Trump administration, when it justified freezing the fuel-efficiency standards. Because of fossil fuel combustion, the earth will heat up by 3.5° by 2100, an internal document admitted; but whether we burn all that gas makes no difference, so we may as well forge ahead – a most specious fatalism, which has belonged to the far-right repertoire for some time.[40] If you are caught in a forest on fire, no single step can bring you to safety. It might require a thousand steps in one running leap, which means that every nationalist denial of the need for action – *our* emissions are infinitesimal, blame China, look at Poland, etcetera – is a fallacy.[41] It might segue into climate nihilism. Yes, global warming exists, and let's burn all the fossil fuels we can get our hands on and enjoy the sight of non-white people going up in smoke.[42] An infinite number of combinations seem possible. As the crisis deepens, the

---

38   Cohen, *States*, 22. Denial can thus, in Cohen's masterful analysis, shift easily between the literal, interpretive and implicatory modes.

39   Mayer, *Dark*, 265.

40   Juliet Eilperin, Brady Dennis and Chris Mooney, 'Trump Administration Sees a 7-Degree Rise in Global Temperatures by 2100', *Washington Post*, 28 September 2018; Branko Marcetic, 'The Trump Administration's Climate Nihilism', *Jacobin*, 4 October 2018; Dana Nuticelli, 'The Trump Administration Has Entered Stage 5 Climate Denial', *Guardian*, 8 October 2018.

41   It is fallacious also because it ignores cumulative and per capita emissions, and because it pretends that science recommends reduction of the *largest* emissions chunks, when in fact it says that *all emissions* have to be brought to zero. A green nationalist making this argument – we won't cut our emissions, someone else should – might have stopped denying the ABCs of climate science. But he denies all the other letters and indeed alphabets that science composes to keep pace with the destabilisation of the climate system.

42   As suggested by Sparrow, *Fascists*, 91.

bandwidth of denial stretches out – not shrinks – all the way to aggressive affirmation.

'One day the world gon' know what you done here'

If continuous reinvestment in business-as-usual engenders denial, so do profits from the past. The climate crisis is peculiar in that it – concretely, biogeochemically – sums up things that have happened between humans over two centuries. In a searching essay, Nancy Tuana brings up the case of the Alabama coal mines. During the Civil War, much of the coal supplying the Confederate armies came from there – slaves were sent from plantations to dig up the fuel – and afterwards, forced labour continued. Thousands of legally free black men were convicted of crimes such as vagrancy and ordered to haul up coal for free. About half of them died within the first four years, their remains incorporated into the ground around the mines. Tuana notes that this episode in the history of the African diaspora and the fossil economy forms the basis of one chapter in Yaa Gyasi's novel *Homegoing*, but she does not cite the central passage:

> And the people on strike broke the line, swarming the few white bosses who were standing guard. They broke the shafts and dumped the coal from the tramcars before breaking those too. H grabbed a white man by the throat and held him over the vast pit. '*One day the world gon' know what you done here*', he said to the man, whose fear was written plainly across his blue eyes, bulging now that H's grip had tightened.[43]

The climate crisis is the one long day when the world learns to *know what you done here*, and rudimentary knowledge of this moment of reckoning has existed since the earliest hour; the UNFCCC speaks of 'common but differentiated responsibility', meaning that humans have caused the problem but *some vastly more*

---

43 Tuana, 'Climate', 15–18; Yaa Gyasi, *Homegoing* (London: Penguin, 2016), 174. Emphasis added.

than others; and that differentiation is, as everyone knew from Rio onwards, inextricably bound up with the history of colonial violence. Taking on full responsibility for preventing climate catastrophe therefore comes with a risk. It holds the white boss over a vast pit crawling with the ghosts from unrepaired injustices. Because the boss has so long turned a blind eye and profited from them, something in him might rise up against the duty to accept responsibility for cumulative emissions, which have, after all, accrued from the invasion of Akka and that of Iraq and countless other acts of constructing material superiority. Where might the reckoning end? The psychic economy could be better maintained by disavowal of the problem, by literal or other tactics. And wrongs denied might easily become wrongs repeated.[44]

'The sudden widening of the eyes, the animal mimicry of innocence' is a signature move of the far right.[45] Palindefence flashes from such eyes. Cohen notes how nationalism positions the perpetrator as victim and vice versa: 'A Serb soldier in 1999 talks about the Battle of Kosovo as if it happened the week before.'[46] Here whites are victims of a timeless aggression, but this type of fantasy has spread beyond the ranks of nationalist warriors. In *Psychoanalyzing the Left and Right after Donald Trump*, Robert Samuels observes that the right – including what remains of mainstream conservatism – excels in imagining its own group as victim, even as it enjoys privileges the putative victimisers cannot come close to. Even the richest billionaires seem able to believe that they are persecuted by governments taxing their wealth.[47] In Europe, again, the far right has an especially strong desire for innocence and victimhood where it was most directly involved in

---

[44] Drawing on Orange, *Climate*, 17, 37, 39, 45. There might, as proposed by Orange, be a particularly intimate link between disavowal of climate change and disavowal of slavery. For a brilliant analysis of the latter, which hints at connections to the former, see Catherine Hall and Daniel Pick, 'Thinking about Denial', *History Workshop Journal* 84 (2017): 1–23.

[45] Ben Lerner, *The Topeka School* (London: Granta, 2019), 142.

[46] Cohen, *States*, 96–7.

[47] Robert Samuels, *Psychoanalyzing the Left and Right after Donald Trump: Conservatism, Liberalism, and Neoliberal Populisms* (Basingstoke: Palgrave Macmillan, 2016), 9–11, 16–17.

the Holocaust. In Germany, Austria, Poland, it is possessed by the idea of the nation as unfairly chastised for the war; the AfD lives to replace atonement with a 'positive self-identification'.[48] For a camp so preoccupied with historical whitewashing and imagined victimhood, acceptance of highly differentiated responsibility for the climate crisis seems a priori excluded. Do you mean to say that *we* have done something wrong? 'History matters, it matters lot', said Santiago Abascal at Covadonga: 'We will never apologise for the works of our elders.'[49]

If comparisons between climate and Holocaust denial could seem insensitive and far-fetched in the first phases of the former, they became unavoidable in the fourth. Here the denialists were sometimes *the very same persons* who had denied or downplayed the Judeocide.[50] Jean-Marie Le Pen, the FPÖ, the SD, the AfD at one point or another engaged, if only with slips of the tongue, in what used to be called Holocaust 'revisionism' as a cognate of climate 'scepticism'. An instructive case is Pat Buchanan. This paleoconservative veteran famously argued that the survivors of the Holocaust must have lied, because 'diesel engines do not emit enough carbon monoxide to kill anybody'.[51] He was also a carrier of the fable of Cultural Marxism and remains, as of this writing, an unbending climate denialist.[52] Denial of the Holocaust and denial of climate change had more than one discursive similarity, including the tropes of evidential unreliability, scientific uncertainty, hoax and conspiracy and the persecution of dissenters, all the way down to the centrality

---

48   Samuel Salzborn, 'Antisemitism in the "Alternative for Germany" Party', *German Politics and Society* 36 (2018): 82–4. Austria: Karin Stoegner, ' "We Are the New Jews!" and "The Jewish Lobby": Antisemitism and the Construction of a National Identity by the Austrian Freedom Party', *Nations and Nationalism* 22 (2016): 484–504. Poland: Christian Davies, 'Under the Railway Line', *London Review of Books* 41, no. 9 (2019): 29–30. Needless to say, this is not to imply any symmetry between German and Polish deeds and fates during the Second World War.

49   González, 'Abascal'.

50   For an excellent piece, see Bernhard Forchtner, 'Climate Change, Holocaust Denial and the "Lies of Our Times" ', *Radical Right Analysis*, 19 December 2018.

51   Adam Peck, 'Why MSNBC Dumped Pat Buchanan: His 10 Most Outrageous Statements', *Think Progress*, 17 February 2012.

52   Rosenberg, 'Donald'; the section on 'Global Warming' at *Patrick J. Buchanan – Official Website*, buchanan.org.

of gases and combustion – 'the gas was too weak to kill', a stock argument of the revisionists.[53]

But the objects were plainly different. In the beginning, global warming was the unintended by-product par excellence, the burning of fossil fuels not initiated for the express purpose of annihilating a people. Cohen has contended that the Holocaust is 'a well-documented set of events that happened in recorded history', whereas the climate crisis 'is a scientific prediction of what is likely to happen in the future'.[54] The latter interpretation was last tenable circa 1988. There is, surely, a difference in temporality: Robert Faurisson and David Irving denied the Holocaust *long after it was over*, while the denialist ISA and the far right denied the climate crisis *as it began to unfold*; only the former, it follows, lacked impact on the course of the events in question. 'No one is being killed as a result of the deniers' lies', writes Deborah Lipstadt in her classic account of Holocaust denial.[55] The same could not be said of climate deniers. Their crime is vast, one more thing *the world gon' know*.

On Feedback Mechanisms

If holding on to business-as-usual generates irrationalities, so does the catastrophe itself. It can drive people crazy. Imagine you are in the loop, see the storms and the droughts, register the unheard-of weather anomalies in your area, receive the news of scientific reports and even worse events farther away. Imagine you keep the open secret inside you for years. It grates on you, wears you down. Nothing seems to ameliorate the situation. The

---

53   Cohen, *States*, 137. Cf. Lipstadt, *Denying*, for example 112–13, 159, 213, 249, 254–5.

54   Stanley Cohen, 'Discussion: Climate Change in a Perverse Culture', in Weintrobe, *Engaging*, 77. For a critique of Cohen's odd views of the climate crisis, see Avi Brisman and Nigel South, 'New "Folk Devils", Denials and Climate Change: Applying the Work of Stanley Cohen to Green Criminology and Environmental Harm', *Critical Criminology* 23 (2015): 449–60.

55   Lipstadt, *Denying*, 3. But plenty forms of denial were, of course, at work in the planning and execution of and connivance with the Holocaust.

forces rushing towards breakdown appear all-powerful, beyond supplication; you cannot see how they could be brought under control. How do you stay sane? You do not: you panic – or, with Adorno and Horkheimer, there is 'panic ready to break out at every moment', because people expect that the world 'will be set on fire' by a 'totality over which they have no control'.[56] If this situation goes on long enough, and if still no breakout seems viable, or if the demands it places on people are extortionate (revolution never being a comfortable undertaking), there might be one option left: affirming the forces of destruction. Life in late capitalism is a life among

> absurdities, the most blatant of which is the threat brought to mankind by the very same technology which was furthered in order to make life easier. Who wants to survive under present conditions is tempted to 'accept' such absurdities, like the verdict of the stars, rather than to penetrate them by thinking which means discomfort.[57]

Note that this does not presuppose any *conscious* knowledge of the catastrophe: it can be suppressed, and precisely for that reason simmer as an indistinct anxiety. Denial is a palliative; it might temporarily manage the anxiety, as the threat out there grows.[58] Eventually the distress must find some outlet. And here the fascist leader comes and offers to channel the insecurities and paranoias away from 'their objective reasons'.[59] The more objective the reasons to be afraid, the more tempting his offer will be; the feelings become overpowering when a destructive status quo appears to be petrified. It cannot be pierced by any simple operation. It has a way of drawing curtains around itself, making it difficult to see what is going on and who is in

---

56   Theodor W. Adorno and Max Horkheimer, *The Dialectic of Enlightenment* (London: Verso, 2008), 29.

57   Adorno, 'The Stars', 57. 'If the status quo is taken for granted and petrified, a much greater effort is needed to see through it than to adjust to it and to obtain at least some gratification through identification with the existent – the focal point of the fascist propaganda.' Adorno, 'Freudian', 134–5.

58   Cohen, *States*, 45.

59   Adorno, 'The Stars', 165.

charge. It cannot be controlled; it is not in control of itself; law-bound and blind, it is a self-driving car without a navigation system.[60] 'Growing masses here are seeking an escape route from the dreadful suffering of our time. This involves much more than filling one's stomach. No, the best of them are seeking an escape from deep anguish of the soul', wrote Zetkin.[61]

Students with pre-test jitters, parachute jumpers, deep-sea fishermen exposed to jeopardy may soothe their dread and recreate predictability by fleeing into superstition. In experiment after experiment, psychologists have showed that when individuals experience a lack of control, they are prone to create 'illusory pattern perception' to restore their sense of control, conspiracy theories being one option.[62] Now, if there is something we can predict about unmitigated climate breakdown, it is that it will intensify the lack of control. To what effect? This question has recently been subjected to initial experiments, with psychologists feeding informants messages about climate change imperilling their well-being. One team tested students in Germany and the UK and found that both reacted by expressing hostility towards out-group others. The climate crisis, the researchers concluded, 'can subtly increase people's general readiness to aggress toward deviant groups' – which would ensure that the source of the malaise remains unaddressed, so that the two processes 'may fuel and catalyse each other'.[63] Another team had Austrians and Argentinians undergo a survey and observed that the former – but, interestingly, not the latter – responded to scientifically accurate information about the climate by reaffirming the greatness of their own ethnic nation. The anxiety was relieved by the in-group closing ranks against non-Austrians who had done

---

60 Adorno, *Elements*, 127; Adorno et al., *The Authoritarian*, 661–5; and see further Paul Leduc Browne, 'Reification and Passivity in the Face of Climate Change', *European Journal of Social Theory* 21 (2018): 435–52.
61 Zetkin, *Fighting*, 60.
62 Jennifer A. Whitson and Adam D. Galinsky, 'Lacking Control Increases Illusory Pattern Perception', *Science* 322 (2008): 115–17.
63 Immo Fritsche, J. Christopher Cohrs, Thomas Kessler and Judith Bauer, 'Global Warming Is Breeding Social Conflict: The Subtle Impact of Climate Change Threat on Authoritarian Tendencies', *Journal of Environmental Psychology* 32 (2012): 6.

nothing to induce it.[64] A sense of control can apparently be re-boosted by charging groups with essences and then derogating or deleting the negatives.[65] Rationalisation of this affective logic could occasionally appear on the far right. Towards the end of the Swedish summer of 2018, one of the leading publicists of the SD argued that voters exerted no influence over the climate, 'but we can have full control over migration', notably by initiating repatriation; Swedes just had to live with the former, while the latter could be dealt with.[66] Some apocalypses are more easily remedied than others.

This, evidently, is the psychology of scapegoating, encountered throughout this book and bound up with the desire for expulsion: in chapter 16 of Leviticus, Aaron is instructed to lay his hands 'on the live goat, and confess over it all the iniquities of the people of Israel, and all their transgressions, all their sins, putting them on the head of the live goat, and sending it away into the wilderness', also identified as 'a barren region'.[67] In Europe, immigrants in general and Muslims in particular have had hands laid on their heads in this manner for a couple of decades by now.[68] The hatred against a racial minority 'cannot be worked off because it can never be fulfilled', with Adorno and Horkheimer; because it carries the weight of a destructive status quo, there can be no satisfaction, only enraged repetition.[69] Has this effect already left an imprint? By the very nature of the thing, the extent to which recent outbreaks of racism have been caused by unconscious anguish related to the state of the planet is anybody's guess. But if the effect exists, we must conclude that it risks becoming more powerful in the near future, insofar as the crisis appears more frightening and unstoppable: there will be a

---

64 Isabella Uhl, Johannes Klackl, Nina Hansen and Eva Jonas, 'Undesirable Effects of Threatening Climate Change Information: A Cross-Cultural Study', *Group Processes and Intergroup Relations* 21 (2018): 513–29.

65 Markus Barth, Torsten Masson, Immo Fritsche and Carolin T. Ziemer, 'Closing Ranks: Ingroup Norm Conformity as a Subtle Response to Threatening Climate Change', *Group Processes and Intergroup Relations* 21 (2018): 498–9.

66 Dick Erixon, 'Agendasättande medier: klimat, klimat, klimat!', *Samtiden*, 13 August 2018.

67 Leviticus 16: 21–2.

68 For a most trenchant analysis, see Emmanuel Terray, 'Headscarf Hysteria', *New Left Review* series 2, no. 26 (2004): 118–27.

69 Adorno and Horkheimer, *Dialectic*, 171.

tremendous amount of energy seeking compensatory control.[70] Living goats are in the enclosure, marked with signs of foreignness and scary globality. If the far right were to be possessed by this spirit, what could hold it back? Nothing in itself: nationalism *is* a slide towards excess.

The Flame for Myself

A case could be made that most symptoms we have inspected here are bound together by the one trait of narcissism. Libidinal investment in self rather than others, or infatuation with one's own grandiosity, is implicated in conspiracy theories in general and new conspiracism in particular: I have reason to believe only myself, plus those who hold up mirrors to my face.[71] 'The answer is me. Me. I talk to myself', Donald Trump responded when asked about his source of information and advice.[72] Contrary to his own belief, he is not one of a kind. There is a type of human being who, Adorno writes, 'ultimately cannot be spoken to, can probably not be reached at all and lives on a kind of narcissistic island' – the profile of the inveterate denialist.[73] What Adorno had before his eyes, of course, was fascist propaganda, which announces that 'the follower, simply through belonging to the in-group, is better, higher and purer than those who are excluded'. Nazism in particular 'increased beyond measure *the collective narcissism*, simply put: national vanity', a concept recently operationalised by a group of scholars led by Agnieszka Golec de Zavala and Aleksandra Cichocka, who argue that some people adore and crave adoration of their own groups rather than themselves as

---

70 Cf. Adorno et al., *The Authoritarian*, 223, 607–8, 619; Marcuse, 'Some', 54; Alvarez, *Unstable*, 87–8, 130.

71 One useful study of the subject in dialogue with Adorno and Marcuse is C. Fred Alford, *Narcissism: Socrates, the Frankfurt School, and Psychoanalytic Theory* (New Haven: Yale University Press, 1988). This work has numerous problems, however: a category of narcissism so inflated that it encompasses nearly all human strivings; an overemphasis on its 'progressive' dimensions; a consequent defence of the pursuit of the domination of nature. But extensive discussion belongs elsewhere.

72 Muirhead and Rosenblaum, *A Lot*, 67.

73 Adorno, 'The Stars', 164.

individuals.[74] Libidinal investment is here moved to the level of the collective, more precisely the nation. Collective narcissists demand constant validation of the greatness of their nation, with real-time and real-world consequences. Using all the data available to social psychologists, Golec de Zavala, Cichocka and their colleagues have demonstrated that people scoring high on this trait supported Trump, voted for Brexit, backed the PiS and believed that Jews and gender studies plotted to devalue the Polish nation.[75] They would probably find the same sublated self-love under any other stone they turned. It is there in the sense of entitlement, the complaint of insufficient recognition, what we have called palindefence, the insistence on giving the nation all the respect and resources it deserves.

The narcissist has a way of behaving with nature. He sees in it a pool reflecting the picture of the great master himself, who does with it as he pleases.[76] There remains to be written an analysis of how the introduction of fossil fuels into the world supercharged ecological narcissism, by allowing the master to disentangle from the cycles and landscapes of the flow and short-circuit the energy through himself, his desires and quest for control. But we have heard British imperialists laud steam for giving them omnipotence and superiority over all others, and in his

---

74   Adorno, 'Freudian', 130; Adorno, *Critical*, 96. Emphasis added. Cf. Alford, *Narcissism*, 63.

75   See for example Agnieszka Golec de Zavala, Aleksandra Cichocka, Roy Eidelson and Nuway Jayawickreme, 'Collective Narcissism and Its Social Consequences', *Journal of Personality and Social Psychology* 97 (2009): 1074–96; Christopher M. Federico and Agnieszka Golec de Zavala, 'Collective Narcissism and the 2016 US Presidential Vote', *Public Opinion Quarterly* 82 (2018): 110–21; Agnieszka Golec de Zavala and Aleksandra Cichocka, 'Collective Narcissism and Anti-semitism in Poland', *Group Processes and Intergroup Relations* 15 (2011): 213–29; Agnieszka Golec de Zavala, Rita Guerra and Cláudia Simão, 'The Relationship between the Brexit Vote and Individual Predictors of Prejudice: Collective Narcissism, Right Wing Authoritarianism, Social Dominance Orientation', *Frontiers in Psychology* 8 (2017): 1–14; Wiktor Soral, Aleksandra Cichocka, Michal Bilewicz and Marta Marchlewska, 'The Collective Conspiracy Mentality in Poland', in Uscinski, *Conspiracy*, 372–84; Marta Marchlewska, Aleksandra Cichocka, Orestis Panayiotou et al., 'Populism as Identity Politics: Perceived In-Group Disadvantage, Collective Narcissism, and Support for Populism', *Social Psychological and Personality Science* 9 (2018): 151–62.

76   Cf. Joseph Dodds, *Psychoanalysis and Ecology at the Edge of Chaos: Complexity Theory, Deleuze/Guattari and Psychoanalysis for a Climate in Crisis* (London: Routledge, 2011), 62; Alford, *Narcissism*, 57, 69, 123.

solitude up on the fuel tank, Marinetti accorded the trait an archetype: 'At last, I break loose and fly freely / over the intoxicating abundance.'[77] Psychoanalysts have shown how late capitalist consumption feeds, and feeds on, the same self-loving self: I buy this commodity 'because I'm worth it', and when I have bought it I have proven my worth – my money really did purchase it – but with such self-referential investment, the cycle cannot come to a rest, as every transaction leaves a taste of incompleteness in the mouth: the narcissist is grandiose and fragile.[78] His self-esteem might be sundered by the smallest setback. He cannot stand a slight. He is bedevilled by the prospect of losing out to others, and precisely this pendulum constitutes narcissism as the psychic accompaniment of capitalist property relations: the mastery over nature ascending to ever greater heights through unsparing competition and shake-out of those not masterly enough.[79]

Now this bourgeois self is being told that he cannot go on like this any longer. He may respond with pharaonic fury.[80] The suggestion of ecological limits is a narcissistic injury of the first order, perhaps comparable only to the Russian Revolution; it would be a wonder if it did not elicit narcissistic rage.[81] Such rage should be most violent among those 'for whom a sense of absolute control over an archaic

---

77 Marinetti, *Selected*, 39.

78 L'Oreal advertisement quoted in Sally Weintrobe, 'The Difficult Problem of Anxiety in Thinking about Climate Change', in Sally Weintrobe (ed.), *Engaging with Climate Change: Psychoanalytic and Interdisciplinary Perspectives* (Hove: Routledge, 2013), 41, and see further this volume, still not accompanied by any similar work. In this regard, the cycle of consumption approaches that of accumulation in limitlessness, arguably precisely by detaching itself from the use-values.

79 This implies that levels of narcissism should be particularly high in countries with a history of few constraints on capitalist property relations (read: the US), that they should increase with continuous capitalist development, and that they should be highest in the capitalist class itself. And for this there is empirical evidence: see for example Paul K. Piff, 'Wealth and the Inflated Self: Class, Entitlement, and Narcissism', *Personality and Social Psychology Bulletin* 40 (2014): 34–43.

80 This is not, as we have seen, the sole possible response: one can also invite Greta to every stage and forum, let her explain how tight the limits are and how much has to be changed, take a selfie with her and then do nothing – that is, do everything just as before with a 3 per cent compound growth rate. There comes a point when this kind of response is a sublimated narcissistic rage.

81 On the Russian Revolution as such an injury, cf. Adorno et al., *The Authoritarian*, 722.

environment is indispensable because the maintenance of self-esteem – and indeed of the self – depends on the unconditional availability' of the material: flying into the rage, they express their 'utter disregard for reasonable limitations'.[82] They don't want to hear about an emergency. They develop a hatred of this reality and the mere thought of it, retreating into optimistic fantasies of endless gratification. Coming closer, objects like climate science and movement are recalcitrant reminders that stoke up the rage, and the same might go for racial others, who by their very difference and existence seem to offend the in-group aggrandised by collective narcissism.[83] If such a rage takes hold, it might start looking for instruments to obliterate those 'objects that belie a subject's expansive sense of self' and, to quote Judith Butler, destroy any 'restrictions imposed on destruction itself'.[84] We have come all the way to the death drive.

---

82   Heinz Kohut, 'Thoughts on Narcissism and Narcissistic Rage', in P. H. Ornstein (ed.), *The Search for the Self*, vol. 2 (New York: International Universities Press, 1972), 645, 639–40; cf. Alford, *Narcissism*, 69–70.

83   For example ibid., 644; Golec de Zavala et al., 'Collective', 1074, 1090–1; Adorno et al., *The Authoritarian*, 147–9, 632–3; Samuels, *Psychoanalyzing*, 19–20, 47–8; Aleksandra Cichocka, 'Understanding Defensive and Secure In-Group Positivity: The Role of Collective Narcissisim', *European Review of Social Psychology* 27 (2016): 294; Earl Gammon, 'Narcissistic Rage and Neoliberal Reproduction', *Global Society* 31 (2017): 524; Hemma Rössler-Schülein, 'The Struggle between Good and Evil: The Concept of the Death Drive from a Kleinian Perspective', in Victor Blüml, Liana Giorgi and Daru Huppert (eds), *Contemporary Perspectives on the Freudian Death Drive: In Theory, Clinical Practice and Culture* (London: Routledge, 2019), 35–7.

84   Gammon, 'Narcissistic', 517; Judith Butler, 'Political Philosophy in Freud: War, Destruction, and the Critical Faculty', in Richard G. T. Gipps and Michael Lacewing (eds), *The Oxford Handbook of Philosophy and Psychoanalysis* (Oxford: Oxford University Press, 2019), 730. The perspective on the death drive we are interested in here is that which considers it primarily directed against others and secondarily against the self – first murder, then suicide (the Nazi trajectory); more Auschwitz, less Nirvana – that is, the perspective of the late Freud, not the Freud of *Beyond the Pleasure Principle*. A useful introduction to these shifts is Victor Blüml, 'The Death Drive: A Brief Genealogy of a Controversial Concept', in Blüml et al., *Contemporary*, 1–22. A study that rather draws on the perspective of *Beyond* to produce illuminating reflections on the self-destructive tendencies in late capitalist digital culture is Benjamin Fong, *Death and Mastery: Psychoanalytic Drive Theory and the Subject of Late Capitalism* (New York: Columbia University Press, 2016). An excellent analysis of the slide from narcissistic aggression to the death drive, focusing on the biographies of Steve Jobs and Anders Breivik, is Harri Virtanen, 'The King of Norway: Negative Individuation, the Hero Myth and Psychopathic Narcissism in Extreme Violence and the Life of Anders Behring Breivik', *Journal of Analytical Psychology* 58 (2013): 657–76.

Of all Sigmund Freud's concepts, this might have produced the most headshakes. Can humans ever have a desire for death, as such, even if it is only the death of others? Surely some kind of pleasure must be the ulterior goal? But if one looks for empirical evidence of the existence of a death drive, there is hardly a better place to start than the writings of, once again, Marinetti and Jünger. At one point, the former declared that he had found the unsurpassable symbol of his project in a story about Japanese merchants producing coal from corpses: 'All gunpowder facilities are working on the production of a new explosive which is deadlier than any other known hitherto. This terrible new compound has carbon from human bones as its principal ingredient', and so the merchants rooted around in battlefields strewn with corpses and recycled them into ammunition: and Marinetti could not imagine anything more entrancing than this death-to-death loop.[85] 'I felt irresistibly drawn to the site of the calamity', Jünger confessed. 'Will to destruction manifested in its purest form by means of machines', he waxed lyrical; everywhere he looked, there were 'gaping craters similar to those of some distant lifeless planet'.[86] It was exactly this *Produktionskrieg* that stimulated Freud to make his scandalous proposition. As the reports from the First World War came in, the way technology enhanced human destructivity led him to suspect that something else was at work, something that could not be explained by the standard desire for pleasure, however disguised or rerouted: it must be a *Todestrieb* – a death drive.[87]

The second impetus for the theory was fascism itself, which has remained the strongest case for those who think this something can have actuality. Here is an ideology that lures the self with a cure for

---

85   Marinetti, *Critical*, 251. Cf. Hewitt, *Fascist*, 157.

86   Jünger, *Storm*, 7; Jünger, *Interwar*, 36, 29. This was the drive of the proto-fascist generation: 'I had always taken particular pleasure in destruction', wrote Jünger's *Freikorps* brother Ernst von Salomon. Theweleit, *Male Fantasies Vol. 2*, 383. Jünger experienced the same exaltation again from watching the 'enormous clouds of smoke' when Paris was bombed in the Second World War. Neaman, *A Dubious*, 148–9.

87   Butler, 'Political', 729–33; Marlène Belilos and Eugénie Lemoine-Luccioni, 'War: A Core Issue for Freud', in Marlène Belilos (ed.) *Freud and War* (London: Karnac, 2016), 66.

narcissistic injury, by giving it 'scope to act out the death drive'.[88] For the German fascists, the injury was double: revolution and a war lost. Masses across Europe had been so traumatised by the war as to break loose from the dominant ideology in packs and move towards revolution. It fell to the fascists to glorify the wasteland, relish and swallow the destruction – the dominant ideology in overdrive – and thereby inspire Freud to give his last word on the subject in *Civilization and Its Discontents*. Here the death drive is derived from narcissism:

> Yet even when it appears without any sexual purpose, in the blindest destructive fury, there is no mistaking the fact that its satisfaction is linked with an extraordinarily high degree of narcissistic enjoyment, in that this satisfaction shows the ego how its old wish for omnipotence can be fulfilled.[89]

And here Freud brings out what we can recognise as an ecological dimension of his theory. The wish for omnipotence is satisfied through 'control over nature', a pursuit in which humans have been so successful that 'they will have no difficulty in exterminating one another, down to the last man'.[90] As controlled as it appears to be, however, nature also becomes the greatest living disproof of the ego's omnipotence. He cannot touch it without breaking the mirror. It reminds him that the source of good lies outside of himself: an outrageous slap in the face. Freud ends with a quote from Goethe's Mephistopheles:

---

88 Virtanen, 'The King of Norway', 672. Virtanen also thinks 'liberal capitalism' can do this. Ibid. Cf. Butler, 'Political', 744; Theodor Adorno, 'Sociology and Psychology – II', *New Left Review* series 1, no. 47 (1968): 88. But Freud is, of course, not necessarily the most reliable guide to politics. 'Uncultivated races and backward strata of the population are already multiplying more rapidly than highly cultivated ones', he complained in one of his letters to Albert Einstein in 1932. Letter included in Belilos, *Freud*, 14. The only people unable to compose such sentences in the interwar period appear to have been the Marxist anti-fascists.

89 Sigmund Freud, *Civilization and Its Discontents* (London: Penguin, 2002 [1930]), 57. The letters to Einstein, however, postdate this book.

90 Ibid., 57, 81.

'For everything that comes into being is worthy of destruction ... So, then, everything you call sin, destruction – in short, evil – is my true element.' As his adversary, the devil himself names not the holy and the good, but nature's power to procreate, to multiply life – in other words, Eros: 'From air, water and earth a thousand germs break forth, in dry, wet, warm and cold! *Had I not reserved the flame for myself, I should have nothing to call my own.*'[91]

Setting living nature on fire, the devil assures himself that some things are his: when his element the flame consumes it.

This theory was formulated, it bears noting, before the bourgeoisie had been told to stop for the sake of the climate. Marcuse radicalised it by suggesting that the death drive is the fully normalised motor of capitalist civilisation, providing 'energy for the continuous alteration, mastery, and exploitation of nature', unfinished work by definition. 'In attacking, splitting, changing, pulverizing things and animals (and, periodically, also men), man extends his dominion over the world and advances to ever richer stages of civilization', while in the very same moves unleashing a surplus of aggression against nature.[92] Some fractions of capital should fit this profile better than others. Some are especially close to the matter of death. Marcuse likewise wrote before the climate crisis, just after the Second World War, with a precocious environmental sensitivity, quoting Baudelaire: 'True civilization does not lie in gas, nor in steam, nor in turntables. It lies in the reduction of the traces of original sin.'[93]

---

91 Sigmund Freud, *Civilization and Its Discontents* (London: Penguin, 2002 [1941]), 59. Emphasis added. One must here clearly distinguish between different spheres of nature, to which the still stock and the devouring fire belong as much as the germinating and the flourishing.

92 Herbert Marcuse, *Eros and Civilization* (Boston: Beacon, 1974 [1955]), 52. Indeed, Marcuse goes so far – and for a Marxist, this really is quite far – as to say that any utility (and by implication profit) is a *by-product* of the fundamental death drive attacking nature (although only, he seems to be saying, in the later stages of capitalist development). Ibid., 87. He here expands on Freud's argument that as much as it is *sui generis*, the death drive comes wrapped up in Eros and does its 'work inconspicuously'. Sigmund Freud, *Beyond the Pleasure Principle* (Peterborough: Broadway, 2011 [1920]), 99. Put differently, destruction for its own sake is *mixed up* with destruction for the sake of profit. Whether the former is primary might be up for debate. cf. note 96.

93 Ibid., 153.

If nature could be identified as the devil's adversary prior to the crisis, in the heat of it, a new flavour of reasserted omnipotence is added to the moment of combustion.[94] Ecological sadism can come into play; to extend Marcuse, it provides energy for continued attacking, splitting, pulverising.[95] We have to 'break through rock walls, mine the depths of the earth, and reach through the ocean floor, to bring every ounce of energy into our homes and commerce and into our lives', cried Trump in his speeches on energy dominance. 'We're loaded! We're richer than all of 'em, richer than all of 'em folks.' 'We are really in the driving seat.'[96] He had a point, as it were. 'Death holds the steering wheel,' Marinetti crowed and drove on.[97]

---

94 'The enemy is the recalcitrant part of an expanded self over which the narcissistically vulnerable person had expected to exercise full control.' Kohut, 'Thoughts', 644.

95 'National Socialism was perhaps able to exploit the death-drives of its followers, but it undoubtedly originated in the very palpable will to live of the most powerful social groups.' Adorno, 'Sociology', 79.

96 Lederer, 'What Is'; Trump, 'Remarks'.

97 Marinetti, *Critical*, 473.

Coda: Rebel for Life

The good news here is that the dominant ideology is showing signs of desperation. It has always purported to improve life; now it is caught in the act of covering for or embracing extinction. For people who have not developed deep investments in the business-as-usual of the capitalist mode of production, it can then be hard to keep faith in it. Why trust an order that cannot stop itself from erasing the foundations for advanced life on earth? It is perhaps significant that the school – for Althusser, the most central ideological state apparatus – was the scene of the young, some would say naïve, demonstration of anger and let-down in the late 2010s. Here the contradictions in the dominant ideology were flagrant and elementary. 'I'm studying for a future that's currently being destroyed'; 'Why should I get educated when politicians don't listen to the educated?'; 'Why go to school if we don't have any future?' These were typical handwritten placards at the strikes – and in Holyrood, the kids held a banner saying 'Capitalism is crisis'; and in Cologne, they unfurled one calling on people to 'Burn capitalism, not coal'.[1] The ecological crisis has a capacity to provoke spontaneous disillusionment with capitalism,

---

1   Kate Buck, 'Take a Look at Some of the Excellent Placards Made by Kids for Climate Change Strike', *Metro*, 15 February 2019; *Guardian*, 'Student Climate Strikes around the World – in Pictures', 24 May 2019.

not the least among a youth that finds itself under its leaden skies: that kind of moment when people might cease responding to standard interpellations. In their perpetual blurring and overlapping – denialism, capitalist climate governance, green nationalism, fossil fascism – the dominant classes and the far right merely demonstrate that they have no real way of dealing with this crisis. Counter-apparatuses have plenty of material to work with here.

The experience of this generation is that fossil capital will not die a natural death.[2] But general interpellations of people who value life will not, of course, be enough to take it down. There has to be some oppositional force mobilised, which is rather easier said than done. Some elements of the working class will, as we have seen, be easier to involve than others. Often overlooked are the racialised parts of the class in Europe, whose members often maintain contacts with their countries of origin and know that an uncle's pistachio fields in Iran have withered, a beach once enjoyed in Jamaica vanished, a best friend's village in Peru swallowed by flood; as Leon Sealey-Huggins observes, they tend to be 'acutely aware of the strange weather befalling the global South', which this South has not itself caused.[3] They are also the prime target of the far right. Yet as of this writing, such people are massively under-represented in European climate movements, which have a whiteness norm of their own to challenge. On the other hand, anti-racist and anti-fascist activists can still occasionally consider climate and ecology some vaguely hippie, middle-class, romantic, uncool red herring. If there is anything this book has sought to demonstrate, it is that the two fronts are merging more fully with every tenth of a degree of warming.

In scientific research on the polarisation between left and right over climate, the policy recommendations tend to be conciliatory. The issue, according to this wisdom, should be communicated in such a way that the right will not bristle at it. The right – apparently including the far right – can be coaxed into caring for the climate if it is told that this is 'a form of patriotism' in line with the quest for

---

2 Paraphrasing Benjamin, *Arcades*, 667.
3 Sealey-Huggins, ' "The Climate" ', 100.

'purity'.[4] The right cannot be won over as long as 'the messenger is perceived as wanting to negatively alter society' – far more effective to stress competition 'for resources or prestige against a common/shared enemy'.[5] This is the road to Christchurch. No climate movement should ever set foot on it. Unfortunately, there seems to be no other option but to find a way to *beat* the far right politically – not humour it, but defeat it.

(This would hold even if the worst comes to pass, as in the kind of future scenario depicted in Lanchester's *The Wall*. He has a network of activists betraying the nation and assisting the Others in breaching the Wall, escaping the digital chips and living on the inside. Race traitors in a warming world, unite: you have nothing to lose but the chains of the Others.)[6]

Moreover, if the far right is in thrall to the death drive and other such urges, nothing could be expected from it other than a fight to the end, as every anti-fascist knows from hard-earned experience. Public Enemy were right when they opened the album *Apocalypse 91 . . . The Enemy Strikes Black* with the line 'The future holds nothing else but confrontation.' Anger and hate cannot be shaken off, as Judith Butler points out; they have to be directed against that which 'imperils the organic persistence of our interconnected lives'.[7] On the way up, conflicts in the elevator escalate.

And as it goes into overdrive, the destructiveness of the dominant ideology becomes plain to see. Rebellion against it really becomes one of and for life itself. Given how powerful the forces of destruction are, the order is tall: can it be accomplished? 'Any doubt', in the words of Daniel Bensaïd, 'bears on the possibility of succeeding, not on the necessity of trying.'[8] Such is now the imperative of a minimum decency.

---

4   Hornsey et al., 'Meta-analyses', 625; Douglas and Sutton, 'Climate', 102.
5   Hoffart and Hodson, 'Green', 47.
6   On the activists behind the Wall, see Lanchester, *The Wall*, for example 191.
7   Butler, 'Political', 747.
8   Daniel Bensaïd, *An Impatient Life: A Memoir* (London: Verso, 2013), 312.

Postscript: A Strange Year in the Elevator

One of the first political casualties of the Covid-19 pandemic was the climate movement. As soon as Europe imposed restrictions, it suspended all activities – school strikes, city occupations, mass actions against infrastructure – and withdrew into digital quarantine. The momentum built up since the summer of 2018 was lost in an instant. Climate fell off the agenda like a rock tumbling into water. By the same token, the far right lost its voice, as neither climate nor immigration preoccupied a public paralysed by fear of this unknown virus scything through populations first in northern Italy and then across the continent.

Nowhere was the change of topic more abrupt than in Germany. Months earlier, the AfD had redefined itself as an anti-climate party and seemingly walked from strength to strength; but in March and April of 2020, the country followed Merkel's instructions to keep a distance, stay at home, wear a mask and reconfigure daily life to slow the contagion – a total, transformative exercise that left no room for other issues. Starved of oxygen, the AfD fell from 15 per cent in the national polls in December to 9 per cent in May, while the Christian Democrats surged by more than 10 percentage points, Merkel exalted as the world's most effective leader in this trial. The same 'rallying around the flag' effect – the nation closing its ranks around the government in a moment of shock – could be observed in much of the continent, the far right sliding in the polls from Sweden to Italy, the Netherlands to Spain.[1] In late spring, Europe thus presented the marvellous spectacle of some aspects of utopia: cities empty of cars, airlines grounded, emissions plunging, the air cleared, wild animals roaming freely and organised racists retreating into the shadows.

If the climate years of 2018–19 were good for the far right, why was the outbreak bad for it? The standard explanation held that demagoguery cannot survive when adherence to scientific expertise makes the difference between life and death. It is obviously incorrect. Like the weather events that forced climate onto the European agenda in the summer of 2018, Covid-19 was a symptom of ecological crisis – more particularly, of deforestation, wildlife trading and livestock industry, the forms of domination of nature that cause zoonotic spillover.[2] In the case of the climate, some dots were connected. Anxiety about an unliveable climate led to a questioning of things like coal and cars (as in Germany, the prime European producer of both). Covid-19, by contrast, was received as a random event, a disease without meaning, no more linked to the operations of global capitalism than an asteroid would have been. Discussions of how companies and consumers pull the supply chains that summon pathogens from their reservoirs were conspicuous by their absence. There were, in short, no profitable privileges to defend from threats of permanent termination. Exactly contrary to the hoary thesis that climate change is a 'post-political' issue, the European far right thrived on the antagonisms it activated, but not on the diagnostics of the virus: *here* was that rare hour when societies came together, distributing sanitizers, avoiding crowding, providing ICU beds and listening to the latest guidelines from the epidemiologists – a brief post-political moment, when apparently neutral skills in disease management set the agenda.[3] Hence the

---

1 For a compendium of reports from this early stage, see Tamir Bar-On and Bàrbara Molas (eds.), *Responses to the Covid-19 Pandemic by the Radical Right: Scapegoating, Conspiracy Theories and New Narratives* (Stuttgart: Ibidem-Verlag, 2020).

2 See Rob Wallace, *Dead Epidemiologists: On the Origins of COVID-19* (New York: Monthly Review Press, 2020); Andreas Malm, *Corona, Climate, Chronic Emergency: War Communism in the Twenty-First Century* (London: Verso, 2020).

3 A re-reading of Erik Swyngedouw's classic essay, 'Apocalypse', will reveal just how wrong he was about climate, and how apt his model was for the moment when Covid-19 broke out.

paradox that the faint preview of a mitigation crisis in 2018–19 served the far right well, while the rather more intense corona crisis of spring 2020 muffled it. But the climate will come back, of course.

Viral Racism Fails

Not that the far right didn't try to drag the pandemic through its funnel. In Spain, streets filled up with demonstrators on 8 March, International Women's Day, a week before the announcement of a national emergency. The cadres of Vox were also out in force, to protest 'gender ideology' and re-elect their leader. In the following days, when the gravity of the situation became clear, the party blamed the swarming feminists for spreading the virus. But then its general secretary, Ortega Smith, tested positive. The entire parliamentary group went into self-isolation; Abascal tested positive too. It turned out that the Vox gathering on 8 March had been the real germ swamp, compelling the party to offer humiliating apologies. Smith tried to save face by tweeting that his 'Spanish antibodies will kick out the damned Chinese virus', but few were impressed.[4] Nor did the demand that 'illegal immigrants' be forced to pay for health services – thereby leaving them to their viral fate – earn Vox much sympathy.[5] The party again picked up the trope of the 'Chinese virus': but it didn't fly. A spate of hate crimes against people of Chinese or East-Asian descent was reported in several European countries; bad as it was, it did not connect with deeper trends. Sinophobia has not been a prominent form of politicised racism in twenty-first century Europe.[6] One can

---

4 20 Minutos, 'Ortega Smith dice que sus "anticuerpos españoles" derrotarán a los "malditos virus chinos"', 20minutos.es, 14 March 2020; and see further Carmen Aguilera-Carnerero and Bàrbara Molas, 'The Spanish Right, Covid-19 and "Socio-Communism"', in Bar-On and Molas, *Responses*, 95–8.

5 Carmen Moraga, 'La extreme derecha trata de expandirse aprovechando la pandemia y saca la artillería contra el Gobierno', *El Diario*, eldiario.se, 24 March 2020.

6 As noticed in the most insightful study of the topic published as of this writing: Jacob Wondreys and Cas Mudde, 'Victims of the Pandemic? European Far-Right Parties and Covid-19', *Nationalities Papers* (2020), online first. There is a slight contrast in this regard between Europe and the US. Sinophobia also had a somewhat prominent role in Italy.

only imagine what would have happened if the pandemic had originated in the heart of the Islamic Middle East, as MERS, the novel coronavirus preceding SARS-CoV-2, did. (Recall Jarosław Kaczyński's rambling on the 'parasites and protozoans' in Muslim bodies.) Now Smith and Abascal became the first on a list of far-right leaders to fall ill, followed by Johnson, Trump and Bolsonaro – none physically or politically killed by the disease, nor boosted or ennobled by it.

Islamophobia could be outright embarrassing. There had long been much indignation in Denmark about the habit of some pious Muslims to abstain from shaking hands, and so the naturalisation ceremony had been rounded out with mandatory handshaking with officials. Therefore, in spring 2020, all naturalisation had to be called off.[7] On the other hand, the far right – from Tories to the FrP, the DF to Le Pen – did its best to assail Muslims and other non-whites for stubbornly socialising and ignoring lockdowns.[8] Such trite scapegoating worked better than elsewhere in Hungary. Orbán did not point his finger at China – to whose government he was close – but seized on the circumstance that the first Covid-19 patient in the country was an Iranian student. A dozen Iranians were swiftly rounded up and deported; having found yet another evil to attribute to Muslims, Orbán declared that he and his government would rule the country by decree. During a state of exception that lasted for three months, Fidesz made sure to reallocate money from municipalities run by the opposition – notably the Greens – and criminalise identification with a sex other than that assigned at birth.[9]

This was not a Hungarian oddity: the exploitation of the pandemic for authoritarian projects unrelated to the causes and spread of a

---

7 Florian Bieber, 'Global Nationalism in Times of the Covid-19 Pandemic', *Nationalities Papers* (2020), online first, 7.

8 For example, Peter Walker and Josh Halliday, 'Tory MP Condemned for Claim BAME People Breaching Lockdown Most', *Guardian*, 31 July 2020; Siv Jensen and Jon Helgheim, 'Ernas berøringsangst', *NRK*, 13 November 2020; Peter Kofod's Facebook page (DF), 5 August 2020; Laura Chanzel, 'France', in Giorgos Katsambekis and Yannis Stavrakakis (eds.), *Populism and the Pandemic: A Collaborative Report* (Thessaloniki: Populismus, 2020), 21.

9 Edit Inotai, 'How Hungary's Orban Blamed Migrants for Coronavirus', *EU Observer*, euobserver.com, 20 March 2020; Kyle Knight and Lydia Gall, 'Hungary Ends Legal Recognition for Transgender and Intersex People', Human Rights Watch, 21 May 2020.

coronavirus. In October, one report listed eighty countries where 'the condition of democracy and human rights has grown worse', an associated blight with its own aetiology. Increased surveillance, the rounding-up of dissidents, a stifled media, executive power grabs, amplified hate campaigns, police violence out of bounds: this was a global story, in which ethnic minorities – such as Muslims in India – bore the brunt.[10] One legacy of Covid-19 could be enhanced authoritarianism in these zones of repression. It set a precedent for coming adaptation crises: when a symptom of ecological collapse breaks out in a population, already exposed minorities may suffer renewed assaults. Europe as a whole saw a relatively mild prefiguration of this scenario. It was most evident in countries with one-party governments of the far right: besides Hungary, Poland.

The PiS corona emergency response chipped away at what remained of the rule of law and freedom of expression. In mid-April, Poland – just like Hungary – recorded its fifth consecutive annual drop on the World Press Freedom Index.[11] But the Polish witch hunt against LGBT people was rather more delirious. In 2019, the PiS began to shift emphasis from race to gender, Kaczyński et al. stressing 'LGBT ideology' – imported, Marxist in character – as a threat to the very existence of the Polish nation. Over this year and the next, roughly one hundred Polish municipalities proclaimed themselves 'LGBT-free zones', covering a third of the country, with a predictable rise in hate crimes, Poland now ranked as the worst country in Europe for queer people.[12] In this still nearly all-white nation, they had now become the domestic enemy number one. Heterosexist apocalypticism dominated the campaign that won Andrzej Duda a second term as president in July 2020. He registered his highest levels of support in the coal areas.

---

10 Sarah Repucci and Amy Slipowitz, *Democracy under Lockdown: The Impact of Covid-19 on the Global Struggle for Freedom*, Freedom House, freedomhouse.org, October 2020.

11 Roman Krakovsky, 'Europe versus Coronavirus – Poland between Reactivity and Opportunism', *Institut Montaigne*, 7 May 2020.

12 See for example Lucy Ash, 'Inside Poland's "LGBT-Free Zones"', BBC News, 20 September 2020.

The Far Right Finds Its Voice

Beyond Poland, however, race was, to paraphrase Stuart Hall, the modality in which the pandemic was lived. The distribution of harm and death mirrored that on the climate front. Half a year into Covid-19, African Americans were nearly three times more likely than whites to contract it and nearly four times as likely to die. A progression of vulnerability exposed them to the disease: sharing crowded homes, living in densely populated neighbourhoods, working in 'essential' jobs where contact could not be avoided – meat plants, hospitals, grocery stores, warehouses – relying on public transport to get to these jobs, carrying the burdens of structural racism – including the siting of refineries and other sources of air pollution – their bodies weighed down by everything from asthma to accelerated ageing, black people would also have the least access to health care. They would wait longer for examinations and receive less adequate treatment than whites.[13] Remarkably, the very same factors primed non-whites for infection in the UK – up to four times higher rates than in white neighbourhoods nearby – as well as in Brazil and Scandinavia. In the latter region, the outbreak struck with disproportionate force against non-white populations and hardest by far against Somali communities, on the absolute bottom of the racial hierarchy – black *and* Muslim – in a pattern that persisted through the year.[14] Climate and corona had one feature in

---

13 An excellent study is Marc A. Garcia, Patricia A. Homan, Catherine Garcia and Tyson H. Brown, 'The Color of Covid-19: Structural Racism and the Disproportionate Impact of the Pandemic on Older Black and Latinx Adults', *Journals of Gerontology: Social Sciences* (2020), online first. Cf. for example Sharrelle Barber, 'Death by Racism', *Lancet Infectious Diseases* 20 (2020): 903; Leonard E. Egede and Rebekah J. Walker, 'Structural Racism, Social Risk Factors, and Covid-19: A Dangerous Convergence for Black Americans', *New England Journal of Medicine* 383 (2020): 1–3; Nancy Krieger, 'Enough: Covid-19, Structural Racism, Police Brutality, Plutocracy, Climate Change – and Time for Health Justice, Democratic Governance, and an Equitable, Sustainable Future', *American Journal of Public Health* 110 (2020): 1620–3.

14 See for example Debora de Souza Santos, Mariane de Oliveira Menezes, Carla Betina Andreucci et al., 'Disproportionate Impact of Coronavirus Disease 2019 (Covid-19) among Pregnant and Postpartum Black Women in Brazil through Structural Racism Lens', *Clinical Infectious Disease* (2020), online first; Dom Phillips, '"Enormous Disparities": Coronavirus Death Rates Expose Brazil's Deep Racial Inequalities', *Guardian*, 9 June 2020; Roger Grosvold, 'Corona sprider sig bland norsk-somaliere: – Vi har ingen å miste', *ABC Nyheter*, 2 April 2020; Freja Sofie

common: a universal danger to humanity propagated through the particular vector of race.

Regarded from this angle, some behaviour on the far right looked familiar. The instincts of Trump and Bolsonaro were, infamously, to deny the existence or severity of Covid-19. The former used the word 'hoax'.[15] He alleged that the virus had been manufactured by China. The Heartland Institute pronounced that 'leftists are stoking Covid-19 panic'.[16] Towards the end of the year, with the second wave in full swing, the denial had hardened in some corners, to the degree that one nurse in South Dakota – the state posting the most cases and deaths in November, after the Republican governor had refused restrictions – testified that patients refused to believe that they had Covid-19, despite seeing positive test results. 'Their last dying words are, "This can't be happening. It's not real."'[17] The escapism echoed in a fresh galaxy of conspiracy theories. There was the theory that Bill Gates had planned it all, or that 5G technology transmitted the virus. There was the theory that SARS-CoV-2 had been concocted in a Chinese laboratory and deliberately let loose on the world (Abascal and Salvini believed this), various theories involving Soros and other Jews – including Jews as hypersusceptible because of their abnormally large nasal cavities holding a glut of droplets – and a French theory of a *coranvirus* (Coran being the French spelling of the Qur'an).[18] One figure close to both RN and the Génération Identitaire spelled out the evidence for the latter: 'The Islamists are delighted:

Madsen and Kenneth Elkjaer, 'Halvdelen av Aarhus' nye smittede har somalisk baggrund: – Vi fik et chok', *TV2 Nyheder*, 4 August 2020; Hasse Burström, Dante Thomsen and Lotta Sima, 'Äldre svenskar med utländsk bakgrund hårdast drabbade av covid-19', *SVT*, 12 December 2020.

15   Lauren Egan, 'Trump Calls Coronavirus Democrats' "New Hoax"', NBC News, nbcnews.com, 29 February 2020.

16   Jim Lakely's Twitter account, 6 May 2020. (Lakely was the vice president of Heartland.)

17   Jodi Doering in Paulina Villegas, 'South Dakota Nurse Says Many Patients Deny the Coronavirus Exists – Right Up until Death', *Washington Post*, 16 November 2020.

18   For example *El Diario*, 'Abascal alienta las teorías de la conspiración sobre el coronavirus y sugiere que fue fabricado por China', 21 October 2020; Matteo Salvini's Twitter account, 25 March 2020; Tamir Bar-On, 'The Oldest Hatred, Conspiracy Theories and the Covid-19 Pandemic', in Bar-On and Molas, *Responses*, 114. For an overview, see Philip Ball and Amy Maxmen, 'Battling the Infodemic', *Nature* 581 (2020): 371–4.

bars closed, women wearing masks, no more shaking hands.'[19] And then we have not yet mentioned the theory that went by the name of QAnon: a cabal of Satan-worshipping paedophiles runs the world, harvests blood from poor children and seeks to undermine Donald Trump, who will one day slay the bastards – a virtual supernova that exploded in 2020 and sucked other theories into its orbit, including those about Covid-19 and climate.[20] With these developments, these trips ever further into the outer space of unreality, it had become nearly impossible to keep pace.

Somewhat closer to the ground, or so it would seem, was the theory that states were needlessly suffocating their citizens with lockdowns. In opposition to the unpleasantness of lockdown was a potential for street mobilisation on the far right. Back in March, the first response by the AfD had been to call for stricter measures and taunt the Merkel government for spending the taxpayers' money on a fake crisis (climate) instead of the real one (corona). Performing a volte-face over a few weeks, the party spent the rest of the year slamming the state for burying the freedom of its citizens. So did tens of thousands who took to the streets in defiance. They were motley crews of anti-vaxxers, mystics, hippies, conservatives, conspiracists (including of the QAnon school), people never engaged in politics and veterans of the hardest right. The *Querdenker* – 'lateral thinkers' – agreed on only one demand: ending the policies for combatting the virus. At the top of their voices, they would claim that the population was being brainwashed; the measures lacked all proportions; there was no virus; there was a virus, but no worse than a flu; there would be forced vaccinations; masks were killing children; a New World Order was under construction. A 'corona dictatorship' had been instituted in Germany, prompting *Querdenker* to liken themselves to victims of the Nazi regime, some taking to wearing yellow Stars of David (that old dream of the German far right: *we* are the real victims). Over them would wave the *Reichsflagge*, a substitute for the

---

19 Damien Rieu quoted in Tristan Berteloot, 'Identitaires, RN: comment l'extrême droite veut tirer profit du coronavirus', *Libération*, 20 March 2020.

20 For the climate component, see for example Sharon Kelly, 'As QAnon Conspiracy Spreads on the Far Right, Climate Science Deniers Jump Onboard', *DeSmog UK*, 27 August 2020.

banned swastika, AfD members marching alongside more radically anti-Semitic squads. The movement scored a symbolic victory when it stormed the stairs of the Reichstag in Berlin on 29 August, and the AfD increasingly appeared – internal conflicts about it notwithstanding – as the parliamentary wing of the street fighters.

Vox likewise switched from demanding that climate funds be earmarked for corona to organising a 'tour for freedom', driving through cities in caravans of SUVs and motorbikes, Spanish flags and pennants on the vehicles, in protest against lockdown.[21] In the UK, Nigel Farage found an old-new calling. 'The idea that the science on Covid-19 is clear is a con', he still maintained in October; weeks later, in response to the second lockdown, he relaunched the Brexit Party as Reform UK.[22] The liberation of the nation from cruel corona policies now took precedence. In Trafalgar Square, Johnson's former weather oracle Piers Corbyn rallied alongside men brandishing the flag of the British Union of Fascists, while in Rome, ultras and neofascists from Forza Nuova clashed with the police.[23] The spring wave of anti-lockdown protests in the US had the appearance of spontaneous eruption, but it was coordinated and nourished by associates of the Koch brothers, the Mercers, the Competitive Enterprise Institute and other big guns.[24]

These were often a carnival of tin foil hats. They seemed less capable of injecting irrationality deep into the body politic than their climate equivalents. Although there were patent similarities – the primacy of economy over science, property over life; the hostility to any sort of change – the protests had no particular corporations or commodities to defend. They targeted general adaptive responses

---

21 For example, Miguel González, 'La protesta de Vox en Madrid acaba en un monumental atasco que incumple las medidas sanitarias', *El País*, 23 May 2020.

22 *Fortune and Freedom with Nigel Farage* (his newsletter), 23 October 2020.

23 See for example Stephen Buranyi, 'How Coronavirus Has Brought Together Conspiracy Theorists and the Far Right', *Guardian*, 4 September 2020; Luca Monaco, 'Roma, Coronavirus. Fascisti, ultrà e negazionisti, anataomia della rivolta contro il coprifuoco', *La Repubblica*, 25 October 2020.

24 Isaac Stanley-Becker and Tony Romm, 'The Anti-Quarantine Protests Seem Spontaneous. But Behind the Scenes, a Powerful Network Is Helping', *Washington Post*, 22 April 2020; Emily Holden, 'US Critics of Stay-at-Home Orders Tied to Fossil Fuel Funding', *Guardian*, 21 May 2020.

undertaken by states, on an expressly temporary basis, in a desperate fight against one symptom of the ecological crisis. None of the scenarios sketched in this book corresponds to that logic. Here were no progressive movements on the march; no proposals to rein in deforestation and other drivers of the problem – as rampant as ever in 2020 – no special material-ecological prerogatives to rescue from enemies. Perhaps the anti-lockdown movements prefigured another form of fascism: a *revolt against adaptation*, in defence of white petty-bourgeois layers constricted or even declassed by it.

But that was not a coherent figure. Where the far right enjoyed unrestricted rule in Europe – Poland and Hungary – the governments neither denied corona nor showed themselves markedly less competent in flattening its curves.[25] Poland was among the first to shut the country down, make masks mandatory and close borders. The border closures then cut up Europe into its national component parts, which some on the far right welcomed as a vindication. 'We are all nationalists now', purred Farage.[26] Here was another opportunity for the far right to expand its pathological normality: making the closures permanent and taking the next logical steps against foreigners (along the lines of the SD's 'next great battle in migration policy'). On the other hand, these closures did not quite meet nationalist desires, as they could be blind to both colour and class. The far right has never wanted people with white skin and high incomes to be locked in place; but in December 2020, to take but one example, British travellers were shut out from European airports in a moment of panic about the new strains emerging from their isles.[27] Sweden Democrats always enjoyed going to Thailand.

And yet the medical nationalism of 2020 might have a long fat tail. So will, more certainly, the vaccine nationalism that made the US and Europe place advance orders for hundreds of millions of doses and thereby clear the shelves of the new medicines, on which the world had banked since the outbreak, leaving poor countries in the

---

25   As argued at length by Wondreys and Mudde, 'Victims'.
26   Nigel Farage, 'Coronavirus Has Shown We Are All Nationalists Now. Does Boris Johnson Realise That?', *Telegraph*, 12 March 2020.
27   For considerations on these contradictions, see Bieber, 'Global'.

global South to wait until maybe 2024 for their turn.[28] What kind of precedent that set for the adaptation crises in the pipeline should be all too obvious.

Sweden Swigs

To the surprise of many, one country deviated from the rest of Europe in the spring of 2020: Sweden. Under the guidance of state epidemiologist Anders Tegnell, the government didn't close any shops or restaurants or schools, satisfying itself with vague recommendations of prudence. SD accused it of sacrificing the elderly and called for the very restrictions the far right in Germany and elsewhere detested. But the initial post-political moment was no less paralysing here: like an addict that can only stay off the drink when his arms are tied behind his backs, Swedish politics had to leave *invandrare* alone for a few months.[29] When the first pandemic wave subsided in the summer, it was immediately back at it. The ills of Swedish society were again as a matter of course imputed to the failure of the non-whites to 'integrate' themselves into said society. By the late summer, the headlines were devoted to incidents of gang crime, and when acolytes of Rasmus Paludan burnt copies of the Qur'an near Rosengård and touched off a night of rioting, the official nation tore its hair out. Vice chief of the Swedish police Mats Löfving offered an explanation: 'Forty clans have arrived in Sweden with the sole purpose of organising criminal activities, building power, earning money', infiltrating corporations and municipalities and the central state apparatus to boot and grooming their children to continue in their path – which was why immigrant neighbourhoods were exploding in violence.[30] The breathtaking conspiracy theory was questioned by lower-level

---

28 Kai Kupferschmidt, '"Vaccine Nationalism" Threatens Global Plan to Distribute Covid-19 Shots Fairly', *Science*, 28 July 2020; Peter Beaumont, 'Scheme to Get Covid Vaccine to Poorer Countries at "High Risk" of Failure', *Guardian*, 16 December 2020.

29 However, the high casualties among Somalis were greeted by the Swedish mainstream, right and far right with responses ranging from callous indifference and blaming of the victims – their deficient culture supposedly causing them to die – to outright glee.

30 SR, 'Polischefen avslöjar: Vi har 40 kriminella klaner i Sverige', 7 September 2020.

cops, who had never seen traces of those forty clans. Nevertheless, the forty clans became accepted reality (the Left Party a lone dissenter), remaining on everyone's lips until the second wave rolled across Sweden.

Then the *laissez-faire* model could no longer be upheld. At long last, Sweden had to put standard restrictions in place, the government reluctantly agreeing that corona liberalism had failed. But Tegnell would still not admit that it had caused the high death rates – 706 per 1 million inhabitants, as against 67 in neighbouring Norway. He had another explanation: 'small groups of *invandrare* have been very much driving' the contagion, and Sweden had more of them than others (14 per cent of the population born outside Europe, as against 8 in Norway). The word that translates as 'driving', *drivande*, has the connotation of active and purposeful propulsion. Facing criticism, Tegnell admitted that it was an 'unfortunate' phrasing – a slip of the tongue – which gave Jimmie Åkesson the chance to defend him for 'having said it as it is'.[31] Once again, the far right had its funnelling and victim-blaming confirmed by a top representative of the state apparatus. The mistakes in the moment of outbreak did not mean that these strategies couldn't work in later stages (including when the pandemic morphs into depression and mass unemployment). More fundamentally, the utter turpitude of the political climate in Sweden appeared bound to hit new lows.

It represented a constantly reaffirmed selection since the summer of 2018: immigration, not global heating, is the menace to be dealt with. New data from 2020 showed the effects on opinion. Out of forty countries polled for the view that climate change is 'very or extremely serious', Sweden ended up in place thirty-eight, some 50 per cent concurring; only in Norway and the Netherlands was the share lower. At the opposite end, with majorities of 90 per cent, stood Chile, Kenya and South Africa. The slumping of these northern European countries towards indifference could only be accounted

---

31 Beri Zangana, 'Tegnell om uttalandet: "Det var olyckligt"', *Aftonbladet*, 4 December 2020; Karin Thurfjell '"Tegnell fick kritik för att han sade som det är"', *Svenska Dagbladet*, 8 December 2020; see also Gusten Holm, 'Norrländska dödstalen visar att Tegnells teori inte stämmer', *Aftonbladet*, 10 December 2020. No individual other than Tegnell, of course, had greater responsibility for devising Sweden's collapsed policy.

for with the syndrome of selection: for anti-immigrant sentiment, Sweden *topped* the European league.[32] Other sentiments barely registered at all. In early December, just as the debate raged over the role of *invandrare* in Covid-19, Oxfam published a report showing that the richest 1 per cent of Swedes caused forty-three tonnes of CO_2 through their annual consumption, as against four for the poorest half. The emissions of the former would, the researchers concluded, need to be cut by 95 per cent in less than a decade – a class war such as the nation has never seen.[33] But the rich need not worry: the uptake of these findings into political forums was zero. Instead, parliament busied itself with taking down every remaining eye of a needle that could lead to asylum in the country, such as 'other need of protection', a residual clause that had hitherto maintained at least a hypothetical chance for refugees from climate disasters. If there was anything exceptional about this nation, it was its breathless ambition to be typical.

France Swings the Truncheon

Similar processes were playing out with rather greater intensity in France. In the lull between the first and second waves, Macron rode forth to assault 'separatism' – mainly targeting the visibility of Muslims in the secular republic – with a new battery of laws. Organisations straying from vacuously defined ideals would be disbanded, Islamic confessional practices subjected to state control, the Ministry of Justice instructed to 'assure a republican presence in every road, every building'.[34] Two weeks later, an eighteen-year-old man of Chechen origin accosted a schoolteacher who had showed

---

32   Simge Andi and James Painter, 'How Much Do People around the World Care about Climate Change? We Surveyed 80,000 People in 40 Countries to Find Out', *The Conversation*, 16 June 2020; Jon Henley and Pamela Duncan, 'European Support for Populist Beliefs Falls, YouGov Survey Suggests', *Guardian*, 26 October 2020.

33   Mira Alestig and Robert Höglund, 'Rika svenskar måste minska utsläppen med 90 procent', *DN*, 17 December 2020.

34   Kim Willsher, 'Macron Outlines New Law to Prevent "Islamist Separatism" in France', *Guardian*, 2 October 2020; cf. Emre Ongün, 'France's Demonization of Muslims Is Getting Worse', *Jacobin*, 22 October 2020.

Muhammad cartoons to his students, pulled out a knife and severed his head. There followed a persecution of Muslim communities unusually broad even by French standards. Armed police raided countless homes to interrogate schoolchildren about their views; some 230 putative radicals were designated for expulsion, seventy-six mosques for shutdown and fifty aid and advocacy groups for dissolution, the prize target being Collectif contre l'islamophobie en France, or CCIF.[35] As its name suggested, CCIF specialised in countering Islamophobia, by publishing reports, providing legal assistance to victims and defending the rights of Muslims in courts – stopping, to mention but one case, the exclusion of mothers in hijab from school trips. Nothing implicated CCIF in the beheading.[36] Pascal Bruckner, however, ran to the defence of the break-up, claiming without a shred of evidence that CCIF and other outlets on the left had 'blood on their hands'.[37] He, at least, had his priorities clear.

Gearing up for another second round against Marine Le Pen in 2022, Macron charged ahead: soon the national assembly approved another legal package, making it an offence to publish photos or films of police in action. But cops could henceforth use drones and facial recognition to prosecute demonstrators. Barely had the ink dried on the law before a film surfaced of three officers breaking into the home of a young black music producer, beating him bloody with truncheons and abusing him with the N-word for no reason other than the colour of his skin. Mass demonstrations thronged cities in denunciation of the carte blanche for police brutality. Forced to beat a retreat, Macron promised to revise the law.

And while all of this was happening, Total, the single largest private corporation in the country, went ahead with plans for massively expanding its fossil gas production from Mozambique to the Arctic. It had the backing of the president. There was little if any controversy about it.[38]

---

35 Adam Nossiter, 'After Teacher's Decapitation, France Unleashes a Broad Crackdown on "the Enemy Within"', *New York Times*, 19 October 2020; Kim Willsher, 'France Cracks Down on 76 Mosques Suspected of "Separatism"', *Guardian*, 3 December 2020.

36 For a superb analysis, see ACTA, 'France's Authoritarian Turn: Islamophobia and the Security State', *RS21*, 1 December 2020.

37 Europe 1, 'Conflands: les "complices" politiques "ont du sang sur les mains", selon Pascal Bruckner', 19 October 2020.

38 Reuters, 'Total Secures $15.8 Billion in Funding for Mozambique Gas Project:

But it would, of course, be a grave error to see this latest instance of the French state flexing its repressive muscles as an intentional deflection from climate crimes. Rather, the causally efficacious background was the cycle of working-class insurrection still rolling through France – uniquely for Europe – in the form of, most recently, the Gilets Jaunes and the strikes against Macron's pension reform. It would also be a mistake to see French (or for that matter Swedish) fascisation as just a setback for the climate. It had a very direct tally of human suffering.

Green Nationalism Mutates

The sudden deglobalisation under Covid-19 provided grist to the mill of French localism. Le Pen, Hervé Juvin and the rest of the RN envisioned a post-pandemic nation that produced and consumed close to home.[39] But their ecology was as inane as ever. When Macron announced a referendum on including climate protection in the constitution and making 'ecocide' a crime – his favoured mode of governing – RN lapsed into well-worn grammar: this initiative could be 'compared to a wind turbine: it's an environmental and political hoax; they think it's beautiful and green, but it actually pollutes our air', raged the young model Julien Odoul from the national leadership. 'Is the climate emergency really the priority of the French people?' he added for good measure. 'There is the migration emergency, the security emergency – why not organise a citizens' committee on these topics?'[40] In Italy, Salvini was equally unable to hold his tongue, lashing out at 'fancy living-room radical-chic environmentalists that have

---

FNB', 8 July 2020; Nabil Wakim, 'La France pourrait soutenir un gigantesque projet gazier dans l'Arctique russe', *Le Monde*, 2 September 2020.

39 See for example Marine Le Pen, 'Allocution de 1er mai', rassemblementnational.fr, 1 May 2020; Nicolas Celnik, 'Localisme, quand le vert vire au bleu marine', *Libération*, 25 June 2020.

40 Odoul in an interview on the LCI channel, posted on his Twitter account, 14 December 2020. He gained a reputation by demanding the ejection of a mother wearing a hijab on a school trip to a regional council: see for example, Jean-Pierre Tenoux, 'Au conseil regional de Bourgogne-Franche-Comté, un responsable du RN aggresse une femme voilée', *Le Monde*, 12 October 2019. For many more relapses into default denial, see the unrivalled study by Lise Benoist, *Green Is the New Brown: Ecology in the Metapolitics of the French Far Right Today*, master's thesis (Lund University, 2020).

never seen a boar or a tree' and asserting that 'we cannot say no to coal, no to petrol, no to methane, no to oil drills – we cannot go around lighting candles.'[41] The banner was held higher by Renaud Camus of the Great Replacement theory. He was convicted of racial insult for the following tweet: 'A box of condoms offered in Africa equals three fewer drowning victims in the Mediterranean, a hundred thousand euros in savings for the CAF [public agency for assisting families], two prison cells freed up and three centimetres of ice sheet saved.' Referred to with the hashtag #icefloetweet by Camus and his supporters, this became a new cause célèbre in the far-right struggle for freedom of expression.[42]

Meanwhile, the limits to green nationalism were laid bare in a European country where climate politics loomed larger: Finland. Popular for its efficient handling of Covid-19, the red-green government could press ahead with its climate ambitions. The main opposition – the Finns – focused on obstructing them. Fuel taxes should be cut not raised, coal and peat dug up not dropped, immigrants sent away and (some) populations shrunk: the standard fare, spiced up in 2020 with violent intrigue. One former member of the Finns struck a current member twenty times in the head with a hammer. The assailant was a founder of KELSU, acronym for 'Nationalist Animal and Nature Protectors', a green nationalist outfit recruiting among the Finns' base. When the investigation into the hammer attack got underway, another plot was uncovered: activists in the same milieu had planned to target and possibly assassinate Maria Ohisalo, minister of the interior representing the Greens. Such far-right violence roiled Finland in 2020, often with a link to the green-brown end of the spectrum. Pentti Linkola passed away in the summer, but a new generation of activists sprung up – such as the collective Suunta, 'Direction', which aimed to forge a true bond between the 'Finnish tribe' and its land – honouring his legacy. Furthest to the right, some Nazis

---

41 Il Metropolitano, 'Salvini: per colpa di ambientalisti da salotto è un casino', ilmetropolitano.it, 14 January 2020; Manlio Lilli, 'L'ossessione di Salvini per l'Ambiente', L'Espresso, 21 January 2019.

42 For example, Franck Johannès, 'L'écrivain Renaud Camus poursuivi pour injure raciale', Le Monde, 26 November 2020.

argued that the future holocaust will only be accepted if it is connected to climate protection.[43]

The Danish government had likewise come to power in a 'climate election' on a mandate to make an actual difference. A new law adopted in 2020 specified a 70 per cent reduction of CO_2 emissions by 2030. How would that goal be attained? Half of it through the deployment of various technologies for capturing and storing carbon – not by avoiding emissions, that is, but by neutralising them. Moreover, the government devised its own 'hockey stick model', in which most of the reductions would be delayed until 2025 and then shoot up as the anticipated carbon-capturing machines rolled out. Unsurprisingly, the plan was savaged by critics – including the government's own council of climate experts – for betting on miracles.[44] But then the social democrats of Denmark drew a line. All new oil and gas exploration in the North Sea was immediately terminated, the next licensing round cancelled. There would be no more hunting for fossil fuels in Danish waters. In an overheating world, this might seem like an unremarkable minimum – not opening *more* fields – but in 2020, it was still a sensational move going against all trends simply by forsaking further accumulation (and state revenues on the order of 2 billion euros). The state did not shut down existing fields – that would have been revolutionary – but allowed companies to keep pumping until they ran empty, maybe in 2050. Its decision had been facilitated by the lukewarm interest of investors due to low oil prices. Yet the climate movement hailed a rare victory. For the first time, a major oil and gas producer – the largest remaining in the EU, measured by barrels produced within its borders – had locked away untouched reservoirs, apparently for the long term, setting an example for the minimum demand elsewhere. From the course followed by the previous right-wing government (see pages 97–8 in this book), it was a U-turn.

---

43 On this latter line of reasoning, see Petri Jääskeläinen, 'Näistä Suomen äärioikeistolaiset unelmoivat keskusteluryhmissään: seksuaalisesta väkivallasta, atleettinatseista – ja hakaristipizzasta', *Seura*, 4 December 2020.

44 See for example Adam Hannestad, 'For første gang præsenterer regeringen et udkast for, hvordan Danmark skal nå sit ambitiøse klimamål. Men eksperter tvivler stærkt på planen', *Politiken*, 29 September 2020.

But it was combined with an escalation of the targeting of non-whites. The same Danish social democrats promulgated even stricter rules for the 'ghettos' – evicting long-time residents, giving the police the right to punish entrants into restricted zones by snatching their jackets, watches and mobile phones – and set up a special register for inhabitants hailing from Muslim countries. 'It is a matter of us against them', said Prime Minister Mette Frederiksen, 'them' being the heavy criminals of the ghettos.[45] This evolution in the ultimate direction of something like ecological fascism had a counterpart in the UK. As post-Brexit prime minister, Johnson ditched denial – the Global Warming Policy Foundation cried betrayal – for a plan to slash emissions by 68 per cent by 2030. How? By building offshore wind farms, centrepiece of his 'Green Industrial Revolution' (a term shamelessly poached from Jeremy Corbyn). Here the former denialist truly turned a page. 'It was offshore wind that puffed the sails of Drake and Raleigh and Nelson, and propelled this country to commercial greatness.' Seven years earlier, he had mocked Labour for promoting a technology that couldn't 'pull the skin off a rice pudding', but now Johnson confessed to the sin of 'forgetting the history of this country'.[46] The British Empire, he should have remembered, was built with wind. By evoking the three heroes, the prime minister hearkened back to the *pre*-fossil empire, when ships sailed the seas with slaves – Drake pioneered slave voyages across the Atlantic; Raleigh paved the way for plantations in Virginia; Nelson professed himself 'a firm friend' of the Caribbean masters and detested the 'cruel doctrine' of abolitionism.[47] This may have been the first time, in the context of the climate crisis, that white nationalist imagery was invested in renewables. Sequels are conceivable, for offshore wind if for no other source. The SD could stir the memory of the Vikings, Vox that of Columbus.

But the Tories had not fully repressed that other part of their past: in November 2020, the first new deep coal mine in three decades was

---

45 Mette Frederiksen, 'Statsministerens åbningstale 2020', regeringen.dk, 6 October 2020.

46 Jon Stone, 'Boris Johnson Condemns His Past Self for "Sneering" at Wind Power', *Independent*, 6 October 2020.

47 Nelson quoted in Michael Taylor, *The Interest: How the British Establishment Resisted the Abolition of Slavery* (London: Penguin, 2020), 71.

given the go-ahead, on the coast of Cumbria. The following month, the legal hurdles against a third runway at Heathrow were cleared (ironically, on the same day a court ruled that the death of a nine-year-old black girl was caused by air pollution from diesel exhaust in her neighbourhood – the first such ruling in the UK).[48] When Johnson oversaw a UK-Africa investment summit, the deals brokered – in the most classical fashion of imperialist capital export – had a preference for fossil fuels, with 90 per cent of energy projects dedicated to oil and gas extraction from Tunisia to Kenya. The black fuel was apparently still the future for capital going to Africa. For actual people coming from that continent, or some other non-white part of the globe, the Tories discussed a range of options: wave machines for capsizing dinghies in the English Channel, or chains or nets for tangling their propellers. Once the migrants had been captured, the question of disposal would arise. One suggestion was to lock them up on oil rigs.[49]

America Burns

George Floyd was murdered in Minneapolis on 25 May 2020. Replayed across millions of screens, this latest atrocity in the endless series inflicted on black people led to a mass uprising against racism such as the US had not experienced since at least the 1960s. If 2019 was the year of the climate movement, 2020 was that of the movement for black lives. Some cross-pollination over the boundary took place, if only on a discursive plane: one investigation revealed that primitive fossil capital, as we have called it in this book, had a predilection for sponsoring US police forces and supplying them with weapons. Companies would spend some of the profits earned from fossil fuels on 'police foundations', raising private money to purchase

---

48 On the latter verdict, see Adam Vaughan, 'Landmark Ruling Says Air Pollution Contributed to Death of 9-Year-Old', *New Scientist*, 16 December 2020.
49 Sascha O'Sullivan, 'Home Office "Looked at Wave Machines" to Stop Migrants in Small Boats Crossing Channel', *The Sun*, 1 October 2020; Jamie Grierson, 'Home Office May Use Nets to Stop Migrant Boats Crossing Channel', *Guardian*, 11 October 2020.

equipment – as though the famously oversized public budgets were insufficient – and some on galas that celebrated the boys in blue. Chevron and Shell were corporate partners of the New Orleans Police & Justice Foundation. Both sponsored the Houston Police Department's mounted patrol. Rick Perry's Energy Transfer chipped in for the Friends of the Dallas Police, while leading coal-burning private utilities donated to foundations in Baltimore and Chicago, Philadelphia and Atlanta. The call to defund the police seemed to harmonise with that to divest from fossil fuels.[50]

And the words George Floyd wheezed while Derek Chauvin choked him to death – 'I can't breathe' – had an ecological ring to them. They spoke to the moment of the pandemic, when the effects of destroyed nature landed with redoubled force on non-white people, and so, on the margins of a global conversation on structural racism, awareness rose of the articulations of race and ecology. Fresh research received publicity: pregnant women exposed to high temperatures and air pollution tend to have stillborn, underweight or premature babies. Black mothers face the greatest risks, because of the concentration of African Americans in 'heat islands' – asphalt and brick soaking up the heat, no greenery to breathe it out – close to industrial zones. Recalling Spike Lee's *Do the Right Thing*, future heatwaves will be worst for them.[51] Segregation predicated on automobility builds the oven and then drives up the heat.

If Black Lives Matter spilled onto the main streets and confronted the police, violently or peacefully, like never before, however, the racial dynamics had the opposite effect on the US climate movement. Its direct actions had always been of the civil and polite type, in which activists practically count on being arrested. That tactic now conclusively manifested itself as a white privilege. While BLM could not but

---

50 Gin Armstrong and Derek Seidman, 'Fossil Fuel Industry Pollutes Black and Brown Communities While Propping Up Racist Policing', *Eyes on the Ties* (Public Accountability Initiative and LittleSis), 27 July 2020.

51 Nina Lakhani, 'Exclusive: New Research Shows How Black and Brown Neighbourhoods Will Be Hit Hardest by Global Heating', *Guardian*, 28 July 2020; Christopher Flavelle, 'Climate Change Tied to Pregnancy Risks, Affecting Black Mothers Most', *New York Times*, 18 June 2020.

be radicalised, the climate movement – still predominantly white in parts, without a militant protocol – shied away from direct action and refocused on parliamentary campaigns, notably the Green New Deal or its closest approximations under Joe Biden. Whether that was a wise choice remains to be seen.

The other side, at any rate, had its own concrete tactic for fusing energy and race: when BLM demonstrations assembled in cities, often blocking an intersection or bridge, counter-protestors would come in their SUVs. They would rev the engines and plough into the crowds. In late September, more than one hundred such incidents had been recorded (eight by police). Near Richmond, Virginia, a member of the Ku Klux Klan rammed his truck into cyclists and pedestrians with BLM placards; in New York, two police SUVs suddenly accelerated into the mass of flesh; in Minneapolis itself, the driver of a tanker truck took aim at thousands on a highway. In Mishawaka, Indiana, a man in an SUV hit the gas – 'we all tried to stop him and were yelling at him, but he just kept going', one demonstrator recalled.[52] The deep symbolism and psychology of this act will now be plain. More lethal than rolling coal, blending the two meanings of the word 'race', the tactic was first adopted in the last years of Obama' administration, against street-blocking rallies protesting the Dakota Access pipeline and the killings of African Americans. As white supremacists chuckled over the clips, they spread the meme 'run them over', later accompanied by 'all lives splatter'.

Riots rarely happen without the friends of status quo faulting 'outside agitators' (the most basic form of the rich man's cognitive mapping). During the Floyd uprising, Trump and his devotees picked 'antifa' for that role. With the methodology of new conspiracism, claims of 'antifa' plotting bedlam multiplied, to the point where heavily armed vigilantes patrolled streets against busloads of 'antifa' that never showed up. The long hot summer eventually set the West Coast aflame. Dried-out and overheated, the forests of California, Oregon and Washington State burnt in wildfires that shattered records for size and

---

52 Demonstrator Trevor Davis quoted in 'Cars Have Hit Protesters More Than 100 Times This Year', *Wall Street Journal*, 25 October 2020.

number and intensity, the smoke dimming the sun and depositing mounds of ash: but in rural Oregon, rumours claimed that *antifa* had organised it all and were coming for more. 'Has anyone seen or heard of 3 guys with Hoodies throwing bottles of gasoline in the Boring golf course?', someone would ask on Facebook. Vigilante checkpoints went up. Emergency operators were overrun with requests for information about the arsonists; sheriffs had to plead for the rumours to stop, but they had, as we have seen, a self-perpetuating logic grounded in structures beyond the Oregon counties.[53]

Adorno also spoke, from the grave, in 2020. A lecture he gave on the German far right in 1967 appeared in an English translation. Among the things he had to say:

> This behaviour is by no means purely psychologically motivated; it also has an objective basis. Someone who is unable to see anything ahead of them and does not want the social foundation to change really has no alternative but, like Richard Wagner's Wotan, to say 'Do you know what Wotan wants? The end.' This person, from the perspective of their own social situation, longs for demise – though not the demise of their own group, as far as possible, the demise of all.[54]

In other words, 'all lives splatter'.

Trump Falls, the Base Stands

But the mixture of virus and wildfires and violence on the streets did not redound to the decisive advantage of the far right. Instead the corona crisis tore the trump card of the incumbent: the appearance of a well-oiled economic machinery. Some argued that in its absence, Trump would have 'flattened the hapless Biden on his road to a second

---

53 For example, Conrad Wilson and Ryan Haas, 'Law Enforcement: Untrue Antifa Rumors Waste Precious Resources for Oregon Fires', *Oregon Public Broadcasting*, opb.org, 11 September 2020; Robin Abcarian, 'The Rumor Mill in Oregon Has Got the Wildfires Figured Out – It's All Antifa's Fault', *Los Angeles Times*, 16 September 2020.

54 Theodor W. Adorno, *Aspects of the New Right-Wing Extremism* (Cambridge: Polity, 2020), 10–11.

term'.[55] If there was a post-political longing for non-partisan know-how in the US, it benefited the candidate cast in that mould. As every discerning commentator noted, however, the election brutally refuted the thesis that Trump was a flukey gust of wind without staying power: he received nine million *more* votes than in 2016. In the four years, uniquely, his approval ratings never exceeded 50 per cent, but, until the very end of his tenure, they stayed within a narrow band of 40–45 per cent of the population, deaf to any bad news about him.[56] This base did not evaporate on the morrow of defeat. Much of it appears, at the time of this writing, to have bought into Trump's latest denial – of the results of a free and fair election – and held fast to its picture of him as legitimate president. This may have denoted a slide away from commitment to democratic norms, which wouldn't be entirely unforeseen. In a survey from early 2020, reported in *Proceedings of the National Academy of Science*, a plurality of Republicans agreed that 'the traditional American way of life is disappearing so fast that we may have to use force to save it'. Few lines could better sum up the politics of fossil fascism. Furthermore, upwards of 40 per cent believed that 'a time will come when patriotic Americans have to take the law into their own hands'. What factor predicted such defection from the normal procedures of democracy? Not lack of education, not rural residence, not conservative values in general, not even adulation of Trump – but 'ethnic antagonism', expressed in demeaning or resentful views of African Americans, Muslims, Latinos and immigrants.[57] If the state would ever act out such attitudes with full force, there would be a crossing over into fascism.

Hence non-Republican post-election debriefings inclined towards the question of the next Trump: will he (or she) be shrewder, slicker, less self-destructive?[58] Is such a character conceivable? For those who

---

55 The brilliant Luke Savage, 'The Trump Era Is Over. Our Leaders Want to Take Us Back to 2015', *Jacobin*, 7 November 2020.

56 See for example Ronald Brownstein, 'Why the Stability of the 2020 Race Promises More Volatility Ahead', CNN, 15 September 2020.

57 Larry M. Bartels, 'Ethnic Antagonism Erodes Republicans' Commitment to Democracy', *PNAS* 117 (2020): 22752–9.

58 For example Zeynep Tufekci, 'America's Next Authoritarian Will Be Much More Competent', *Atlantic*, 6 November 2020; Bhaskar Sunkara, 'Brace Yourselves. The Next Donald Trump Could Be Much Worse', *Guardian*, 10 November 2020.

wish to peer into the mental state of the base on the way into the 2020s, a useful work is *American Crusade: Our Fight to Stay Free* by Iraq veteran and Fox host Pete Hegseth, who poses on the cover with a flag and a machine gun tattoo on his right bicep and a gaze so stern it borders on the comic. In an exceedingly mediocre fashion, his book popularises many of the notions we have encountered – the evils of the left (Ilhan Omar in particular); the Muslim demographic conquest of Europe; palindefence from the seventh century via Vienna to present-day Minnesota; the dangers of feminism; the inexhaustible wonders of capitalism; unrepentant, categorical denial of climate change; the joys of 'driving a gas-guzzling SUV' and the pride in American inventions such as the aeroplane – or what must now be considered a bundle of ideas of stupendous durability and uniformity. The main novelty in Hegseth might be his explicit rejection of democracy and undisguised relish for armed violence. The flaps inform the reader that 'Pete and his wife, Jennifer, have seven kids, all future Crusaders'. What does he have to offer them? 'Crusaders, welcome to the Warring Twenties!' is his *cri de coeur*.[59]

The Wealth Is Seized

Trump used his last months in the White House to orchestrate one final saturnalia of deregulation, under the maxim 'seize the wealth while there is time': auctioning leases for drilling in the Arctic National Wildlife Refuge; opening Tongass Alaska's National Forest – the world's largest intact temperate rainforest – to loggers; relieving oil and gas producers of the duty to fix methane leaks. Whether these initiatives would survive the transition to Biden remained unclear. But every piece of sold-off land and tonne of emitted gas added to the challenge of changing course. In the pandemic, the Trump regime found a pretext to further aggravate the damages: in March, the EPA announced that it wouldn't hassle power plants and factories any more for violating air and water pollution standards. Billions of

---

59 Pete Hegseth, *American Crusade: Our Fight to Stay Free* (New York: Hachette, 2020). SUV, 311; 'Warring Twenties', 15; rejection of democracy, 50–3.

dollars from corona aid packages – tax reliefs, forgivable loans, federal bond buybacks – then found their way into the coffers of Peabody, ExxonMobil, Koch Industries and a second tier of lesser known fossil fuel producers, drowning hopes that the crisis would spontaneously euthanise the sector.[60] But there were other parts of the world where it had something to that effect. The coal fields of Poland were among the more important.

In the early summer, stocks of unsold coal piled up in the yards of Silesia. An unusually warm winter had reduced demand, while cheaper coal imported from far afield – Australia, Colombia, Mozambique – had eaten into the market. Then the industry came down with Covid-19. Labouring underground in close proximity, their lungs already worn by soot, miners made up one-tenth of the cases in the country and the region of Silesia half, earning it a reputation as the 'Wuhan of Poland'. An effusion of online hate against the miners ensued. The PiS government reacted by temporarily closing a dozen mines, fanning fears that – despite all the rhetoric about 'standing on coal' – the days of the industry were numbered. With an election to win, the PiS still tried to burnish its image as the guardian angel of the miners, paying them 100 per cent of their wages on furlough, Duda handing out bread rolls, lignite mines receiving licenses for extension: but other realities were closing in.

Brussels applied a mix of sticks and carrots to make the PiS give up on coal. Poland would ostracise itself if it persisted, or be entitled to more than a quarter of the Just Transition Mechanism funds – part of the Green Deal touted by the Union and, after dogged resistance, finally signed by the PiS leaders in late July. Having accepted the Deal, the government suddenly unveiled a new energy strategy that would cut coal from 77 per cent of the nation's consumption to between 11 and 28 over two decades. Offshore and onshore wind, solar and nuclear would take over. Staring death in the face, miners responded with wildcat strikes. By the autumn, an unstable compromise between unions and the PiS envisioned a

---

60 On this crisis management plan, see for example Lisa Friedman, 'E.P.A., Citing Coronavirus, Drastically Relaxes Rules for Polluters', *New York Times*, 28 March 2020; Fiona Harvey, 'US Fossil Fuel Giants Set for a Coronavirus Bailout Bonanza', *Guardian*, 12 May 2020.

gradual closure of the industry, ending in 2049.[61] Environmentalists welcomed a step in their direction – although another three decades of coal combustion, one decade more than in Germany, were hard to swallow – while miners came away with a deepened sense of victimisation. Covid-19 may have been the nail in the coffin of the PiS iteration of coal nationalism. If so, who will scoop up the emotional fallout?

If the EU could be perceived as an angel of death for Polish coal, it certainly didn't satisfy hopes about the pandemic as the 'once-in-a-generation' opportunity to break with business-as-usual.[62] The European Central Bank (ECB) led the efforts to pump oxygen into capital accumulation. Purchasing bonds and other assets, it injected billions of euros into companies – Shell and Total, utilities running coal-fired power plants, auto producers like Volkswagen, Daimler, BMW, Renault. Marxist state theory would predict that, absent anti-capitalist forces besieging them, state apparatuses would indeed seek to resuscitate accumulation across the board, and in proportion to the fossil character of capital, they would come to its rescue. But in late spring, the ECB asset portfolio was *more* biased towards fossil fuels than the market in general.[63] By November, the countries that had chosen to pour the most in billions into such ventures were the US, the UK, Germany and France (with China contributing mere scraps).[64] So much for talk of a green recovery.

Germany marked the year of Covid-19 with a trifecta of expansion: a brand-new coal-fired power plant in North Rhine-Westphalia, a mega-airport in Berlin, an extension of the autobahn through an ancient oak forest in Hesse. Norway lifted tax burdens from oil and gas and aviation. Sweden resurrected its airline from near-death and devoted twice as much money to climate destruction as to anything that could be labelled green business, while Australia went for broke

---

[61] For example Barbara Oksińska, 'Polskie kopalnie węgla pofedrują do 2049 r.', *Parkiet*, 26 September 2020.

[62] Piers Forster quoted in Damian Carrington, 'Covid-19 Lockdown Will Have "Negligible" Impact on Climate Crisis – Study', *Guardian*, 7 August 2020.

[63] Paul Schreiber, *Quantitative Easing and Climate: The ECB's Dirty Secret*, Reclaim Finance, 18 May 2020; cf. Greenpeace, *Bankrolling the Climate Crisis*, June 2020.

[64] SEI, IISD, ODI, E3G and UNEP, *The Production Gap Report: 2020 Special Report*, productiongap.org, 3 December 2020.

with a 'gas-fired recovery' – very much as usual, capitalist states across the global North *doubling down* on fossil fuels. A modest fraction of the trillions of dollars in corona packages would have sufficed to initiate a transition consistent with a 1.5°C pathway. Instead, the countries determining the fate of the planet readied for an increase in fossil fuel production, by an average of 2 per cent per year in the decade ahead.[65] That means so much more material for the far right to defend.

(This formidable inertia had a pendant in the department of neo-optimism, which in 2020 struck back against the wave of the previous year with two bestsellers: *False Alarm* by Bjorn Lomborg and *Apocalypse Never* by Michael Shellenberger.[66] Both accepted trend and attribution and denied impact. Global warming is not very bad. CO_2 makes the world greener, and so on, ad infinitum; while Oregon burned, these two volumes topped the 'climate change' section on Amazon, *The Facts* with Mark Steyn having dropped to thirteenth place. It was 2020, the hottest year on record, and bourgeois denialism had still not gone out of fashion.)

Developments in the Amazon then appeared rather unexceptional. One cannot deny that Bolsonaro had a point when he indignantly retorted to European diplomats that he treated natural riches just like they did. The wildfires in the 2020 season were greater in number, larger, even more cataclysmic than in the previous year, but received a sliver of the attention from a world absorbed by the virus. Beyond the Amazon, the Pantanal – the world's largest tropical wetland area, home to indigenous peoples and a treasure house of endangered species – lost nearly a third of its surface to burning. Scientists feared that the ecosystem would never recover. Like Trump, Bolsonaro knew how to seize the opportunity of the pandemic, scaling back monitoring and firefighting patrols even further, so that

---

[65] Marina Andrijevic, Carl-Friedrich Schleussner, Matthew J. Gidden et al., 'Covid-19 Recovery Funds Dwarf Clean Energy Investment Needs', *Science* 370 (2020): 298–300; SEI et al., *The Production*.

[66] Bjorn Lomborg, *False Alarm: How Climate Change Panic Costs Us Trillions, Hurts the Poor, and Fails to Fix the Planet* (New York: Basic, 2020); Michael Shellenberger, *Apocalypse Never: Why Environmental Alarmism Hurts Us All* (New York: HarperCollins, 2020).

ranchers could have their way. When occasionally pressed on the matter, he blamed the inferno on the indigenous peoples.[67]

The Pendulum Swings

But on the whole, 2020 was a poor year for the far right, by its own recent standards of success. The significance of losing the gravitational pull from the White House can hardly be exaggerated. It might have closed the conjuncture that began with the Brexit referendum and Trump. European parties received the results with sad silence (broken by claims of fraud), some of them embroiled in their own internal crises: having dropped out of the government in January, the FrP fell in the polls until it reached the lowest levels in three decades, triggering a civil war inside the party. The national leadership sought to bring the Oslo confederacy of hardcore xenophobes and denialists to heel.[68] There were rumours of a new party further to the right, á la Hard Line of Paludan. The FvD was heavily shaken after *Het Parool* (the newspaper formed by the Dutch resistance in 1941) revealed endemic Nazism in its youth wing. App groups and Instagram accounts were filled up with SS regalia, theorising about Jewish networks forcing women into pornography and the music played by Brenton Tarrant on his way to the Al Noor mosque.[69] The position of Baudet was unsettled. A splinter group of MPs set up a new party, JA21 (among the priorities: ending climate policies in the Netherlands). The remainder of the FvD focused on the fight against lockdowns and vaccines and moved closer to the scene of spiritual,

---

67 For example Tom Phillips, 'Amazon Deforestation Surges to 12-Year-High under Bolsonaro', *Guardian*, 30 November 2020; Emiliano Rodríguez Mega, '"Apocalyptic" Fires Are Ravaging a Rare Tropical Wetland', *Nature* 586 (2020): 20–1; Jenny Gonzales, 'As Fire Season Ends, Brazil Cited for Failed Amazon and Pantanal Policies', *Mongabay*, mongabay.com; and for the global political economy of the inferno, Amazon Watch, *Complicity in Destruction III: How Global Corporations Enable Violations of Indigenous People's Rights in the Amazon*, amazonwatch.org, October 2020.

68 For example Pedja Kalajdzic, Astrid Randen, Martin H. W. Zondag and Camilla Helen Heiervang, 'Oslo Frp-leiar ekskludert av partiet – Oslo Frp under administrasjon', *NRK*, 15 December 2020.

69 Harm Ede Botje and Mischa Cohen, 'Nazifoto's geen bezwaar bij de jongeren van Forum voor Democratie', *Het Parool*, 21 November 2020.

vegan conspiracists, while continuing to harass the traitors who signed climate agreements and took in immigrants. Over in Austria, the disgraced FPÖ was in free fall. The ability of far-right formations to self-destruct should not be written off.

But a hydra has many heads: when the Lega declined in 2020, there soared the Fratelli d'Italia, or 'Brothers of Italy', derived from the neofascist movement of the postwar era, whose leader bypassed the prime minister in popularity after a year of devastating pandemic in the country. The DF fell; the New Bourgeoisie rose. Bolsonaro stumbled without falling, while the Finns ended the year as the largest party in the polls – uneven outcomes, putting an end to hopes that Covid-19 would kill the far right. On average, the initial losses from the spring were recouped in less than half a year.[70] The year 2020 was bad for the far right because it *didn't grow* as it had for years (much as with emissions). It was too early to tell, at the time of this writing, whether this constituted a turning point. If so, the anti-climate politics of the far right would have reached its historical peak of influence around 1°C of global warming – that is, just at the moment when action to avoid 1.5°C was most exigent. That would now be the best-case scenario.

The victory of Biden signified a swing of the pendulum back towards capitalist climate governance. Might it crash through its wall, under the combined pressure of compounding disasters and billowing movements? Biden won 'as the embodiment of American business-as-usual', the compromise fetishist, the anti-Sanders candidate who rebuffed the Green New Deal and promised a gathering of New York capitalists that 'nothing would fundamentally change' under his watch.[71] But he might be a guy to blow with the winds, if they are strong enough. In the weeks before his inauguration, he had already made key nominations – to the EPA, the Department of the Interior, various climate policy posts – upsetting enough to appeal to the climate movement and rile up the right in equal measure. If this would be a replay of the Obama years, there would also be a reaction.

---

70 See Wondreys and Mudde, 'Victims'.
71 Jeff Sparrow, 'Trumpism Will Persist Until We Rekindle Faith in People's Ability to Reshape the World', *Guardian*, 13 November 2020; Igor Derysh, 'Joe Biden to Rich Donors: "Nothing Would Fundamentally Change" If He's Elected', *Salon*, 19 June 2020.

If Biden would go further in irritating or even hectoring primitive fossil capital, the reaction would presumably be stronger. All experience from the past three decades indicates that as long as such capital exists, it will resist its own abolition: in this regard, it does bear comparison with slaveholding capital. To flee from the fight would be to surrender beforehand.

The Zetkin Collective
late December 2020

Acknowledgements

Andreas Malm and Ståle Holgersen gratefully acknowledge Formas grant 2018-01702, which made work on this book possible. Agnes Andersson Djurfeldt, head of the Department of Human Geography at Lund University, has very generously supported this project from the start. Troy Vettese read an earlier draft and gave extremely valuable comments (special thanks to Octavia Butler). Richard Seymour helped to improve the manuscript in crucial respects. Thanks to Fabrício Belzoff, Santiago Gorostiza, Torsten Krause, Esther Leslie, Marcus Mohall, Laura Siggelkow. Much gratitude is due to Brian Baughan for essential polishing of the manuscript. Thanks to the participants at the anti-fascist climate camp in Leipziger Land in 2019 and at the Political Ecologies of the Far Right conference. Last but not least, thanks to the wider community of CPS, Culture, Power and Sustainability, the master's programme in human ecology at Lund University, where this project began.

Index

Abascal, Santiago 11, 156, 225, 297, 383, 494
Abbey, Edward 142–3
adaptation crises 241–7, 247–9, 252, 515, 521
Adorno, Theodor 294, 301, 305, 306, 311–13, 335, 383, 435, 447–8, 455–6, 482, 485, 496, 498, 499, 532
advertising 371–2, 378, 419
Africa 71, 363
African Americans 368–72, 516
   car workers 365–7
   cars and car ownership 364–5, 374, 388–9
Afro-Brazilians 207
Åkesson, Jimmie 46, 50, 65–6, 94, 307, 379, 522
Al Noor mosque massacre, Christchurch 150–3
Ali, Ayaan Hirsi 58–9
Allen, Irma xv
Alles Leben ist Kampf (film) 464–5
al-Qaida 54
Alternative for Germany (AfD) 4–8, 35, 40n3, 83–4, 89, 152, 178, 187, 223–4, 307, 451–2, 472
   climate denial 35–6, 481
   and coal industry 99–101
   and covid-19 pandemic 511–12, 518–19
   defence of car ownership 379, 380–1
   gender ideology 392–3
   green nationalism 167–70
   Grundsatzprogramm 5
   and immigration 52–3
   keywords 224
   'Learning from European History' posters 52–3
   opposition to wind power 87
   party motto 6
   and Thunberg 309
   truth-telling crusade 6–7
Althusser, Louis 14, 322–3, 324–5, 381–2, 383, 485, 507
Alvarez, Alex 274
Alvim, Robery 224–5
Amazon 339–42
   deforestation 208–11, 217–18, 341
   fires 217–18, 310–11, 397, 489, 537–8
America Alone (Steyn) 55–6
American Crusade (Hegseth) 534
America First Committee 440
America First Energy Conference 185–6
American Academy of Political and Social Science 358–9
American Enterprise Institute 32, 58–9
American Legislative Exchange Council 34
American Petroleum Institute 18, 24–5, 27
American War (El Akkad) 247–9, 251, 270
Amoco 18
An Inconvenient Truth (film) 32

Anderson, Perry 285–7
Anglo-American exceptionalism 37
anti-communism 22
anti-feminism 391–2
anti-Semitism 114–15, 312, 439
antifa 531–2
apocalypticism 50–1, 53
Araújo, Ernesto 214–15, 219, 304–5
Arctic, disappearance of ice from 194
asylum-seekers 41, 159, 162–3
Atlanta 369–71
attention deflection 199–206
Auschwitz 436–8
Australia xv, 57–8, 490, 536–7
Austria 84, 90, 172, 178, 259, 476
authoritarianism 475
Avanti! (magazine) 400, 416
aviation 416–18, 423–4, 440–1, 446
Axess (magazine) 55

Balbo, Italo 416–17
Baldwin, James 329
Banks, Arron 362
Bannon, Steve 188
Bardella, Jordan 136, 155
Bastié, Eugénie 138n22
Batten, Gerard 72, 80
Battistoni, Alyssa x
Baudet, Thierry 9–10, 36, 53, 302, 306
Bawer, Bruce 58
Beck, Glenn 281, 295
Benegal, Salil D. 201
Benjamin, Walter 400, 407, 444–5, 447
Bensaïd, Daniel 509
Bernhardt, David 185
Biden, Joe 531, 539–40
Bild (newspaper) 117
biodiversity 472–3
black extremists 69
Black Lives Matter 315–16, 529–32
blackness 252, 328
Błaszczak, Mariusz 111–12, 112
Bloch, Ernst 457–8n188, 470, 474
blood and soil 460–9, 474
Bobako, Monika 454–5
Bolsonaro, Carlos 212–13
Bolsonaro, Eduardo 213, 309
Bolsonaro, Flavio 213
Bolsonaro, Jair 206–7, 292, 310, 361, 397, 537–8, 539
  appointments 213–15
  climate denial 212–13
  and covid-19 pandemic 514, 517
  and Cultural Marxism 304–5
  and deforestation 209–11, 217–18

  gender ideology 391–2
  immigration policies 216
  mitigation funding cut 219
  non-white land seizure 216–17
  racism 211, 216, 340–2
Bolton, John 194
Bonapartism 226, 236
border fortifications 174, 471–2
Bosch, Carl 431
Bosnian genocide 263–7, 269, 273–4
bourgeois optimism 486–8
Bourke-White, Margaret 365
BP 18, 26, 28, 31, 32
Brazil xv, 206–19, 260–1
  Amazon fires 217–18, 310–11, 397, 489, 537–8
  cattle ranching 341
  climatic weight 208
  and Cultural Marxism 304–5
  deforestation 208–11, 217–18, 341
  energy sector 207
  frontier racism 339–42
  gender ideology 391–2
  Green Revolution 208
  immigration policies 216
  Landless Workers' Movement 210
  mitigation funding cut 219
  non-white land seizure 216–17
  oil and gas industry 207–8, 219
  premonitions of fascism 224–5
  racism 211, 216
Brazilian proposal, the 211–12
Breitbart 302, 306–7
Breitbart, Andrew 302
Breivik, Anders 61, 125, 129, 130, 150, 237, 246, 272–3, 273–4, 303–4
Brexit 69–81, 362–3, 450
  and general election, 2019 77–80
  and immigration 69–71, 76–7
  and Islamophobia 69–71
  leave campaign 71–7
  referendum 71–2
Brexit Party 73, 519
Britain First 80
Britannia Unchained 75–6, 81
British Empire 343–63, 362–3, 443
British National Party 80
Brooks, Mo 281–2
Bruckner, Pascal 59–61, 394, 524
Brulle, Robert 33
Buchanan, Pat 494
Bush, George 396–8
Bush, George W. 55, 196
Business for Innovative Climate and Energy Policy 29

Index

Butler, Judith 509
Butler, Octavia, *Parable of the Sower* 250
By the Feet of Men (Price) 250

California 490
Camus, Renaud 51, 526
capital accumulation 483, 536
capitalist classes 482
capitalist climate governance 28–30, 33, 34, 83, 85, 130, 131, 174, 189, 212, 484
capitalist mode of production 291–2, 296, 481–4, 507
Caplan, Jane 234–5
car industry 91–2, 93, 193, 241, 365–7, 374
carbon sequestration, forests 208
carbon vitalism 20–1, 74, 109, 112, 117, 187, 193–4
Cardoso, Fernando Henrique 212
cars and car ownership 363–81, 447–8, 456
  advertising 371–2, 378
  African Americans 364–5, 374, 388–9
  China 388
  defence of 379–81
  emissions 363
  Europe 374–81
  and gender 393
  Germany 418–21, 425–6
  Italy 406–7, 414–16
  non-emitting fuel 363–4
  racial status 364–5
  significance 385–6
  SUVs 372–4, 378, 381, 385, 386
  Sweden 376–8, 379–80
  United States of America 363, 364–74, 388
  and whiteness 381–6
Carter Hett, Benjamin 233
Carvalho, Olavo de 213, 217, 261, 304–5
Castello Branco, Roberto 207–8
Chanin, James 319
Chapoutot, Johann 461, 467–8
Charlotte, North Carolina 303
chemtrails 292–3
Chevron 530
China 202–3, 316, 346, 388
Christchurch, Al Noor mosque massacre 150–3
Chrysler 28
Cichocka, Aleksandra 499–500
class xv, 458–9, 482
class bias 22–3
class consciousness 293, 453, 457

class struggle 450–9
class war 411
clean coal 107, 110
climate breakdown, reality of xiii
climate change
  as non-problem 6
  and skin colour 315–22
climate crisis, responsibility 492–5
climate delayers 130–1
climate denial ix, 3–38, 139, 170, 511–13
  Austria 10–11
  Bolsonaro 212–13
  comparison with Holocaust denial 494–5
  conservative white men 295–6
  contradictions 481
  crisis 24–35
  drivers 490–1
  faith 4
  feedback mechanisms 495–9
  first phase of innocence 281–4
  fourth phase 296–8
  future of 489–92
  Germany 4–8, 36–7
  green nationalism 171–4
  implicatory 480–6
  irrationality 482–3
  liberal ideology 486–8
  literal 479, 480, 489
  Netherlands, the 8–10
  origins 12–24, 285
  periodisation 38
  push to far right 284
  and responsibility 492–5
  revenge of 35–8
  Spain 11–12
  spread 3–12
  status 12
  taxonomy 479–80
  temporality 495
  Trump 299–300
climate fiction 247, 247–53
climate hoax claims 9–10, 21–2, 72–3, 159, 201, 282–3, 298, 391
climate hysteria 161, 165, 310
Climate Intelligence Foundation (Netherlands) 10
climate justice 177–8
climate neutrality 117
climate profiteering 193
climate refugees 42, 177–9
climate refusal 396
climate religion 11–12
climate research community, communication viii

climate scepticism 3–4
climate science vii–viii, 3, 70
climate shocks 241, 252
climate stabilisation 241
Climate Surprise conference 56–7
climate system
  anthropogenic interference 13
  models 172–4
  volatility xv–xvi
Climategate 283–4, 288
Cliteur, Paul 10, 302
CO2 Coalition 56–7
CO2 emissions xvi, 3, 154
  American 24, 316
  atmospheric impacts 24–6
  atmospheric rate 26
  as a godsend 5, 9–10
  China 316
  European Union 117–18
  Germany 98
  Norway 118–19, 122
  peat 95
  Poland 118
  responsibility 316
  Saudi Arabia 122
  United Kingdom 316
coal exit commission', Germany 99–100
coal industry
  denial machine 19
  employment 277
  Germany 98–101, 277, 456
  and the Holocaust 435–8
  hydrogenation 430–4
  Nazi Germany 429–34
  Poland 105–7, 110–11, 457, 535–6
  United Kingdom 96–7
  United States of America 192, 195, 277
coal nationalism, Poland 116–17
coal workers 277
Coates, Ta-Nehisi 200–1, 204
cognitive mapping 290–2, 296
Cohen, Stanley 479–80, 480–1, 484, 490–1, 493, 495
Collectif Contre L'Islamophobie en France 524
commodity circulation 333–4
Communism 306
Competitive Enterprise Institute 19–21, 31, 32, 38, 75, 184, 219, 284, 287–8, 361
concentration camps 266, 435–8
conservationism 139
conspiracy theories 279–300, 497
  absence of observable evidence 281
  accept no coincidences 284

chemtrails 292–3
Climategate 283–4, 288
cognitive mapping 290–2
conspiratorial network 280–1
credibility establishment 298–9
Cultural Marxism 300–13
defence of the social system 293–6
Earth Day 287–8
features 285
the hidden hand 280–1
immune to falsification 281
left climate conspiracy theory 283–92
new conspiracism 298–300, 305–6, 313
role 293–4
status 289–90
and Thunberg 309
Continental Resources 182
control, sense of 497–8
Conway, Erik M. 285
COP15 43, 44, 212, 320
COP21 34, 109, 134, 196–7, 212
COP24, Katowice 110–11, 117, 186, 215
COP25 219, 307
Copenhagen 43
Corbyn, Jeremy 78
Corbyn, Piers 70, 519
corporate funding 23
cosmopolitanism 278–9
counterjihad movement 59, 258–61
Covid-19 pandemic xvi, 511–13
  anti-lockdown movements 518–20
  Chinese virus trope 513
  exploitation for authoritarian projects 514–15
  France 523–5
  hoax claims 517
  medical nationalism 520–1
  non-white infection rates 516–17
  and Poland 535–6
  and racism 513–15, 516–21
  and Sweden 521–3
Cox, Jo, shooting of 80, 128
Craig, David 75
Crawfurd,, John 354
Crusades, the 258
Crusius, Patrick Wood 471
Cruz, Ted 56
Cultural Marxism 9, 300–13, 494
Culture, Law of 324–5
Cummings, Dominic 80
Cyclone Idai 153–4

Dagens Nyheter (newspaper) 63
Daggett, Cara xii, 242, 393, 394–5, 396

Index

Dakota Access pipeline 190, 198
Danish People's Party 43–5, 86, 97–8, 162–5
Dark Mountain Project 146–7, 148
Darré, Richard 463–4
Darwish, Mahmoud 360–1
David Horowitz Freedom Center 58–9
de Benoist, Alain 137, 258
death drive 502–6, 509
decarbonisation xvi, 8
deep state, the 302
deforestation 208–11, 215n114, 217–18, 341
deglobalisation 525
degrowth movement 137–8
deindustrialisation 90
de-islamisation 89
Delingpole, James 306–7, 308–9
denial
  drivers 490–1
  taxonomy 479–80
denial machine 13–14, 18–24, 26–8, 35–8
Denmark 43–5, 86–7, 97–8, 477, 487
  climate elections, 2019 164–6, 527
  green nationalism 162–6, 527–8
  islamophobia 514
  non-Western population 163–4
Detroit 365–7, 384, 389
DeVaughn, William 389
Dewey, John 449
Di-Aping, Lumumba 320
Dias, Tereza Cristina 210
DiCaprio, Leonardo, ignites Amazon fires 310–11
differential mobilities 372
Donors Capital Fund 32, 61
DonorsTrust 32
Dorling, Danny 362–3
d'Ornano, Mireille 133
Dörre, Klaus 451–2, 453–4, 455
Du Bois, W. E. B. 359–60
Duda, Andrzej 111, 117, 515, 535
DuPont 28

Earth Day 287–8
Earth First! 144
Earth Hour 56–7
Ebell, Myron 184, 188, 219, 284
Echeverría, Bolívar 390
ecological fascism 475–7, 528
ecological sadism 396–8, 506
ecologie intégrale 138–9
Ecologism 59–61
ecology 469–77

biodiversity 473
border fortifications and 471–2
and Nazism 469–70
patriotic 134, 137, 139, 174, 474–5
and self-interest 473–4
tactics 474–5
economic self-sufficiency 86
Economist (magazine) 196
eco-populist voodoo 7
eco-Shiites 311
eco-socialism 477
education, lack of 282
Egypt 350, 389
Ehrlich, Paul 140
El Akkad, Omar, *American War* 247–9, 251, 270
El Paso killer, the 471
elevator metaphor xiii–xiv
Eley, Geoff 232, 235, 236, 238–9, 242
Elliott, Matthew 75
Emerson, Ralph Waldo 356–8, 442
emissions cuts 239, 322
Ende Gelände 7, 99
Energy Agency 482
Energy and Man symposium 24
'Energy' commercial 19–21
energy dominance 191–2
energy efficiency 94
energy independence 183, 190–2, 388
energy stock 274–9
English Defence League 80, 258
environmental degradation, and racism 337–8
Environmental Protection Agency 184, 188–9, 193, 534, 539
environmentalism
  American 139–46
  British 146–9
Equinor 483
escalator metaphor xiii
Esposito, Fernando 417
ethnic cleansing 67, 176
ethnocultural segregation 168
Ethnological Society of London 353–4
ethnonationalism 37, 38, 39–40, 67–8, 199, 227
ethnopluralism 94, 124
Eurabia 47–50, 55, 61, 178, 257, 306, 388
Europäisches Institut für Klima und Energie 187
Europe
  car industry 374
  cars and car ownership 374–81
  Islamisation threat 39–54, 60, 67–8, 93

secessionist automobility 375–81
European Central Bank 536
European integration 448
European Parliament 390–1, 472
European Union 92, 115, 196
  CO2 emissions 117–18
  migrant quotas 104
existential threat 51, 227–8, 246
Extinction Rebellion 7, 79, 245–6, 307
extraction, definition 277–8
Exxon 13, 18, 19, 25–6, 27
ExxonMobil 28, 31, 32, 38, 54, 58, 190, 279, 280, 484

Facebook 303
faith 4
fake news 298
Fanaticism of the Apocalypse, The (Bruckner) 59–61
Fanon, Frantz xii, 317, 336–7, 340, 389–90
far right, the
  advance of ix
  climate change denial ix
  and the natural environment x
  rise of ix–x, xvi
  struggle against xvi
Far Right Today, The (Mudde) ix–x
Farage, Nigel 72, 73, 76, 78, 155–6, 519, 520
Farrar, Frederic 353–4
Fasci di Combattimento 399–400
fascisation 164, 251, 452–5
fascism xiv–xv
  American 439–41
  basic strands 405
  becomes avantgarde of fossil capital 399–407
  causes of 231–5
  classical xii, xiv, 225–6, 271, 272, 405–6, 438, 449–50, 459
  definition 227–39, 274
  following 233
  keywords 224
  liberal theory of 457–8
  machine fetishism 400–7, 442–50
  Marxist theories of 452–5
  myth of 255–7
  and nature 460–9
  non-contemporaneity of 457–8n188
  outside the Axis 438–41
  premonitions of 223–6
  racism 235
  as real historical force 229–30, 231–9, 273

as set of ideas 230–1
stirring power 228
usage 226–7
fatalism 491
Faurisson, Robert 495
feedback mechanisms 495–9
feminism 306, 403
Ferguson, Stephen C., III 332
Ferrara, Peter 192
Fiat 415–16
Fidesz party 87, 90, 91–3, 166, 171, 253, 261-2, 309, 392
Finland 87–8, 95–6, 161–2, 380, 526–7 503
  climate elections, 2019 156–61, 161, 395–6
First Opium War 346
First World War 231, 234, 403, 407–8, 503
Five Star Movement 176
Floyd, George, murder of 529–30
Foča genocide 263–4
Forbes (magazine) 291–2, 293
Ford, Henry 439, 440
Ford Motor Company 18, 28
Foreman, Dave 144–6, 149, 155
forests, carbon sequestration 208
fortress mentalities 242
Forum for Democracy (Netherlands) 9–10
fossil capital
  as class fraction 16–17
  defence of 90–101
  definition 15
  denial machine 13–15, 18–24, 26–8, 35–8
  existential crisis 17–18
  political force 17
  primitive 16–18, 20–1, 95–101, 181, 195–6, 205, 280
  primitive accumulation 16
  profit generation 15
  structure 15–18
  Trump and 188–94, 195–7, 205
fossil fascism xii, 457–8
  adaptation crises 241–7, 247–9, 252
  causes of 231–5
  in climate fiction 247–53
  danger of 313
  definition 239, 274
  keywords 224
  mitigation crisis 239–41, 243–4, 245–6, 247–9, 251, 253
  premonitions of 223–6
  scenarios 239–47
  stirring power 228
fossil fuels

Index

energy stock 274–5
imports 94
irrationality 482–3
need to shed 3
reduction planning 12–13
Trump and xi
Fox Business 197
Fox News 33, 55, 308
fracking 96, 96–7, 183, 195
France 41–2, 120, 133–9, 151, 474–5
   climate protection referendum 525
   and covid-19 pandemic 517–18, 523–5
   Islamophobia 523–5
   police brutality 524
   working-class insurrection 524–5
Frankenberg, Ruth 331, 332, 332–3
Frankfurt School 300–2, 306, 390, 392, 447, 458
Frankfurter Allgemeine Zeitung (newspaper) 223–4
Franklin, Benjamin 327
Fransson, Josef 379
Fratelli d'Italia 539
Frederiksen, Mette 164, 528
Freedom Party (Austria) 10–11, 259
free-market, the 22, 34, 38, 54
Freiberg, Kjell-Børge 127, 128
Freud, Sigmund 502n85, 503–6
Fridays for Future 7, 149–50
Frohardt-Lane, Sarah 365–7
Front National (later Rassemblement National) 40, 42, 120, 133–5
frontier racism 338–43
FrontPage Magazine (magazine) 44, 58–9
fuel-efficiency standards 491
fugitive flow 275
funnel methodology 53–4
Futurist Manifesto 400–7

gas extraction 97–8
gatedness 242, 270
Gates, Bill 486
Gates of Vienna website 259
Gauland, Alexander 5–6, 36, 52, 223–4
gender xv, 391–6
gender ideology 391–3
General Motors 26, 28
Génération Identitaire 258–9
genetics 326–7
genocide 263–7, 273
George C. Marshall Institute 13, 56
Georgia Highway Contractors Association 370
Germany 36–7, 40n3, 83–4, 116, 473, 474–5
asylum applicants 111
aviation 423–4
cars and car ownership 379, 380–1, 418–21, 425–6, 456
chemical industry 429–31, 432, 441
climate denial 4–8, 36–7
CO2 emissions 98
coal exit commission 99–100
coal industry 98–101, 277, 429–34, 456
coalition government 5
and covid-19 pandemic 511–12, 518–19
emissions 8
Freikorps 278
green nationalism 167–70
hydrogenation 446
machine fetishism 407–14, 442–3
mitigation crisis 241
nationalism 100
and nature 460–9
Nature Protection Law (Nazi) 465–6
Nazi 418–38, 442–3, 445, 457–8, 460–9
Nazi political economy 424–6
Nazi techno-racism 427–9
opposition to wind power 87
racial divides 37
renewable energy 275
renewable energy sector employment 277
resentment against the US 425–6
rise of Nazism 412–13
road development 421–3, 445, 446–7
the *Volk* 411–12, 420–1, 424, 426, 428–9, 460–1
Weimar 410–12, 421–2, 425, 455
Gilroy, Paul 369, 385, 389
Global Climate Coalition 13, 18, 19, 26, 28, 29, 31, 37, 38
Global Environmental Change (journal) 36–7
global financial crisis, 2008 168
Global Warming Policy Foundation 74, 79, 487
globalisation 134, 278
globalism 148, 213, 262, 279
globalist elite, the 223–4
Golden Dawn xvi
Golec de Zavala, Agnieszka 499–500
Google 34
Górak-Sosnowska, Katarzyna 113–14
Gore, Al 32, 287n85, 309
Gorgon, HMS 344

Göring, Hermann 434
Gramsci, Antonio 22, 137n19, 235, 300, 303
Great Replacement, the 258, 306
Green Industrial Revolution 457
green nationalism 139–49, 306–7, 461, 470–1, 525–9
  balance sheet 171–5
  central European 166–70
  climate denial 171–4
  climate system models 172–4
  definition 154–6
  Denmark 162–6, 527–8
  Finland 156–61, 161–2, 526–7
  Germany 167–70
  Italy 175–9
  practice 149–54
  theory 139–49
Green New Deal 457
green technology 10
greenhouse effect 12, 24–6
greenhouse gas emissions, human attributed 3
Greenpeace 144, 210
Greens 7
greenwashing 30, 484
Griffin, Roger 227–9, 230, 231, 231–2, 235, 237, 270, 272–3
Guardian (newspaper0 34–5, 147–8, 293
gunboat diplomacy 343–6

Hage, Ghassan 318, 337–8
Hagen, Carl I. 119, 124, 125
Haimbuchner, Manfred 90
Hall, Stuart 71, 328, 334, 516
Halla-aho, Jussi 156–7, 158, 161, 304
Hamm, Harold 182–4, 195, 197, 388
Hanebrink, Paul 271
Hansen, James 3, 12, 27, 252–3, 316–17
Happer. William 193–4
Hardin, Garrett 140–2, 144, 155
Heartland Institute 22, 31, 32, 38, 56, 59, 111, 187, 192, 192–3, 282, 517
Hegseth, Pete, *American Crusade* 534
Helmer, Roger 72–3, 74, 86, 96, 380
Henderson, Jason 371
Henwood, Doug 195–6, 204
Herf, Jeffrey 412–13, 442, 447
Heritage Foundation 32, 59
hidden hand, the 280–1
Hilse, Karsten 6, 8, 99, 100
Himmler, Heinrich 460, 465
Hitler, Adolf 224, 232, 233, 243, 272, 409, 430, 433, 434–5, 443, 445, 465
  assault on nature 468–9

aviation 423–4
  cars and car ownership 418–21, 425–6
  political economy 424–6
  resentment against the US 425–6
  road development 421–3
Höcke, Björn 167–70, 392
Holgersen, Ståle xv
Holocaust, the 435–8
Holocaust denial 4, 494–5
Holocaust scenario 242–3
Hopkins, Katie 69, 80
Horkheimer, Max xiii, 496, 498
Horowitz, David 58–9
Huber, Matthew 371–2
Huhtasaari, Laura 157
Hultgren, John 139–40
human descent groups 326–7
human species, biological degeneration 135–6
Hungary 87, 89, 90–3, 94, 166–7, 261–2, 309–10, 342
Hurricane Marie 310
Hutchinson, Sikivu 369
hyper-segregation 377–8

ideological state apparatus 13–15, 53, 131
  denialist 18–24, 26–8, 30–1, 35–8, 59, 185–8, 279, 480
  parallel 29–30
  raw material 333
ideology 333–5
  definition 381–2
  Marxist theories of 22–3
IG Farben 430–1, 433, 435–6, 441
Ikard, Frank 24–5
Il Populista 177–8
immigration 61–2, 141, 145–6, 148, 158–61, 484–5, 522–3
  Brazil 216
  and Brexit 69–71, 76–7
  call to halt 40–1
  as funnel issue 41
  hostility towards 39–54
  illegal 58
  Italian policy 176–9
  preoccupation with 54
  prohibition 168
  repatriation 66–7
  and scapegoating 498–9
  zero tolerance 41
Immonen, Olli 158, 304
implicatory denial 480–6
India xv, 353, 363
indifference 318–19, 335

Index 551

indigenous peoples 339–41
inferiorization 332, 335–43
Information Council on the Environment 19, 28, 38
infrastructure construction ban 239
Inhofe, James 21–2, 184, 282–3, 295, 298
Inikori, Joseph 384
Institut für Staatspolitik 167–8, 168–9
Institute of Economic Affairs 74, 79, 80–1
Institute of Public Affairs 57–8
integral ecology 138–9
integration 63n72
Intergovernmental Panel on Climate Change viii–ix, 3, 13
  reports 25, 28, 32, 186–7
International Center for Western Values 53
International Socialist Party 284–5
interpellation 322–8, 332, 335, 382–3, 508
interpretive denial 480
intersectionality 303
invandrares 324–6
invasion myths 256–74
Investor's Business Daily (newspaper) 283
Iran Hostage Crisis 387–8
Iranian Revolution 387
irrationality 482–6, 495–6
Irving, David 495
ISIS 69
Islam 48, 60
  Johnson on 70–1
  as permanent war 257–8
Islamic places, targeting 88–9
Islamic terrorism 184
Islamisation threat 266, 272
  Europe 39–54, 60, 67–8, 93
  stealth 124
Islamofascism 227
Islamophobia 44–8, 54–62, 258, 388, 523–5
  and Brexit 69–71
  and covid-19 pandemic 514–15
  Poland 112–14, 116, 454–5
Italy 175–9, 259–60, 399–407, 414–18, 445, 525–6, 539

Jacobs, Benjamin 436–8
Jacques, Peter J. 4, 23
Jameson, Fredric 290–1
Jardina, Ashley 200, 322
Jensen, Siv 42–3, 43, 119, 124, 126, 127
Jews 113, 177–8, 224, 231, 304, 439–40, 461, 462

Holocaust scenario 242–3
imagery 312
Judeo-Bolshevism 271–2, 301, 425, 426, 434
and nature 462–3
subversive 92–3
Jobbik 166–7
John Birch Society 203
Johnson, Antonia Ax:son 55, 308
Johnson, Boris 70–1, 77–81, 96–7, 487, 514, 528–9
Jomshof, Richard 46–7, 67, 379
Judeo-Bolshevism 271–2, 301, 425, 426, 434
Jünger, Ernst 407–14, 447, 503
Juvin, Hervé 135–7, 476, 525

Kaczyński, Jarosław 103–4, 104–5, 111, 112, 116
Karadžić, Radovan 266
Karlsson, Mattias 68, 256–7, 269, 304
Katowice 110–11, 117, 186, 215
Kelly, R. 389
Kingsnorth, Paul 146–9, 155
Kinnunen, Martin 64, 94
Kjærsgaard, Pia 43, 165
Klein, Naomi 292, 338, 489
Koch, Charles 31, 185, 195, 203, 441, 491
Koch, David 31, 185, 195, 203, 441, 491
Koch, Fred 203, 441
Kosovo Polje, Battle of 264–5
Kraft, Rainer 6–7, 8
Kukla, Rebecca 323–4, 325–6, 333, 381–2
Kurdi, Alan 103
Kuswa, Kevin Douglas 372
Kyoto Protocol 29–30, 43

Laatsch, Harald 152
Laird, Macgregor 349–50, 442
Lajos, Kepli 166–7
Lämsä, Tapio 158
Lanchester, John, *The Wall* 249–50, 251, 270, 509
Lander, Richard and John 348–9
Landless Workers' Movement, Brazil 210
late fascism 253
Latouche, Serge 137–8
Lawson, Nigel 74, 115
Le Bon, Gustave 356
Le Monde (newspaper) 42
Le Pen, Jean-Marie 35, 40, 41–2, 134, 135, 178, 361, 494
Le Pen, Marine 86, 89, 133, 134–5, 136–7, 151, 174, 225, 524
Lee, Spike 530

left climate conspiracy theory 283–92
Lega Nord 175–9, 259–60, 268, 539
Lemos, Anabela 154
Lennon, John 301
Les Patriotes 137
Lewis, David Levering 269
Ley, Robert 424
LGBT rights 392
liberal ideology 486–8
Lien, Tord 127
Lilico, Andrew 319
Limburg, Michael 187n26
limits 449–50, 501–2
Lind, William S. 301, 303, 306
Lindbergh, Charles 439–41
Lindzen, Richard 27
Linkola, Pentti 159–60
Lipsitz, George 386
Lipstadt, Deborah 495
Listhaug, Sylvi 127, 130
literal denial 479, 480, 489
Littman, Gisèle 47–8
Löfving, Mats 521
Lomborg, Bjørn 487, 537
London bombings, 2005 70–1
London City Airport 315–16
Löwenthal, Richard 311
lubrication 347–8
Lukács, Georg 22, 300

Maastricht Treaty 448
machine fetishism
  critique of 444–5
  fascism 442–50
  Germany 407–14
  Italy 400–7
  Mussolini 414–18
  Nazi Germany 418–38
Macierewicz, Antoni 104, 105–6
McIntyre, Lee 296
Macron, Emmanuel 135, 524–5, 525
Malmö 377–8
Malthusianism 140, 155, 245–6
Mandel, Ernest 459n190
Marcuse, Herbert 447n159, 505–6
Maréchal-Le Pen, Marion 134, 138
Marinetti, Filippo Tommaso 400–7, 408, 409, 418, 459, 488, 501, 503, 506
Martineau, Harriet 353, 382, 442
Marx, Karl 15–16, 17, 226, 229, 288–9, 321
Marxism 404–5
  Cultural 300–13
  theories of fascism 452–5
masculinity 393–5

mass appeal 37
May, Theresa 97
Mayer, Jane 203
Mechanics' Magazine (magazine) 346
medical nationalism 520–1
mega-emitters 212
Mercer, Robert 194
Merchant, Carolyn 393
Merkel, Angela 99, 511–12
Messerschmidt, Morten 44
Mészáros, Lörinc 92
methane emissions 189–90
Mexico 191
Miéville, China 396
migrant policy 174
migrants and migration 103–5, 252
Miliband, Ed 320
Miliband, Ralph 482
Milošević, Slobodan 265–6, 269
minarets, hatred of 88–9
misogyny 403
Mitchell, Don 373, 385–6, 386
mitigation 3, 85, 88, 90, 211, 292, 474
mitigation crisis 239–41, 243–4, 245–6, 247–9, 251, 253
Mladić, Ratko 266, 267
Mobil 21–2, 26
mobile combustion, acts of 342–3
Morano, Marc 31, 186, 215
Mordaunt, Terence 79n130
Mos, Richard de 8
Mosley, Oswald 438–9
Mourão, General Hamilton 215
Mozambique 153–4
Mudde, Cas ix–x, 485
Muhammad Ali 343–6
Muhammad cartoon crisis 54
Muirhead, Russell 297–8, 298, 300
multiculturalism 157, 306
Murdoch, Rupert 33
Murer, Philippe 134
Murray, Robert E. 182, 188, 192–3, 195
Murray Energy 31, 182
Muslim oil crisis 387–8
Muslims
  American ban 182
  call to halt immigration 40–1
  and covid-19 pandemic 514, 517–18
  demonisation 44
  deportation 49–50
  genocide 263–7
  Holocaust scenario 242–3
  in France 523–5
  majority threat 43–6, 55–6, 88, 112, 150–3, 158–61

Index

oil mastery 387–8
population growth 47, 51, 55–6, 158–61
re-primitivization threat 56–7
subversive 92–3
threat 39–54, 54–62
threat to white women 51–3, 112, 114
Mussolini, Benito 232, 399–400, 414–18, 439, 445

Nanterre conference 42, 178
Napier, Charles 344, 345, 345–6
narcissism 499–502, 504
nation, defence of 256–74
National Mining Association 31
nationalism 39–40, 84, 100, 166–70, 275, 383–4, 404–5
   green 139–79, 162, 461, 470–1
   petro 124–31
natural resources, duty to use 109–10
nature 137–8, 139–40, 155, 336
   destruction of 402
   domination of 335–43, 504–5
   and fascism 460–9
   law-making authority 461–2
   love for xv
   Nazi assault on 466–9
Nature Protection Law (Nazi Germany) 465–6
Nazi ventriloquism 225–6
Nazism 243, 271, 412–13
   assault on nature 466–9
   concentration camps 435–8
   green window dressing 469–70
   and nature 460–9
   political economy 424–6
   sterilisation programmes 464–5
   techno-racism 427–9, 442–3
   violence 443
Neo (magazine) 54–5
neofascism 226
Nerbrand, Sofia 54–5
Netherlands, the 8–10, 83, 89, 97, 206
Neumann, Franz 431–3, 434
New Bourgeois 165
new conspiracism 298–300, 305–6, 313, 531
New York (magazine) 34
New York Times (newspaper) 12–13, 21–2, 192–3, 219, 395, 439
New Zealand 355
   Al Noor mosque massacre 150–3
Niger river 348–50, 443
nigger climate accords 204
Nonnenbruch, Fritz 427–9

Norgaard, Kari Marie 480
North Sea Agreement 97–8
North Sea oil 97
Norway xv, 42–3, 61, 206, 480, 484, 536, 538–9
   cars and car ownership 380
   climate policy 125
   climate scepticism 119
   CO2 emissions 118–19, 122
   and covid-19 pandemic 522
   denialism 125–6
   energy ministers 127–9
   migration policies 129–30
   Ministry of Petroleum and Energy 119, 122
   mitigation crisis 241
   oil and gas industry 118–31
   petro-nationalism 124–31
   reliance on oil 120–1
   self-image 121–4
   welfare state 120
Norwegian Oil and Gas 123
Norwegian Progress Party 42–3, 119–21, 124–9, 131–2, 380, 538–9
Notre Dame cathedral fire 281
Nouvelle Droite, the 41–2, 137
Nouvelle Écologie 133, 137, 472
Nye Borgerlige 165

Obama, Barack 33, 189–90, 196–7, 198, 200–1, 203, 204, 212
Ocasio-Cortez, Alexandria 130–1
Odoul, Julien 525
oil and gas industry 96, 386–8
   American production 192
   Brazil 207–8, 219
   Muslim mastery 387–8
   Norway 118–31
   prices 386, 387
oil embargo, 1973 47–8, 386–7
Oil Fund 120–1
Opperman, Romy 317
Orbán, Viktor 87, 90–1, 92, 261–2, 514
Oreskes, Naomi 285
originality, lack of 54
Ost, David 454
Ottoman Empire 259, 261–2, 261–3, 264–5, 343
Oxfam 523
Oxford Handbook of the Radical Right x, 40

Painter, Nell Irvin 356–8
Palestine 343–6
palindefence, myth of 256–74

fascist character 271–4
ideological trick 268–9
interpellative power 269–70
purported historical depth 267
palingenesis, myth of 255–7
palingenetic ultranationalism 228, 229
palm oil trade 347–50
Palmerston, Lord 343, 345
Paludan, Rasmus 165–6, 178–9, 521
Parable of the Sower (Butler) 250
Paris Climate Accord 34, 70, 77, 109, 134
  goals 85
  opposition to 83–6, 175–9, 213
  US withdrawal 134–5
Party for Freedom (Netherlands) 8–10, 10, 83, 89, 97
Patriarchy 394
patriotic ecology 134, 137, 139, 174, 474–5
patriotism 508–9
Paxton, Robert 230, 232, 235, 237
Peabody 26
peat 95–6, 161
PEGIDA, 167
Pence, Mike 185
Perry, Rick 191–2
Perussuomalaiset 83, 87–8, 95–6, 156–61
Peterson, Jordan 303, 306, 307, 393–4
Petrobras 207–8, 219, 483
petroleum 386–8
petroleum advertisements 371–2
petro-nationalism 124–31
Pinker, Steven 487–8
Pittsburgh synagogue shootings 304
Poland xv, 103–18, 131–2, 206, 435–6
  anti-Semitism 114–15
  asylum applicants 111
  CO2 emissions 118
  coal industry 105–7, 110–11, 457, 535–6
  coal nationalism 116–17
  COP24, Katowice 110–11, 117, 186, 215
  and covid-19 pandemic 515, 520, 535–6
  defence ideology 263
  diversity 112–13
  Electromobility Development Plan 107
  fascisation 454–5
  foreign capital penetration 115–16
  immigration policy 111–12
  Islamophobia 112–14, 116, 454–5
  Jewish population 113
  migrant threat 103–5, 241
  Muslim population 113
  renewable energy 108, 275
  view of the climate problem 108–10
polarisation 508–9
Polish Defence League 114
political correctness 157, 306
political influence 36
pollution xv, 50–1
Pompeo, Mike 185, 194
Popper, Karl 449
Popular Science Monthly (magazine) 358
population ecology 169
population growth and overpopulation 140–1, 145, 158–61, 169, 172, 172–3, 184, 215n114, 245–6
population policy, ecologically sustainable 169
populism 65
Porsche, Ferdinand 421
postfascism 226, 253
post-truth 296–7, 313
Poulantzas, Nicos 205–6, 234, 383, 445
poverty 60
power relations 383
powerlessness, 456
Price, Grant, *By the Feet of Men* 250
primitive accumulation 17
problem-solving ingenuity ix
Proceedings of the National Academy of Science 533
proletarian consciousness 456
protectionism 155
Protocols of the Elders of Zion 139
Pruitt, Scott 184, 188
psychotic thinking 313
public transport 365–6
Pulido, Laura 202, 319–20
Puricelli, Piero 415

QAnon 518
quasifascism 226
Querdenker 518–19

race and racism xv, 244, 318, 322, 404–5, 462
  as biological units 327
  Bolsonaro 211, 216
  classification 327–8
  and covid-19 pandemic 513–15, 516–17
  and environmental degradation 337–8
  fascism 235
  frontier 338–43
  and genetics 326–7

Index

ideology 334
ontological status 331–2
petrification 329–30
primacy over class 458–9
steam power and 343–63
Trump 199–204
white superiority claims 350–9
see also skin colour
race prejudice 336–8
race riots, Detroit, 1943 365–6
racial divides 37
racial identities 329–30
racial others 244–5
racialisation 201, 330–2
Rackete, Carola 178
Rahmstorf, Stefan 3n1
Rassemblement National 135–7, 155, 171, 225, 463, 525
rationality viii–ix, 482, 489
Raymond, Lee 27
reactionary modernism 442–3
reality 485
  adjustment of 489–90
  break with 484
reconquista 260–1
REDD+ 122
Rees-Mogg, Jacob 72–3, 81, 96
refugee crisis 114, 176–9, 319, 330
refugees x, 71–7, 158–9, 162
remigration 66–7, 168
renewable energy 86, 100–1, 176, 205, 240, 275–6
  employment 277
  flow 274
  need for internationally integrated grids 276–7
  opposition to 299–300
  Poland 108, 275
  Trump mockery of 186
  waterpower 276
  wind power 86, 86–9, 108, 147, 275, 299–300
repatriation 66–7
re-primitivization threat 56–7, 61
Republika Srpska 264
responsibility 211–12, 316–17, 492–5
reverse discrimination 303n126
Ridley, Matt 487
Riley, Dylan 204
road development 367–9, 414–18, 421–3, 445–6, 446–7
Robertson, Corbin, Jr. 203
Robertson, Thomas 140, 245
Robinson, Tommy 80
Rockström, Johan 206

rolling coal 394–5
Rosalía, Fuck Vox tweet 278–9
Rosenberg, Arthur 228–9, 231, 235
Rosenblum, Nancy L. 297–8, 298, 300
Rosling, Hans 486–7
Ross, Edward A. 358–9, 442
Ross, Wilbur 185
Rydgren, Jens 40

Said, Edward 45–5, 387
Salles, Ricardo 213–14, 219, 311
Salvini, Matteo 175–9, 225, 260, 525–6
Samuels, Robert 493
Sato, Makiko 316–17
Saudi Arabia 122
Saull, Richard 238
Scandinavia 42–8
scapegoating 498–9
school strikes 7, 149–50, 153, 164, 307–10, 507–8
Science (journal) 141
science, war on 27–8
scientific expertise, adherence to 512
scientific methodology 4
sea level rise 281–2
Sealey-Huggins, Leon 319, 508
secessionist automobility 371–4, 375–81, 385–6
Second World War 234, 266, 267, 365–7, 385, 433–5, 505
segregation 168, 368–72, 375–81, 385–6
self-sufficiency 134, 139
Serbs 263–7, 269
Sessions, Jeff 187
sharia law 266
Shearman, David 475–6
Shell 18, 26, 32–3, 58, 530
Shellenberger, Michael 537
Sheller, Mimi 372
Shiite ecologist activism 311
Shriver, Lionel 246, 246–7
Sierra Club 144
Simms, Brendan 425
Singer, Fred 27, 35, 43, 282, 285, 295
Sinophobia 513
skin colour 315–91
  blackness 252, 328
  and car ownership 363–81
  and climate change 315–22
  and genetics 326–7
  and immigrant status 330
  and inferiorization 332, 335–43
  negation of whiteness 329–30
  non-white 327–8, 329–30
  and temperature rise 318–21

Trump and xi
whiteness xi, 116, 204, 313, 322, 322–35, 336, 356–9, 390
see also race and racism
slaves and slavery 336, 348, 384, 492
Smith, Joseph Wayne 475–6
Smith, Ortega 513–14
Smith, Richard 74
Snyder, Timothy 239, 242–3
social artefacts 331
social Darwinism 473
social sadism 396
socialism 284–5, 291, 455
socialist regulation 10
Söder, Björn 67, 308
Solberg, Erna 127, 130
Soral, Alain 139
Soros, George 92–3, 114, 213, 262, 306, 307–8, 310, 392
South Africa 69
sovereignty 84, 85–6, 135
Søviknes, Terje 127–8
Spackman, Barbara 405–6
Spain 94, 260, 262, 270n46, 342, 513–14
climate denial 11–12, 489–90
Spectator (magazine) 71, 79–80, 246, 246–7, 308
Sproat, Gilbert Malcolm 352–3
Srebrenica massacre 266–7, 273
standard of living 60
Standard Oil 441
stealth Islamisation 124
steam power 343–63, 384–5, 385
stereotypy 312
sterilisation programmes 464–5
Stern Report 32
Steyn, Mark 55–8, 61, 62, 151, 258, 361
stock myth, the 274–9
Strache, Heinz-Christian 11, 51–2
Stram Kurs 165–6
summer, 2018 62–9
Sun (newspaper) 69
Suomen Sisu 158
supranational treaties 84
Sussman, Brian 288–9, 298
sustainability 168
sustainable dystopia 477
Sweden 46–8, 50, 54–5, 83, 89, 94–5, 149–50, 234, 256–7, 450–1
asylum applicants 111
cars and car ownership 376–8, 379–80
and covid-19 pandemic 521–3
election, 2018 64–9
gender ideology 392
invandrare 521–3

Remigration Agency 66–7
secessionist automobility 376–8
summer, 2018 62–9
Syrian refugees 330
wildfires 310, 321
Sweden Democrats 46–8, 50, 64–9, 83, 89, 94–5, 304, 310, 321, 376–8, 390–1
Swedish Meteorological Institute 64
Szijjártó, Péter 91
Szyszko, Beata 105, 109–10
Szyszko, Jan 108–9

tactics, diversity of 30–1
Tarrant, Brenton 150–3, 155, 273–4, 538
tax and taxation 65–6
TaxPayers' Alliance 75, 79
Taylor, Jared 179
technology, fetishization of
critique of 444–5
fascism 442–50
techno-racism 356–9, 427–9, 442–3
Tegnell, Anders 521–3, 522
Telegraph (newspaper) 70, 283, 319
Teller, Edward 24
temperature rise 154
1°C threshold crossed 489
above 1.5°C viii–ix
natural fluctuations 27
and population growth 172–3
responsibility 211–12, 316–17
secular trend 3
and skin colour 316–17, 318–21
summer, 2018 62
Terre et Peuple 138–9
terrorism 184
Texaco 18, 28
Thatcher, Margaret 71, 96, 287n85
Theweleit, Klaus 278, 411
think tanks 18, 31, 35, 38, 59, 187
Thomson, John Turnbull 355
Thornton, Bruce 58
Thunberg, Greta 149–50, 307–10, 397–8
Thyssen, Fritz 431
Tillerson, Rex 185
Tomlinson, Sally 362–3
Tooze, Adam 425
Total 524–5
total mobilisation 408–14, 426, 433
Trans-Amazonian Highway 340
Traverso, Ernesto 443
trend denial 6, 11, 19–22, 23, 27, 35, 42–3, 54–5, 66, 108–9, 187, 288
Triumph of the Will (film) 419, 423, 442
Trotsky, Leon 230, 231, 236

Trudeau, Justin 483
True Finns, the 83, 395–6
   black man cartoon vii–viii, ix
Trump, Donald xvi, 64, 252, 277, 310,
     387–8, 397, 440, 450, 488, 499
   administration 181–99
   approval ratings 533
   attention deflection 199–206
   and Black Lives Matter demonstrations 531
   cabinet 184–5
   Clean Power Plan 188–9
   climate denial x–xi, 299–300, 490, 491
   and covid-19 pandemic 514, 517
   denialist ISA 185–8
   election victory 38
   electoral defeat, 2020 532–4
   energy advisors 182–4
   energy dominance doctrine 191–2
   first week in office 181–2
   and fossil capital 188–94, 195–7, 205
   fossil fuels and xi
   last months in office 534–5
   and methane emissions 189–90
   misjudgement 204–5
   Muslim ban 182
   and Obama 200–1
   presidency x–xi
   President's Council on Climate Change 193–4
   racism 199–204
   rhetorical excrescences 202
   secretary of energy 184–5
   whiteness and xi
Tuana, Nancy 492
Twitter 298
Tybring-Gjedde, Christian 124, 129

ultranationalism 227, 228, 229, 245, 274
unemployment 41
Union of Concerned Scientists 32
United Kingdom 83–4
   Brexit 69–81
   Brexit referendum 71–2
   CO_2 emissions 316
   coal industry 96–7
   conservative party 71–2
   and covid-19 pandemic 516, 519–20
   environmentalism 146–9
   fascism in 438–9
   Green Industrial Revolution 528
   green nationalism 528–9
   immigration 148–9
   Islamophobia 69–71
   Middle Eastern oil fields 435
   North Sea oil 97
United Kingdom Independence Party 71–3, 76, 80, 83–4, 86
   climate denial 171–2
   and coal industry 96, 96–7
   general election, 2019 77–80
United Nations Framework Convention on Climate Change 13, 493
United States of America
   attention deflection 199–206
   Black Lives Matter demonstrations 529–32
   car industry 365–7
   cars and car ownership 363, 364–74, 385, 388–9
   Chinese threat 202–3
   Civil War 492
   Clean Power Plan 188–9, 195
   CO_2 emissions 24, 316
   coal industry 19, 192, 195, 277
   and covid-19 pandemic 517
   Dakota Access pipeline 190, 198
   denialist ISA 185–8
   Department of Energy 191–2
   Detroit race riots, 1943 365–6, 384
   drought, 1988 12–13
   energy dominance 191–2
   energy independence 183, 190–2
   environmentalism 139–46
   fascism in 439–41
   highway construction 367–9
   Hitler's resentment against 425–6
   Make America Great Again myth 255–6
   mitigation crisis 241
   Muslim ban 182
   oil and gas production 192
   Paris Agreement withdrawal 134–5
   police foundations 530
   presidential election, 2020 532–4, 539–40
   President's Council on Climate Change 193–4
   racism 199–204
   renewable energy 275
   renewable energy sector employment 277
   segregation 368–72
   SUVs 372–4, 385
   Tea Party movement 200
   Trump administration 181–99
   Trump presidency 199–206
   Trumps last months in office 534–5
   Warren County protests 143–4
   white supremacy 188

whiteness 368
wildfires 531–2
universal humanity 321
US Chamber of Commerce 18

Vancouver Island 352–3, 360, 384
victim blaming 154, 522
victimisation 265
Vienna 259, 263, 269–70, 274
vivisection 465–6
Voegler, Albert 431
Volk, the 411–12, 420–1, 424, 426, 428–9, 460–1
Völkischer Beobachter (newspaper) 412, 413, 462
Vox party 11–12, 156, 225, 260, 262, 270n46, 278–9, 307, 380, 489–90, 513–14, 519

Waddington, Lorna 272
Wall, The (Lanchester) 249–50, 251, 270, 509
Wall Street Journal (newspaper) 33, 55, 206, 283
war 407–9
War on Terror 258
Warren County, North Carolina 143–4
Waszczykowski, Witold 117
waterpower 276
Webber, Jeffery 216
welfare state 120
Western Fuels Association 19
Westminster, 55 Tufton Street 73–6, 80
Wheeler, Andrew 184, 188, 199
white burners 326
white flight 369
white genocide 69, 151
white pride 9
white privilege 199–204, 332, 334, 386
white superiority, claims of 350–9
white supremacy 58, 158, 188, 362

whiteness 116, 204, 313, 322, 336, 368
  and car ownership 381–6
  constitution of 329
  construction of 356–9
  definition 332–3
  fossil 322–35
  historical production of 333
  interpellations 382–3
  mestizo version 390
  negation of 329–30
  possessive investment in 386
  as talisman 328–9
  Trump and xi
wilderness 352–3, 355, 460
Wilders, Geert 9, 10–11, 51, 97
Williams, Michelle Hale 41
Williamson, Jeffrey G. 350
wind power 275, 299–300, 528
  opposition to 86, 86–9, 108, 147
Winter, Susanne 10–11, 51
women
  association with nature and animality 393
  attitudes towards 354, 403, 412
  Muslim threat to 51–3, 112, 114
working-class history 450–9
world, the, state of 3
World Bank 177
World Business Council for Sustainable Development 29, 32–3
World Climate Declaration 10
wSieci (magazine) 112
WWF 210

Ye'or, Bat 47–50, 53, 257–8

Zeller, Thomas 423
Zetkin, Clara xvi, 226, 228, 229, 231, 233, 454, 459, 497
Zinke, Ryan 185
zoonotic spillover 512